SOIL BIOCHEMISTRY
Volume 9

BOOKS IN SOILS, PLANTS, AND THE ENVIRONMENT

Soil Biochemistry, Volume 1, edited by A. D. McLaren and G. H. Peterson
Soil Biochemistry, Volume 2, edited by A. D. McLaren and J. Skujinš
Soil Biochemistry, Volume 3, edited by E. A. Paul and A. D. McLaren
Soil Biochemistry, Volume 4, edited by E. A. Paul and A. D. McLaren
Soil Biochemistry, Volume 5, edited by E. A. Paul and J. N. Ladd
Soil Biochemistry, Volume 6, edited by Jean-Marc Bollag and G. Stotzky
Soil Biochemistry, Volume 7, edited by G. Stotzky and Jean-Marc Bollag
Soil Biochemistry, Volume 8, edited by Jean-Marc Bollag and G. Stotzky
Soil Biochemistry, Volume 9, edited by G. Stotzky and Jean-Marc Bollag

Organic Chemicals in the Soil Environment, Volumes 1 and 2, edited by C. A. I. Goring and J. W. Hamaker
Humic Substances in the Environment, M. Schnitzer and S. U. Khan
Microbial Life in the Soil: An Introduction, T. Hattori
Principles of Soil Chemistry, Kim H. Tan
Soil Analysis: Instrumental Techniques and Related Procedures, edited by Keith A. Smith
Soil Reclamation Processes: Microbiological Analyses and Applications, edited by Robert L. Tate III and Donald A. Klein
Symbiotic Nitrogen Fixation Technology, edited by Gerald H. Elkan
Soil-Water Interactions: Mechanisms and Applications, Shingo Iwata and Toshio Tabuchi with Benno P. Warkentin
Soil Analysis: Modern Instrumental Techniques, Second Edition, edited by Keith A. Smith
Soil Analysis: Physical Methods, edited by Keith A. Smith and Chris E. Mullins
Growth and Mineral Nutrition of Field Crops, N. K. Fageria, V. C. Baligar, and Charles Allan Jones
Semiarid Lands and Deserts: Soil Resource and Reclamation, edited by J. Skujinš
Plant Roots: The Hidden Half, edited by Yoav Waisel, Amram Eshel, and Uzi Kafkafi
Plant Biochemical Regulators, edited by Harold W. Gausman
Maximizing Crop Yields, N. K. Fageria
Transgenic Plants: Fundamentals and Applications, edited by Andrew Hiatt
Soil Microbial Ecology: Applications in Agricultural and Environmental Management, edited by F. Blaine Metting, Jr.
Principles of Soil Chemistry: Second Edition, Kim H. Tan

Water Flow in Soils, edited by Tsuyoshi Miyazaki
Handbook of Plant and Crop Stress, edited by Mohammad Pessarakli
Genetic Improvement of Field Crops, edited by Gustavo A. Slafer
Agricultural Field Experiments: Design and Analysis, Roger G. Petersen
Environmental Soil Science, Kim H. Tan
Mechanisms of Plant Growth and Improved Productivity: Modern Approaches, edited by Amarjit S. Basra
Selenium in the Environment, edited by W. T. Frankenberger, Jr., and Sally Benson
Plant–Environment Interactions, edited by Robert E. Wilkinson
Handbook of Plant and Crop Physiology, edited by Mohammad Pessarakli
Handbook of Phytoalexin Metabolism and Action, edited by M. Daniel and R. P. Purkayastha
Soil–Water Interactions: Mechanisms and Applications, Second Edition, Revised and Expanded, Shingo Iwata, Toshio Tabuchi, and Benno P. Warkentin
Stored-Grain Ecosystems, edited by Digvir S. Jayas, Noel D. G. White, and William E. Muir
Agrochemicals from Natural Products, edited by C. R. A. Godfrey
Seed Development and Germination, edited by Jaime Kigel and Gad Galili
Nitrogen Fertilization in the Environment, edited by Peter Edward Bacon
Phytohormones in Soils: Microbial Production and Function, W. T. Frankenberger, Jr., and Muhammad Arshad
Handbook of Weed Management Systems, edited by Albert E. Smith
Soil Sampling, Preparation, and Analysis, Kim H. Tan
Soil Erosion, Conservation, and Rehabilitation, edited by Menachem Agassi
Plant Roots: The Hidden Half, Second Edition, Revised and Expanded, edited by Yoav Waisel, Amram Eshel, and Uzi Kafkafi
Photoassimilate Distribution in Plants and Crops: Source–Sink Relationships, edited by Eli Zamski and Arthur A. Schaffer
Mass Spectrometry of Soils, edited by Thomas W. Boutton and Shinichi Yamasaki

Additional Vclumes in Preparation

Handbook of Photosynthesis, edited by Mohammad Pessarakli

Fauna in Soil Ecosystems: Recycling Processes, Nutrient Fluxes, and Agricultural Production, edited by G. Benckiser

Chemical and Isotopic Groundwater Hydrology: The Applied Approach, Second Edition, Revised and Expanded, Emanuel Mazor

SOIL BIOCHEMISTRY

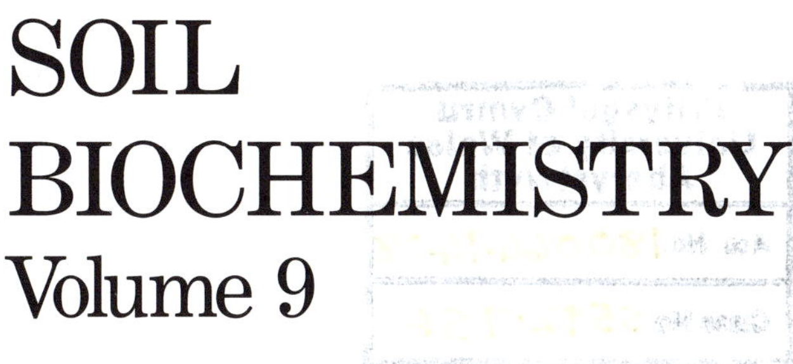

Volume 9

edited by

G. STOTZKY
Department of Biology
New York University
New York, New York

JEAN-MARC BOLLAG
Center for Bioremediation and Detoxification
Environmental Resources Research Institute
The Pennsylvania State University
University Park, Pennsylvania

Marcel Dekker, Inc. New York • Basel • Hong Kong

ISBN: 0-8247-9441-9

The publisher offers discounts on this book when ordered in bulk quantities. For more information, write to Special Sales/Professional Marketing at the address below.

This book is printed on acid-free paper.

MARCEL DEKKER, INC.
270 Madison Avenue, New York, New York 10016

Current printing (last digit):
10 9 8 7 6 5 4 3 2 1

PRINTED IN THE UNITED STATES OF AMERICA

Preface

The quality of the environment continues to be a major scientific and societal concern. Pollution of soil, water, and the atmosphere proceeds essentially unabated, and natural ecosystems and efficient agricultural production remain at risk. Some parts of the world now produce more food and fiber than ever before, but this high level of productivity may be only temporary, as soils worldwide are increasingly serving as repositories for a variety of anthropogenic wastes, most of which are toxic to microbes, plants, and animals. Moreover, soils in many parts of the world, especially where human habitation is dense, are insufficiently productive to sustain the indigenous populations. Some of these soils are being further stressed by their use as dumps for toxic wastes.

One promising approach for remediation of such pollution is the use of microorganisms and their products. Unfortunately, despite a few reports of successful bioremediation in the last few years, progress has fallen short of original expectations. This disappointing record results from our failure to understand adequately the physicochemical and biological variables of soil environments that influence the establishment, activity, ecology, and population dynamics of microorganisms within them. This applies especially to those microorganisms, whether indigenous or introduced, with the capacity to remediate pollutants. A major purpose of *Soil Biochemistry* is to provide up-to-date reviews of these factors and their influences on a spectrum of basic biological phenomena in soil, most of which have a biochemical base and affect the quality of soil.

Similar to its eight predecessors, this volume of *Soil Biochemistry* provides basic information on factors that affect the quality of soil and on methods to measure the effects of soil management and bioremediation. Without quantitative benchmarks, applications of the results of basic research to in situ conditions have limited relevance. This is especially true in the present worldwide economic and political climate that discourages the funding of basic research without a demonstrable "bottom-line" payoff. However, this "bottom line" will not be achieved in the short run or sustained in the long run without the accumulated results of basic research—a knowledge base that may often be of limited relevance to an immediate payoff.

Consequently, several chapters in this volume discuss the influence of microorganisms and their enzymes on soil quality, including the suppression of plant diseases caused by soil-borne pathogens. Several chapters discuss constraints by the physicochemical and biological characteristics of the soil environment on these influences and how they can be modified by management practices. One chapter deals with the effects of these characteristics on in situ bioremediation. Another chapter evaluates methods for determining the microbial populations in the rhizosphere; another discusses the movement of bacteria in soil; and another is a critical analysis of methods that measure metabolic activity in soil and of the relevance of these methods to the "real world." The application and relevance of fractals to the biology and biochemistry of soil is discussed in another chapter.

In this volume of *Soil Biochemistry*, we continue the tradition of providing a range of topics rather than focusing on a single theme. There are many single-theme book series, and these are obviously valuable. However, we believe that a multi-theme approach better serves the interests of our readers. The soil environment is complex, and the diversity of advances in its study requires a multi-theme treatment.

We express our gratitude to the authors who have contributed to this and past volumes, as well as to the editorial staff of Marcel Dekker, Inc., who have been instrumental in publishing this work.

G. Stotzky
Jean-Marc Bollag

Contents

Contributors

Claude Alabouvette Laboratoire de Flore Pathogène du Sol, Institut National de la Recherche Agronomique, Dijon, France

Jay Scott Angle Department of Agronomy, University of Maryland, College Park, Maryland

Richard J. F. Bewley Dames & Moore, Manchester, England

Jean-Marc Bollag Center for Bioremediation and Detoxification, Environmental Resources Research Institute, The Pennsylvania State University, University Park, Pennsylvania

Jerzy Dec Center for Bioremediation and Detoxification, Environmental Resources Research Institute, The Pennsylvania State University, University Park, Pennsylvania

Lloyd F. Elliot National Forage Seed Production Research Center, Agricultural Research Service, U.S. Department of Agriculture, Corvallis, Oregon

Ralph C. Foster Division of Soils, Commonwealth Scientific and Industrial Research Organization, Glen Osmond, South Australia, Australia

Joel V. Gagliardi Department of Agronomy, University of Maryland, College Park, Maryland

Liliana Gianfreda Dipartimento di Scienze Chimico-Agrarie, Università degli Studi di Napoli Federico II, Naples, Italy

Heinrich Hoeper Laboratoire de Flore Pathogène du Sol, Institut National de la Recherche Agronomique, Dijon, France

Jeffrey N. Ladd Division of Soils, Commonwealth Scientific and Industrial Research Organization, Glen Osmond, South Australia, Australia

Philippe Lemanceau Laboratoire de Flore Pathogène du Sol, Institut National de la Recherche Agronomique, Dijon, France

Morris A. Levin University of Maryland Biotechnology Institute—Columbus Center, Baltimore, Maryland

J. M. Lynch School of Biological Sciences, University of Surrey, Guildford, England

Eugene L. Madsen Division of Biological Sciences, Cornell University, Ithaca, New York

Pertti J. Martikainen Department of Environmental Microbiology, National Public Health Institute, Kuopio, Finland

Maria S. McIntosh Department of Agronomy, University of Maryland, College Park, Maryland

Stephanie L. Murphy Department of Environmental Sciences, Cook College, Rutgers University, New Brunswick, New Jersey

Paolo Nannipieri Department of Soil Science and Plant Nutrition, Università degli Studi, Florence, Italy

J. Malcolm Oades Department of Soil Science, University of Adelaide, Glen Osmond, South Australia, Australia

R. I. Papendick Agricultural Research Service, U.S. Department of Agriculture, Washington State University, Pullman, Washington

Pacifico Ruggiero Istituto di Chimica Agraria, Università degli Studi di Bari, Bari, Italy

Nicola Senesi Istituto di Chimica Agraria, Università degli Studi di Bari, Bari, Italy

Christian Steinberg Laboratoire de Flore Pathogène du Sol, Institut National de la Recherche Agronomique, Dijon, France

Robert L. Tate III Department of Environmental Sciences, Cook College, Rutgers University, New Brunswick, New Jersey

1

The Microbial Component of Soil Quality

Lloyd F. Elliott U.S. Department of Agriculture, Corvallis, Oregon

J. M. Lynch University of Surrey, Guildford, England

R. I. Papendick U.S. Department of Agriculture, Washington State University, Pullman, Washington

I. INTRODUCTION

Soil quality has been defined as the capacity of soil to produce healthy and nutritious crops, resist erosion, and reduce the impact of environmental stresses on plants [1,2]. The need for preservation and improvement of soil quality is recognized worldwide. Declines in soil quality are evidenced by severe soil erosion, increased soil crusting and compaction, reduced water infiltration, adverse pH changes, contamination by pollutants, salinization, and a reduced capability of crop production, all of which can be linked to poor agricultural practices. The substitution of pesticides, fertilizers, and energy (in the form of intensive tillage) for crop rotations and the use of soil as a waste repository will not, in the long term, support sustainable agriculture and will result in the loss of soil and environmental quality. These approaches to crop production have required a reassessment of long-term nonrenewable resources and environmental considerations. For example, Babich and Stotzky [3] called for a Clean Soil Act in 1985. It is increasingly evident that inputs (fertilizers, pesticides, and energy inputs for tillage) can mask the degradation of soil quality in terms of productivity, but they cannot remedy the degradation because many of these inputs promote erosion, damage the environment, and are nonrenewable and energy dependent. Intensive cropping without inputs such as fertilizers can also be nonproductive and damage the resource base. This is evident in areas such as the Middle East, which have been tilled for centuries without fertilizer additions and

where much erosion has occurred through lack of crop cover. Harvesting of residues in these areas has also contributed to the problem.

Soil quality does not depend just on the physical and chemical properties of soil, but it is very closely linked to the biological properties of soil. Many soil properties affect soil quality, and most are influenced by microbiological processes. Soil properties affected by the size and composition of the microbial biomass include water holding capacity, infiltration rate, crusting, erodibility, aggregate stability, susceptability to compaction, nutrient cycling, available nitrogen, nutrient capacity, and soil organic matter content. Presently, however, methods for measuring soil quality are not well defined.

Several studies have shown the effect of microorganisms on soil structure. Martin and Waksman [4] and Martin [5] demonstrated the importance of microorganisms and their by-products to soil structure and aggregate stability. The degree of microbial stabilization of soil and volcanic ash during the decomposition of straw was shown by Lynch and Elliott [6]. They showed that excess nitrogen could inhibit soil aggregation by decreasing polysaccharide production during decomposition of residues [7]. The production of polysaccharides probably also influences infiltration, crusting, and erodibility. Rapid advances are now being made in understanding soil structure/soil biota interrelationships [8]. However, the influence of microorganisms on soil aggregation and related processes is only one facet that affects soil quality.

Recently emphasis has been placed on defining and developing sustainable agricultural systems. An important component of sustainable agriculture is soil quality, because a healthy, functioning soil is fundamental to sustainable crop and livestock production and a healthy environment. The biological component of soil quality is very important because it has a major effect on the processes and properties previously mentioned. Moreover, loss of soil quality components that are controlled by biological properties and processes can be remedied through the use of crop rotations, cover crops, and green manures, utilization of organic wastes, and effective management of crop residues. However, it is likely that remediation will be slow and the rate will depend on the degree of degradation.

The first substantive information showing the value of green manures and crop rotations on soil quality was collected during studies comparing soil properties of an organic farm with an adjacent conventional farm in eastern Washington state. Bolton et al. [9] showed the organic farm had significantly higher levels of urease, phosphatase, dehydrogenase, and soil organic matter content (1.14 and 0.91% organic carbon, respectively) and a significantly higher microbial biomass content in two out of three samplings. Reganold et al. [10] studied the same farms and found that the organically farmed soil had a significantly higher polysaccharide content, a lower modulus of rupture, and the thickness of the topsoil was almost 16 cm greater than in the soil that had been conventionally farmed. This large difference in depth of the topsoil was attributed to greater soil erosion from the conventionally farmed soil during the period from 1948 to 1985. It was estimated that at existing soil erosion rates, all topsoil would be lost from the conventional farm in 50 years, and crop yields then would be half or less of the present yields, even with high inputs of fertilizers and pesticides.

It was suggested that the lower rate of erosion on the organic farm resulted from a greater number of crops in the rotation and the use of green manures. The benefits of crop rotations continue to be emphasized [11–13]. With the decline in the quality of the soil on the conventional farm, it would not be possible to build the size of the microbial biomass and nutrient status to its previous level without remediation because of the loss of water holding capacity, which is related to soil organic matter content. The biological components would also be adversely affected because degradation of the physical properties could impede gaseous exchange and reduce water availability.

More recently, Reganold et al. [14] compared the physical, biological, and chemical soil properties and the profitability of biodynamic and conventional farms in New Zealand. They found that the biodynamic farms had better soil quality and were just as financially viable on a per hectare basis as adjacent conventional farms. These studies further show how different farming systems can dramatically affect soil quality. Of the changes observed, decreases in soil organic matter content and microbial biomass and increases in erodobility are strong indicators of decreases in soil quality [15].

There is a positive correlation between soil organic matter content and soil microbiological activity [16–18]. Soil organic matter tends to be greater in cropping systems utilizing crop rotations than in continuous cropping and in systems utilizing animal manures rather than commercial fertilizers [16,19].

Sod or forage legume-based rotations reduce soil loss in two main ways. First, soil loss by erosion during the meadow sequence is negligible after establishment because of the characteristic dense vegetative cover and rooting of these crops. Second, when the sod or legume is cultivated, the residual effects of these plantings improve infiltration and leave the soil less erodible for several years thereafter [20]. With conventional tillage and planting systems, soil losses from soils cropped to maize following meadow ranged from 14 to 68% of the corresponding losses from soils cropped to maize without meadow in the rotation [20]. The residual effect of meadow in reducing soil loss with intensive tillage may continue, though on a declining level, for 2 to 3 years and, hence, would carry through the entire small-grain and row-cropping sequence for many of the rotations used by organic farmers. For example, the annual soil loss from a 4-year rotation of wheat, hay, and two years of conventionally planted maize averages about one-third of that from conventionally planted continuous maize [20]. This indicates that the effect of the hay in reducing soil loss carries through the two succeeding years of maize.

Microbial activity in soil can be altered greatly by the addition of heavy metals and persistent organic compounds, especially xenobiotics [3,21]. Pollutant input can be no greater than the ability of the soil to ameliorate the input. Activities such as mineralization of carbon, nitrogen, phosphorus, and sulfur, N_2 fixation, and soil respiration are reduced by heavy metal pollution. Enzyme activity in soil is also sensitive to heavy metals. Decomposition of plant litter, nutrient cycling, and organic matter turnover are all sensitive to heavy metal contamination. Disruption to these processes would greatly change the ecology and productivity of a system and would likely result in lower soil quality and lower productivity. The effect of heavy metals in soils on the microflora and plants appears to be highly sensitive to changes in pH [22]. Decreases in quality caused by pollutants, such as heavy metals and persistent toxic organic compounds, are extremely difficult to reverse and must be stopped and remedied before the situation becomes severe.

Obviously, indicators that are sensitive to increases and decreases in soil quality are needed. Changes in microbiological characteristics will probably be among the most sensitive indicators. For example, bacterial numbers in the top 7.5 cm of soil were lowest in a rotation of fallow wheat, intermediate in fallow wheat-wheat, higher in continuous wheat fertilized with nitrogen and phosphorus, and highest in continuous wheat fertilized with only phosphorus. Microbial biomass carbon in these four rotations was 180, 226, 217, and 357 kg/ha, and the erodible fraction in the top 15 cm of soil was inversely related to the amount of trash conserved [23]. They concluded that land degradation and the decline in amount and quality of organic matter caused by frequent summer fallowing can be arrested by more frequent cropping. Removal of residues by burning appears to reduce soil organic matter, reduce the amount of water-stable aggregates, and increase peak discharge rates of runoff [24–27].

II. QUANTIFYING SOIL QUALITY

Soil quality cannot yet be defined in quantitative terms. However, it should be possible eventually to define soil quality quantitatively by the appropriate integration of specific quantitative terms, so that the effects of management and cropping systems on soil quality can be determined. Soil quality is a dynamic character, and any significant indicators must be sensitive to small changes in key soil properties. Consequently, as improved management schemes are applied, it should be possible to rapidly detect changes in soil quality. However, tools to detect the impact of changes in management schemes are needed. Measurements to define changes in soil quality are proposed in Doran et al. [15].

A model to describe soil quality should probably include initially most of the biological, physical, and chemical measurements of soil that can be made.

An appropriate model should integrate these characteristics into a hierarchical progression to provide quantification of soil quality. It should then be possible to determine those characteristics that must be included to accurately describe changes in soil quality. In this manner, key descriptors can be sorted out. This differentiation is important for technology transfer because many geographic areas may have only the instrumentation and means necessary for a general description. Coleman et al. [28] have developed ecosystem-level process models describing carbon budgets for the decomposition of buried and surface *Secale* litter in no-tillage and conventional tillage systems. With selective inhibitors, they have described the contribution of the various soil biotic groups to carbon flow and have been able to provide carbon conversion rates. They described the effects of tillage and residue management on soil structure. These effects could be included in the model to determine management favorable to soil structure and potentially soil quality. Other aspects of soil quality, such as soil-borne disease relationships, beneficial rhizosphere relationships, N_2 fixation, etc., could be included in the model. This approach is complicated, but once the model is completed it can be greatly simplified in most cases.

An approach to measuring soil quality was developed at a meeting in the United States by the Soil Quality Working Group [29]. A framework for characterizing soil quality was provided and broken into five levels (Table 1). An example was provided for approaching the major issue of soil quality-crop productivity. The index should be defined for a specific agroecosystem (e.g., dryland wheat, paddy rice) and local weather/climatic conditions. The productivity index should be a function of absolute yield, spatial and temporal variability of yield, and resource use efficiency. The meso-level indices of soil quality assigned to productivity were fertility, hydrology, toxic factors, and temperature. Components and measurable properties for these indices are presented in Table 2.

Another example was provided for addressing a second soil quality issue, i.e., environmental. The target or agroecosystem regions could be synonymous with those for the productivity issue. The working group used the United States as the target area and subdivided environment into three concerns (meso-levels): water quality (including surface and ground water), soil contamination, and air quality (including dust and gases). For this analysis, the group attempted to include several soil quality indices that would bridge environmental issues with soil properties. Two possible indices were soil tilth and soil contamination. Soil properties were given the main focus of attention, and these are presented in Table 3. An approach such as this should aid in the development of methods for predicting changes in soil quality.

Singh et al. [30] developed a method for predicting a soil tilth index. Five parameters are measured and incorporated into an index that could provide a component of soil quality. A critical requirement for assessing soil quality mea-

Table 1 An Example of an Operational Framework for Characterizing Soil Quality

Level	Component			
Target area	Temperate	Humid tropics	Semiarid tropics	Savanna
Soil quality issues	Productivity	Environment	Health	
Meso-level soil quality index	Tilth	Fertility	Nutrition	Toxicants
Chemical				
First-order soil properties	Electrical conductivity (salinity), pH Site-specific toxicities (toxic elements, pesticides, organics)			
Physical				
	Water status (0.03–1.5 MPa) Infiltration rate Soil depth, horizons		Water holding capacity Soil texture Soil bulk density	
Biological				
	Soil organic matter Soil microbial biomass qCO_2 (resp./biom. ratio) Mineralizable N (labile)		Labile organic C Soil respiration Vegetation growth and cover Abundance of "key" earthworms	
Second-order soil properties	Aggregate stability (turbidimetric dispersable clay cation-anion-exchange capacity numbers and diversity of "key" soil fauna and flora)			

surements is being able to overcome variability. Smith et al. [31] developed a multivariate kriging method for integrating multiple measurements into a single index. This approach will prove valuable for assembling data for a predictive effort.

III. COMPONENTS OF SOIL QUALITY

There are many components of varying sensitivity and importance to soil quality that encompass a broad range of soil physical, chemical, and biological properties. However, the key is to identify those components that rapidly respond to

Table 2 Some Examples of Meso-Level Soil Quality Indexes and Components and Measurable Properties for Characterizing Soil Productivity

Component	Measurable properties
A. Fertility component	
a. Nutrient holding capacity	Cation exchange capacity
b. Nutrient levels	Anion exchange capacity
c. Nutrient transformation	Total and available amounts of nutrients
	Dentrification rates
	C mineralization and respiration
	P and K cycling and fixation of P and K
	Microbial biomass/total carbon
B. Hydrology component	
a. Infiltration/runoff	Exchangeable sodium, aggregate stability, crusting, infiltration rate, runoff rate, electrolytes, surface cover, plant emergence
b. Available water holding capacity	Soil depth, moisture release curves, organic matter content
c. Hydraulic conducivity	Saturated and unsaturated conductivities
d. Depth to restrictive barrier	Water table depth, aluminum subsoil, bedrock, pans
e. Landscape position	Slope, aspect, trafficability
C. Toxicity	
Nonanthropogenic	Salinity, pH, aluminum, manganese, selenium
Anthropogenic	Salinity, pH, pesticides and pesticide degradation products, industrial wastes
D. Temperature	
Radiation	Albedo, solar, net, long-wave, UV
Soil temperature	Surface temperature, thermal conductivity

changes in soil quality. Finally, the minimum components or measurements that can be used to describe changes in soil quality must be determined.

The most critical factor in determining soil quality is probably the productivity of crops grown on the soil, both in terms of the biomass they produce and the flow of nutrients at the root-soil interface. Thus, root and shoot biomass and soil carbon:nitrogen and carbon:phosphorus ratios are good indicators of soil quality.

Many scientists agree that organic matter content is the best single indicator of soil quality and the main constituent that differentiates topsoil from subsoil. Soil organic matter is unique in that it is a part of the physical, chemical, and biological properties of soil. Loss of organic matter results in loss of soil tilth

Table 3 Examples of Soil Properties for Characterizing the Environmental Issue of Soil Quality

Soil properties	Chemical	Physical	Biological
First order	pH Salinity Cation exchange capacity Organic matter Site-specific toxicities a. Heavy metals b. Toxic organics c. Nitrate d. Radioactivity	Infiltration Available water Soil depth	Vegetative cover Microbial biomass Labile organic carbon Labile organic nitrogen
Second order		Water-stable aggregates Dispersible clay Bulk density	Key invertebrates Earthworms

and productivity. Organic matter serves as the major granulating agent in soil and provides structural stability and optimum air and water relations [32]. Presently it is not possible to measure subtle changes in organic matter; however, it is possible to assess the readily available fraction (that which can be rapidly decomposed) with respiration techniques.

Organic matter and its management have a strong effect on soil aggregation. Lynch and Elliott [6] showed that degraded straw significantly increased the aggregate stability of Palouse silt loam and volcanic ash. However, soil organic matter content cannot be separated from ease of decomposition and from the soil microbial biomass. In fact, a soil organic matter decomposibility measurement, such as respiration, the mineralization potential of nitrogen, phosphorus, or sulfur, size and/or composition (diversity) of the microbial biomass, or a plant vigor index, may be more meaningful than the total organic matter content of soil. For example, a virgin soil with an organic matter content of 2% is probably of higher quality than an intensively farmed soil with a monoculture and an organic matter content of 3%. The largest, most readily decomposed fraction of the soil organic matter is likely the most important.

Doran et al. [33] proposed a list of basic soil properties that should be indicators of soil quality (Table 4). The indicators were chosen to be definitive of soil quality and also to be basic enough so they can be measured almost anywhere without the need for complicated and expensive apparatus. These indicators could be fundamental for developing predictions of changes in soil quality.

Table 4 Proposed Soil Physical, Chemical, and Biological Characteristics to be Included as Basic Indicators in Screening Soil Quality

Soil characteristic	Relationship to soil condition and function; rationale as a priority measurement
Physical	
Texture	Retention and transport of water and chemicals; modeling use, soil erosion and variability estimate
Depth of soil, topsoil and rooting*	Estimate of productivity potential and erosion; normalizes landscape and geographic variation
Soil bulk density (SBD) and infiltration*	Potential for leaching, productivity, and erosivity; SBD needed to adjust analyses to volumetric basis
Water holding capacity* Water retention characteristic	Related to water retention, transport, and erosivity; available water, calculated from SBD, texture, organic matter content
Chemical	
Total organic C	Defines soil fertility, stability, and erosion and N extent; use in process models and for site normalization
pH	Defines plant and microbial activity thresholds; essential to process modeling
Electrical conductivity	Defines plant and microbial activity thresholds; presently lacking in most process models
Extractable, N, P, and K	Plant available nutrients and potential for N loss; productivity and environment quality indicators
Biological	
Microbial biomass	Microbial catalytic potential and repository for C and N; modeling, early warning of management effects on organic matter content
Potentially mineralizable N (anaerobic incubation)	Soil productivity and N-supplying potential; process modeling, surrogate indicator of biomass
Soil respiration, water content, and temperature*	Microbial activity measure (in some cases, plants); process modeling, estimate of biomass activity

*Measurements made in field for varying row orientation and management conditions.
Source: Refs. 15 and 33.

IV. SOIL PHYSICAL FACTORS

Soil physical factors of greatest importance to soil quality include water holding capacity, aeration, crusting, color, temperature, strength, resistance to compaction, and porosity. Although it is not strictly a soil physical property, soil organic matter will directly affect most soil physical factors. None of these factors is an independent variable; a change in one will usually result in a change in others. For example, a high soil water content may result in lowered soil temperatures, a deficiency of oxygen for plants and microorganisms, and changes in soil strength.

The physical factors of soil are influenced, to a large extent, by the structure or aggregation of the primary particles. The characteristic of soil structure that is most important to plants is the pore size distribution (relative numbers of pores in different size classes), which results from aggregation and aggregate stability (the hardness of the individual aggregates and their ability to hold their shape upon wetting or to reform upon drying). Many physical problems of soils are related to poor structure. Good structure is needed to maintain adequate aeration and ready infiltration and to provide an optimal root environment [32]. Each of these properties is strongly affected by the content and quality of soil organic matter [34]. Soil crusting also appears to be very sensitive to the status of soil organic matter. Soil crusting also is greatly affected by chemical relations, as discussed by Sumner and Miller [35].

Logsdon et al. [36] studied the effect of farming practices on water infiltration. They suggested that infiltration rates were somewhat higher for soils on alternative farms that used rotations and green manures for fertility and pest control than on conventional farms. They found that bulk densities were somewhat lower and the volume of large pores was slightly higher in soils of the alternative farming system. Jordahl and Karlen [37] found that alternative farming practices, when compared with conventional practices, resulted in greater water stability of soil aggregates, higher soil organic matter content, and lower bulk density. They concluded that the practices mainly responsible for these differences were rotations, including oat and hay crops; ridge tillage; and additions of 45 Mg/ha of a mixture of animal manure and municipal sludge during the first 3 years of each 5-year rotation.

Arshad and Coen [38] proposed some physical and chemical properties that may be indicative of soil quality. These are soil depth to a root restricting layer, available water holding capacity, bulk density/penetration resistance, hydraulic conductivity, aggregate stability, organic matter content, nutrient availability/retention capacity, pH, and where appropriate, electrical conductivity and exchangeable sodium.

Soil water is probably the most important physical factor affecting the physical condition of soil. Soil water is not only essential for life processes of plants and microorganisms, but it also has many interacting components that, individually or in combination, affect biological systems and their activities. These effects may arise from the potential energy of the water per se or from the indirect effects of water potential or water content on such factors as gas or solute diffusion and soil strength. The amount of water a soil will hold is also affected by the soil organic matter content.

V. SOIL CHEMICAL FACTORS

Soil chemical properties are associated with the organic matter content of soil [39], although some properties are associated with the mineral composition of the soil. Availability of an element to plants and microorganisms is generally associated with organic matter content, microbial activity, and pH. The nitrogen content of soils and the ability of a soil to recycle nitrogen is largely dependent on the soil organic matter content and associated microbial biomass [40]. Cultivation has the greatest effect on the availability of soil nitrogen and associated nutrients [39,41]. Cultivation results in a loss of soil organic matter and in a decrease in soil nitrogen and nitrogen-cycling capacity [11]. The efficiency of crop utilization of native soil nitrogen with intensive tillage is very low [41]. Reduced tillage will conserve the soil organic matter content and its nutrient storage and cycling capability [11,42]. The value of crop residues on the content of soil nitrogen and organic matter was demonstrated by Smith et al. [43]. Nutrients in tropical soils recycle rapidly because of higher temperatures, and in some cases, higher moisture in these zones [42]. An approach for evaluating chemical indicators of soil quality was presented by Karlen and Stott [44].

VI. SOIL BIOLOGICAL FACTORS

If erosion is not greater than the rate of soil formation, the major factor that regulates soil quality is its biological properties. These biological properties include size and diversity of the microbial, micro- and macrofaunal biomass, enzyme quantities and activities, mineralizable nitrogen, carbon, phosphorus, sulfur, etc., respiration, and soil organic matter content. The measurement and utility of some of these properties were summarized by Parkinson and Coleman [45] and Horwath and Paul [46]. Immunological and molecular techniques for studying microbial populations in soil were summarized by Sayler et al. [47] and Carter and Lynch [48]. The correlation of these properties with tillage, residue management, crop rotations, and chemical inputs will be necessary for assessing soil quality.

Recent studies on reduced-input farming have shown benefits for improved soil quality (reduced input involves reduced inputs of mineral fertilizers and biocides, less intensive tillage, and integrated use of manures and mineral fertilizers). Hassink et al. [49] reported that the size and activity of the microbial biomass were greater in the top 25 cm of soil from a reduced-input farming system than from a conventional system during a 3-year study. They compared microbial communities of the fields based on activity indices, a multivariate procedure, and diversity indices [50]. They found that the composition of the rhizosphere populations changed more than those in the bulk soils, and the diversity index of the fungal population in the rhizosphere of the conventional field decreased more than that of the reduced-input field. The diversity of the bacterial and fungal populations decreased in July and August, with the largest decrease occurring in the rhizosphere of the conventional field [50]. When conventional and integrated (integrated differed from conventional in reduction of inputs of fertilizer nitrogen and pesticides and in reduction of tillage) farming plots were sampled, the total biomass of soil microorganisms averaged 690 kg C/ha and 907 kg C/ha for the conventional and integrated farming plots, respectively [51]. After 10 years, a tillage rotation study showed that microbial biomass, total carbon, and total nitrogen were highest under no-till surface soils (0 to 5 cm). Rotation alone had little effect on the size of the microbial biomass [52]. Studies have also provided evidence that the size of the soil microbial biomass is a function of the climate [52–54]. Thus, it may be difficult to extrapolate microbial biomass results across climatic zones.

It should be recognized that many soil microorganisms (probably about 90%) are not culturable, and therefore diversity is difficult to assess. Nonculturable organisms, as presently defined, could have a significant role in soil quality. Currently there is no means to assess their effect, if any. A recent approach to study diversity has been with an ecophysiological index (F. A. A. M. De Leij, J. M. Whipps, and J. M. Lynch, unpublished), where diversity is classified according to bacterial growth sites and limited to the spectrum of γ and κ strategies in a population. Cook [55] suggested that soil-borne diseases can also be used as an indicator of soil quality. For example, crop rotation reduces the inoculum potential of soil-borne pathogens, thus improving soil quality. Deleterious rhizobacteria may also become a problem when crop rotations are decreased [56,57].

Visser and Parkinson [58] reviewed the literature and concluded that ecosystem (process)-level studies may be the best approach for rapidly assessing changes in soil quality. Ecosystem-level studies could include carbon cycling, soil respiration, microbial biomass carbon, nitrogen cycling, and nutrient leaching. Coleman et al. [28] provided convincing evidence for ecosystem-level studies for assessing soil quality, but they were not convinced that measurements of soil enzyme activity would correlate with soil quality. Domsch et al. [59] proposed an interesting concept for assessing the effects of agrochemicals on ecosystems. They proposed following organisms of high sensitivity, which included nitrifiers, rhizobia, and actinomycetes, and organic matter degradation rate and nitrification. For organisms of medium sensitivity, they proposed algal populations, total bacterial and fungal populations, soil respiration, denitrification, and ammonification. They identified organisms of low sensitivity as *Azotobacter* populations, general microorganisms, ammonifiers, and aerobic N_2-fixing capacity. Although organisms and processes may be associated with groups different from those they identified, the concept may be useful for identifying changes in soil quality.

The beneficial effects of crop residues and accompanying microbial activity on soil stability have been extensively studied [4–6,8,9,34]. There appear to be three mechanisms whereby microbes stabilize soil. Physical binding, such as by hyphae around and through soil particles, attraction between microbes and soil by physical means or sticky material on the cells, and the production of extracellular materials, such as polysaccharides [34]. It was shown that if nitrogen was limited during the decomposition of crop residues, polysaccharide production would be encouraged, resulting in a greater potential for stabilizing soil. Experiments conducted with decomposing wheat straw of different nitrogen contents showed this to be true [2]. These studies point out the value of surface-managed, low-nitrogen crop residues for increasing soil resistance to erosion and for decreasing soil crusting. Improved microbial activity benefits water infiltration, resistance to erosion, and crop growth, and reduces crusting. Crop rotation also beneficially affects some of these properties. Bolton and Weber [60] studied the effect of cropping systems on aggregation of the top 10 cm of soil in Ontario, Canada. They found soil aggregation decreased in the order of bluegrass, alfalfa-brome, oats, and continuous corn. Many investigations have shown the positive effect of a grass sod in the rotation [20].

Studies have also shown the benefit of a legume in the rotation for increasing soil stability, resistance to erosion, and crop productivity [11,61]. Elliott et al. [11] included several legumes in their rotation. Although red clover produced the smallest amount of plant biomass and fixed nitrogen of the legumes tested, winter wheat following the red clover outyielded winter wheat following the other legumes tested, such as Austrian winter peas. The effect of the red clover was greater than that of the added fixed nitrogen and the amount of plant biomass. Studies are needed to determine more adequately the mechanisms of how legumes benefit a rotation.

A portion of the beneficial effects of rotations and the use of green manures probably results from the interruption of cycles of soil-borne diseases and, possibly, of deleterious rhizobacteria [57]. It is tempting to postulate that these practices change biodiversity in soil (e.g., that there are shifts in the microbiota to the disadvantage of harmful organisms) for the benefit of soil quality. This does not mean that biodiversity is increased; in fact, evidence indicates this is not true [62]. However, the balance of microorganisms is apparently changed. Altering residue management and tillage also changes these relations [28,63]. No-till cropping favors fungi in the surface soil, whereas conventional tillage favors bacteria. Thus, an increase in organic matter content in the soil surface through management of crop residues and reduced tillage may even be of greater benefit if crop rotations are maintained. The benefit would result from the increase in soil microbial biomass and possibly enzyme activity [64,65]. As the microbial biomass of soil increases, the challenge is to ensure that beneficial organisms are favored.

There are some interesting new data on biodiversity of higher plants that were developed by Tilman and Downing [66] during a long-term study of grasslands. They found that the primary productivity of more diverse plant communities is more resistant to and recovers more fully from drought. Their data suggested that as a species was lost from a grassland, there was a greater negative impact on drought resistance. These studies indicate that preservation of plant biodiversity in grasslands is essential to the maintenance of stable productivity in ecosystems. This principle may extend to soil quality.

A major obstacle to defining soil quality has been the lack of sensitive approaches for measuring temporal changes in the size, activities, and composition of the soil microbiota. If these temporal characteristics can be measured with precision, such as with microbial biomass techniques, measuring mRNA, using reporter genes, or other quantitative approaches, then it may be possible to link them with properties such as erodibility (stability), crusting ability, nutrient cycling capacity, water infiltration, and energy of wetting.

Immunological and molecular techniques may offer approaches for measuring temporal soil changes associated with soil quality. Carter and Lynch [48] summarized these techniques for studying the dynamics of microbial populations in soil and the following points emerged. Antibodies have been used successfully to track bacteria in soils; this technique has been much more difficult with fungi, with specificity being a problem. Immunological techniques, such as monoclonal antibodies, can be quite useful in some situations. Molecular techniques offer promise for following specific populations and potential activities. DNA probes have been used successfully to follow populations. RNA probes have been developed to determine the potential for a process, and mRNA probes may be capable of following biological activities. Reporter genes and transcriptional fusions offer promise for tracking organisms, especially those in the rhizosphere. These approaches might be used to assess soil quality by determining the presence of and potential for activity by deleterious and beneficial bacteria. These could be very sensitive indicators of soil quality.

The importance of the meso- and macrofauna to soil quality must be included [28,63,66,67]. These organisms greatly affect the nutrient status, structure, and processes in soil. They have a major role in ecosystem processes and should provide indications of soil quality.

VII. IMPROVING OR MAINTAINING SOIL QUALITY

Soil quality is a qualitative concept. However, new integrative techniques are becoming available that may allow definition of the process. Soil quality must be a "moving target." As new management techniques and approaches are developed, it should be possible to improve soil quality. The approach for achieving maximal crop yields must be changed. It is highly unlikely that maximal crop yields will be sustainable or promote soil quality. With current technology, these yields are possible only with high inputs of fertilizer, energy, pesticides, and irrigation. If irrigation is unavailable, there is a likelihood that pesticides and fertilizers will be overapplied. All of these practices result in environmental perturbations that can have many adverse long-term consequences.

Karlen et al. [68] suggested that the physical, chemical, and biological processes and interactions within soil govern indicators of soil quality. They also stated that the biological processes provide the resiliency and buffering capacity to ameliorate stress. They pointed out that conservation tillage, cover crops, and rotations may improve soil quality and that practices, such as alley or narrow-strip cropping, may also be beneficial.

Measuring and defining soil quality will probably require two approaches: one for policy makers and one for researchers. Each will require a hierarchical approach but with much different levels of precision. One approach, e.g., the Prevention Index approach described by Rodale Press [69], may be appropriate for the policy decision-making process. In this approach, 21 questions were developed that were presumed to be related to the general health of an individual. By polling the population, an attempt was made to develop a determination of an individual's health by the degree to which the person practiced the presumed health-promoting behavior. This approach may be sharpened as the research approach provides routes to do so. With regard to the prediction of soil quality, the index could be composed of a series of questions and answers concerning, for example, climate, soil type, soil color, crop production, pollution problems, and erosion rate. Using the hierarchical approach, relevant questions would be identified and others would be eliminated.

The research approach would involve two or three sites dedicated to a variety of integrated farming systems. All possible physical, chemical, and biological measurements of the soils should be recorded. The biological measurements, especially, will be very important. These quantitative terms can be carefully integrated to provide a quantitative index of soil quality. The following is an example of a possible progression to quantification.

First order: Soil productivity, tilth, and crop quality.

Second order: Climate, microflora, microfauna, and mesofaunal diversity, tillage resistance, compactibility, disease incidence, and crop vigor.

Third order: Soil physical, chemical, and biological properties (soil type, nutrient content, soil organic matter content, number of disease-causing propagules).

This progression merely illustrates the approach, and more measurements would probably be required. It would rapidly become apparent which measurements could be discarded and which were essential. A similar approach has been suggested by Larson and Pierce [70]; however, they did not include the biological aspects of soil quality, which are critical to the definition.

An integrated team research approach is also important. It is also critical that there is public involvement in the development of the model. This involvement will provide checks and balances to questions such as, "Is the approach achievable? Does it make sense? and Will it be accepted by the public?" These questions are simple, but they could prevent the introduction of an approach for improving soil quality that would be unacceptable to society. There is much that can be done to improve soil quality, but this will not be accomplished until the predictive tools are developed.

VIII. RESEARCH NEEDS

Several parameters of biological properties must be developed. These include microbial biomass, general composition of the soil microflora, microfauna, and mesofauna, turnover rates of different classes of chemical substrates, microbiota associated with root surfaces, soil enzyme activity, and biological effects on soil stability, infiltration, crusting, nutrient cycling relationships, soil organic matter content and activity, and soil-borne diseases. These are only several obvious measurements.

The effect of several cultural practices on these properties should be assessed. Some of these include the effect of the addition of organic materials (crop residues, sludge, contaminants), tillage practices and residue management, soil fracturing and deep fertilizer placement (such as phosphorus on eroded or degraded soils), the effects of legumes and grasses in the cropping system, differential effects of forage and seed legumes, if any, and the effects of types and rates of fertilizers and pesticides. Crop quality (such as nutritional value and chemical composition) must also be measured so that it can be factored into the soil quality equation.

When these interactions are defined, it should be possible to determine approaches for maintaining or increasing soil quality, as judged by biological indicators. It should also be possible to determine the sustainability of a cropping system. Having produced definitions and word models, it should then be possible to construct mathematical models that would allow, for example, prediction of the long-term capacity of a soil to sustain its quality.

REFERENCES

1. Papendick, R. I., and J. F. Parr. 1992. Soil quality—the key to sustainable agriculture. Amer. J. Alter. Agric. 7:2–3.

2. Elliott, L. F., and J. M. Lynch. 1994. Biodiversity and soil resilience, p. 353–364. In D. J. Greenland and I. Szabolcs (eds.), Soil Resilience and Sustainable Land Use. CAB International, Wallingford, United Kingdom.
3. Babich, H., and G. Stotzky. 1985. Heavy metal toxicity to microbe-mediated ecologic processes: A review and potential application to regulatory policies. Environ. Res. 36:111–137.
4. Martin, J. P., and S. A. Waksman. 1940. Influence of microorganisms on soil aggregation and erosion I. Soil Sci. 50:29–47.
5. Martin, J. P. 1971. Decomposition and binding action of polysaccharide in soil. Soil Biol. Biochem. 3:33–41.
6. Lynch, J. M., and L. F. Elliott. 1983. Aggregate stabilization of volcanic ash and soil during microbial degradation of straw. Appl. Environ. Microbiol. 45:1398–1401.
7. Elliott, L. F., and J. M. Lynch. 1984. The effect of available carbon and nitrogen in straw on soil and ash aggregation and acetic acid production. Plant Soil 78:335–343.
8. Brussaard, L., and M. J. Kooistra. 1993. Soil Structure/Soil Biota Interrelationships. Geoderma 56:1–648; 57:1–181.
9. Bolton, H., Jr., L. F. Elliott, R. I. Papendick, and D. F. Bezdicek. 1985. Soil microbial biomass and selected soil enzyme activities: Effect of fertilization and cropping practices on soil microbial biomass. Soil Biol. Biochem. 17:297–302.
10. Reganold, J. P., L. F. Elliott, and P. W. Unger. 1987. Long-term effects of organic and conventional farming on soil erosion. Nature 330:370–372.
11. Elliott, L. F., R. I. Papendick, and D. F. Bezdicek. 1987. Cropping practices using legumes with conservation tillage and soil benefits, p. 81–89. In J. F. Power (ed.), The Role of Legumes in Conservation Tillage Systems. Soil Conservation Society of America, Ankeny, IA.
12. Collins, H. P., P. E. Rasmussen, and C. L. Douglas, Jr. 1992. Crop rotation and residue management effects on soil carbon and microbial dynamics. Soil Sci. Soc. Amer. J. 56:783–788.
13. Rovira, A. D. 1994. The effect of farming practices on the soil biota, p. 81–87. In C. E. Pankhurst, B. M. Doube, V. V. S. R. Gupta, and P. R. Grace (eds.), Soil Biota Management in Sustainable Farming Systems. CSIRO, East Melbourne, Victoria, Australia.
14. Reganold, J. P., A. S. Palmer, J. C. Lockhart, and A. N. Macgregor. 1993. Soil quality and financial performance of biodynamic and conventional farms in New Zealand. Science 260:344–349.
15. Doran, J. W., D. C. Coleman, D. F. Bezdicek, and B. A. Stewart (eds.). 1994. Defining Soil Quality for a Sustainable Environment. American Society of Agronomy, Madison, WI.
16. Martyniuk, S., and G. H. Wagner. 1978. Quantitative and qualitative examination of soil microflora associated with different management systems. Soil Sci. 125:343–350.
17. Power, J. F., and J. W. Doran. 1984. Nitrogen use in organic farming. p. 585–598. In R. D. Hauck (ed.), Nitrogen in Crop Production. American Society of Agronony, Madison, WI.
18. Schnurer, J., M. Carholm, and T. Rosswall. 1985. Microbial biomass and activity in an agricultural soil with different organic matter contents. Soil Biol. Biochem. 17:611–618.

19. Sahs, W., and G. Lesoing. 1985. Crop rotations and manure versus agricultural chemicals in dryland grain production. J. Soil Water Conser. 40:511–516.
20. Stewart, B. A., D. A. Woolhiser, W. H. Wischmeier, J. H. Caro, and M. H. Frere. 1976. Control of Water Pollution from Cropland, Vol. II. U.S. Department of Agriculture and Environmental Protection Agency, Washington, D.C.
21. Hicks, R. J., G. Stotzky, and P. van Voris. 1990. Review and evaluation of the effects of xenobiotic chemicals on microorganisms in soil. Adv. Appl. Microbiol. 35:195–253.
22. Chaney, R. L., J. A. Ryan. 1993. Heavy metals and toxic organic pollutants in MSW-composts: Research results on phytoavailability, bioavailability, fate, etc. p. 451–506. In Harry A. J. Hoitink and Harold M. Keener (eds.), Science and Engineering of Composting: Design, Environmental, Microbiological and Utilization Aspects. Renaissance Publications, Ohio State University, Columbus.
23. Biederbeck, W. V., C. A. Campbell, and R. P. Zentner. 1984. Effect of crop rotation and fertilization on some biological properties of a loam in southwestern Saskatchewan. Can. J. Soil Sci. 64:355–367.
24. Dormarr, J. F., U. J. Pittman, and E. D. Spratt. 1979. Burning crop residues: Effect on selected soil characteristics and long-term wheat yields. Can. J. Soil Sci. 59:79–86.
25. Marston, D. 1978. The effect of two stubble management practices on runoff characteristics in arable black soils, p. 130-133. In Proceedings of the Technical Conference of the Institute of Australian Engineers, Toowoomba, Queensland, Australia.
26. Unger, P. W., R. R. Allen, and J. J. Parker. 1973. Cultural practices for irrigated winter wheat production. Soil Sci Soc. Amer. Proc. 37:437–442.
27. Rasmussen, R. E., H. P. Collins, and R. W. Smiley. 1989. Long-term management effects on soil productivity and crop yield in semi-arid regions of eastern Oregon. Oregon Agricultural Experiment Station Bulletin 675. Oregon State University and USDA Agricultural Research Service, Pendleton.
28. Coleman, D. C., P. F. Hendrix, M. H. Beare, D. A. Crossley, S. Hu, and P. J. van Vliet. 1994. The impacts of management and biota on nutrient dynamics and soil structure in sub-tropical agroecosystems: Impacts on detritus food webs, p. 133–143. In C. E. Pankhurst, B. M. Doube, V. V. S. R. Gupta, and P. R. Grace (eds.), Soil Biota Management in Sustainable Farming Systems. CSIRO, East Melbourne, Victoria, Australia.
29. Papendick, R. I. 1991. Report on the International Conference on the Assessment and Monitoring of Soil Quality. Washington State University Press, Pullman, WA.
30. Singh, K. K., T. S. Colvin, D. C. Erbach, and A. Q. Mughal. 1992. Tilth index: An approach to quantifying soil tilth. Tran. Amer. Soc. Agricul. Eng. 35:1777–1785.
31. Smith, J. L., J. J. Halvorson, and R. I. Papendick. 1993. Using multiple-variable indicator kriging for evaluating soil quality. Soil Sci. Soc. Amer. J. 57:743–749.
32. Papendick, R. I., and L. F. Elliott. 1983. Soil physical factors that affect plant health, p. 168–179. In T. Kommendahl and P. H. Williams (eds.), Challenging Problems in Plant Health. American Phytopathology Society, St. Paul MI.
33. Doran, J. W., M. Sarrantonio, and R. Janke. 1994. Strategies to Promote Soil Quality and Health, p. 230–247. In C. D. Pankhurst, B. M. Doube, V. V. S. R. Gupta, and

P. R. Grace (eds.), Soil Biota Management in Sustainable Farming Systems. CSIRO, East Melbourne, Victoria, Australia.

34. Lynch, J. M., and E. Bragg. 1985. Micro-organisms and soil aggregate stability. In Advances in Soil Science, vol. 2, p. 133–170. Springer-Verlag, New York.
35. Sumner, M. E., and W. P. Miller. 1992. Soil crusting in relation to global soil degradation. Amer. J. Altern. Agricul. 7:56–62.
36. Logsdon, S. D., J. K. Radke, and D. L. Karlen. 1993. Comparison of alternative farming systems. I. Infiltration techniques. Amer. J. Altern. Agricul. 8:15–20.
37. Jordahl, J. L., and D. L. Karlen. 1993. Comparison of alternative farming systems. III. Soil aggregate stability. Amer J. Altern. Agricul. 8:27–33.
38. Arshad, M. A., and G. M. Coen. 1992. Characterization of soil quality: Physical and chemical criteria. Amer. J. Altern. Agricul. 7:25–30.
39. Stevenson, F. J. 1982. Humus Chemistry. John Wiley & Sons, New York.
40. Haider, K. 1992. Problems related to the humification processes in soils of temperate climates, p. 55–94. In G. Stotzky and J. M. Bollag (eds.), Soil Biochemistry. Marcel Dekker, New York.
41. Campbell, C. A., E. A. Paul, and W. B. McGill. 1976. Effect of cultivation and cropping on the amounts and forms of soil N, p. 7–101. In Western Canadian Nitrogen Symposium Proceedings, Calgary, Alberta.
42. Smith, J. L., and L. F. Elliott. 1990. Tillage and residue management effects on soil organic matter dynamics in semiarid regions, p. 69–88. In Advances in Soil Science. Springer-Verlag, New York.
43. Smith, V. T., L. C. Wheeting, and S. C. Vandecaveye. 1946. Effects of organic residues and nitrogen fertilizers on a semiarid soil. Soil Sci. 61:393–410.
44. Karlen, D. L., and D. E. Stott. 1994. A framework for evaluating physical and chemical indicators of soil quality, p. 53–72. In J. W. Doran, D. C. Coleman, D. F. Bezdicek, and B. A. Stewart (eds.), Defining Soil Quality for a Sustainable Environment. SSSA Special Publication No. 35. American Society for Agronomy, Madison, WI.
45. Parkinson, D., and D. C. Coleman. 1991. Methods for assessing soil microbial populations, activity and biomass. Microbial communities, activity and biomass. Agricul. Ecosys. Environ. 34:3–33.
46. Horwath, W. R., and E. A. Paul. 1994. Microbial biomass, p. 753–773. In R. W. Weaver, J. S. Ande, and P. S. Bottomley (eds.), Methods of Soil Analysis, part II, Microbiological and Biochemical Properties. SSSA Book Series 5. American Society for Agronomy, Madison, WI.
47. Sayler, G. S., K. N. Nikbakht, J. T. Fleming, and J. Packard. 1992. Applications of molecular techniques to soil biochemistry, p. 131–172. In G. Stotzky and J.-M. Bollag (eds.), *Soil Biochemistry*, Vol. 7, Marcel Dekker, New York.
48. Carter, J. P., and J. M. Lynch. 1993. Immunological and molecular techniques for studying the dynamics of microbial populations and communities in soil, p. 249–272. In J.-M. Bollag and G. Stotzky (eds.), Soil Biochemistry, Vol. 8, Marcel Dekker, New York.
49. Hassink, J., G. Lebbink, and J. A. van Veen. 1991. Microbial biomass and activity of a reclaimed-polder soil under a conventional or a reduced-input farming system. Soil Biol. Biochem. 23:507–513.

50. Hassink, J., J. H. Oude Voshaar, E. H. Nijhuis, and J. A. van Veen. 1991. Dynamics of the microbial populations of a reclaimed-polder soil under a conventional and a reduced-input farming system. Soil Biol. Biochem. 23:515–524.
51. Brussaard, L., L. A. Bouwman, M. Geurs, J. Hassink, and K. B. Zwart. 1990. Biomass, composition and temporal dynamics of soil organisms of a silt loam soil under conventional and integrated management. Netherlands J. Agricul. Sci. 38:283–302.
52. Granatstein, D. M., D. F. Bezdicek, V. L. Cochran, L. F. Elliott, and J. L. Hammel. 1987. Effect of long-term tillage and rotation practices on soil microbial biomass, carbon, and nitrogen. Biol. Fert. Soils 5:265–278.
53. Insam, H. 1990. Are the soil microbial biomass and basal respiration governed by the climatic regime? Soil Biol. Biochem. 22:525–532.
54. Insam, H., D. Parkinson, and K. H. Domsch. 1989. Influence of microclimate on soil microbial biomass. Soil Biol. Biochem. 21:211–221.
55. Cook, R. J. 1994. Introduction of soil organisms to control root diseases, p. 13–22. In C. E. Pankhurst, B. M. Doube, V. V. S. R. Gupta, and P. R. Grace (eds.), Soil Biota Management in Sustainable Farming Systems. CSIRO, East Melbourne, Victoria, Australia.
56. Elliott, L. F., and R. E. Wildung. 1992. What biotechnology means for soil and water conservation. J. Soil Water Conserv. 47:17–20.
57. Rovira, A. D., L. F. Elliott, and R. J. Cook. 1990. The impact of cropping systems on rhizosphere organisms affecting plant health, p. 389–436. In J. M. Lynch, (ed.), The Rhizosphere. John Willey & Sons, Chichester.
58. Visser, S., and D. Parkinson. 1992. Soil biological criteria as indicators of soil quality: Soil microorganisms. Amer. J. Altern. Agricul. 7:33–37.
59. Domsch, K. H., G. Jagnow, and T. H. Anderson. 1983. An ecological concept for the assessment of side-effects of agrochemicals on soil microorganisms. Residue Rev. 86:65–105.
60. Bolton, E. F., and L. R. Webber. 1952. Effect of cropping systems on the aggregation of a Brookston clay soil at three depths. Sci. Agricul. 32:555–558.
61. Bruce, R. R., S. R. Wilkinson, and G. W. Langdale. 1987. Legume effects on soil erosion and productivity, p. 127–138. In J. F. Power (ed.), The Role of Legumes in Conservation Tillage Systems. Soil Conservation Society of America. Ankeny, IA.
62. Pankhurst, C. E., and J. M. Lynch. 1994. The role of the soil biota in sustainable agriculture, p. 3–9. In C. E. Pankhurst, B. M. Doube, V. V. S. R. Gupta, and P. R. Grace (eds.), Soil Biota Management in Sustainable Farming Systems. CSIRO, East Melbourne, Victoria, Australia.
63. Gupta, V. V. S. R. 1994. The impact of soil and crop management practices on the dynamics of soil micro-fauna and meso-fauna, p. 107–124. In C. E. Pankhurst, B. M. Doube, V. V. S. R. Gupta, and P. R. Grace (eds.), Soil Biota Management in Sustainable Farming Systems, CSIRO, East Melbourne, Victoria, Australia.
64. Doran, J. W. 1980a. Microbial changes associated with residue management and reduced tillage. Soil Sci. Soc. Amer. J. 44:518–524.
65. Doran, J. W. 1980b. Soil microbial and biochemical changes associated with reduced tillage. Soil Sci. Soc. Amer. J. 44:765–771.

66. Tilman, D., and J. A. Downing. 1994. Biodiversity and stability in grasslands. Nature 367:363–365.
67. Stork, N. E., and P. Eggleton. 1992. Invertebrates as determinants and indicators of soil quality. Amer. J. Altern. Agricul. 7:38–47.
68. Karlen, D. L., N. S. Nash, and P. W. Unger. 1992. Soil and crop management effects on soil quality indicators. Amer. J. Altern. Agricul. 7:48–55.
69. The Prevention Index. 1992. Rodale Press, Emmaus, PA.
70. Larson, W. E., and F. J. Pierce. 1994. Dynamics of soil quality as a measure of sustainable management, p. 37–51. In J. W. Doran, D. C. Cokman, D. F. Bezdicek, and B. A. Stewart (eds.), Defining Soil Quality for a Sustainable Environment. SSSA Special Publication. American Society for Agronomy. Madison, WI.

2
Soil Structure and Biological Activity

Jeffrey N. Ladd and Ralph C. Foster Commonwealth Scientific and Industrial Research Organization, Glen Osmond, South Australia, Australia

Paolo Nannipieri Università degli Studi, Florence, Italy

J. Malcolm Oades University of Adelaide, Glen Osmond, South Australia, Australia

I. INTRODUCTION

Almost 30 years ago, in the Preface to the first volume of the book series *Soil Biochemistry*, McLaren and Peterson [1] reflected on the complexity of the soil matrix as a medium for biological and biochemical activity. Not only does the soil offer a great diversity of decomposer organisms and substrates, but the organisms and their enzymes function in spatially heterogeneous environments, a "spectrum of microenvironments" characterized by soil matrix induced variations in substrate concentrations, water availability, and gradients in pH and redox potential.

Earlier studies had established that soil organic matter decomposed on average relatively slowly, despite substantial proportions of the organic matter being composed of carbohydrates and (peptide-linked) amino acids. Soil organic matter was considered to be mainly of microbial origin as a result of the growth and turnover of decomposer organisms, and there was conjecture on the nature and importance of physical mechanisms of protection and stabilization of organic residues within the soil matrix.

A focus on the influence of soil structure on biological activity has been maintained through studies of (1) the properties of macro- and microaggregates, (2) the behavior of substrates, enzymes, and microorganisms adsorbed on surfaces, (3) the dynamics of populations of soil biota and the linking of predator-prey interactions with soil biological activities, and (4) simulation modeling of carbon and nitrogen turnover through decomposer organisms, which collectively have been measured and modeled as a small labile organic pool, i.e., the microbial biomass. Recent reviews testify to the continued widespread interest in soil microenvironments and the mechanisms by which processes of decomposition and turnover are physically constrained within soil [2–13].

In considering biological activity in relation to the properties of the soil matrix, the following view has been taken.

1. Substrates in soil are predominantly chemically complex. Fresh plant and animal residues are comprised of varying proportions of particulate and soluble components, which determine the extent to which such residues are initially rapidly decomposed upon entering soil.
2. The majority of substrates in soils are particulate and can only be colonized by organisms on their outer surfaces. Thus, particle size and surface area of the substrates control their rates of decomposition. Coating of substrate surfaces with inorganic materials confers on particulate substrates various degrees of protection from decomposition.
3. The soil microbiota are the major agents of decomposition; the activities of phyllosphere and rhizosphere microbiota, and of soil enzymes, are also significant at times and under suitable conditions. In addition to rhizospheres, decomposing organic particles represent concentrations of biological activity in the soil matrix. These "hot spots" of activity are remote from each other with respect to the vast surfaces present in soils, and for microorganisms, they represent oases in an otherwise barren landscape.
4. Microbial biomass is the major organic product of substrate metabolism, and biomass turnover is the major mechanism for the accumulation of dead microbial residues.
5. The rates of decomposition of carbon and nitrogen of microbial biomass and products are influenced by their location in the soil. During the early phases of decomposition, biomass associated with particulate residues is less protected than biomass formed from the metabolism of soluble substrates that have diffused into the soil matrix. Clay soils have a greater capacity for the protection of biomass within the soil matrix than do sands.
6. Interactions of microbial metabolites and products with clay surfaces stabilize the organic materials. Particle-size fractionation of dispersed soils commonly reveals a concentration of soil organic matter and microbial biomass in fine silt-size/coarse clay-size materials. Clay fractions isolated from soils may be artifacts that contain both very old and modern carbon.

7. There is no simple single measure of soil structure. In addition, soil structure is dynamic. In many soils, the dynamics of organic matter turnover and soil structure are inexorably linked.

In this chapter, an attempt is made to unravel some of these links by proposing various conceptual approaches to an understanding of soil structure and biological activity across wide ranges of scale and time.

II. SOIL STRUCTURE: DEFINITION AND DESCRIPTION

A. Definition and Properties

The simplest generic definition of soil structure is the size, shape, and arrangement of particles in soils. This simultaneously defines the size, shape, and arrangement of pores in soils, which represent the space in which all the biological activity in soils takes place. It includes the continuity and connectivity of pores, which is so important for air/water relations in soil and, consequently, for biological activity from root growth, hyphal growth, and the mobility of a range of organisms.

Soil structure is not static and changes with water status as the soil shrinks on drying and swells on wetting. The structure is altered by various forms of management. Large pores shrink under compression or compaction caused by vehicles or animals, particularly ungulates. Large pores may be created by cultivation, but they are also infilled by particles released from aggregates by external and internal sources of energy, such as cultivation and rapid wetting, respectively.

Biological activity is involved in both the stabilization and degradation of soil structure. A well-structured soil encourages biological activity, which helps to maintain favorable structural conditions.

Discussions of soil structure should include references to scale, because structure applies across some nine orders of magnitude, and some means of dealing with this vast scale are necessary to ensure consistent terminology and to avoid fruitless arguments. Waters and Oades [14] prepared an overview of the relations of particles, aggregates, pores, and biota with scale (Figure 1).

The various particles described under Particles in Figure 1 are the result of a complete disaggregation of soils after oxidation of organic materials, including organisms, and dispersion of clay particles, aided by saturation of the soil with sodium. While this represents a standard procedure for determining the particle size distribution of a soil, it may give a false impression with respect to the way that soils behave naturally. (Texturing of soils by hand may be closer to reality than the classical "mechanical analysis.")

The degree to which these various particles are aggregated in soils is outlined under Aggregations in Figure 1. In some soils there appears to be an

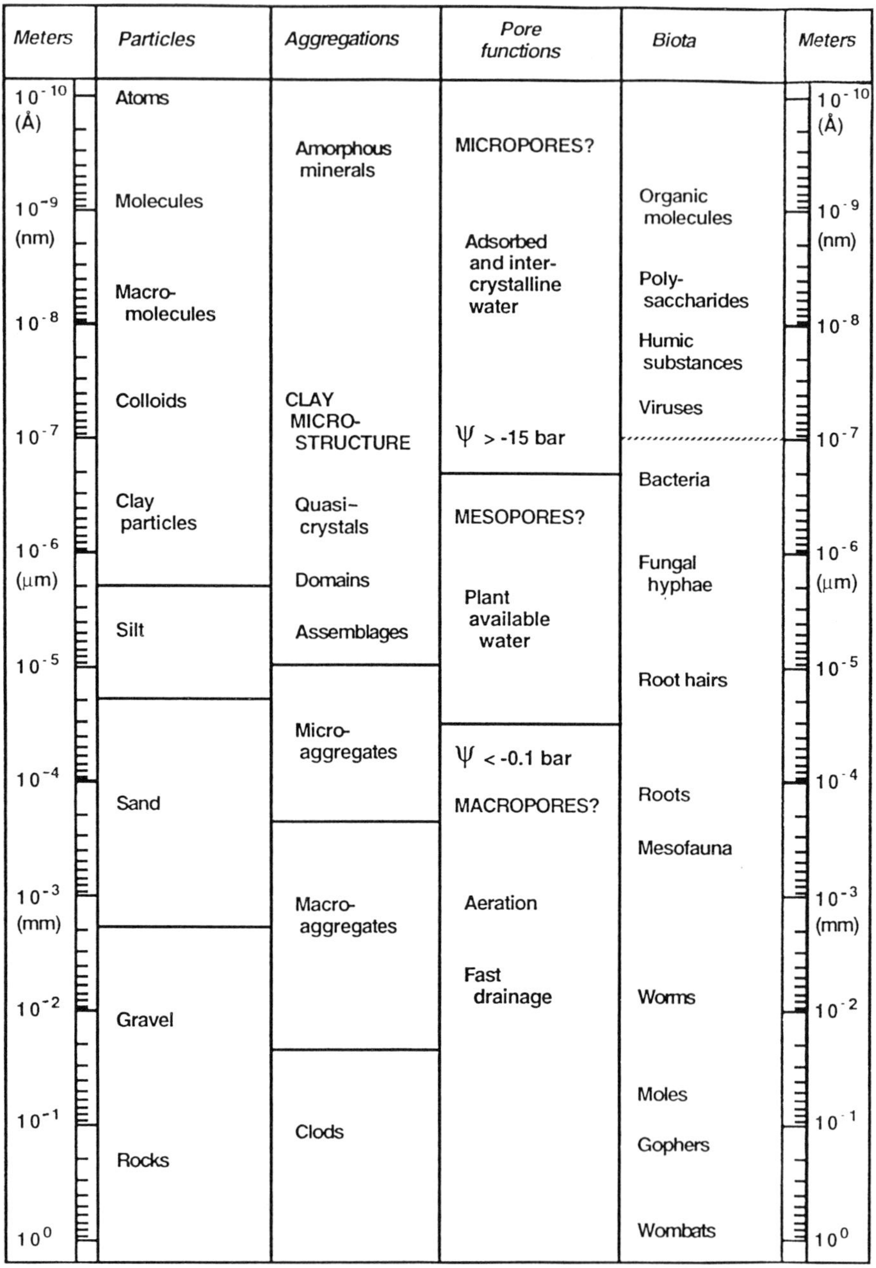

Meters	Particles	Aggregations	Pore functions	Biota	Meters
10^{-10} (Å)	Atoms	Amorphous minerals	MICROPORES?		10^{-10} (Å)
10^{-9} (nm)	Molecules		Adsorbed and inter-crystalline water	Organic molecules	10^{-9} (nm)
	Macro-molecules			Poly-saccharides	
10^{-8}				Humic substances	10^{-8}
	Colloids	CLAY MICRO-STRUCTURE	ψ > -15 bar	Viruses	
10^{-7}	Clay particles	Quasi-crystals	MESOPORES?	Bacteria	10^{-7}
10^{-6} (μm)		Domains	Plant available water	Fungal hyphae	10^{-6} (μm)
	Silt	Assemblages			
10^{-5}		Micro-aggregates		Root hairs	10^{-5}
			ψ < -0.1 bar		
10^{-4}	Sand		MACROPORES?	Roots	10^{-4}
				Mesofauna	
10^{-3} (mm)		Macro-aggregates	Aeration		10^{-3} (mm)
			Fast drainage	Worms	
10^{-2}	Gravel				10^{-2}
				Moles	
10^{-1}	Rocks	Clods		Gophers	10^{-1}
10^{0}				Wombats	10^{0}

Figure 1 Scale in soil structure (from Ref. 14).

aggregate hierarchy, which in the secondary and tertiary stages is related to soil biology. Proposals by Edwards and Bremner [15] were developed by Tisdall and Oades [16], who showed that a separation of micro- and macroaggregates could be made at a size of 250 μm in diameter. Microaggregates were shown to be relatively stable, because they are held together by a range of mechanisms, the most important of which are polysaccharide-based glues (sometimes referred to as mucigel or mucilages) and "calcium bridges." These mucilages are known to be produced by plant roots, by some fungal hyphae, and by some bacteria as extracellular slimes, which help attach organisms to surfaces and attach fine clay particles to the surfaces of the mucilages. This is done in the presence of a supply of calcium ions, which maintain clays in a stable (flocculated) state and crosslink organic functional groups, such as carboxyl and hydroxyl groups, with clay surfaces. Calcium ions may also crosslink functional groups on biopolymers and render them less soluble and, thus, more stable with respect to decomposition. Sodium has the opposite effect and renders organic biopolymers more soluble and all colloidal materials, both organic and inorganic, more dispersible. Mucilages appear to be ubiquitous in soils that are biologically active, and they have been demonstrated elegantly by various metal-binding techniques in preparation for transmission electron microscopy (see Section II. C).

In soils under grassland or where annual organic inputs are large, the microaggregates are bound together by a network of roots and hyphae. The roots and hyphae form a network of materials with extracellular mucilages on their surfaces and act as a "sticky string bag" that stabilizes aggregates with diameters of a few millimeters (macroaggregates) [17].

Macroaggregates (>250 μm) and microaggregates (<250 μm) are readily separated and measured by various wet-sieving techniques, and it has been demonstrated many times that under grassland, the proportion of macroaggregates stable to rapid wetting increases with time. This leads to greater macroporosity and a range of benefits for plant growth and crop production. Under frequent cultivation and monoculture, the networks of roots and hyphae are decomposed, and the proportion of the soil present as water-stable macroaggregates decreases. Thus, aggregate stability in the size range of a few millimeters is dependent on soil management. Figure 2 shows the general trends obtained for water-stable aggregation as controlled by the organic carbon content of soil, which was controlled by the years of pasture in various pasture/wheat rotations in a mediterranean environment.

The term slaking is used for the breakdown of macroaggregates caused by rapid wetting, which leads to the formation of microaggregates but not to the discrete particles listed under Particles in Figure 1. Slaking is easily demonstrated when a dry aggregate is placed in good quality water. Two phenomena are obvious: the aggregate appears to crumble into a heap of smaller aggregates, and simultaneously, bubbles of air can be seen escaping from the disintegrating

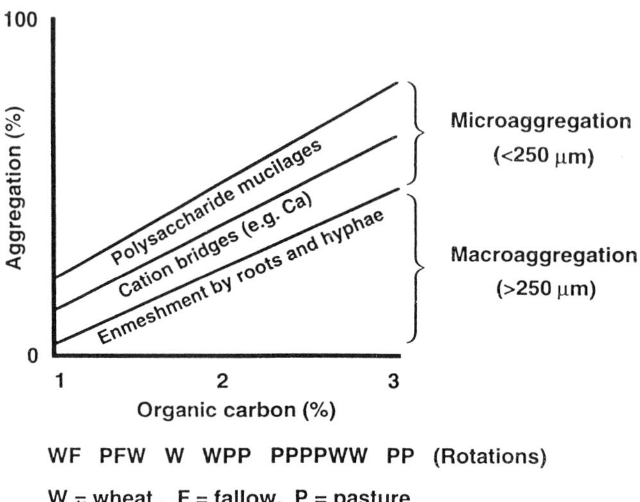

Figure 2 Effect of carbon content and crop rotation on aggregation in an Alfisol (from Ref. 16).

aggregate. The instability is caused by rapid entry of water into pores that do not traverse the aggregate. This traps air, which escapes vigorously when the pressure created in the pores exceeds the strength of the aggregate. Simultaneously, pressures created by unequal swelling within the wetted aggregate also aid the crumbling process. Factors that slow the wetting process help prevent slaking [18].

The stability of microaggregates prevents the complete disaggregation of soils and, in cultivated, cropping situations, allows soils to be managed with care to avoid runoff of rainfall and erosion. However, microaggregates will also eventually become unstable if organic matter reserves are severely depleted. Under these conditions, microaggregates may disperse and release clay particles under the impact of rainfall and rapid wetting.

Similarly, the development of sodicity (i.e., the exchange of calcium by sodium), even to a small extent, e.g., an exchangeable sodium percentage of less than 10, will lead to some clay dispersion and instability of microaggregates. Generally the development of sodicity and the instability of structural hierarchy are expected to accelerate decomposition processes by exposing protected organic debris and by rendering organic materials more dispersible and more soluble. Similar principles would apply to the dispersion of colonies of microorganisms.

What may be referred to as clay microstructure or the first stage of aggregation of clay particles is probably not influenced by biological activities. Simpli-

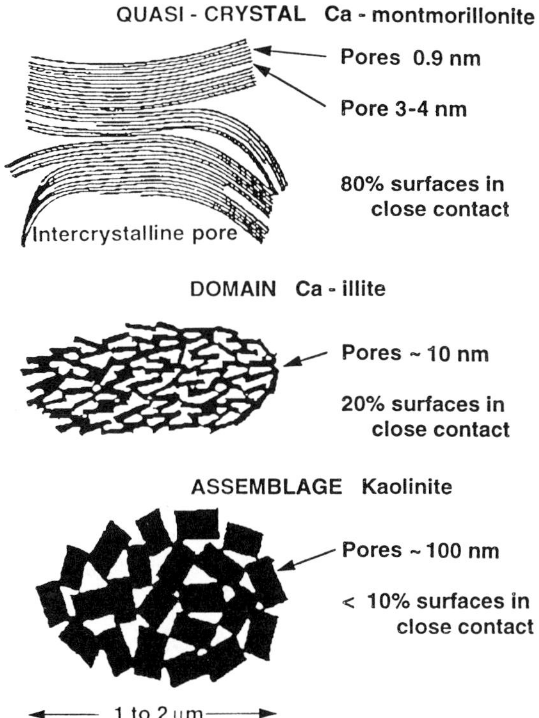

Figure 3 Packing of clay particles in primary aggregates (from Ref. 19).

fied schematic representations of clay microstructure (Figure 3) indicate an increasing degree of ordered packing that is controlled by the morphology of individual clay particles or platelets [19]. These clay aggregates are stable entities in soils, and even in laboratory situations, it is difficult to separate the individual plates from each other, i.e., to disperse completely the individual clay particles.

Bearing in mind the net negative charge on clay particles and the adsorptive capacities of clay surfaces, it seems unlikely that modern organic compounds will penetrate these units, although they are involved in exchange reactions with inorganic ions. Thus, the smallest pores accessible to organic materials in soils are probably those created between the clay aggregates. Packets of clay plates with mucilage between the packets are evident in transmission electron micrographs of soils containing smectites [20].

The location of substrates and organisms in and around aggregates of various sizes is discussed later (see Sections II. C and II. D).

B. Formation, Stabilization, and Degradation of Structure

Structure is dynamic and changes with the water content of soil; and because structural stability is related to organic materials, it changes with season in many soils. Oades [13] suggested that studies of soil structure and biota should recognize three major groups of soils—sands, loams, and clays—and that these distinctions should be based on the shrink-swell capacity of the soil [coefficient of linear extensibility, (COLE)] rather than on a traditional particle-size distribution. The major factor involved in the formation of soil aggregation is the swelling and shrinking of soil materials during wetting-drying cycles. On drying of a uniform soil material with a significant COLE, a series of cracks develop at regular intervals. Further wet-dry cycles may accentuate the process, and eventually secondary and tertiary cracks develop leading to a three-dimensional crack arrangement that defines soil aggregates. The forces involved in wetting and drying of soils are large and, in loams and clays, are the major factor in the creation of structure and aggregation.

Plants are involved indirectly in aggregate formation in that they help to dry the soil at many sites, thus creating areas of weakness (wet areas) and strength (dry areas) throughout the soil. Soils containing bushy root systems generally exhibit a high degree of aggregation as a result of site-specific drying and wetting of the soil. In sands, aggregation of particles cannot occur by this mechanism; the effect is maximal in clay soils. These changes in structure are, thus, dominantly abiotic and are controlled by the water regime.

Biota are strongly involved in stabilizing aggregates, both micro- and macroaggregates. At the microaggregate level, stabilization is brought about largely by mucilages around hyphae, bacterial colonies, and roots, which glue together microaggregates. At the larger scale, roots and hyphae help enmesh the microaggregates in larger compound units. The relative importance of these processes in the three classes of soil is shown in Table 1.

C. Location and Characterization of Substrates

Microscopy and Histochemistry

Environmental scanning electron miscroscopy (ESEM) has the advantage that uncoated, moist soils can be examined. When applied to soil aggregates this technique shows that aggregates are frequently interconnected by soil organic matter in the form of roots, fungi, and tissue fragments [21,22]. Roots may be bonded to aggregates by their external gels [23], by the penetration of laterals or root hairs into aggregates, or by fungi, especially mycorrhizal fungi [16,24].

Transmission electron microscopy (TEM) of sections shows the interrelationships between the various kinds of soil organic matter and the mineral and biological components of microaggregates [5,21,22,25,26]. The most obvious organic deposits are remnants of plant tissues. These are both numerous and

Table 1 Biotic and Abiotic Influences on Soil Structure

	Sand (<15% clay)	Loam (15–35% clay)	Clay (>35% clay)
Shrink-swell capacity	Minimum	Important	Maximum →
Abiotic aggregate formation	Minimum	Important	Maximum →
Biotic influences	Aggregate stabilization and degradation	Aggregate stabilization and degradation	Minimal (biopores?)

extensive in surface soils (1 to 5 cm deep), especially in aggregates from undisturbed (never plowed) sites [21]. Aggregates from these soils are largely composed of organic residues in which the cellular ultrastructure is still recognizable [22]. The characteristic multilaminate structure of the cell walls is often still apparent [27], and they do not stain strongly with OsO_4. Since carbohydrates do not react with OsO_4 [28], this indicates that there is still a lot of carbohydrate present. Some fresh material displays plasmadesmata. Root fragments (Figure 4) and three-dimensional sheets of cell wall material closely bind nearby minerals, especially clay (Figures 5,6), and link other more distant (>20–30 μm) aggregate components [5].

Bacteria are often largely confined to the periphery of these remnants of plant tissues. The lack of access of microorganisms to the interior of tissues appears to delay their decomposition, as such large deposits of tissue are seldom observed in the aggregates from tilled sites [21,22]. Cellular debris in microaggregates from long-term (>100 years) tilled sites often consists of fragments of wall from individual cells rather than of pieces of tissue, and these fragments stain densely with OsO_4, suggesting that the carbohydrate had been lysed by enzymes from nearby microorganisms to leave more recalcitrant fractions of the cell walls.

Most soils contain organic deposits with no recognizable substructure (Figure 6) and which become highly electron-dense as their size decreases until it is impossible to determine their origin [29]. Such amorphous deposits are common in worm casts [30]. Treatment of cell wall fractions and pure organics with heavy metals commonly used as stains in electron microscopy suggest that such highly electron-dense materials are rich in lignin residues, certain amino acids, and polyphenolics [28]. Inasmuch as polyphenols inhibit many enzymes, these densely stained deposits probably constitute part of the chemically stable fraction of the soil organic matter.

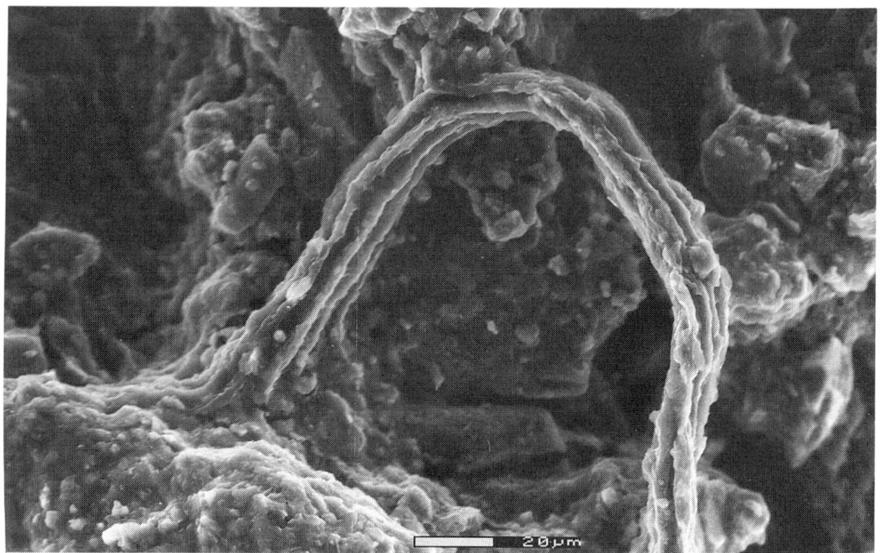

Figure 4 Environmental scanning electron micrograph of an aggregate surface showing a root fragment linking together four microaggregates.

Microaggregates isolated by centrifugation from gently dispersed soil often have an organic matter core, which consists of either amorphous polysaccharide gel or plant cell wall debris [12,22,26]. In electron probe microanalysis, these gels produce strong peaks for osmium, uranium, and lead, the heavy metals used to stain sections for TEM. All soil organic matter tested, even root surface mucigels near the root apex, give peaks also for aluminum and silicon. These amorphous gels, reactive to the ruthenium red-osmium tetroxide (Ru/Os) reagent [29], are similar to those found at root surfaces or secreted by soil microorganisms as sheaths of extracellular polysaccharide and provide short-range bonding (1 to 3 μm) between soil constituents. Cell wall fragments also form the organic core of other microaggregates. Such mineral coated organics are unlikely to be accessed by decomposer microorganisms and may constitute part of the physically protected soil organic matter.

These clay cutans that enclose organic materials may consist of a few layers of clay platelets [20], which exhibit face-to-face orientation at the surface of the organic fragments. However, it is possible that such oriented clays on organic materials, as observed in isolated microaggregates, are an artifact of the preparation technique. When ultrathin sections of whole aggregates are examined by TEM, clay platelets enclosing organic fragments are generally not well oriented face-to-face at the organic surface (Figure 7).

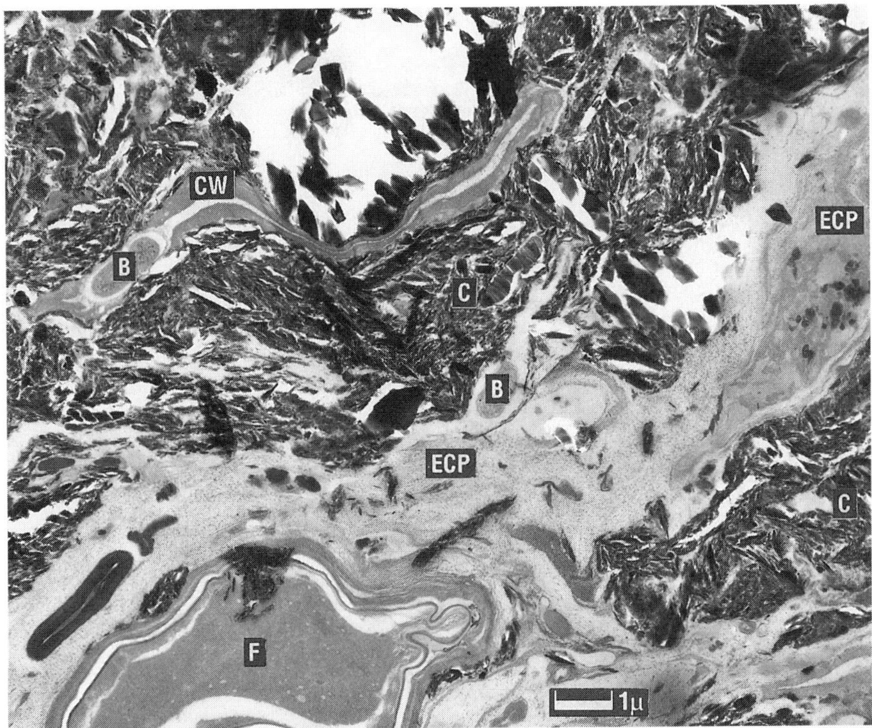

Figure 5 Transmission electron micrograph of an ultrathin section of a soil aggregate showing clay microaggregates (C) linked by ruthenium/osmium-stained, extracellular polysaccharide (ECP) of fungal hyphae (F). The wall of a collapsed cell (CW) is being lysed by a bacterium (B).

Chemical Characterization of Substrates

All natural organic materials are potentially mineralizable. If this were not the case, there would be accumulations in soils of specific organic chemicals, which would be readily observable in old soils. Existing information indicates that even the organic matter at depth (which has mean residence times measured in thousands of years), has similar organic chemistry to more recent organic carbon (E. M. M. Fongers, personal communication).

The development and applications of cross-polarization, magic-angle spinning, ^{13}C nuclear magnetic resonance spectroscopy (CPMAS ^{13}C NMR) in the last decade have yielded total chemical signatures of complex mixtures of biopolymers, without recourse to chemical pretreatments or extractions. Thus, although there are still questions about the quantitative nature of NMR spectra,

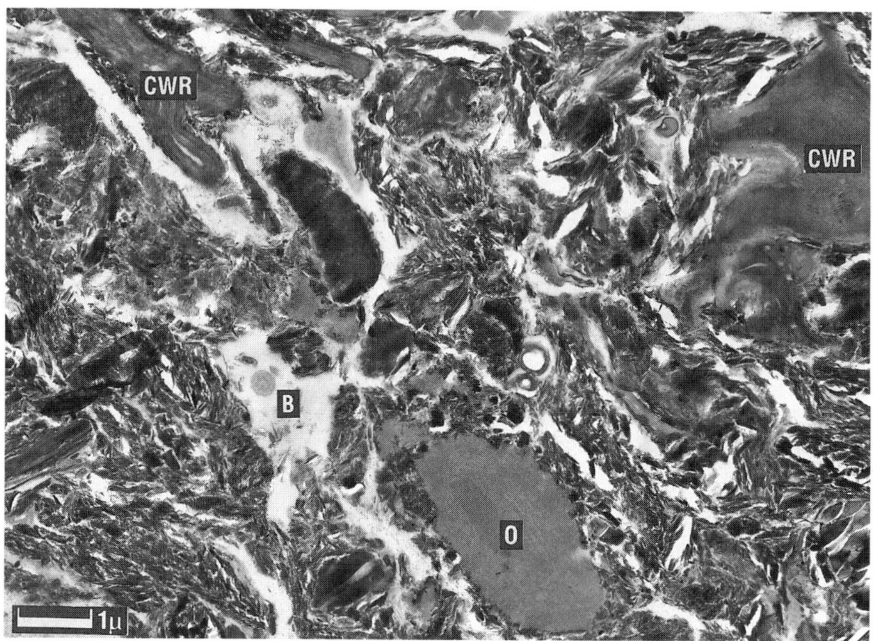

Figure 6 Transmission electron micrograph of an ultrathin section of a soil aggregate showing clay (C), bacteria (B), cell wall remnants (CWR), and amorphous, highly humified organic matter (O).

the problems of artifact formation during interactions of soils with reactive chemicals are avoided. Consequently it is now feasible to make gross comparisons of the total organic chemistry of different soils, of soils under different management regimes, and of different fractions of soils obtained by physical or chemical methods.

Studies of the dynamics of carbon in soils involving extractive chemical procedures have not provided much useful information, and in the last decade, fractionations of organic matter in soils by physical methods have become popular. There are several reviews that cover the methodology and reasons for physical fraction procedures. Oades and Ladd [3] suggested that approaches to fractionation of soils should attempt to isolate pools or fractions that were biologically meaningful. Turchenek and Oades [32] described studies of organomineral associations using sedimentation and density techniques, and these approaches have been reviewed in detail by Oades [33], Stevenson and Elliott [34], and Christensen [9,35]. The major problem associated with studies of soil organic materials in fractions obtained from soils by physical processes is

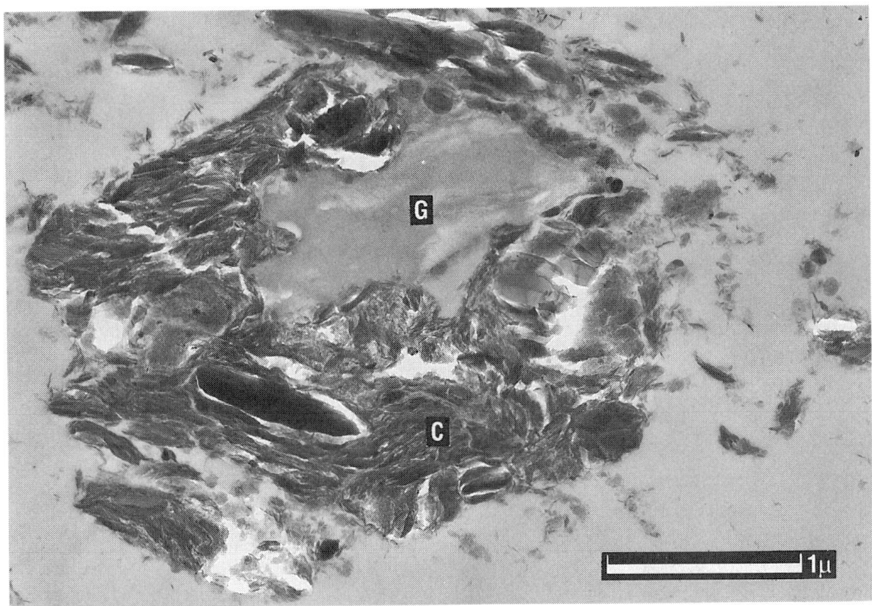

Figure 7 Transmission electron micrograph of an ultrathin section of a soil aggregate showing ruthenium/osmium-stained gel (G) enclosed in clay (C).

the degree of disaggregation of the soil before particles are separated by sedimentation procedures using water or various heavy liquids.

The chemistry of soil organic matter begins with plant materials, which are eventually converted to microbial biomass and the products of microbial turnover. Microbial cells and cell debris and products are intimately associated mainly with clay fractions (see Section II.D), which provide most of the surface available for interactions with microorganisms. In contrast, plant debris is mainly in large pieces not associated with the mineral fraction and, therefore, is of a density close to 1 Mg/m^3. Substantial portions of plant debris can be separated from the rest of the soil organic matter by flotation on water; by sieving out of larger, lighter particles; and by winnowing from sandy soils.

Although plant debris free from inorganic materials can be readily separated from soil, further separations require the input of energy to yield fractions that are chemically different. Ultrasonic energy has often been used to promote disaggregation of soils, but ball mills, macerators, and other means can be used. The difficulty with all these procedures is to standardize the methods and to compare the energy inputs with other methods.

Soils have different structural stabilities and react differently to disaggregating procedures. For example, some coarse-textured and slightly acidic soils,

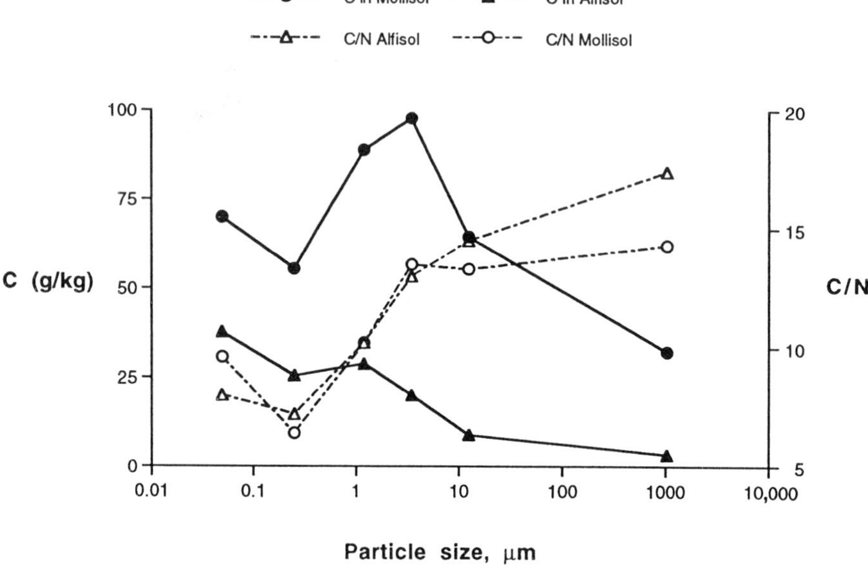

Figure 8 Particle size distributions of an Alfisol and a Mollisol after a 5-min treatment with ultrasonic energy (from Ref. 32).

such as Xeralfs, are usually dispersed thoroughly using ultrasonic energy. Other heavier textured, neutral soils, such as Mollisols, are more difficult to disperse, and even after several minutes of treatment using ultrasonic energy, particles in the size range from about 1 to 20 μm remain aggregated. The differences between the two soil orders are illustrated in Figure 8, which shows that in an Alfisol, the carbon content is greatest in the finest fraction, but that in a Mollisol, the carbon content is highest in particles in the size range of about 1 to 20 μm in diameter. In both soils, the carbon:nitrogen (C:N) ratios are about 16 in the coarse fractions and 8 or less in the clay-size fractions.

Differences in the stability of soils are also illustrated in diagrams that show particle size distributions with increasing inputs of energy for disruption (Figure 9). The graphs illustrate the concept of aggregate hierarchy in an Alfisol and a Mollisol, but not in an Oxisol. In the Alfisol and the Mollisol, larger aggregates were broken down into smaller aggregates with little release of clay particles. The Oxisol was the most stable soil and resisted disaggregation even after 16 h of shaking. Only small amounts of smaller aggregates were produced after 16 h of shaking, but the input of ultrasonic energy showed that the Oxisol contained more than 60% clay.

The chemistry of the organic materials in this range of particle-size fractions varies systematically from coarse to fine particles and, in general terms, reflects

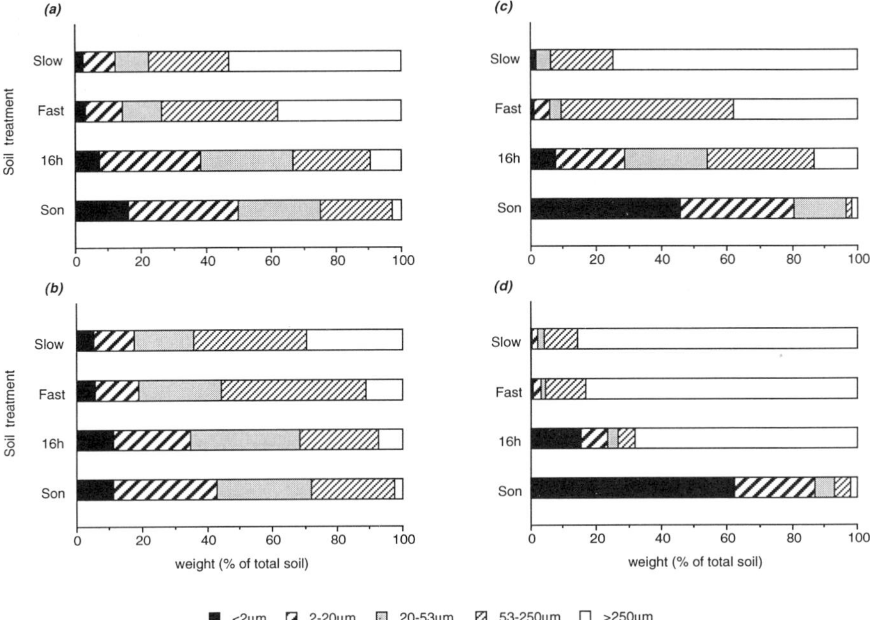

Figure 9 Particle size distributions of an (a) Alfisol (cropped), (b) Alfisol (pasture), (c) Mollisol, and (d) Oxisol using increasing intensity of aggregate disruption; slow—slow wetting under suction, fast—wetting by immersion in water, 16h—16h end-over-end shaking, son—5 min ultrasonic treatment (from Ref. 8).

a change from a chemistry dominated by plant materials to the chemistry of microbial biomass and metabolites. In general, coarse fragments are dominated by plant carbohydrates, mainly cellulose; silt-size particles show significant amounts of aromatic carbon; and clay-size fractions are rich in aliphatic components. Baldock et al. [36] suggested that the ratio of carbohydrate to polymethylene materials could be used to assess the degree of decomposition of organic materials and as a measure of their degree of bioavailability. Polymethylene materials contain $-(CH_2)n$-, where n can be 1 to more than 30, and include all alkyl chains. Baldock et al. [36] proposed a simplified model to describe, in chemical terms, the oxidative decomposition of plant materials based on data from a Mollisol (Figure 10).

Two important points need consideration in studies of the organic materials in different particle-size fractions. Firstly, the C:N ratios of particulate organic material in soils are relatively low. Plant materials added to soils usually have C:N ratios exceeding 30. However, most "light fractions" or coarse plant debris isolated from soils have already undergone a two or three-fold concentration of nitrogen during the early stages of decomposition. Secondly, the finer particle-

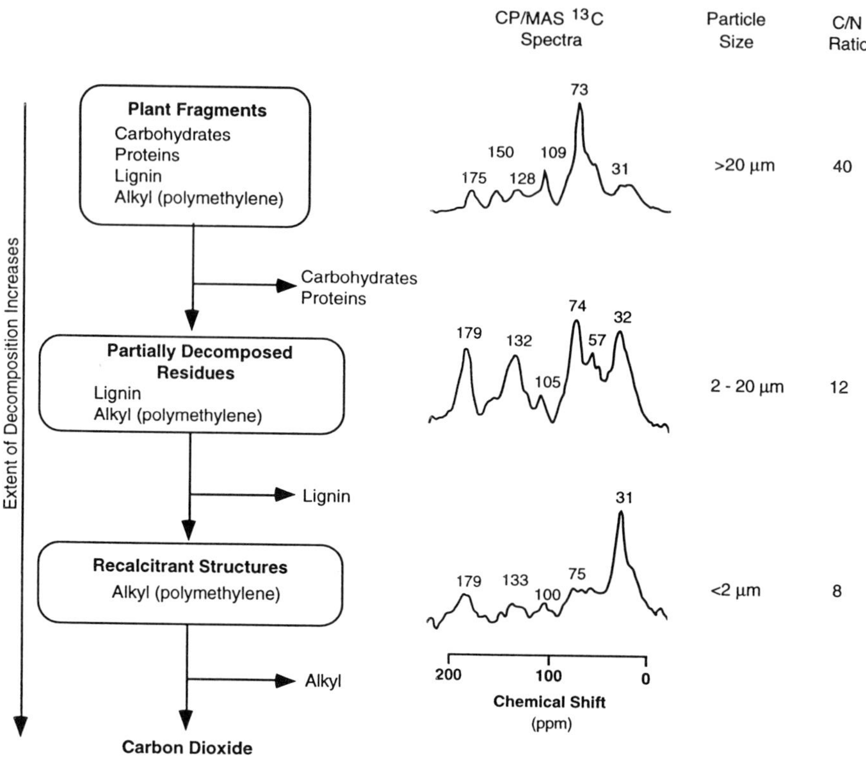

Figure 10 A schematic model of oxidative degradation of organic materials in soils (from Ref. 36).

size fractions with high surface areas may contain artifacts associated with the disaggregation procedure. For example, disruption of organisms by energetic dispersion procedures may release soluble materials that could readily associate with clay surfaces exposed for the first time to modern biological chemicals. Thus, clay materials in clay microstructures and even in microaggregates that may have been associated with organic materials for millenia may become further associated with modern carbon compounds of similar chemistry during the disaggregation process. The modern carbon will mask the old carbon in respect to ^{14}C dating procedures.

Jocteur Monrozier et al. [26] fractionated an Alfisol and a Vertisol after partial dispersion of the soils by gentle mechanical disruption of suspensions in water [37]. Size fractions included coarse and fine sand (>250 and 50 to 250 μm), coarse and fine silt (20 to 50 μm and 2 to 20 μm), and coarse and fine clay (0.1 to 2 μm and <0.1μm). The sand-size fractions contained no stable microag-

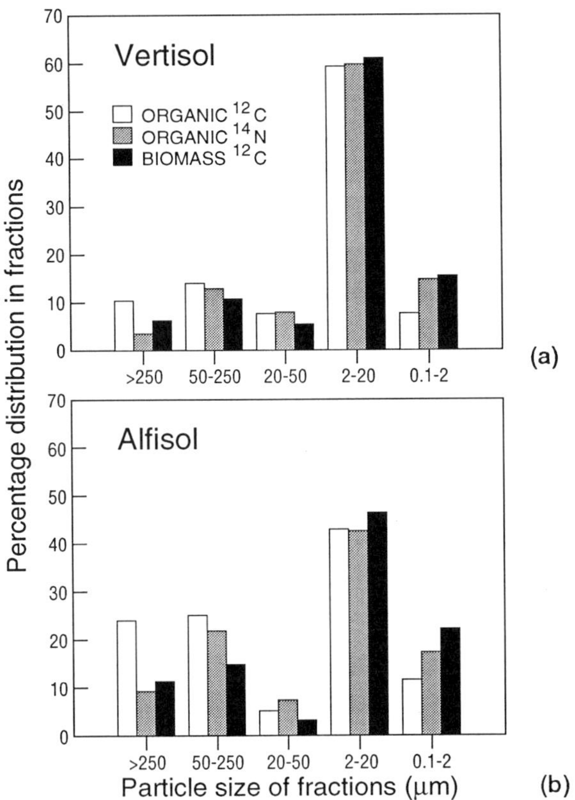

Figure 11 Percentage distribution of organic matter C and N biomass C in particle size fractions of (a) a Vertisol and (b) an Alfisol (from Ref. 26).

gregates. Microaggregates in the silt-size fractions accounted for about 10% of the clay of the Alfisol and nearly 60% of the clay of the Vertisol. Recoveries after fractionation of total organic carbon and nitrogen and of biomass carbon, generally exceeded 90%, indicating that artifacts resulting from cell destruction were minimal. The organic carbon and nitrogen were bimodally distributed, with most (40 to 60%) being associated with the fine silt-size (2 to 20 μm) material and a lesser amount with the fine sand-size (50 to 250 μm) fraction (Figure 11) [38]. The C:N ratios of fractions from both soils decreased with decreasing particle size, to about 6–8:1 in the clay-size (0.1 to 2μm) fraction. The C:N ratio of the >250-μm-size fraction ranged from about 15:1 for the Alfisol to about 45:1 for the Vertisol, indicating a more recent addition of plant residues to the Vertisol than to the Alfisol.

D. Location and Characterization of Biota

Microscopy

Because most soil bacteria are <0.5 μm in diameter [39], they are difficult to detect in situ by light microscopy against a background of opaque minerals and organic debris when sections of soil are prepared by usual methods [40]. Also, it has been difficult, on purely morphological grounds, to distinguish between live bacteria, which contribute to the biomass, and dead bacteria, which do not.

Some of these problems have been solved by the use of fluorescence microscopy. In the past this technique has been limited by the reaction of many fluorescent stains (e.g., acridine orange) with particulate organic debris, and by rapidly fading images under UV. However, by using fluorescein isothiocyanate (FITC) staining, Lawrence and Germida [41] have been able to detect live bacteria on elemental sulfur grains incorporated into soil, and Calcofluor has been used successfully to locate bacteria in soil sections [42]. Morgan et al. [43] have used Calcofluor and fluorescein diacetate to specifically detect soil fungi, and by combining fluorescence microscopy with automated image analysis, they have estimated fungal biomass.

These studies have markedly advanced and complemented research utilizing higher resolution, electron microscopy for locating microorganisms in situ. Sample preparation is simpler for light microscopy, and much larger, more representative volumes of soil are used than with electron microscopy. Moreover, automated image analysis obviates the tedium and subjectivity of manual recognition and counting techniques.

Microorganisms, although present in large numbers in topsoils (10^6 to 10^{10} colony forming units (CFUs)/cm^3), are neither uniformly nor randomly distributed, but, as revealed by TEM of soil sections, are clumped near or within cellular residues or in micropores accessible to soluble soil organics. Such sites must also provide compatible pH, water potential, and oxygen concentrations [5,44].

Bacteria and actinomycetes are concentrated in rhizospheres, fecal pellets, and plant and animal debris. Many microaggregates have an organic core [12,26], which not only binds components of the microaggregate together, but also provide substrates for the microbiota within. In energy-rich microhabitats, such as fecal deposits and rhizospheres, bacterial populations may rise to 10^{10} to 10^{12} bacterial cells/cm^3 [45] and individual cells are larger than in the bulk soil [46]. Also, numerous small colonies of many different kinds of actinomycetes and bacteria are observed where substrates are plentiful [23,47]. TEM of sections of aggregates suggests that in the bulk soil each microhabitat tends to be colonized by one type of microorganism [5].

Many soil bacteria produce extracellular polysaccharides (ECP), so that both colonies and individual cells isolated in micropores are often enclosed in a polymer sheath that, in the rhizosphere, can be distinguished ultrastructurally and histochemically from root surface ECP [47]. Microbial ECP can become coated with clay particles so that bacterial colonies often form the core of microaggregates [10,48].

Bacteria in the interior of clay dominated microaggregates are small, gram positive, and devoid of polysaccharide and of granules of polyhydroxybutyrate (PHB) and polyphosphate (PP) [5]. These organisms are often associated with polyphenol-rich materials that become electron-dense after staining with heavy metals [22]. In contrast, bacteria near the surface of microaggregates are large, gram negative, and often contain granules of PHB and PP. These observations suggest that cells near the surface of microaggregates are able to intercept most, and perhaps all, of the soluble organic substrates that diffuse within the soil matrix, thereby effectively screening the readily decomposable substrates from microorganisms in the interiors of microaggregates [49].

Although bacteria occur in micropores near the surface of aggregates, they seldom occur on the surface, except in sites rich in organic matter [22]. Conversely, fungi frequently occur in the pores between the aggregates in bulk soil, with occasional branches into adjacent microaggregates. Fungi penetrate the microaggregates mechanically by inserting wedge-shaped hyphae between the various components. These hyphae later vacuolate, swell, and push the soil minerals and organic matter aside [22]. Macroaggregates are often composed of microaggregates that are linked together in this way by hyphae, especially those of mycorrhizal fungi [24]. Fungi are also common in the rhizosphere as commensals, symbionts (mycorrhizae), and root pathogens [50].

Many soil fungi also secrete ECP which attach their hyphae to the mineral surfaces [51,52]. In some voids that contain fungi, most of the space not occupied by the hyphae is filled with Ru/Os-reactive ECP [5,22], which serve to glue the parts of the aggregate together. Many mycorrhizal fungi secrete extensive coats of gel (Figure 5) and support colonies of gel-secreting bacteria [52]. These microbial gels may act as glues that hold aggregates together from within. In annual cropping ecosystems, death of the crop leads to the death of the soil fungi that the crop has supported, and aggregates must be recolonized de novo each year. Failure to recolonize, combined with decomposition of gels from previous crops by resident microorganisms, may lead to net losses of organic matter and of fine clay no longer stabilized by gel [21,22]. In contrast, in permanent pastures and natural ecosystems where some plants remain alive all year, extensive annual recolonization of aggregates by fungi is unnecessary, and gel production maintains aggregate stability.

Many soil bacteria also secrete extensive coats of Ru/Os-reactive ECP, which bind nearby minerals to the surfaces of both individual cells and microcolonies [20,48]. In micropores that are no longer accessible to recolonizing bacteria, the ECP may persist after the death of the microorganisms that secreted them, continuing to link various parts of the microaggregates [29]. The content of microbial ECP is highly correlated with aggregate stability [53].

Protozoa accumulate in regions rich in bacteria, actinomycetes, and fungi, such as beneath plant litter and in rhizospheres where the root hairs and the outer rhizodermal cells are undergoing autolysis, i.e., at sites of abundant energy supply for the growth of the microbiota and their predators [54].

Biota and Biomass of Soil Fractions

Interest in soil structure and in the mechanisms by which biological activities are physically constrained within the soil matrix has stimulated research on methods of aggregate dispersion and on the nature and distribution of biota and biomass in fractions differing in particle size and density. Hattori [6] showed that organisms can be separated into two broadly distinct groups, according to their ease of dispersion after repeated, and gentle agitation of soil aggregates in sterile water. Organisms in suspension, after settling of partly dispersed soil, were assumed to have been previously adsorbed to the outer surfaces of aggregates or within pores of relatively large (>6 μm) pore-neck diameters. Organisms in suspension after ultrasonic dispersion of repeatedly washed aggregates were assumed to have been previously located within relatively narrow pores of water-stable aggregates.

Differences in the number (viable counts on selective media), type, and behavioral responses of cells to soil amendments were consistent with expectations of cells in microhabitats as influenced by pore size and location. Protozoa and fungi were preferentially located in the outer pores of aggregates, whereas bacteria existed predominantly in the inner pores. Bacteria in the outer areas were more responsive to amendment with soluble substrates and to air drying of the soils, were affected sooner by soil fumigation, and were more sensitive to bactericides than were bacteria located internally. Predation was predominantly of bacteria located in the outer areas of aggregates, as demonstrated in a sterilized soil to which protozoa (*Colpoda* sp.) and bacteria had been added.

Kanazawa and Filip [55] utilized a more energetic procedure of soil fractionation than Hattori [6], but they also demonstrated both qualitative and quantitative differences in the distribution of the soil microbiota and of biomass in the fractions. By combining wet sieving and sedimentation of a blended soil suspension, they obtained a silt/clay-size fraction (<50 μm diameter) and a series of coarser size fractions that contained light materials of relatively high organic matter content that was attributed mainly to plant debris in different stages of decay. Fungi, actinomycetes, and bacteria (copiotrophic and oligotrophic) and total biomass carbon were most concentrated in the light fractions. Concentrations in the light fractions decreased with the particle size of the sand/coarse silt components from which they were prepared, consistent with trends of decreasing

C:N ratios with decreasing particle size and with progressively advancing stages of residue decomposition. Cell numbers and microbial biomass were most concentrated in the <50 μm-size components of the particle-size fractions.

Application of the Brückert [37] method enabled Monrozier et al. [26] to separate soils into a range of particle-size components without appreciable destruction of the microbial biomass (see Section II. C). As also found for organic carbon and nitrogen, biomass carbon was bimodally distributed among the size fractions of two topsoils, with the greatest amounts located in the fine silt-size fraction (47 to 61% of recovered biomass carbon) and lesser amounts in the fine sand-size components (11 to 15%) (Figure 11). All the biomass carbon of the coarse and fine sand-size fractions was associated with plant fragments, which were readily separated by flotation from inorganic material (J.-L. Chotte, personal communication).

Chotte et al. [56] have compared the distribution of biomass in the size fractions of a Vertisol that was partly dispersed by the Brückert technique and by a gentler, briefer procedure. The coarse sand-size fraction (>250 μm) of the Vertisol prepared by the latter method contained organic residues and sand particles coated with a clay matrix, and about one-third of the soil microbial biomass. Application of the Brückert technique removed the aggregated clay from the >250-μm fraction, freed the organic residues from the fraction, and transferred biomass to finer-size fractions, as was also reported by Monrozier et al. [26].

Other studies have utilized more energetic procedures than the Brückert technique for soil dispersion before particle-size fractionation [57–59], sometimes resulting in low recoveries of biomass carbon. Recovered biomass carbon was located mainly with the (fine) silt-size fractions, as found for gently dispersed soils, or in increased proportions in the coarse clay-size fractions that accompanied increased dispersion of microaggregates.

E. Location and Characterization of Enzymes

Microscopy

There are two methodologies for specifically locating enzymes in biological tissues: ultrahistochemistry and immunolabeling [60].

Cytochemical/histochemical methods for locating enzymes in cells and tissues have been used for more than 40 years, and standard texts for preparing biological tissues for electron microscopy provide methods for almost all groups of enzymes [e.g., 61,62]. Soil samples must be fixed sufficiently well to kill microorganisms and fine roots, to prevent de novo synthesis of target enzymes and leakage of existing intracellular enzymes into the nearby soil as cells die, and to preserve enough cytoplasmic ultrastructure (e.g., ribosomes) to enable the distinction between living and dead cells. The fixation techniques must not, however, substantially inhibit the activity of the target enzyme.

The required balance between adequate fixation and adequate enzyme activity often requires delicate adjustment of the concentration and pH of the fixative and the temperature and duration of fixation. For example, many methods in standard texts have been developed for mammalian tissues, and enzymes of the soil biota may exhibit optimal activity at >20°C below recommended temperatures. Also, short initial fixation times (<1h) are adequate for plant and animal tissues, but for bacteria enclosed in clays it may be necessary to fix overnight at low temperature and low concentration to get adequate fixation of cells. Furthermore, whereas some enzymes (e.g., acid phosphatase) retain most of their activity after fixation, others (e.g., succinic dehydrogenase) are so sensitive that it is necessary to fix the soil sample after the cytochemical reaction is completed.

After fixation, soil samples are incubated in a medium containing a substrate specific for the target enzyme and a heavy metal-trapping agent that produces an electron-dense complex with a product of the enzyme reaction. Some heavy metals complex nonspecifically with many organic and inorganic components. Adequate controls are, therefore, essential. These include heat treatment, omission of the substrate or cofactors (e.g., ATP, Mg^{2+}), and the use of specific inhibitors [63].

The soil samples are then refixed for up to 3 days, counterstained in OsO_4, dehydrated, embedded in plastic, and ultrathin sections are prepared. Unlike adjacent sections of tissues of plants and animals, adjacent soil samples differ in microstructure, so control and experimental samples are often not directly comparable. In practice, a number of easily recognizable sites are examined (e.g., near root surfaces, hyphae, bacterial colonies, particular minerals, etc.) and comparisons are made with similar sites in sections from the controls. Usually some sections contain plant or animal tissues or microorganisms (e.g., fungi) known to contain the target enzyme, and these cells provide an internal check on the adequacy of the procedures. They also indicate the type of electron-dense product to search for in the extracellular matrix. Any similar electron-dense deposits present in experimental samples but absent from control samples are taken to indicate the location of the enzyme. The specificity of the reaction can be checked by electron probe microanalysis.

The ultracytochemical tests used to locate enzymes are limited in regions where the soil components are naturally electron-dense (e.g., minerals) or react with OsO_4 (e.g., humified organic matter) or with the heavy metal component of the enzyme-specific medium. Because of these reactions, the locations of extracellular enzymes in soils can be determined unequivocally only in electron-transparent materials, such as microbial and root ECP, fragments of cell walls, and microbial membranes.

Despite these limitations, enzymes such as acid phosphatase [23,29], succinic dehydrogenase, peroxidase, and catalase [64] have been detected in soil

Figure 12 Scanning electron micrograph of a root surface in the zone of cortical cell death. The dead cells (acid phosphatase-negative) are dark and the live cells (acid phosphatase-positive) are bright.

microorganisms in situ in rhizospheres and aggregates. Thus, acid phosphatase has been detected in roots (Figure 12), mycorrhizae [63,65], soil microorganisms [25,29], and fragments of microbial membranes as small as 7×20 nm.

Immunocytochemical techniques require that the target enzyme is isolated and purified and that an antibody is prepared against it. The antibody is isolated, purified, and labeled with colloidal gold. The gold-antibody complex is either infiltrated into the soil or applied to ultrathin sections. This method has proved effective in investigations of animal and plant tissues [63], and of bacteria cultured in the laboratory [66], but is far less so when applied to soils. The gold tag becomes nonspecifically attached to various soil components in the absence of the antibody and when the gold-labeled antibody is applied to heated specimens, or to samples treated with an enzyme-specific inhibitor.

Enzyme Activity of Soil

Enzymes in soil are believed to occur in live and dead cells and tissues, in cell and tissue fragments, in soil solution as free enzymes, and as immobilized enzymes as the result of their binding or entrapment by clays or humic colloids [67–71]. The occurrence of enzymes with intact cells and cell debris has been demonstrated for a few specific cases by TEM of soil aggregates (see the previous subsection, Microscopy). Soil dehydrogenases are assumed to function only intracellularly, based on the consideration of their cofactor requirements. The complete loss of dehydrogenase activity in soils fumigated with chloroform to destroy the microbial biomass supports the assumption that dehydrogenases are exclusively associated with live, intact cells [72]. Free enzymes are believed to be short-lived in soils, being vulnerable to degradation and to immobilization on the surfaces of soil colloids. Immobilization may protect enzymes against degradation and denaturation, usually at the cost of some loss of activity. Thus, immobilized enzymes may accumulate in soils. However, methodological difficulties have prevented attempts to confirm their presence by electron microcopy, and there is no direct evidence for the occurrence in soils of clay-enzyme complexes per se [71,73].

By implication rather than by clear definition, the term *soil enzymes* may denote enzymes in any of the states of occurrence in soils, referred to above. Burns [68] proposed an approach to permit an appraisal of the individual contributions of an enzyme in different states of occurrence in soil, to total activity. However, little has been further achieved in this respect. Not even a partitioning between activities of intra- and extracellular enzymes in soil can be determined unequivocally. Whereas the activity of a stabilized extracellular enzyme in soil is independent of current cell growth, the activity of an intracellular enzyme is affected by both cell growth and enzyme induction and repression mechanisms [71]. The use of chemical compounds (e.g., toluene) or physical agents (e.g., high-energy ionizing radiation) to discriminate between the two types of activities has provided unsatisfactory results [67,69].

Most assays of soil enzymes have been designed to determine total potential activity against a specific substrate, irrespective of the source and the state of occurrence of the enzyme in soil, and under incubation conditions that eliminate or minimize cell proliferation. Assays have been devised for a wide range of soil enzymes that are representative of all enzyme classes. Oxidoreductases, transferases, and hydrolases have been the most studied because of their role in organic matter degradation and release of nutrients. Dehydrogenase activity has been used as a general index of biological activity because of the involvement of dehydrogenases in oxidative energy-transfer sequences.

Simple and accurate assays of enzymes in soil utilize suspensions of soil, incubated at controlled temperatures with substrates in sufficient concentrations to approach zero-order kinetics, and buffered at optimal pH. The assay period must be sufficiently brief to avoid microbial growth and to minimize further metabolism of the products of the enzyme reaction on which the assay is based. When assays involve colorimetic determination of a specific reaction product, the efficient extraction of the product must not result in coextraction of interfering amounts of colored organic matter [71]. Ideal assay conditions are rarely met.

The potential activities of enzymes in soil, as assayed under controlled conditions of temperature, pH, and high and uniform concentrations of substrate (usually artificial), are not easily related to their actual activities in soil. In soil, natural substrates are present mainly in rate-limiting concentrations and are metabolized under fluctuating environmental conditions [68,70,74]. Thus, the contributions of soil enzymes to the processes of organic matter decay and nutrient release have not been assessed. In some cases, their involvement may be substantial, e.g., hydrolysis by soil ureases of urea applied as fertilizer.

Free and Immobilized Enzymes

The preparation in vitro and the properties of clay-enzyme complexes have been described [3,75]. Protein binding on clay is principally by hydrogen bonding and is influenced by protein properties (isoelectric point, number of binding sites, solubility, concentration, shape and size of molecules) and by clay type (surface area and charge, the nature of saturating cations). Clay minerals in soil have a mixture of exchangeable cations and may be coated, at least partially, with mixtures of polymeric oxides and hydroxides of iron, aluminium, and manganese. Covering of the surface of montmorillonite with different amounts of OH-A1 species reduced the activity of adsorbed urease [76], suggesting that studies conducted with homoionic pure clay minerals may not reflect conditions in situ in soil. Nevertheless, Fusi et al. [77] reported that the amounts of the proteins, catalase, and β-lactoglobulin, bound to calcium-montmorillonite and other clays coated with polymeric $Fe(OH)_3$ were, respectively, no greater than the amounts bound to the pure uncoated clays.

Synthetic humic-enzyme complexes also have been prepared by reacting enzyme with polymerizing phenols; enzyme immobilization occurs during the oxidative coupling of quinones, which are enzymatically or abiotically produced from the respective phenols [78,79]. Enzymes immobilized in these complexes have been regarded as being analogous to enzymes immobilized in soil by bonding to organic colloids. The synthetic humic-enzymes, prepared by adding the enzyme during the formation of quinones, were more resistant to thermal and proteolytic degradation than those prepared by adding enzyme before the formation of quinones from added phenols or at the end of the polymerization process [80]. The oxidative coupling of phenolic constituents and entrapment of enzymes during humification has been suggested as the main factor in enzyme stabilization in soil [79].

A better understanding of the properties of naturally occurring humus-enzyme complexes has resulted from their extraction from soil [73]. Extraction of enzymes in adequate yields must be accomplished by procedures that minimize the release of enzymes from lysed microorganisms, the adsorption or entrapment of enzymes by coextracted organic colloids, and the generation of activities from reactions between the extractant and inorganic components (e.g., extractants containing citrate ions form complexes with manganese that exhibit laccase-like activity [81]).

A range of conventional buffers have been used as extractants of enzymes from soil [73,82]. Pyrophosphate buffer, a common extractant of soil organic matter, was also a relatively effective extractant of humic ureases and immobilized proteases. The latter was shown not to be the result of the formation of an artifact during extraction [73,83,84].

Studies of partially purified enzymes from soil extracts have resulted in different but not mutually exclusive views of the nature of soil humus-enzyme complexes. Mayaudon [85] reported that extracted enzymes of fungal origin were stabilized through their bonding to partially decomposed lipopolysaccharides of bacterial origin. The enzymes, enveloped by the polysaccharide, were suggested to be immobilized in soil by their bonding through calcium to humic colloid surfaces.

According to Burns et al. [86] and Nannipieri et al. [87,88], enzymes, such as ureases and phosphatases, are immobilized in organo-mineral complexes in such a manner that pores in the organic network surrounding the enzymes permit the passage of small molecules of substrate and product, but not those of large molecules such as proteolytic enzymes. This structure confers a measure of stabilization on the enzyme while allowing retention of enzyme activity. A corollary of this hypothesis is that an enzyme normally active against substrates of high molecular weight, e.g., a protease active against a protein substrate, would lose that activity if immobilized and protected in a similar organic complex, but may well retain activities toward substrates of low molecular weight, such as peptides [67,89].

The kinetics of enzymes in soil differ from those of enzymes in solution. In heterogeneous soil microenvironments, enzymatic rates are regulated by the pH [90] and humidity [91] at substrate-enzyme interfaces. The responses to temperature of enzymes in soils follow similar but not necessarily identical patterns to those of enzymes in solution [71].

The kinetics of enzymes immobilized on clays or on charged organic colloids are influenced by the nature of the matrix surrounding the enzyme. For example, the surface pH values of negatively charged clays and organic colloids are more acidic than the bulk solution, causing an apparent shift toward the alkaline region of the optimal pH of clay- and colloid-bound enzymes [67,71]. Thus, the pH-activity profiles of trypsin-humic acid analogs were displaced toward more alkaline regions compared with that of free trypsin, due to the polyanionic humic acid moiety concentrating hydrogen ions in the microenvironment of the active site of the bound enzyme [78]. Similar effects on the pH-activity profiles of other water-insoluble derivatives of trypsin have been described [92]. Furthermore, the Michaelis constant of the trypsin enzyme entrapped within a polyanionic matrix was about 30 times lower than that of the soluble free enzyme [92]. The effect was the result of the polyanionic carrier increasing the concentration of positively charged substrate molecules within the immediate vicinity of the entrapped enzyme. Similar effects have been postulated for enzymes entrapped by humic molecules extracted from soil [73]. For example, Nannipieri et al. (93) showed that the K_m values of proteases extracted from soil and active on the positively charged substrate, benzoylarginine amide, were lower than the proteases purified from microbial culture. The effect was attributed to the increased concentrations of substrate molecules within the vicinity of the active sites of the extracted soil proteases as a result of the association of the enzymes with humic compounds.

Enzymes Active on Particulate Substrates

The classical Michaelis-Menten equation

$$v = \frac{V_{max} S}{K_m + S}$$

is valid for a soluble enzyme reacting with soluble substrate(s) in non-rate-limiting concentrations. When the substrate is insoluble, the enzyme (E) is in excess and the relationship between activity and substrate/enzyme concentrations is described [94] by

$$v = \frac{V_{max} S}{K_m + E}$$

This equation can be approximated to $V = kE^n$, where n and k are constants; k is the proportion of enzyme distributed between the solution and the surface of the insoluble substrate. Thus, enzyme activity is proportional to the surface area of the substrate and to the amount of enzyme interaction per unit surface area of substrate. The interaction of an enzyme with an insoluble substrate depends on (1) the concentration and properties (size, charge) of the enzyme, (2) the nature of the substrate (e.g., crystalline cellulose is less accessible to enzymes than amorphous cellulose), (3) the presence of other constituents (e.g., the presence of lignin affects cellulose degradation), and (4) the environmental conditions (e.g., pH, salt concentration, temperature).

Enzymes of Soil Fractions

Information on the distribution and properties of enzymes in soil fractions is scarce and fragmented and analyses of enzyme kinetics of the soil fractions, although soundly-based, have done little to resolve the question of the states of occurrence of enzymes in soil. Methodologies for soil dispersion and fractionation into components ranging in their particle size and, in some cases, particle density have differed between studies, thereby making direct comparisons difficult.

Substantial proportions (up to 70 to 76%) of total invertase and amylase activities of some grassland soils were recovered in a light fraction after dispersion of dried soils in an organic liquid of high specific gravity (s.g. 2.06) [95]. Part of the recovered activities were attributable to enzymes of plant residues in the light fraction. Rojo et al. [96] reported that soil phosphatase activity was mainly located in the 100 to 200-μm size fraction and, on the basis of other evidence, attributed the activity to residues of plant roots and associated microorganisms.

Kanazawa and Filip [55] found that most of the activities of β-glucosidase, β-acetylglucosaminidase, and proteinase of a soil were located with plant residues in different stages of decomposition. The enzyme activities of the fractions decreased with increased stage of decomposition of the residues and were directly correlated with microbial numbers and biomass of the fractions.

Particle-size fractions prepared from soils by gentle sieving, with or without prior mechanical dispersion, differed in their enzymic activities [97,98]. The activities of catalase, urease, dehydrogenases, and proteases were each mainly associated with aggregates that were 200 to 2000 μm in diameter when soils were sieved, but they were mainly located with finer-size materials (<50 μm) as the energy of dispersion was increased. In none of the above studies did soil fractionation distinguish between plant and microbial sources of the assayed enzymes.

Gupta and Germida [99] reported that the proportional decreases in microbial biomass and enzyme activities of a soil following long-term cultivation were greater, collectively, for macroaggregate fractions than for microaggregate fractions. However, the distributions of phosphatase and sulfatase activities among macro- and microaggregate-size fractions differed from each other and from that of microbial biomass carbon, nitrogen, and sulfur. Sequi et al. [100] showed that cultivation decreased urease and phosphatase activities and soil porosity in the pore size range of 30 to 200 μm in diameter.

As discussed above, the kinetic properties of enzymes bound to or entrapped within solid matrices differ from those of enzymes free in solution and are influenced by the properties of the solid matrix. Nevertheless, Tabatabai [101] found that the K_m values of urease in sand-, silt-, and clay-size fractions were similar, suggesting similar "average" states of occurrence of the enzyme in the fractions, despite presumed differences in the proportions present as urease-clay complexes. In contrast, other studies [98] have demonstrated that urease of different particle-size fractions differed in its pH optimum and its resistance to thermal denaturation. These differences in properties were attributed to different states of association of extracellular urease with soil colloids.

III. DECOMPOSITION OF ORGANIC RESIDUES IN SOIL

Plant residues are chemically complex organic substrates that enter soil annually. The properties of the substrates and soil that influence the rates of residue decomposition have been amply documented [102,103].

A. Influence of Residue Properties

Early studies by Waksman [102], and confirmed by many others [104–107], have demonstrated a direct relation between the proportion of water-soluble components in the residues and the extent of the initial decomposition of the residues in soil. Simulation models of the cycling of substrate carbon through the soil biota have acknowledged the influence of substrate complexity on decomposition rates. Relatively high values for decay rates and efficiencies of utilization have been assigned for the decomposition of the water-soluble constituents [108,109].

An understanding of the net rates of decomposition of plant residues in the short to medium term (weeks to years) requires consideration not only of the rates of decomposition of the soluble and particulate prime substrates, but also of the opportunities for stabilization of the microbial biomass and turnover products. Decomposer organisms located with particulate organic residues are probably more vulnerable to predation and their carbon is more available for further breakdown than those organisms that metabolize soluble substrates that diffuse within soil pores, which imposes greater constraints on predator movement and activity.

This concept was explored by Ladd et al. [109] to explain the marked differences in the behavior of ^{14}C in legume residues and glucose. Decomposition of legume leaf ^{14}C (about 35% of which was soluble in hot water) in a clay soil was initially far less extensive than that of soluble glucose ^{14}C, but exceeded that of glucose ^{14}C after several months of incubation. Whereas the ^{14}C in microbial residues derived from glucose was rapidly stabilized in the highly aggregated soil matrix, the particulate plant residues plus some microbial residues decomposed at a higher net average rate during the later incubation period. The ratio of biomass ^{14}C to ^{14}C evolved as CO_2 in the soil amended with the plant residue was much less than in the glucose-amended soil, indicating a faster turnover of biomass ^{14}C (and perhaps a lower efficiency overall of substrate use) in the former treatment.

Although it has generally been assumed that some particulate organic components of plant residues constitute energy sources for decomposers relatively early in the process of decomposition, recent information, particularly from the use of $\delta^{13}C$ values, has also indicated considerable longevity for significant proportions of the particulate fraction. A number of workers have utilized changes in the abundance of ^{12}C and ^{13}C in soils where there have been changes in vegetation from C-3 to C-4 plants (or the reverse) to determine the persistence of various soil organic fractions [110–116]. In all cases, the particulate fraction persisted for decades, and even after 50 years, significant proportions of particulate material from a previous vegetation remained in the soil.

B. Influence Of Soil Properties

The generally observed direct relation between the soil organic matter and clay content, as recorded by Jenny [117] for regions of similar climate and vegetation and management regimes, has underpinned early conclusions that clay retards the rate of decay of soil organic matter. Laboratory studies on the effects of clay on organic substrates have generally supported this conclusion (reviewed by Jenkinson [103]).

The use of ^{14}C-labeled substrates has enabled comparisons of the rate of decomposition of substrates, simple and complex, as known inputs into soils of varying clay content and type and incubated under laboratory and field conditions for short (weeks) and medium (years) terms. Many studies [118–127] have demonstrated that the percentage retention of applied substrate ^{14}C as total residual organic ^{14}C is greater the higher the soil clay content.

Evidence suggested that the clay effect resulted from an enhancement within the soil matrix of the stabilization of product(s) of rapid substrate decomposition rather than of the prime substrates themselves, as (1) the effect was achieved with simple, soluble substrates, such as glucose, which completely disappeared within the first few days of incubation [124,128], (2) the effect was more readily demonstrable with ground rather than unground plant residue substrates (P. Sørensen, personal communication), and (3) soils of higher clay content accumulated higher concentrations of substrate-derived, microbial biomass carbon, an effect that was already measurable early in the incubations [122,127].

Thus, the clay-related effect is established during the initial rapid phase of metabolism of primary substrates and turnover of decomposer cells and cell products. The effect may be apparent within days, in the case of some laboratory experiments, and is demonstrable with both simple and complex substrates. The effect persists for years. For example, Jenkinson [119] and Ladd [127] demonstrated that under field conditions, the greater percentage retention of organic ^{14}C from the decomposition of ^{14}C-labeled plant residues in soils of relatively high clay content was clearly established at the time of the first sampling (within a few weeks or months of plant residue incorporation), and remained evident for 6 to 10 years thereafter, when the experiments concluded. After the initial phase of rapid decomposition, the first-order decay rates of the more resistant plant components plus microbial metabolic products were relatively independent of soil clay content (Figure 13).

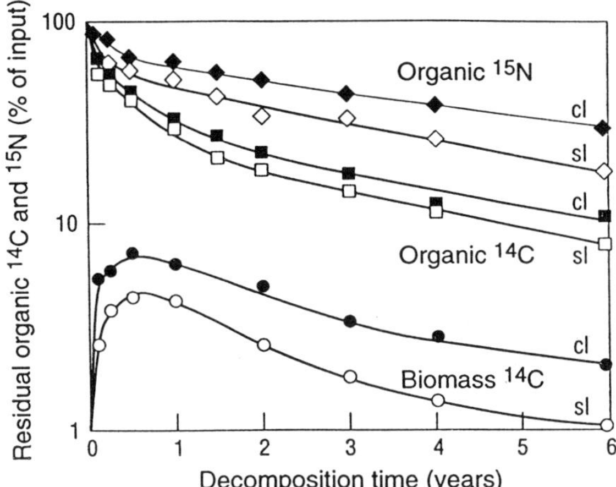

Figure 13 Residual organic ^{14}C and ^{15}N and biomass ^{14}C in a clay soil (cl) and a sandy loam (sl) accompanying the decomposition of isotope-labeled legume tops under field conditions (from Ref. 38).

Amato and Ladd [126] incubated 23 soils with clay contents ranging from 2 to 73% with ^{14}C-glucose for 44 weeks and demonstrated that clay content was directly correlated with concentrations of total residual organic ^{14}C and microbial biomass ^{14}C. Similar relationships were evident for 11 soils with neutral to alkaline pH that were incubated for 66 weeks with ^{14}C-labeled plant residues. The differences between soils in the amounts of total residual organic ^{14}C were statistically accounted for by the differences in the amounts of biomass ^{14}C. Thus, the amounts of nonbiomass ^{14}C, i.e., the calculated differences between the amounts of total residual organic ^{14}C and biomass ^{14}C, ranged from 13 to 18% of input ^{14}C in the glucose-incubated soils, and from 17 to 26% of input ^{14}C in the 11 soils with neutral to alkaline pH that were incubated with plant residues. However, neither the clay content nor the cation exchange capacity (CEC) of the soils was correlated with the amount of nonbiomass ^{14}C from the decomposition of the glucose or plant residues.

Substituting the CEC of the soils for clay content only marginally improved the correlation with the amount of biomass ^{14}C and total residual organic ^{14}C, indicating that clay type may not have been an important determinant of the extent of mineralization of substrate ^{14}C. A mineralogical analysis of the 23 soils was not undertaken, although the clay of some of the soils was known to be predominantly smectitic in character, and of others, to be predominantly kaolinitic (e.g., Ref. 26). The apparent absence of a substantial effect of clay type on the extent of decomposition of the organic substrates in the soils of

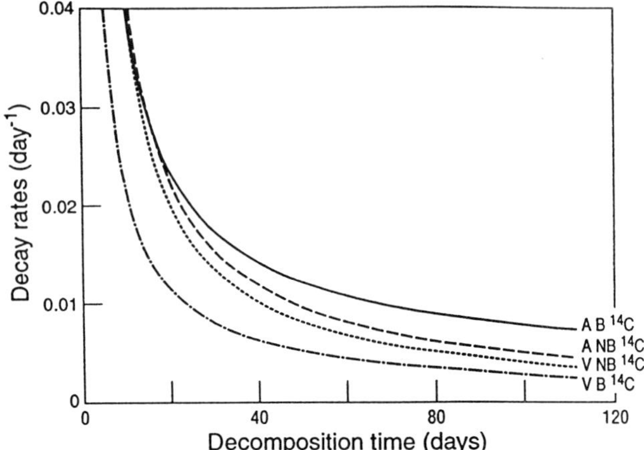

Figure 14 Decay rates of biomass ^{14}C (B ^{14}C) and nonbiomass ^{14}C (NB ^{14}C) in an Alfisol (A) and a Vertisol (V) calculated using an average substrate utilization efficiency of 35% (from Ref. 12).

different clay content differed from the results of other studies on the effects of added clay on metabolic activities, as extensively reviewed by Stotzky [3]. However, as Stotzky pointed out, the influence of different clay minerals on heterotrophic metabolism was very variable, ranging from an enhancement, to no effect, to a reduction, and depended inter alia on the concentration of clay and the nature of the specific substrate being decomposed.

Ladd et al. [128] also showed for a pair of soils of contrasting texture amended with ^{14}C glucose and incubated for 112 days that the gross first-order rates of decay of biomass ^{14}C were about 2.0 to 2.8 times faster in the sandy loam than in the clay soil. In contrast, gross rates of decay of nonbiomass ^{14}C were only 1.0 to 1.2 times faster in the soil of lighter texture (Figure 14). The same average efficiency of substrate utilization (35%) was assumed for both soils.

The results of Amato and Ladd [126] and Ladd et al. [128] suggested that the decreased mineralization of added substrate carbon in soils of increasing clay content resulted from the enhanced stabilization or protection of substrate carbon in the microbial biomass. This conclusion does not exclude mechanisms of stabilization of nonbiomass constituents which also involve their interactions with clays (see Section IV. D). However, the extent to which such mechanisms may have operated was not measurably related to the clay content of the soils in the wide range tested by Amato and Ladd [126]. Christensen [9], in reviewing studies on the distribution of organic matter in different particle-size fractions, demonstrated that both native soil organic matter and ^{14}C-labeled products of

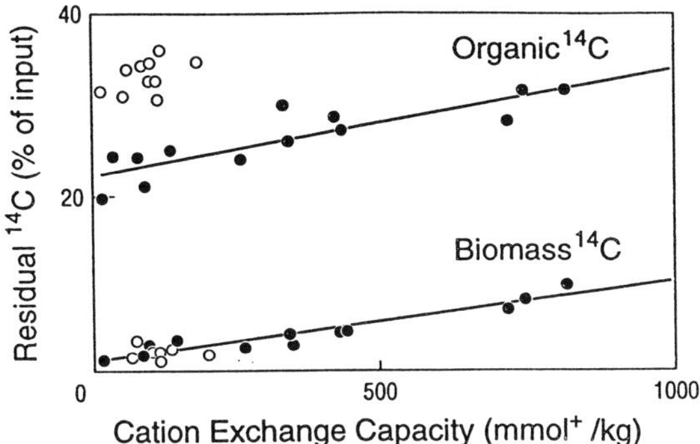

Figure 15 Relationship between residual organic ^{14}C and biomass ^{14}C and soil cation exchange capacity of ^{14}C legume-amended soils: o, acidic soils; •, neutral to alkaline soils (from Ref. 126).

the metabolism of specific substrates were associated mainly with clay fractions. Significantly the clay-size and silt-size components of light-textured soils contained higher concentrations of organic matter than the respective fractions in heavy-textured soils, suggesting that small amounts of clay in soils were effective in stabilizing organic residues.

The influence of soil clay content on carbon turnover processes has been accommodated in several simulation models, although the precise manner of treatment has varied. In the Century model [129], higher clay plus silt contents slow the rate of decay of the designated "active" soil carbon pool (which encompasses biomass carbon and other labile soil components), and lower its efficiency of conversion to the "slow" soil carbon pool. In the Rothamsted models [130,131], clay content has no influence on pool decay rates, but the efficiency of conversion of all plant and soil pools to biomass plus metabolic products is arbitrarily set to increase in common with an increase in clay content. Modified versions [109,124,125] of the van Veen model allow not only for increasing efficiencies of substrate utilization in soils of increasing clay content, but also for protection of increasing proportions of biomass newly formed from substrate decomposition.

Not all experimental data support the conclusion that the decomposition of organic substrates is less extensive in soils of high clay content. For example, Gregorich et al. [132] incubated soils ranging in clay content from 18 to 38% with ^{14}C-glucose for up to 90 days, at which time, neither total residual organic ^{14}C nor biomass ^{14}C was correlated with the amount of clay. Although Amato and Ladd [126] found that analogous relations were highly significant ($P < 0.001$), as determined by residual organic ^{14}C for glucose-incubated soils of a wider range (2 to 73%) of clay content, the correlations were much weaker for those soils lying within the clay range of 12 to 45%.

Others factors, perhaps of a biological nature, may under some circumstances override clay or clay-related effects. Differences in the properties of plant residues and soils may encourage the growth and activity of primary decomposer microorganisms that differ in their efficiencies of substrate utilization and vulnerability to predation. For example, fungi utilize substrates with higher efficiencies for cell synthesis than do bacteria, although variations between species occur [133,134]. Also, using soils amended with ^{14}C-labeled plant materials, Amato and Ladd [126] determined that soils of mildly acidic pH (6.0 to 6.7) retained more total organic ^{14}C, even after 66 weeks incubation, than did neutral to alkaline coils of similar CEC (Figure 15). Plant residues may decompose less extensively in mildly acidic soils, possibly because of enhanced stability of particulate remainders of the residues, resulting from their reaction with soil inorganic constituents. (see Section IV. D).

IV. CONSTRAINTS ON BIOLOGICAL ACTIVITY

A. Soil Disturbance and Biological Activity

Several lines of evidence indicate that some organic materials persist in soils as the result of the physical isolation of the organic substrates from the appropriate decomposer organisms and enzymes. For example, organic residues of clearly cellular origin (tissue fragments, polysaccharides) are commonly revealed at sites within the internal ultrastructure of aggregates by TEM (Figures 5 and 6) (see Section II. C). Chemically defined substrates (carbohydrates), relatively stable in soil in situ, are readily decomposed after extraction from and readdition to soil and incubation [135]. Constraints on biological activities may be reduced as the result of changes in soil structure and hence substrate availability (see Sections II.B and II.C). The increased biological activities (e.g., rates of carbon and nitrogen mineralization, nitrification) that are caused by soil disturbance or fluctuating soil moisture conditions result, in part, from enhanced opportunities (albeit temporary) for enzyme-substrate interaction [136–149].

Decomposable organic substrates, made available after intermittent drying and remoistening of soils, are derived partly from the death of a portion of the soil biota, and partly from the nonliving soil organic matter [140,142]. The susceptibility of soil organisms to drying and remoistening treatments depends on the type of organism [145], their physiological state [139,143], and their location in the soil fabric [144]. Various studies have shown that drying and rewetting of soils decrease microbial biomass by about 15 to 30% [143,146,147, 149]. Van Gestel et al. [149] demonstrated with soils subjected to various sequences of fumigation, drying, and storage that (1) desiccation reduced biomass carbon in a sandy loam and a clay soil by 30% and 26%, respectively, and (2) fumigation incubation of soils to destroy the majority (86 to 92%) of the biomass carbon had little effect on the amounts of the flush of carbon and nitrogen mineralized when the soils were subsequently dried, rewetted, and incubated. They concluded that cells killed by soil desiccation were only a minor source of carbon and nitrogen mineralized after soil rewetting and incubation.

Hassink [150] measured carbon and nitrogen mineralization in undisturbed and sieved soils and used the gains in mineralization rates in sieved soils to estimate the amounts of physically protected organic matter in soils of contrasting clay content. Clay type was not reported. Small pores (<1 to 2 μm) accounted for higher proportions of the total pore volume in clay and loam soils compared with sandy soils. Flushes of carbon, and especially of nitrogen, mineralized were greater in the heavier textured soils. Fine sieving appeared to expose a physically protected fraction of the organic matter of a C:N ratio smaller than that of the total soil organic matter.

B. Microbial and Enzymic Diversity

Plant residues, freshly incorporated into topsoils, are quickly colonized by a variety of microorganisms [23,151]. Tester [152] has implicated microbial cells of the introduced plant residues themselves as the main agents of enzymic attack during the first stages of residue decomposition, whereas soil organisms are more involved later in the decomposition of more recalcitrant components of the residues.

Under suitable conditions of pH, temperature, and moisture content, plant residues are extensively decomposed within a few months [103]. Complete heterotrophic degradation of complex organic residues requires that decomposers either individually possess a broad biochemical capability (e.g., to degrade cellulose, pectins, hemicelluloses, lignins, etc.) or have the capacity to function within a community of microbial species, each with specific capabilities for degradation of these polymers [153]. The latter option appears to be the more likely. Evidence suggests that as decomposition proceeds (at progressively slower rates), some specialization of function among microorganisms occurs. For example, different morphological types of bacteria have been observed in highly lignified and carbohydrate-rich regions of cell wall remnants [151]. Studies on the enzymology of ligno cellulose degradation have revealed species differences in biochemical capabilities (see Section IV.C).

Although the diverse microbiota of topsoils provide for a range of integrated enzymic activities that allow extensive decay of complex substrates, opportunities also arise for stabilization of organic materials because of their location within the soil matrix. Individual microniches in aggregates frequently appear to be colonized by only one type of microorganism. In such cases, complete decomposition of any accompanying organic residues must await the disruption of the aggregates and the exposure of the materials to more diverse populations of microorganisms.

The benefits of a wide enzymic capability possessed by consortia of different microbial species extend to the decomposition of secondary metabolites and toxic by-products and the provision of cofactors and growth stimulants.

C. Activity and Persistence of Extracellular Enzymes

To be active against particulate substrates, extracellular enzymes must diffuse from the producer organisms to the substrates. The shorter the distance for enzyme diffusion, the greater the likelihood of enzyme-substrate reaction and effective utilization of the products by the producer cells. In soil, extracellular enzymes can be denatured or degraded, or they can be immobilized by adsorption or other bonding mechanisms to inorganic and organic components. Thus, free, extracellular enzymes are likely to have an ephemeral existence in soils and continued activity by free enzymes would depend on the availability of energy for de novo enzyme synthesis [70]. In contrast, immobilized enzymes are probably more able to persist as potentially active entities in soils, although they may be constrained in their activities on substrates of high molecular weight and in their opportunities for reaction with particulate substrates. Thus, immobilized enzymes may not be able to fulfill in soil the potential role envisaged for them by Burns [69], namely, to release from substrates of large molecular weight the inducer molecules required to trigger microbial synthesis of appropriate extracellular enzymes for substrate degradation.

The decomposition of plant residues is mediated primarily by extracellular enzymes whose activities are influenced by the physical and chemical nature of the substrates and the enzymes, as well as by environmental factors. Cellulose is the most abundant polymeric constituent of plant material, and it is more easily degradable than lignin [154], with which it is often associated. Haider [155] has recently reviewed the biochemical pathways of the degradation of cellulose and lignin in soils.

Cellulose is degraded by the concurrent action of extracellular cellulases, a group comprising three distinctive hydrolases: (1) exoglucanases or cellobiohydrolases (β-1,4-exoglucanases), which release glucose or cellobiose from the nonreducing ends of cellulose; (2) endoglucanases (β-1,4-endoglucanases), which randomly cleave internal glucosidic linkages; and (3) β-glucosidase, which hydrolyses cellobiose and soluble oligosaccharides to glucose [156]. Multiple forms of the same cellulases exist. For example, in *Trichoderma reseei*, endoglucanases of relatively high molecular weight are generally more active against insoluble substrates, whereas smaller molecules of endoglucanases produced after proteolytic cleavage of the parent enzyme appear to be more effective against the shorter-chain and water-soluble substrates produced during cellulose hydrolysis [157]. Some microorganisms, e.g., the white rot fungus, *Sporotrichum pulverulentum*, degrade cellulose by using oxidative enzymes and products of lignin degradation. The reaction sequences are complex and involve extracellular attack by several enzymes of both cellulose and lignin and their by-products [157]. The primary metabolism of wood polysaccharides continues until fungal growth becomes nitrogen limited, when secondary metabolism of lignin occurs [158,159].

The interactions of cellulases and cellulose are mainly affected by the surface area and crystallinity of the substrate [160]. However, preferential interaction of endoglucanases occurs during early hydrolysis, and of exoglucanases later, as more free end groups of substrate are exposed [161]. Lignin also binds cellulases, preferentially immobilizing endoglucanases as more lignin is exposed during the degradation of lignocellulose [162,163].

The various cellulases are differently inhibited by degradation products [164]. Studies on the degradation of lignocellulose by *T. viride* have supported the view that exo- and endocellulases (which directly attacked the cellulose component) are less inhibited by phenols produced by the oxidative degradation of lignin than is β-glucosidase. β-glucosidase remains associated with the fungus and is, therefore, spatially separated from the site of lignocellulose degradation. Inhibition of β-glucosidase by phenols (and by cellobiose from cellulose degradation) requires that the inhibitors diffuse to and are taken up by the fungus.

The enzymatic degradation of cellulose is influenced by the crystallinity, the molecular conformation, and the steric rigidity of the cellulose and by the cellulose pore size, which affects the accessibility of the hydrolases [153,160]. Cellulase isomers of molecular mass of about 12,500 daltons are more effective than are those of about 50,000 daltons in the early stages of degradation [153]. Reaction rates are also influenced by the nature of the internal surfaces within substrate pores. Most of the enzymatic hydrolysis occurs within gross capillaries of smaller surface areas.

Pure cellulose is a relatively easily degradable substrate. Its natural association with other compounds, principally lignin but also with hemicelluloses (branched heteropolymers of xylose, galactose, mannose, arabinose, glucose, and uronic acids), physically confer protection on the cellulose component against attack by cellulases [160].

D. Accessibility of Substrate to Enzymes and Organisms

There are at least two reasons for the persistence over decades of particulate organic matter in soils. The first relates to the chemistry of these plant materials, which is mainly lignin-encrusted cellulose as solid particles with limited surface area, as discussed above. Secondly, the intrinsic stability of such material is enhanced when the particles are encrusted with clay materials, which become associated with organisms and the metabolic products derived from the particulate substrate. The protective shell created around the particulate material constrains the diversity of associated organisms and their biochemical capabilities (see Section IV.B) and limits access of enzymes from external organisms to the encrusted substrate. Continued decomposition will depend on the removal of the enveloping clay-organic shell or, in the case of a particulate organic core within an aggregate, on the fracturing of the aggregate.

The process of encrustation of particulate organic material applies at the scale of microaggregates. In macroaggregates (>250 μm in diameter), it is common to see fragments of roots encrusted to some extent with inorganic materials. In smaller aggregates, elongate shapes can be observed, the result of partial or complete encapsulation of cellular remnants, probably the lignified skeletons of plant tissue. Such aggregates cannot be found when soils are subjected to vigorous dispersion procedures but may be found after slaking during rapid wetting.

Based on systematic scanning electron microscopy, Waters and Oades [14] suggested a sequence from encrusted roots to encrusted particulate organic matter at a scale of about 100 μm, to smaller aggregates showing cavities thought to be left after complete decomposition of particulate materials, i.e., comparable to root channels. The principle also applies at smaller scales (see Section II. C).

Visual and experimental evidence has linked decomposition with soil aggregation. A conceptual model that describes the simultaneous dynamics of microaggregates and their organic cores has been proposed by Golchin et al. [165]. The data which form the basis of the model were obtained by examination of density fractions obtained from soils before and after disruption of aggregates by ultrasonic dispersion (Figure 16). The free light fraction was floated from the soil using sodium polytungstate solution, of a density 1.6 Mg/m^3. During decomposition of this light fraction material in soil, there is an interaction between clay materials and microorganisms and metabolites formed from the primary substrate to form a stable aggregate with an organic core. These aggregates resist ultrasonic disruption and exhibit the characteristics shown in Figure 16. With time, decomposition of the organic core continues, albeit slowly, and the polysaccharide-based gels responsible for the stability of the aggregates are also decomposed to some extent. The aggregate is no longer totally stable and, after ultrasonic treatment of the soil, organic-rich material is found in a lighter fraction with a higher C:N ratio. In cases where decomposition of the core is even more extensive and the polysaccharide mucilages are completely decomposed, the core, on disruption, is separated at a density <2.0 Mg/m^3. The residual core material has a high C:N ratio, and has high contents of aromatic and aliphatic components.

The high-density components represent the clay-organic associations that develop as the primary substrate is decomposed. They show a chemical composition similar to clay-size fractions, with significant contents of polymethylene components, proteins, and some carbohydrates (presumably all of microbial

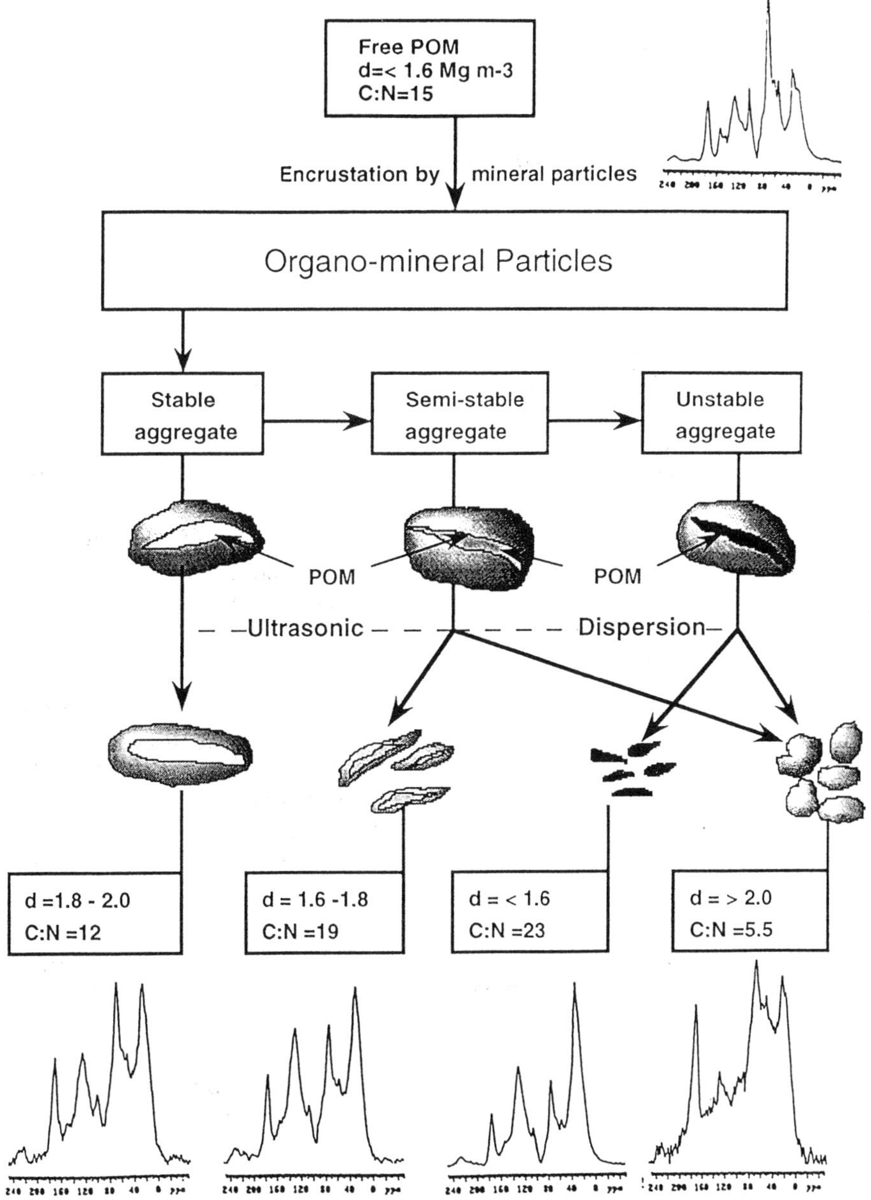

Figure 16 Role of particulate organic matter in the dynamics of soil aggregation (from Ref. 165).

origin). The cores obtained from the unstable aggregates have been shown to have persisted for decades, by using $\delta^{13}C$ procedures after changes in vegetation from C-3 to C-4 plants.

E. Predator-Prey Interactions

Predator-prey relationships and their impact on carbon and nutrient turnover in soil and the rhizosphere have been the subject of numerous studies [54,166–172], as have the specific effects of soil texture and structure (habitable pore space) on predator-prey interactions [11,44,173–184]. A simulation model of carbon turnover [109] emphasizes the differences in the decay rates of microbial biomass carbon depending on the location of decomposer cells with particulate organic residues or with (principally) silt-size clay aggregates. The operation of the model requires that the soil matrix confers a measure of protection on the microbiota from predation, the level of protection being a characteristic of the soil and being directly related to the clay content of the soil.

The various mechanisms by which the soil microbiota and their predators are able to coexist in soils have been reviewed by Alexander [2]. These mechanisms include location of the soil microbiota in pores of sufficiently small neck diameters to prevent access of larger predator organisms, and threshold or critical prey densities (cell numbers or biomass per unit of soil pore volume) below which too little energy is obtained by the predator to maintain its active search for food.

Both mechanisms are supported directly by experimental evidence and implicate soil structure and clay content as determinants of survivor numbers and biomass. The appeal of the second mechanism is that the microbiota may survive in relatively large soil pores from which predators are not excluded, and are able to respond quickly to the additions of substrates, even insoluble materials. Rapid growth of microbiota to above-threshold levels in large pores ensures, in turn, rapid responses in predator numbers and activities and subsequently a return to a protected level of biomass, which in part is comprised of newly formed cells at the expense of cells formerly protected before substrate addition.

The principle of microbial survival by virtue of their location in soil pores from which predators are excluded has been tested in several ways. Heijnen et al. [185], Rattray et al. [186], and Wright et al. [187] found that the rates of survival in soils of introduced bacteria were greater when the organisms were inoculated into relatively dry soils. Inoculation of drier soils favored movement by mass flow of the introduced bacteria to sites within pores of relatively small neck diameters, thus affording them greater protection from attack by larger size predators. White et al. [188] also have demonstrated that cells of bioluminescence-marked *Pseudomonas fluorescens*, when introduced into pores of smaller neck diameter, were less sensitive to chloroform fumigation.

Further support for the "predator exclusion" principle of microbial survival was provided by Heijnen et al. [185,189,190], who showed that the survival in soil of introduced bacterial cells was improved by the addition of bentonite clay. Amendment of soil with bentonite increased the number of pores of neck diameters <6 μm. Premixing of bentonite with the inoculum was more effective for bacterial survival than mixing bentonite with the soil before inoculation.

The rationale for survival of introduced organisms according to their pore-size location can be extended to decomposer organisms growing in situ on substrates introduced to different locations within the soil matrix. Killham et al. [49] introduced ^{14}C-glucose into soil pores of two size classes (nominally of neck diameters of <6 μm, and 6 to 30 μm) of a well-aggregated Vertisol. The decomposition of glucose itself was unaffected by its pore-size location, but the turnover of decomposer biomass ^{14}C located in pores of the smaller neck diameter was slower. These observations were consistent with the hypothesis that cells formed in pores of small pore necks were protected by virtue of the exclusion of larger microfaunal predators.

Precise estimates of critical pore neck size have been made from studies of the survival of cells of specific bacterial species introduced to soils. However, given the diversity of predators and prey involved in the processes of carbon and nutrient turnover in soils, critical pore neck diameter(s) necessary to constrain biological activities would be anticipated to range widely within and between soils.

The principle of microbial survival by virtue of the existence of threshold densities of prey numbers below which predators cannot maintain active grazing has been derived from direct experimentation on the relationships between prey densities and predator growth rates [2]. Even in suspensions without soil, large numbers of protozoa do not eliminate prey organisms entirely [191]. Microbial survival is further enhanced by the presence of soil or of clays or of colloidal soil organic matter. For those added bacteria that do survive in soil, their numbers appear to be at or near some steady-state level [2].

F. Organic Residues of Soil Fractions

Medium-Term Incubations with Isotope-Labeled Substrates

Although the potential for the production of artifacts is acknowledged in any fractionation procedure, analysis of the behavior of biomass and nonbiomass carbon and nitrogen of particle-size fractions over time and in relation to other properties of the fractions has aided in the interpretation of the dynamics of carbon and nitrogen turnover in whole soil aggregates. Appraisal of qualitative and quantitative differences between fractions has been favored by the use of mild techniques for (partial) dispersion of soils with minimal loss of microbial biomass. For example, partial dispersion of soils by the Brückert technique [37] and particle-size fractionation have been carried out with soils amended and

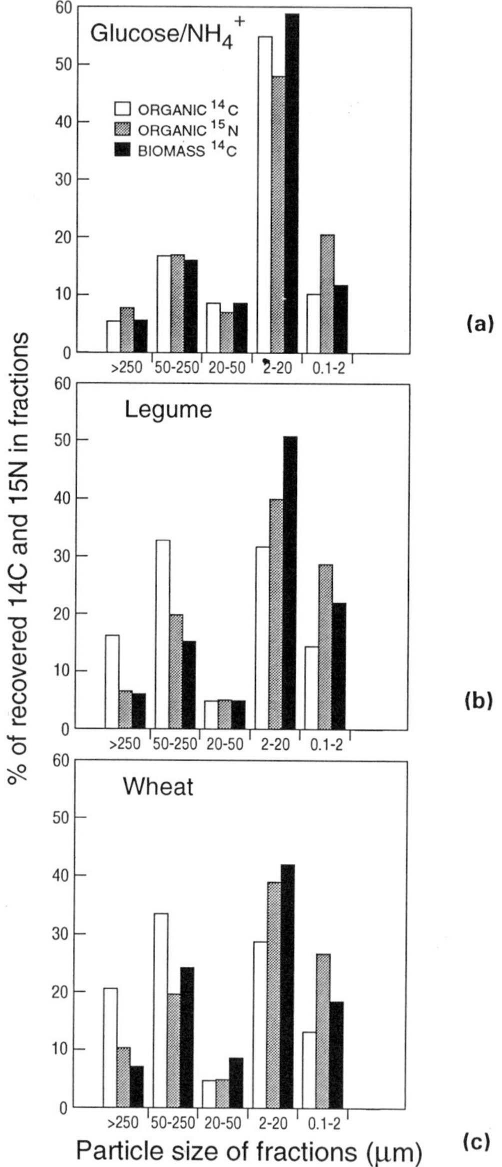

Figure 17 Percentage distribution of organic ^{14}C and ^{15}N and biomass ^{14}C in particle-size fractions of a Vertisol incubated with ^{14}C-, ^{15}N-labeled (a) glucose/NH_4^+ for 3 days, (b) legume tops for 49 days, (c) wheat straw for 49 days. (Adapted from J. N. Ladd, M. van Gestel, L. Jocteur Monrozier, and M. Amato. Soil Biol. Biochem, submitted for publication.)

incubated with either ^{14}C-glucose/^{15}N-NH_4 or ^{14}C, ^{15}N-labeled plant residues (J. N. Ladd, M. van Gestel, L. Jocteur Monrozier, and M. Amato, unpublished data). Irrespective of soil, substrate amendment, and incubation time, isotope-labeled organic carbon and nitrogen and biomass carbon were bimodally distributed (Figure 17). Most of the residual ^{14}C and ^{15}N was recovered in the fine silt-size fraction (2 to 20 μm) and a lesser peak of labeled residues was accounted for in the fine sand-size fraction (50 to 250 μm) as was found also for native carbon and nitrogen of unamended soils (compare Figures 17 and 11).

More specific comparisons between the amounts of organic carbon and nitrogen (labeled and unlabeled) in the fractions with time revealed that

1. The percentage distributions of recovered biomass ^{14}C and total biomass carbon were broadly similar over the entire size range of the fractions. An exception to the general observations was the relatively greater percentage decrease with time of biomass ^{14}C than total biomass carbon associated with the coarse sand-size fraction. Nevertheless, the factors that determined the distribution within the soil matrix of the great majority of biomass ^{14}C that accumulated after metabolism and turnover of specific substrate amendments also appeared to determine the distribution of native biomass carbon. Soil factors that affect microbial survival appeared to be of overriding importance in determining concentrations of microbial biomass. The direct relationship observed between concentrations of biomass ^{14}C and total biomass carbon, for a range of particle-size fractions from a given soil, was also observed for a range of unfractionated soils of varying clay contents incubated with ^{14}C-labeled glucose or legume residues and assayed 44 and 66 weeks after cessation of active metabolism [126].
2. Although total organic ^{14}C and ^{15}N and biomass ^{14}C were, in general, similarly distributed in soil particle-size fractions in the Vertisol amended with legume and wheat materials, the fine sand-size fraction (50 to 250 μm) accumulated disproportionately higher amounts of total organic ^{14}C than of biomass ^{14}C and to a lesser extent of organic ^{15}N. The results indicated that even after 6 years of decomposition (when only about 10 to 15% of input ^{14}C remained in the total soil), partly decomposed plant residues accounted for a significant minority of the residual ^{14}C (see Section IV. D).
3. The coarse clay-size fraction (0.1 to 2 μm) of both glucose- and plant residue-amended soils accumulated disproportionately higher amounts of organic ^{15}N than of biomass ^{14}C in the short to moderate term. Such differences disappeared on longer (6 years) incubation. The results indicated that those inorganic soil components that dispersed readily as clay-size materials preferentially became associated with exocellular nitrogenous organics formed during early periods of active metabolism.

Such clay-bound organic ^{15}N components were either subsequently metabolized during longer incubation or the soil microaggregates in which they were bound were less readily dispersed. Alternatively, the results may have stemmed from the formation of artifacts resulting from adsorption of exocellular metabolites on clays during soil dispersion before fractionation. There was, however, no evidence to suggest that the exocellular nitrogenous organics were themselves artifacts of the dispersion technique. High recoveries of biomass ^{14}C after fractionation minimized the probability that cell destruction during fraction preparation caused the release of the nitrogenous organics.

Interpretation of net changes in the amounts of biomass carbon or organic carbon and nitrogen of fractions with time remains superficial without further knowledge of the biological and physicochemical mechanisms of carbon and nitrogen transfers between fractions, and without accommodation of the effects with time of a changing ease of soil dispersion. In the fractionation studies discussed above, the trend was for a greater percentage decrease in the biomass ^{14}C that was formed and located with sand-size fractions (collectively >50 μm) than for biomass ^{14}C located with silt plus clay-size components. Biomass ^{14}C associated with clay microaggregates (2 to 20 μm) of a Vertisol appeared to be no more stabilized than biomass ^{14}C in the dispersed clay-size (0.1 to 2 μm) fraction. Total biomass carbon of the undispersed soils changed far less with time than did biomass ^{14}C, and net changes in biomass carbon of the >50 μm and <50 μm fractions appeared to be heavily influenced by small changes in the distribution of fraction weights at different sampling times.

Persistent Organic Residues of Soil Fractions

The use of ^{14}C dating to determine mean residence times for carbon in soils and soil fractions has not led to unequivocal conclusions. Organic matter in some soils, e.g., Mollisols, has been shown to have persisted for thousands of years, whereas in other soils more influenced by modern carbon inputs, organic matter has been dated as modern. One conclusion that can be drawn (with relatively few exceptions, such as with Spodosols) is that the age of organic matter increases with depth to several meters. At 1 m depth, the mean residence time of organic matter for a range of soils is several thousand years [33].

Recent investigations using CPMAS ^{13}C NMR of organic materials to depths of about 1 m in an Alfisol found little change in the chemistry of organic materials with depth. Thus, the greater age of organic matter at depth cannot be attributed to a particular group of organic chemicals. The simple explanation for the apparent changes of mean residence time with depth is that there is "old" carbon associated with clay surfaces throughout the soil, and that "age" profiles are simply a dilution of this older carbon by modern carbon from current biological activities. This explanation cannot be proven. Indeed, there are a number of studies that indicate that the organic matter of coarser size fractions up to about 20 μm in diameter are older than that of other fractions [9,33]. Nevertheless, interpretations of such data still can rest on the variable extent of dilution of "old" organic matter with modern carbon. Stable aggregates with "old" internal organic materials may not be influenced by the modern carbon cycle, whereas clay materials not in aggregates may be associated with modern carbon. Any association of "old" carbon with the clay would be masked by ^{14}C from modern carbon.

Some support for this explanation is given by the work of Skjemstad et al. [192], who have shown that the organic material on the outside of microaggregates was modern, presumably because external surfaces interact with organics from the modern carbon cycle, whereas that inside the microaggregates was much older. The differentiation was made using high-energy, UV oxidation of organic materials. The technique preferentially oxidizes organics located on external surfaces, and organic matter inside aggregates is relatively protected. Possible complications resulting from the creation of artifacts during fractionation procedures are recognized.

There remains the possibility that specific chemicals may be so difficult to decompose that they survive pedologic processes and accumulate in soils in significant amounts. Two classes of compound may be suggested. One is charcoal, which is present in a range of soils and sediments and which may cause complications with respect to carbon dating. Charcoal is believed to be inert and not available to microorganisms. Charcoal is chemically complex, but it may represent significant portions of the aromatic carbon in soils under grasslands and cannot readily be distinguished from the aromatic components of humic compounds [192]. A second group of compounds that may persist in soils are alkyl nitriles, produced by protozoa. These compounds have been described by Largeau [193] from marine systems, but they could also be present in clays and account for the low C:N ratios and large alkyl peaks as shown by CPMAS ^{13}C NMR.

REFERENCES

1. McLaren, A. D., and G. H. Peterson. 1967. Introduction to the biochemistry of terrestrial soils, p. 1–13. *In* A.D. McLaren and G. H. Peterson (eds.), Soil biochemistry. Marcel Dekker, New York.
2. Alexander, M. 1981. Why microbial predators and parasites do not eliminate their prey and hosts. Ann. Rev. Microbiol. 35:113–133.
3. Stotzky, G. 1986. Influence of soil mineral colloids on metabolic processes, growth, adhesion, and ecology of microbes and viruses, p. 305–428. *In* M. Huang and M. Schnitzer (eds.), Interactions of soil minerals with natural organics and microbes. Special Publication No. 17. Soil Science Society of America, Madison, WI.

4. Nedwell, D. B., and T.R.G. Gray. 1987. Soils and sediments as matrices for microbial growth, p. 21–54. *In* M. Fletcher, T. R. G. Gray, and J. G. Jones (eds.), Ecology of microbial communities. Cambridge University Press, Cambridge.
5. Foster, R. C. 1988. Microenvironments of soil microorganisms. Biol. Fertil. Soils 6: 189–203.
6. Hattori, T. 1988. Soil aggregates as microhabitats of microorganisms. Rep. Inst. Agr. Res. Tohoku Univ. 37:23–36.
7. van Veen, J. A. and P. J. Kuikman. 1990. Soil structural aspects of decomposition of organic matter by microorganisms. Biogeochemistry 11:213–233.
8. Oades, J. M., and A. G. Waters. 1991. Aggregate hierarchy in soils. Austr. J. Soil Res. 29:815–828.
9. Christensen, B. T. 1992. Physical fractionation of soil and organic matter in primary particle size and density separates, p. 1–90. *In* B. A. Stewart (ed.), Advances in soil science, Vol. 20. Springer-Verlag, New York.
10. Robert, M., and C. Chenu. 1992. Interactions between soil minerals and microorganisms, p. 307–404. *In* G. Stotzky and J.-M. Bollag (eds.), Soil biochemistry, Vol. 7. Marcel Dekker, New York.
11. Juma, N. G. 1993. Interrelationships between soil structure/texture, soil biota/soil organic matter and crop production. Geoderma 57:3–30.
12. Ladd, J. N., R. C. Foster, and J. O. Skjemstad. 1993. Soil structure: carbon and nitrogen metabolism. Geoderma 56:401–434.
13. Oades, J. M. 1993. The role of biology in the formation, stabilization and degradation of soil structure. Geoderma 56:377–400.
14. Waters, A. G., and J. M. Oades. 1991. Organic matter in water-stable aggregates, p. 163–174. *In* W. S. Wilson (ed.), Advances in soil organic matter research: the impact on agriculture and the environment. Royal Society of Chemistry, Cambridge.
15. Edwards, A. P., and J. M. Bremner. 1967. Dispersion of soil particles by sonic vibration. J. Soil Sci. 18:47–63.
16. Tisdall, J. M., and J. M. Oades. 1982. Organic matter and water-stable aggregates in soils. J. Soil Sci. 33:141–163.
17. Oades, J. M. 1984. Soil organic matter and structural stability: mechanisms and implications for management. Plant Soil 76:319–337.
18. Emerson, W. W. 1977. Physical properties and structure, p. 78–104. *In* J. S. Russell and E. L. Greacen (eds.), Soil factors in crop production in a semi-arid environment. University of Queensland Press, St. Lucia, Queensland.
19. Oades, J. M. 1987. Associations of colloidal materials in soils. Trans. XIII Congress of ISSS, Hamburg, 1986, VI:660–674.
20. Emerson, W. W., R. C. Foster, and J. M. Oades. 1986. Organo-mineral complexes in relation to soil aggregation and structure, p. 521–548. *In* P. M. Huang and M. Schnitzer (eds.), Interactions of soil minerals with natural organics and microbes. Special Publication No. 17. Soil Science Society of America, Madison, WI.
21. Churchman, G. J., and R. C. Foster. 1994. The role of clay minerals in the maintenance of soil structure. Trans. 15th Congress of ISSS, Acapulco, Mexico, 1994, 8a: 17–34.
22. Foster, R. C. 1994. Microorganisms and soil aggregates, p. 144–155. *In* C. E. Pankhurst, B. M. Doube, V. V. S. R. Gupta, and P. R. Grace (eds.), Soil biota.

Management in sustainable farming systems. CSIRO, East Melbourne, Victoria, Australia.

23. Foster, R. C., A. D. Rovira, and T. W. Cock. 1983. Ultrastructure of the root-soil interface. American Phytopathological Society, St. Paul, MN.
24. Tisdall J. M., and J. M. Oades. 1980. The management of ryegrass to stabilize aggregates of a red-brown earth. Austr. J. Soil Res. 18:415–422.
25. Foster, R. C. 1985. *In situ* localization of organic matter in soils. Quaestiones Entomologicae 21:609–633.
26. Jocteur Monrozier, L., J. N. Ladd, R. W. Fitzpatrick, R. C. Foster, and M. Raupach. 1991. Components and microbial biomass content of size fractions in soils of contrasting aggregation. Geoderma 49:37–62.
27. Foster, R. C. 1981. The ultrastructure and histochemistry of the rhizosphere. New Phytol. 88:263–273.
28. Bland, D. E., R. C. Foster, and A. F. Logan. 1971. The mechanism of permanganate and osmium tetroxide fixation and the distribution of lignin in the cell wall of *Pinus radiata*. Holzforschung 25:137–143.
29. Foster, R. C., and J. K. Martin. 1981. *In situ* analysis of soil components of biological origin, p. 75–110. *In* E. A. Paul and J. N. Ladd (eds.), Soil biochemistry, Vol. 5. Marcel Dekker, New York.
30. Lee, K. E., and R. C. Foster. Soil fauna and soil structure. Austr. J. Soil Res. 29: 745–775.
31. Oades, J. M., and J. N. Ladd. 1977. Biochemical properties: carbon and nitrogen metabolism, p. 127–160. *In* J. S. Russell and E. L. Greacen (eds.), Soil factors in crop production in a semi-arid environment. University of Queensland Press, St. Lucia, Queensland.
32. Turchenek, L. W., and J. M. Oades. 1979. Fractionation of organo-mineral complexes by sedimentation and density techniques. Geoderma 21:311–343.
33. Oades, J. M. 1989. An introduction to organic matter in mineral soils, p. 89–159. *In* J. B. Dixon and S. B. Weed (eds.), Minerals in soil environments. Soil Science Society of America, Madison, WI.
34. Stevenson, F. J., and E. T. Elliott. 1989. Methodologies for assessing the quantity and quality of soil organic matter, p. 173–179. *In* D. C. Coleman, J. M. Oades, and G. Uehara (eds.), Dynamics of soil organic matter in tropical ecosystems. NifTAL Project, University of Hawaii, Manoa, HI.
35. Christensen, B. T. 1995. Carbon in primary and secondary organomineral complexes. *In* B. A. Stewart (ed.), Advances in soil science. Springer-Verlag, New York.
36. Baldock, J. M., J. M. Oades, A. G. Waters, X. Peng, A. M. Vassallo, and M. A. Wilson. 1992. Aspects of the chemical structure of soil organic materials as revealed by solid-state ^{13}C NMR spectroscopy. Biogeochemistry 16:1–42.
37. Brückert, T. S. 1979. Analyse des complexes organo-minéreaux des sols. *In* D. Duchaufour and B. Souchier (eds.), Pédologie, 2: Constituants et propriétiés du sol. Masson, Paris.
38. Ladd, J. N., M. Amato, L. Jocteur Monrozier, and M. van Gestel. 1990. Soil microhabitats and carbon and nitrogen metabolism. Trans. 14th Congress of ISSS, Kyoto 1990, III:82–87.

39. Bae, H. C., E. H. Cota-Robles, and L. E. Casida. 1972. Microflora of soil as viewed with transmission electron microscopy. Appl. Microbiol. 23:637–648.
40. Murphy, C. P. 1986. Thin section preparation of soils and sediments. Academic Publishers, Berkhamsted, U.K.
41. Lawrence, J. T., and J. J. Germida. 1991. Microbial and chemical characteristics of elemental sulfur beads in agricultural soils. Soil Biol. Biochem. 23:617–622.
42. Postma, J., and H. J. Altemuller. 1990. Bacteria in thin soil sections stained with the fluorescent brightener Calcofluor white M2R. Soil Biol. Biochem. 22:89–96
43. Morgan, P., C. J. Cooper, N. S. Battersby, S. A. Lee, G. T. Lewis, T. M. Machin, S. C. Graham, and R. S. Watkinson. 1991. Automatic image analysis method to determine fungal biomass in soils and solid matrices. Soil Biol. Biochem. 23:609–616.
44. Hattori, T., and R. Hattori. 1976. The physical environment in soil microbiology. An attempt to extend principles of microbiology to soil microorganisms. CRC Crit. Rev. Microbiol. 4:423–461.
45. Ladd, J. N., and R. C. Foster. 1988. Role of soil microflora in nitrogen turnover, p. 113–144. *In* J. R. Wilson (ed.), Advances in nitrogen cycling in agricultural ecosystems. CAB International, Wallingford, U.K.
46. Foster, R. C. 1985. The biology of the rhizosphere, p. 75–79. In C. A. Parker, A. D. Rovira, K. J. Moore, P. T. W. Wong, and J. F. Kollmorgen (eds.), Ecology and management of soil-borne plant pathogens. American Phytopathology Society, St. Paul, MN.
47. Foster, R. C., and A. D. Rovira. 1978. The ultrastructure of the rhizosphere of *Trifolium subterraneum* L., p. 278–289. In M. Loutit (ed.), Proceedings of the 1st International Symposium on Microbial Ecology. Springer-Verlag, Berlin.
48. Dormaar J. F., and R. C. Foster. 1991. Nascent aggregates in the rhizosphere of perennial ryegrass (*Lolium perenne* L.). Can. J. Soil Sci. 71:465–474.
49. Killham, K., M. Amato, and J. N. Ladd. 1993. Effect of substrate location in soil and soil pore-water regime on carbon turnover. Soil Biol. Biochem. 25:57–62.
50. McGee, P. A., S. E. Smith, and F. A. Smith. 1989. Research directions on the structure and function of the plant-microbe interfaces. Austr. J. Plant Physiol. 16:1–3.
51. Chenu, C. 1989. Influence of a fungal polysaccharide, scleroglucan on clay microstructure. Soil Biol. Biochem. 21:299–305.
52. Foster, R. C. 1981. Mycelial strands of *Pinus radiata* D. Don: ultrastructure and histochemistry. New Phytol. 88:705–712.
53. Lynch, J. M., and E. Bragg, 1985. Microorganisms and soil aggregate stability, p. 133–171. *In* B. A. Stewart (ed.), Advances in soil science, Vol. 2. Springer-Verlag, New York.
54. Foster, R. C., and J. F. Dormaar. 1991. Bacteria-grazing amoebae *in situ* in the rhizosphere. Biol. Fertil. Soils 11:83–87.
55. Kanazawa, S., and Z. Filip. 1986. Distribution of microorganisms, total biomass, and enzyme activities in different particles of brown soil. Microb. Ecol. 12:205–215.
56. Chotte, J.-L., L. Jocteur Monrozier, G. Villemin, and A. Albrecht. 1993. Soil microhabitats and the importance of the fractionation method, p. 39–45. *In* K. Mulongoy and R. Merckx (eds.), Soil organic matter dynamics and sustainability of tropical agriculture. Wiley-Sayce, New York.

57. Ladd J. N., M. Amato, and J. W. Parsons. 1977. Studies of nitrogen immobilization and mineralization in calcareous soils. III. Concentration and distribution of nitrogen derived from the soil biomass, p. 301–310. In Soil organic matter studies, Vol. I. IAEA, Vienna.
58. Amato, M., and J. N. Ladd. 1980. Studies of nitrogen immobilization and mineralization in calcareous soils. V. Formation and distribution of isotope-labeled biomass during decomposition of ^{14}C- and ^{15}N-labeled plant material. Soil Biol. Biochem. 12:405–411.
59. Ahmed, M., and J. M. Oades. 1984. Distribution of organic matter and adenosine triphosphate after fractionation of soils by physical procedures. Soil Biol. Biochem. 16:465–470.
60. Lemoine, M. C., V. Gianinazzi-Pearson, S. Gianinazzi, and C. J. Straker. 1992. Occurrence and expression of acid phosphatase of *Hymenoscyphus ericae* (Reid) Korf and Kerman in isolation or associated with plant roots. Mycorrhiza 1:137–146.
61. Hayat, M. A. 1989. Principles and techniques of electron microscopy. Biological Applications. CRC Inc., Boca Raton, FL.
62. Lewis, P. R., and Knight, D. P. 1977. Staining methods in electron microscopy. North Holland Publishing Co., Amsterdam.
63. Gianinazzi-Pearson, V., S. E. Smith, S. Gianinazzi, and F. A. Smith. 1990. Enzymatic studies on the metabolism of vesicular-arbuscular mycorrhizas. New Phytol. 117:61–74.
64. Heritage, A. D., and R. C. Foster. 1984. Catalase and sulfur in the rice rhizosphere: An ultrastructural, histochemical demonstration of a symbiotic relationship. Microb. Ecol. 10:115–121.
65. Smith, S. E., and F. A. Smith. 1990. Structure and function of the interfaces in biotrophic symbioses as they relate to nutrient transport. New Phytol. 114:2–38.
66. Rohde, M., H. Gerberding, T. Mund, and G-W. Kohring. 1988. Immunoelectron microscopic localization of bacterial enzymes: pre- and post-embedding labelling techniques on resin-embedded samples, p. 175–210. *In* F. Meyer (ed.), Methods in microbiology, Vol. 20. Academic Press, London.
67. Ladd, J. N. 1978. Origin and range of enzymes in soils, p. 51–96. *In* R. G. Burns (ed.), Soil enzymes. Academic Press, London.
68. Burns, R. G. 1978. Enzyme activity in soil: some theoretical and practical considerations. p. 205–196. *In* R. G. Burns (ed.), Soil enzymes. Academic Press, London.
69. Burns, R. G. 1982. Enzyme activity in soil: location and possible role in microbial ecology. Soil Biol. Biochem. 14:423–427.
70. Ladd, J. N. 1985. Soil enzymes, p. 175–221. *In* D. Vaughan and R. E. Malcolm (eds.), Soil organic matter and biological activity. Martinus Nijhoff/Dr. W. Junk Publishers, Dordrecht, Netherlands.
71. Nannipieri, P. 1995. Enzyme activity. *In* C. W. Finkl (ed.), The encyclopedia of soil science. Chapman and Hall, New York.
72. Amato, M., and J. N. Ladd 1988. Assay for microbial biomass based on ninhydrin-reactive nitrogen in extracts of fumigated soils. Soil Biol. Biochem. 20:107–114.
73. Nannipieri, P., P. Fusi, and P. Sequi. 1995. Humus and enzyme activity. *In* A. Piccolo (ed.), Humic substances in terrestrial ecosystems. Elsevier Applied Science, New York.

74. Nannipieri, P. 1994. The potential use of soil enzymes as indicators of productivity, sustainability and pollution, p. 238–244. *In* C. E. Pankhurst, B. M. Doube, V. V. S. R. Gupta and P. R. Grace (eds.), Soil biota. Management in sustainable farming systems. CSIRO, East Melbourne, Victoria, Australia.
75. Boyd, S. A., and M. M. Mortland. 1990. Enzyme interactions with clays and clay-organic matter complexes, p. 1–28. In J.-M. Bollag and G. Stotzky (eds.), Soil biochemistry, Vol. 6. Marcel Dekker, New York.
76. Gianfreda, L., M. A. Rao, and A. Violante. 1992. Adsorption, activity and kinetic properties of urease on montmorillonite, aluminium hydroxide and $Al(OH)_X$-montmorillonite complexes. Soil Biol. Biochem. 24:51–58.
77. Fusi, P., G. Ristori, L. Calamai, and G. Stotzky. 1989. Adsorption and binding of protein on "clean" and "dirty" (coated with Fe oxyhydroxides) montmorillonite, illite and kaolinite. Soil Biol. Biochem. 21:911–920.
78. Rowell, M. J., J. N. Ladd, and E. A. Paul. 1973. Enzymatically active complexes of proteases and humic acid analogues. Soil Biol. Biochem. 5:699–703.
79. Burns, R. G. 1986. Interactions of enzymes with soil minerals and organic colloids, p. 429–431. *In* P. M. Huang and M. Schnitzer (eds.), Interactions of soil minerals with natural organics and microbes. SSSA Special Publication No. 17. Soil Science Society of America, Madison, WI.
80. Sarkar, J. M., and R. G. Burns. 1984. Synthesis and properties of β-D-glucosidase-phenolic copolymers as analogues of soil humic-enzyme complexes. Soil Biol. Biochem. 16:619–625.
81. Leonowicz, A., and J.-M. Bollag. 1987. Laccases in soil and the feasibility of their extractions. Soil Biol. Biochem. 19:237–242.
82. Tabatabai, M. A., and M. Fu. 1991. Extraction of enzymes from soils, p. 197–227. *In* G. Stotzky, and J.-M. Bollag (eds.), Soil biochemistry, Vol. 7. Marcel Dekker, New York.
83. Nannipieri, P., S. Cervelli, and F. Pedrazzini. 1973. Concerning the extraction of enzymatically active organic matter from soil. Experientia 31:513–514.
84. Nannipieri, P., B. Ceccanti, S. Cervelli, and P. Sequi. 1974. Use of 0.1 M pyrophosphate to extract urease from podzol. Soil Biol. Biochem. 6:359–362.
85. Mayaudon, J. 1986. The role of carbohydrates in the free enzymes in soil, p. 263–309. *In* C. H. Fuchsman (ed.), Peat and water. Elsevier Applied Science, New York.
86. Burns, R. G., A. H. Pukite, and A. D. McLaren. 1972. Concerning the location and persistence of soil urease. Soil Sci. Soc. Am. J. 36:308–311.
87. Nannipieri, P., B. Ceccanti, S. Cervelli, and P. Sequi. 1978. Stability and kinetic properties of humus-urease complexes. Soil Biol. Biochem. 10:143–147.
88. Nannipieri, P., B. Ceccanti, and D. Bianchi. 1988. Characterization of humus-phosphatase complexes extracted from soil. Soil Biol. Biochem. 20:683–691.
89. Ladd, J. N., and J. H. A. Butler. 1975. Humus-enzyme systems and synthetic organic polymer-enzyme analogs, p. 143–194. *In* E. A. Paul and A. D. McLaren (eds.), Soil biochemistry, Vol. 4. Marcel Dekker, New York.
90. McLaren, A. D., and G. V. F. Searman. 1968. Concerning the surface pH of clays. Soil Sci. Soc. Am. Proc. 32:127.
91. Skujins, J. J., and A. D. McLaren. 1967. Enzyme reaction rates at limited water activities. Science 158:1569-1570.

92. Goldstein, L., Y. Levin, and E. Katchalski. 1964. A water-insoluble polyanionic derivative of trypsin. II. Effect of the polyelectrolyte carrier on the kinetic behavior of the bound trypsin. Biochemistry 3:1913–1919.
93. Nannipieri, P., B. Ceccanti, S. Cervelli, and C. Conti. 1982. Hydrolases extracted from soil: kinetic parameters of several enzymes catalysing the same reaction. Soil Biol. Biochem. 14:429–432.
94. McLaren, A. D., and J. J. Skujins. 1971. Trends in the biochemistry of terrestrial soils, p. 1–15. *In* A. D. McLaren and J. J. Skujins (eds.), Soil biochemistry, Vol. 2. Marcel Dekker, New York.
95. Ross, D J. 1975. Studies on a climosequence of soils in tussock grassland. 7. Distribution of invertase and amylase activities in soil fractions. New Zeal. J. Sci. 18: 527–534.
96. Rojo, M. J., S. G. Carcedo, and M. P. Mateos. 1990. Distribution and characterization of phosphatase and organic phosphorus in soil fractions. Soil Biol. Biochem. 22:169–174.
97. Mateos, M. P., and S. G. Carcedo. 1985. Effect of fractionation on location of enzyme activities in soil structural units. Biol. Fertil. Soils 1:153–159.
98. Mateos, M. P., and S. G. Carcedo. 1987. Effect of fractionation on the enzymatic state and behaviour of enzyme activities in different structural soil units. Biol. Fertil. Soils 3:151–154.
99. Gupta, V. V. S. R., and J. J. Germida. 1988. Distribution of microbial biomass and its activity in different soil aggregate size classes as affected by cultivation. Soil Biol. Biochem. 20:777–786.
100. Sequi, P., G. Cercignani, M. de Nobili, and M. Paglia. 1985. A positive trend among two soil enzyme activities and a range of soil porosity under zero and conventional tillage. Soil Biol. Biochem. 17:255–256.
101. Tabatabai, M. A. 1973. Michaelis constant of urease in soils and soil fractions. Soil Sci. Soc. Am. Proc. 37:707–710.
102. Waksman, S. A. 1952. Decomposition of plant and animal residues in soils and in composts, p. 95–123. *In* Soil microbiology. John Wiley & Sons, New York.
103. Jenkinson, D. S. 1981. The fate of plant and animal residues in soil, p. 505–561. *In* D. J. Greenland and M. H. B. Hayes (eds.), The chemistry of soil processes. John Wiley & Sons, New York.
104. Jenkinson, D. S. 1965. Studies on the decomposition of plant material in soil. I. Losses of carbon ^{14}C labeled rye grass incubated with soil in the field. J. Soil Sci. 16:104–115.
105. Amato, M., R. B. Jackson, J. H. A. Butler, and J. N. Ladd. 1984. Decomposition of plant material in Australian soils. II. Residual organic ^{14}C and ^{15}N from legume plant parts decomposing under field and laboratory conditions. Austr. J. Soil Res. 22:331–341.
106. Reinertsen, S. A., L. F. Elliott, V. L. Cochran, and G. S. Campbell. 1984. Role of available carbon and nitrogen in determining the rate of wheat straw decomposition. Soil Biol. Biochem. 16:459–464.
107. Collins, H. P., L. F. Elliott., and R. I. Papendick. 1990. Wheat straw decomposition and changes in decomposability during field exposure. Soil Sci. Soc. Am. J. 54: 1013–1016.

108. Paul, E. A., and J. A. van Veen. 1978. The use of tracers to determine the dynamic nature of organic matter. Trans. 11th Congress of ISSS, Edmonton, 1978, 3:61–102.

109. Ladd, J. N., M. Amato, P. R. Grace, and J. A. van Veen. 1995. Simulation of ^{14}C turnover through the microbial biomass in soils incubated with ^{14}C-labeled plant residues. Soil Biol. Biochem. 27:777–783.

110. Cerri, C., C. Feller, J. Balesdent, R. Victoria, and A. Plenecassagne. 1985. Application du tracage isotopique naturel en ^{13}C, a létude de la dynamique de la matiere organique dans les sols. C. R. Acad. Sc. Paris, Serie II, 300:423–428.

111. Balesdent, J., A. Mariotti, and B. Guillet. 1987. Natural ^{13}C abundance as a tracer for studies of soil organic matter dynamics. Soil Biol. Biochem. 19:25–30.

112. Balesdent, J., G. H. Wagner, and A. Mariotti. 1988. Soil organic matter turnover in long-term field experiments as revealed by carbon-13 natural abundance. Soil Sci. Soc. Am. J. 52:118–124.

113. Skjemstad, J. O., R. P. Le Feuvre, and R. E. Prebble. 1990. Turnover of soil organic matter under pasture as determined by ^{13}C natural abundance. Austr. J. Soil Res. 28: 267–276.

114. Bonde, R. A., B. T. Christensen, and C. C. Cerri. 1992. Dynamics of soil organic matter as reflected by natural ^{13}C abundance in particle size fractions of forested and cultivated oxisols organic matter. Soil Biol. Biochem. 23:275–277.

115. Cambardella, C. A., and E. T. Elliott. 1993. Carbon and nitrogen distribution in aggregates from cultivated and native grassland soils. Soil Sci. Soc. Am. J. 57: 1071–1076.

116. Golchin, A., J. M. Oades, J. O. Skjemstad, and P. Clarke. 1994. Structural and dynamic properties of soil organic matter as reflected by ^{13}C natural abundance, pyrolysis mass spectrometry and solid-state ^{13}C NMR spectroscopy in density fractions of an oxisol under forest and pasture. Austr. J. Soil Res. 33:59–76.

117. Jenny, H. 1941. Factors of soil formation. McGraw-Hill, New York.

118. Sørensen, L. H. 1975. The influence of clay on the rate of decay of amino acid metabolites synthesized in soils during decomposition of cellulose. Soil Biol. Biochem. 7:171–177.

119. Jenkinson, D. S. 1977. Studies on the decomposition of plant material in soil. V. The effects of plant cover and soil type on the loss of carbon from ^{14}C labeled rye grass decomposing under field conditions. J. Soil Sci. 28:424–434.

120. Ladd, J. N., J. M. Oades, and M. Amato. 1981. Microbial biomass formed from ^{14}C, ^{15}N-labeled plant materials decomposing in soils in the field. Soil Biol. Biochem. 13:119–126.

121. Sørensen, L. H. 1981. Carbon-nitrogen relationships during the humification of cellulose in soils containing different amounts of clay. Soil Biol. Biochem. 13:313–321.

122. Ladd, J. N., M. Amato, and J. M. Oades. 1985. Decomposition of plant material in Australian soils III. Residual organic and biomass C and N from isotope-labeled legume material and soil organic matter, decomposing under field conditions. Austr. J. Soil Res. 23:603–611.

123. Merckx, R., A. den Hartog, and J. A. van Veen. 1985. Turnover of root-derived material and related microbial biomass formation in soils of different texture. Soil Biol. Biochem. 17:565–569.

124. van Veen, J. A., J. N. Ladd, and M. Amato. 1985. Turnover of carbon and nitrogen through the microbial biomass in a sandy loam and a clay soil incubated with [^{14}C(U)] glucose and [^{15}N]$(NH_4)_2SO_4$ under different moisture regimes. Soil Biol. Biochem. 17:747–756.

125. van Veen, J. A., J. N. Ladd, J. K. Martin, and M. Amato. 1987. Turnover of carbon, nitrogen and phosphorus through the microbial biomass in soils incubated with ^{14}C-, ^{15}N- and ^{32}P-labeled bacterial cells. Soil Boil. Biochem. 19:559–565.

126. Amato, M., and J. N. Ladd. 1992. Decomposition of ^{14}C-labeled glucose and legume material in soils: Properties influencing the accumulation of organic residue C and microbial biomass C. Soil Biol. Biochem. 24:455–464.

127. Ladd J. N. 1989. The role of the soil microflora in the degradation of organic matter, p. 169–174. *In* T. Hattori, Y. Ishida, Y. Maruyama, R. Y. Morita, and A. Uchida (eds.), Recent advances in microbial ecology. Japan Scientific Societies Press, Tokyo.

128. Ladd, J. N., L. Jocteur Monrozier, and M. Amato. 1992. Carbon turnover and nitrogen transformations in an Alfisol and Vertisol amended with [U-^{14}C] glucose and [^{15}N] ammonium sulfate. Soil Biol. Biochem. 24:359–371.

129. Parton, W. J., D. S. Schimel, C. V. Cole, and D. S. Ojima. 1987. Analysis of factors controlling soil organic matter levels in Great Plains grasslands. Soil Sci. Soc. Am. J. 51:1173–1179.

130. Jenkinson, D. S., P. B. S. Hart, J. H. Rayner, and L. C. Parry. 1987. Modelling the turnover of organic matter in long-term experiments at Rothamsted. INTECOL Bull. 15:1–8.

131. Jenkinson, D. S. 1990. The turnover of organic carbon and nitrogen in soil. Phil. Trans. R. Soc. Lond. B. 329:361–368.

132. Gregorich, E. G., R. P. Voroney, and R. G. Kachanoski. 1991. Turnover of carbon through the microbial biomass in soils with different textures. Soil Biol. Biochem. 23:799–805.

133. Alexander, M. 1977. Introduction to soil microbiology. John Wiley & Sons, New York.

134. Sakamoto, K., and Y. Oba. 1994. Effect of fungal to bacterial biomass ratio on the relationship between CO_2 evolution and total soil microbial biomass. Biol. Fertil. Soils 17:39–44.

135. Cheshire, M. V., M. P. Greaves, and C. M. Mundie. 1974. Decomposition of soil polysaccharide. J. Soil Sci. 25:483–498.

136. Adu, J. K., and J. M. Oades. 1978. Physical factors affecting decomposition of organic materials in soil aggregates. Soil Biol. Biochem. 10:109–115.

137. Rovira, A. D., and E. L. Greacen. 1957. The effect of aggregate disruption on the activity of microorganisms in soil. Austr. J. Agric. Res. 8:659–673.

138. Stevenson, I. L. 1956. Some observations on the microbial activity in remoistened air-dried soils. Plant Soil 8:170–182.

139. Birch, H. F. 1960. Nitrification in soils after different periods of dryness. Plant Soil 12:81–96.

140. Soulides, D. A., and F. E. Allison. 1961. Effect of drying and freezing of soils on carbon dioxide production, available mineral nutrients, aggregation and bacterial population. Soil Sci. 91:291–298.

141. Sørensen, L. H. 1974. Rate of decomposition of organic matter in soil as influenced by repeated drying-rewetting and repeated additions of organic matter. Soil Biol. Biochem. 6:287–292.
142. Jenkinson, D. S. 1966. Studies in the decomposition of plant material in soil. II. Partial sterilization of soil and the soil biomass. J. Soil Sci. 17:280–302.
143. Bottner, P. 1985. Response of microbial biomass to alternate moist and dry conditions in a soil incubated with ^{14}C- and ^{15}N- labeled plant material. Soil Biol. Biochem. 17:329–337.
144. Hattori, T. 1973. The microenvironment of microbes in the soil, p. 263–312. *In* Microbial life in soil: an introduction. Marcel Dekker, New York.
145. Chen, M., and M. Alexander. 1973. Survival of soil bacteria during prolonged desiccation. Soil Biol. Biochem. 5:213–221.
146. Amato, M., and J. N. Ladd. 1980. Studies of nitrogen immobilization and mineralization in calcareous soils. V. Formation and distribution of isotope-labeled biomass during decomposition of ^{14}C and ^{15}N-labeled plant material. Soil Biol. Biochem. 20: 107–114.
147. Sørensen, L. H. 1983. Size and persistence of the microbial biomass formed during the humification of glucose, hemicellulose, cellulose, and straw in soils containing different amounts of clay. Plant Soil 75:121–130.
148. Kieft, L. T., E. Soroker, and M. K. Firestone. 1987. Microbial biomass response to a rapid increase in water potential when dry soil is rewetted. Soil Biol. Biochem. 19: 119–126.
149. van Gestel, M., J. N. Ladd, and M. Amato. 1991. Carbon and nitrogen mineralization from two soils of contrasting texture and microaggregate stability: influence of sequential fumigation, drying and storage. Soil Biol. Biochem. 23:313–322.
150. Hassink, J. 1992. Effects of soil texture and structure on carbon and nitrogen mineralization in grassland soils. Biol. Fertil. Soils 14:126–134.
151. Foster, R. C., and A. D. Rovira. 1976. Ultrastructure of the wheat rhizosphere. New Phytol. 76:343–352.
152. Tester, C. F. 1988. Role of soil and residue microorganisms in determining the extent of residue decomposition in soil. Soil Biol. Biochem. 20:915–919.
153. Burns, R. G. 1982. Carbon mineralization by mixed cultures, p. 475–543. *In* A. T. Bull and J. H. Slater (eds.), Microbial interactions and communities, Vol. 1. Academic Press, London.
154. Adler, E. 1997. Lignin chemistry—past, present and future. Wood Sci. Technol. 11: 169–218.
155. Haider, K. 1992. Problems related to the humification processes in soils of temperate climates, p. 55–94. *In* G. Stotzky and J.-M. Bollag (eds.), Soil biochemistry, Vol. 7. Marcel Dekker, New York.
156. Eriksson, K.-E., and T. M. Wood. 1985. Biodegradation of cellulose, p. 469–503. *In* T. Higuchi (ed.), Biosynthesis and biodegradation of wood components. Academic Press, London.
157. Eriksson, K.-E., R. A. Blanchette, and P. Ander. 1990. Microbial and enzymatic degradation of wood and wood components. Springer-Verlag, Berlin.
158. Kirk, T. K. 1980. Physiology of lignin metabolism by white-rot fungi, p. 51–63. *In* T. K. Kirk, T. Higuchi, and H. M. Cheng (eds.), Lignin biodegradation: microbiology, chemistry, and potential applications, Vol. 2. CRC Press, Boca Raton, FL.

159. Jeffries, T. W., S. Choi, and T. K. Kirk. 1981. Nutritional regulation of lignin degradation by *Phanerochaete chrysosporium*. App. Environ. Microbiol. 52:251–254.
160. Marsden, W. L., and P. P. Gray. 1986. Enzymatic hyrdrolysis of cellulose in lignocellulosic materials. CRC Crit. Rev. Biotechnol. 3:235–276.
161. Castanon, M., and C. R. Wilke. 1980. Adsorption and recovery of cellulases during hydrolysis of newspaper. Biotechnol. Bioengin. 22:1037–1053.
162. Sinsabaugh, R. L., and A. E. Linkins. 1988. Adsorption of cellulase components by leaf litter. Soil Biol. Biochem. 20:927–932.
163. Sinsabaugh, R. L., and A. E. Linkins. 1989. Cellulase mobility in decomposing leaf litter. Soil Biol. Biochem. 21:205–209.
164. Sinsabaugh, R. L., and A. E. Linkins. 1987. Inhibition of the *Trichoderma viride* cellulase complex by leaf litter extracts. Soil Biol. Biochem. 19:719–725.
165. Golchin, A., J. M. Oades, J. O. Skjemstad, and P. Clarke. 1994. Soil structure and carbon cycling. Austr. J. Soil Res. 32:1043–1068.
166. Clarholm, M. 1981. Protozoan grazing of bacteria in soil—impact and importance. Microb. Ecol. 7:343–350.
167. Clarholm, M. 1985. Interactions of bacteria, protozoa and plants leading to mineralization of soil nitrogen. Soil Biol. Biochem. 17:181–187.
168. Acea, M. J., and M. Alexander. 1988. Growth and survival of bacteria introduced into carbon-amended soil. Soil Biol. Biochem. 20:703–709.
169. Freckman, D. W. 1988. Bacterivorous nematodes and organic matter decomposition. Agricul. Ecosyst. Environ. 24:195–217.
170. Gupta, V. V. S. R., and J. J. Germida. 1989. Influence of bacterial-amoebal interactions on sulfur transformations in soil. Soil Biol. Biochem. 21:921–930.
171. Kuikman, P. J., A. G. Jansen, J. A. van Veen, and A. J. B. Zehnder. 1990. Protozoan predation and the turnover of soil carbon and nitrogen in the presence of plants. Biol. Fertil. Soils 10:22–28.
172. Kuikman, P. J., A. G. Jansen, and J. A. van Veen. 1991. ^{15}N-Mineralization from bacteria by protozoan grazing at different soil moisture regimes. Soil Biol. Biochem. 23:193–200.
173. Elliott, E. T., R. V. Anderson, D. C. Coleman, and C. V. Cole. 1980. Habitable pore space and microbial trophic interactions. Oikos 35:327–335.
174. Vargas, R., and T. Hattori. 1986. Protozoan predation of bacterial cells in soil aggregates. FEMS Microbiol. Ecol. 38:233–242.
175. Bamforth, S. S. 1988. Interactions between protozoa and other organisms. Agricul. Ecosyst. Environ. 24:229–234.
176. Darbyshire, J. F., B. S. Griffiths, M. S. Davidson, and W. J. McHardy. 1989. Ciliate distribution amongst soil aggregates. Rev. Ecol. Biol. Sol. 26:47–56.
177. Kuikman, P. J., J. D. van Elsas, A. G. Jansen, S. L. G. E. Burgers, and J. A. van Veen. 1990. Population dynamics and activity of bacteria and protozoa in relation to their spatial distribution in soil. Soil Biol. Biochem. 22:1063–1073.
178. Rutherford, P. M., and N. G. Juma. 1992. Influence of texture on habitable pore space and bacterial-protozoan populations in soil. Biol. Fertil. Soils 12:221–227.
179. Rutherford, P. M., and N. G. Juma. 1992. Influence of soil texture on protozoa-induced mineralization of bacterial carbon and nitrogen. Can. J. Soil Sci. 72:183–200.

180. Darbyshire, J. F., S. J. Chapman, M. V. Cheshire, J. H. Gauld, W. J. McHardy, E. Paterson, and D. Vaughan. 1993. Methods for the study of interrelationships between microorganisms and soil structure. Geoderma 56:3–23.
181. England, L. S., H. Lee, and J. T. Trevors. 1993. Bacterial survival in soil: Effect of clays and protozoa. Soil Biol. Biochem. 25:525–531.
182. Hassink, J., L. A. Bouwman, K. B. Zwart, and L. Brussaard. 1993. Relationships between habitable pore space, soil biota and mineralization rates in grassland soils. Soil Biol. Biochem. 25:47–55.
183. Hassink, J., L. A. Bouwman, K. B. Zwart, J. Bloem, and L. Brussaard. 1993. Relationships between soil texture, physical protection of organic matter, soil biota, and C and N mineralization in grassland soils. Geoderma 57:105–128.
184. Hattori, R., and T. Hattori. 1993. Soil aggregates as microcosms of bacteria-protozoa biota. Geoderma 56:493–501.
185. Heijnen, C. E., J. Postma, and J. A. van Veen. 1990. The significance of artificially formed and originally present protective microniches for the survival of introduced bacteria in soil. Trans. 14th Congress of ISSS, Kyoto, 1990, III:88–93.
186. Rattray, E. A. S., J. I. Prosser, L. A. Glover, and K. Killham. 1992. Matric potential in relation to survival and activity of a genetically modified microbial inoculum in soil. Soil Biol. Biochem. 24:421–425.
187. Wright, D. A., K. Killham, L. A. Glover, and J. I. Prosser. 1993. The effect of location in soil on protozoal grazing of a genetically modified bacterial inoculum. Geoderma 56:633–640.
188. White, D., D. A. Wright, L. A. Glover, J. I. Prosser, D. Atkinson, and K. Killham. 1994. A partial chloroform-fumigation technique to characterise the spatial location of bacteria introduced into soil. Biol. Fertil. Soils 17:191–195.
189. Heijnen, C. E., C. H. Hok-A-Hin, and J. A. van Veen. 1992. Improvements to the use of bentonite clay as a protective agent, increasing survival levels of bacteria introduced into soil. Soil Biol. Biochem. 24:533–538.
190. Heijnen, C. E., C. Chenu, and M. Robert. 1993. Micro-morphological studies on clay-amended and unamended loamy sand, relating survival of introduced bacteria and soil structure. Geoderma 56:195–207.
191. Danso, S. K. A., S. O. Keya, and M. Alexander. 1975. Protozoa and the decline of *Rhizobium* populations added to soil. Can. J. Microbiol. 21:884–895.
192. Skjemstad, J. O., J. M. Oades, J. A. Taylor, P. Clarke, and S. G. McClure. 1995. The chemistry and nature of protected carbon in soils. Austr. J. Soil Res.
193. Largeau, C., S. Derenne, and P. Metzger. 1994. The formation of refractory organic matter from biological precursors. *In* J. Lupp (ed.), The role of non-living organic matter in the earth's carbon cycle. Dahlem Workshop Report LS16, Berlin, Germany.

3

Soil as a Catalytic System

Pacifico Ruggiero Università degli Studi di Bari, Bari, Italy

Jerzy Dec and Jean-Marc Bollag The Pennsylvania State University, University Park, Pennsylvania

I. INTRODUCTION

Soil is in a constant state of disequilibrium as the result of the multitude of simultaneous transformation reactions that involve both natural and xenobiotic organic compounds. In agricultural soils, the equilibrium is also disturbed by continuous modification of the soil environment by fertilization, crop rotation, and other agronomic practices. Various biotic (enzymes, and free-living components such as bacteria, fungi, algae, microfauna) and abiotic (clay minerals, oxides, hydroxides, noncrystalline components, organic matter) components of the soil may be involved in the same transformation reaction. Therefore, it is difficult to determine the importance of a specific process. In addition, interactions among the various components increase the complexity of the soil environment.

The determination of the rates and extent of abiotic and biotic transformations is a prerequisite for the rational use of soil and for the elimination of toxic compounds that have accumulated in soil as a result of intended and unintended human activities. Soils and the environmental conditions that influence rates of transformation may vary significantly from point to point in situ. The multiplicity of factors that affect transformation processes requires a joint interdisciplinary effort by basic and applied scientists. The major goal is to develop suitable methods of handling soil to meet current demands for food, energy, and fiber, and to decompose toxic chemicals.

Studies with laboratory model systems have strong appeal because conditions can be controlled and the results are usually reproducible. However, these systems may not reflect the actual field conditions and the data can seldom be extrapolated. Nevertheless, considering the complexity of soil systems, it is almost impossible to conduct field studies on a particular transformation reaction in isolation. Moreover, field experiments usually leave unanswered the question of whether reaction products are formed via biological or abiotic pathways. The relevance of laboratory data to the prediction of transformation reactions in soil in situ is best examined under controlled conditions that are designed to mimic natural microenvironments. Such experiments enable rapid screening of transformation processes, facilitate control of reaction conditions, and yield mechanistic information. Conclusions drawn from controlled microcosm studies help build the conceptual framework for reactions that may occur in more complex natural systems.

The plethora of transformation reactions that occur in soil involve primarily enzymes. Therefore, it is not surprising that enzymatic processes are one of the most important subjects of model studies. Enzymes as soil catalysts are influenced by a variety of natural or anthropogenic factors that considerably modify their performance and properties. Knowledge of the nature of these factors is essential for understanding the role of enzymes in soil processes.

In an attempt to summarize the current state of knowledge on soil enzymes, the general principles of catalysis will be discussed and the concept of soil as a catalytic system introduced. Subsequently, a survey of major transformation reactions that occur in soil will be presented and, based on the outlined background, the actual discussion of enzymes as soil catalysts will follow.

II. GENERAL PRINCIPLES OF CATALYSIS

The free energy change (ΔG) is a measure of the amount of useful energy that may be derived from a particular reaction. The driving force for any chemical reaction is the difference in free energy between the reactants and the end products. This parameter indicates how far a reaction mixture is from its equilibrium composition, but it provides no information about the rate of reaction.

Soil constitutes a dynamic and thermodynamically "open" system, in which matter and energy are freely exchanged with the surrounding environment. Even a simple reaction that occurs in soil may involve a series of steps, each of which has a different rate. As a result, a large part of the free energy can be accumulated and used for the cycling of macro- and micronutrients. In each step of a chemical reaction, only molecules with sufficient energy and of a suitable stereochemical state will be transformed. Chemical reaction rates are controlled by such factors as the concentration and nature of the reactants, temperature, presence of catalysts, and radiation. In general, the higher the concentration of reactants and the temperature, the more rapid the reaction. However, the range of temperatures in soil is too narrow to affect critically the rate of chemical reactions. Therefore, the observed enhancement of the reaction rates in soil is primarily the result of the presence of catalysts.

Reactions in soil can be mediated by a multitude of ions and molecules, as well as by crystalline and noncrystalline surfaces. These catalytic materials are effective at concentrations much lower than those of the reactants, although some soil enzymes and inorganic catalysts can be present in high concentrations relative to their substrates.

According to the transition state theory developed by Eyring [1], a reaction will occur only when a reactant molecule acquires sufficient internal energy to reach an activated state in which there is a high probability that an existing bond will be broken or a new one formed. The transition state complex contains a higher free energy and, hence, a lower stability than that of the reactants, and it then dissociates to yield products whose free energy is less than that of the activated complex. The rate of any chemical reaction is proportional to the number of activated molecules, and, therefore, depends strongly on the activation energy, which is defined as the energy required for the formation of one mole of the transition state complex.

Figure 1 depicts an energy barrier that must be overcome before reactants can be transformed to final products. A catalyst increases the rate of reaching the equilibrium of a reaction by providing an alternative route from reactants to products—one with a lower activation energy barrier. As the reaction proceeds, the concentration of products continuously increases while that of reactants decreases. The catalyst participates in the mechanism of the reaction but is subsequently regenerated to its original state and may be reused. However, the life of a catalyst is limited because it tends to be modified by the incorporation of impurities or secondary reaction products.

Catalysis can be homogeneous or heterogeneous. It is homogeneous when the reaction occurs in only one phase. Many organic reactions in liquid phases involve a catalyst that is an enzyme or simply a proton or metal ion with multiple oxidation states, such as copper, iron, or manganese. Catalysis is considered heterogeneous when the catalyst is present in a phase different from that of the reactants and products, and the chemical change occurs at the interface between phases, usually solid-liquid. Both homogeneous and heterogeneous catalysis may occur in soil. The presence of a diffuse liquid phase, in which many organic and inorganic solutes are solubilized, and of a colloidal fraction, which offers a large surface area for interactions between reactants, provides enormous potential for catalytic effects.

The mechanism of heterogeneous catalysis involves the following processes: (1) adsorption of the reactants on the surface of the catalyst; (2) chemical reactions of the activated forms on the surface; and (3) desorption of the reaction products and their diffusion into the homogeneous phase. Any of these processes

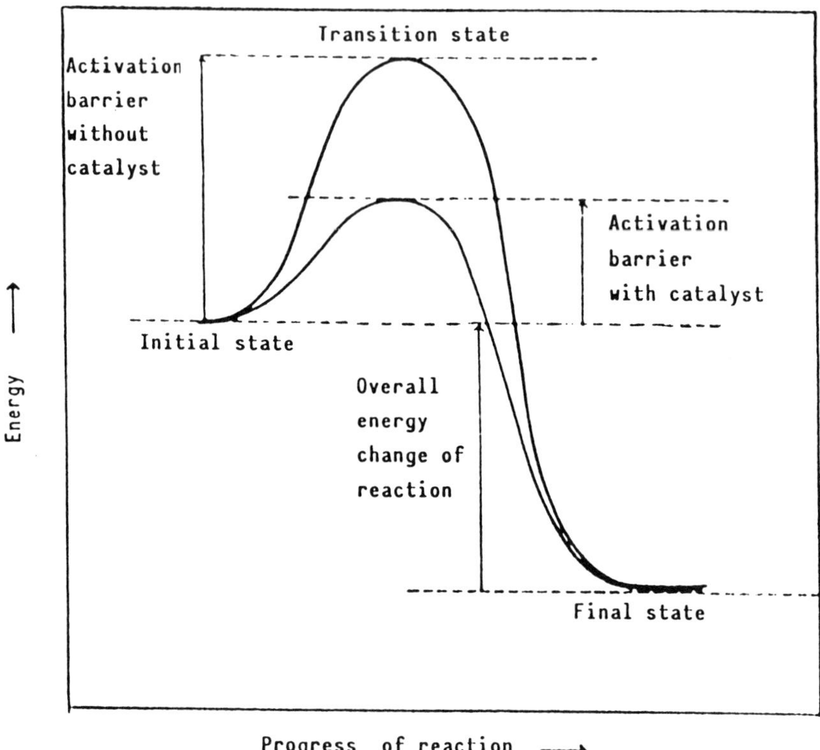

Figure 1 Free-energy graph showing the energy of activation of an uncatalyzed and a catalyzed reaction.

may be rate controlling. Processes (1) and (3) are diffusion phenomena and are fundamentally governed by the laws of diffusion. The interphase effects will cause additional changes in the reaction rates, which during homogenous catalysis, are affected only by temperature and the concentration of the reactants. As a result, the mechanisms of heterogeneous catalysis will be more obscure than those of one-phase, homogeneous catalysis. The kinetics of a heterogenous reaction will be determined by the slowest of the processes that contribute to the mechanism of catalysis.

III. CATALYSIS IN SOIL

Soil is a complex, interacting mixture of components such as clay minerals, oxides and hydroxides of various elements, organic matter, organomineral complexes, ions, free radicals, and molecules in the excited state after exposure to light. All these components abiotically promote a variety of reactions between organic and inorganic molecules that contribute to natural environmental processes and are responsible for the circulation of macro- and micronutrients. Free electrons are also present and available for involvement in redox reactions [2]. The direction of electron flow depends upon the relative affinities (i.e., redox potentials) of the reactants. Theoretically, oxidation and reduction are balanced, and no free electrons should remain at the end of a reaction. In soil, however, small amounts of extra free electrons are present as a result of continuous shifts in the equilibria of reactions.

By definition, a true catalyst cannot be altered by the reaction it catalyzes. However, some soil components that seem to mediate a reaction behave more as reactants than as catalysts. For instance, rapid chemical oxidation of certain phenolic acids in the soil was shown to be coupled with a reduction of iron and manganese oxides [3,4]. According to McBride [5], Fe^{2+}, generated by the oxidation of catechol and hydroquinone by Fe(III) oxides, was rapidly reoxidized by O_2, suggesting that Fe(III) oxides were true catalysts of the oxidation reaction. A similar reduction of manganese oxides by aromatic and nonaromatic compounds was determined [6,7]. Humic-like phenolic polymers were obtained through the catalytic action of manganese, iron, aluminum, and silicon oxides, and of clay minerals, such as nontronite, bentonite, and kaolinite. The metal oxides and clay minerals that bring about the oxidation reaction may function as Lewis acids, by accepting electrons from monomeric phenols, or as electron transfer agents and complexation substances [8–10].

The biotic component of soil provides another source of catalysts. Much of the extensive research on biological processes is related to the metabolic activities of microorganisms. This chapter focuses on the role of extracellularly active enzymes produced by microorganisms, as well as by other sources. All enzymes are proteins that often need additional components, such as metal ions or small organic molecules, to exhibit catalytic activity. Some enzymes are active only within intact cells; others also function extracellularly. In contrast to many inorganic catalysts, enzymes are highly efficient. They accelerate reactions by factors ranging from 10^8 to 10^{20}. For example, the half-life of urea at 25°C is about 32 years, but in the presence of urease, the half-life is reduced to 10^{-4} sec. The number of substrate molecules transformed per enzyme molecule per minute (i.e., the turnover number) is normally above 10^3, but in some cases, it may be greater than 10^6.

With respect to their specificity, enzymes are classified into six groups: oxidoreductases, transferases, hydrolases, lyases, isomerases, and ligases. In general, enzymes act on a limited number of substrates that show similar structural features (substrate specificity), with the exception of hydrolytic enzymes, which react with a large number of substrates. Stereospecificity enables enzymes to recognize molecules that are sterically nonequivalent.

The catalytic activity of enzymes depends on a very narrow number of ions and/or molecules, which must be smaller than the apoprotein part of the enzyme. All enzyme-catalyzed intracellular reactions are subject to cellular control; the regulation of enzyme synthesis and of enzyme activity is the basis of the functional organization of cellular metabolism.

The critical factor in enzymatic catalysis is the structure of the transition state, which is far more complex than in the case of inorganic catalysts. Proteins undergo substantial changes in their tertiary conformation before, during, and after reaction, and many details of the molecular mechanisms of catalysis are unknown. Only a small portion of the enzyme—the active site—is in direct contact with the substrate. Other sections of the protein molecule function indirectly by maintaining the active site in the suitable configuration for efficient and specific catalysis.

IV. BIOTIC AND ABIOTIC REACTIONS

It is difficult to distinguish among the factors that contribute to transformation reactions in soil, as a specific compound can be transformed by various pathways. Moreover, biotic and abiotic reactions may show similarities in reaction mechanisms and in the products obtained. In general, reactions catalyzed by enzymes display greater specificities than abiotic reactions. The major biological and nonbiological reactions that contribute to transformations of various natural compounds and xenobiotics in soil are hydrolysis, oxidation-reduction, generation of free radicals, condensation and polymerization, and phototransformation. Any reaction in these general categories may be mediated by specific catalysts working in one-phase or heterogeneous systems.

A. Hydrolysis

Hydrolytic reactions are responsible for the transformation of many important classes of compounds, including amides, carbamates, esters, nitriles, thiocarbamates, alkyl and aryl halides, epoxides, imines, silanols, and organometallic compounds. Hydrolytic transformations in soil are affected by a wide range of factors such as moisture, temperature, pH, organic matter content, and ionic strength.

Biochemical transformation can be mediated by extracellular enzymes originating from microorganisms. Therefore, conditions that favor microbiological growth usually enhance biochemical transformation of naturally occurring and synthetic molecules. Commonly occurring enzymes, such as esterases, amidases, phosphatases, and proteases, catalyze the hydrolysis of the respective chemical bonds. Reaction rates are strongly influenced by the chemical properties of the substrates.

Abiotic mechanisms of degradation involve principally sorption-catalyzed hydrolysis of chemical linkages, and they may be related to the presence on clay surfaces of acidic sites of either the Brönsted or the Lewis type. Brönsted acidity occurs in acid-activated clays when Al^{3+} and Fe^{3+} ions are released from octahedral sites in the clay layers and become acid centers in the interlayer space. Clay acidity may also result from the dissociation of water molecules of the hydrated cations associated with clay and the polarizing effect of the cations on the water molecules. Acidity increases with decreasing water content and increasing charge of the cation. The Brönsted acidity of hydrated cations on clay surfaces may be illustrated as follows:

$$[M(OH_2)_n]^{m+} \rightarrow [M(OH_2)_{n-1}OH]^{(m-1)+} + H^+$$

The proton is immediately available for participation in hydrolysis: it withdraws electrons, thus weakening the bond to be broken. Positively charged metal ions and polyhydroxy aluminum may also be located at the edges of clays, thereby introducing potential Lewis-type centers that reduce electron density at a site of the substrate subjected to bond breaking or nucleophilic attack.

B. Oxidation-Reduction

Oxidation-reduction or redox reactions involve transfer of electrons from one reactant to another. When available, O_2 functions as the electron acceptor during oxidation of organic compounds. In the absence of O_2, other compounds, such as nitrate, sulfate, and various oxides of iron, manganese, and copper, may serve as electron acceptors. In the absence of catalysts, most redox reactions are extremely slow.

Clays and humic substances are integral components of the heterogeneous catalytic system of soil. Clays containing transition metal cations can catalyze the oxidation of aromatic molecules [11]. Humic and fulvic acids can bind certain metals and form stable complexes with different redox potentials. Electron spin resonance (ESR) spectra of such complexes suggest that Fe^{3+} and Cu^{2+} are strongly bound and protected by humic substances in the inner-sphere complexes; therefore, they are considerably resistant to reduction [12].

In addition to intact microbial cells, extracellular enzymes in soil contribute to the transformation of naturally occurring and xenobiotic compounds. Enzymes involved in redox reactions are oxidoreductases, including tyrosinase, laccase, peroxidase, nitrogen reductases, dehydrogenases. Copper, iron, or molybdenum ions function as catalytic centers of these enzymes, and oxygen or hydrogen peroxide act as electron acceptors.

The main oxidative reactions include hydroxylation and cleavage of aromatic and heterocyclic rings; hydroxylation and dealkylation of aromatic side chains; cleavage of ether linkages; epoxidation of carbon-carbon double bonds; oxidative coupling, in which two alkyl or aryl groups connect after oxidation by an enzyme or a suitable chemical reagent, yielding dimerized or polymerized products; and sulfoxidation reactions, in which organic sulfides (thioethers) and sulfites are oxidized to the corresponding sulfoxides and sulfates.

The major reduction reactions are reduction of certain metals; denitrification by nitrate reductase; transformation of nitro groups to amines; saturation of double bonds; reduction of aldehydes and ketones; reduction of sulfoxides; and reductive dehalogenation and deamination, in which a carbon-halogen or carbon-nitrogen bond is broken and hydrogen replaces the halogen or NH_2 group substituent.

C. Formation of Free Radicals

Free radicals are highly reactive chemical species that contain an unpaired electron. In soil, free radicals can exist for long periods and at relatively high concentrations. Steelink and Tollin [13] analyzed the origin and role of stable free radicals in humic substances. Experiments using ESR demonstrated that the concentration of free radicals in humic substances was highly dependent on a number of factors, including pH, redox potential, irradiation, hydrolysis, alkylation, and temperature [12]. Nonhumic organic compounds such as lignins, tannins, and pigments, all containing quinone and phenolic groups, formed stable free radicals by an electron transfer reaction between the two partners in a donor-acceptor pair [13]. Other substituted aromatic chemicals, including pesticides, underwent oxidation reactions that involved the formation of stable free radicals through the removal of a proton and an electron from the molecule [14].

Free radicals can be generated both biologically and abiologically. The abiological genesis is influenced by pH, irradiation, and redox potential. Formation of free radicals was observed during interaction between humic substances and some herbicides, wherein radical charge-transfer complexes were formed [12]. Free radical formation was enhanced by the presence of clay minerals saturated with iron, copper, and other trace metals [10,14,15] capable of accepting a single electron from organic substrates. Generation of alkyl radicals was also observed during the reductive dechlorination of halogenated aliphatic compounds, which was effected by the presence of transition metals [16].

According to Sjoblad and Bollag [17], the biological origin of free radicals in soil is the result of the activity of phenoloxidase enzymes, such as laccase and peroxidase, that provoked the oxidation of suitable phenols by two or more steps. The initial step involved the production of an aryloxy radical, the stability of which depended on the structure of the phenolic substrate [17]. Humic substances gave rise to semiquinone radicals in the same way. Further steps involved a coupling or condensation of the free radicals, disproportionation in two molecules with paired electrons, or a radical reduction. An alternative reaction route for the organic free radicals consists of radicals interacting with metal ions by forming complexes that may or may not retain unpaired electrons.

Tyrosinase, a metalloenzyme frequently associated with laccase and peroxidase, caused an analogous oxidation reaction of phenols. However, in contrast to the two other enzymes, it transferred electrons in pairs without the formation of free radicals [18].

The formation of hydroxyl radicals in soil ($HO^{\cdot}$) is an important source of free radicals. The possible involvement of $HO^{\cdot}$ radicals in the oxidative coupling reaction of phenols by a soil extract was hypothesized by Suflita et al. [19]. The genesis of the $HO^{\cdot}$ radical can be related to the generation of the anionic superoxide radical $O^{\cdot}{}_2{}^-$, promoted by electron transfer from photoexcited molecules, such as humic substances, to molecular oxygen [20]. In water, $O^{\cdot}{}_2{}^-$ is partly protonated to its conjugate acid, H-O-O$^{\cdot}$ ($pK_a = 4.8$), and is likely to be disproportionated to oxygen and hydrogen peroxide by the microbial enzyme superoxide dismutase [21]:

$$\text{H-O-O}^{\cdot} + O^{\cdot}{}_2{}^- + H^+ \rightarrow H_2O_2 + O_2$$

The mild oxidizing action of hydrogen peroxide is dismutated by a catalase or manganese oxide, at near neutral or high pH, to O_2 and H_2O. Also, peroxidase may reduce H_2O_2 to H_2O. More importantly, H_2O_2 probably reacts with certain metal catalysts, such as reduced iron, to form the hydroxyl radical, $HO^{\cdot}$, the most powerful free radical oxidant known [22]:

$$Fe^{2+} + H_2O_2 \rightarrow HO^{\cdot} + HO^- + Fe^{3+}$$

Limited research has been conducted on the presence and role of superoxide anion and hydroxyl radicals in soils [19,20,23]. These free radicals have a very short lifetime, as they react readily with one another and with transition metals and, albeit more slowly, with many organic compounds. As a result, their concentration becomes very low and difficult to detect.

In conclusion, free radicals in soil appear to have important roles. The enhanced interest in free radical reactions is the result of their active role in various chain processes. Reactions of interest include polymerization and depolymerization of humic substances, participation as photosensitizers in photochemical reactions of environmental importance, transformation of toxic pollutants, complexation interactions with metal ions, and physiological effects on microorganisms and higher plants [12,17,24].

D. Condensation and Polymerization

Synthetic processes are responsible for the incorporation of carbon and nitrogen into soil organic matter. Phenoloxidases, a class of microbial enzymes, have the principal role in humification. Phenoloxidases have been intensively investigated for their ability to catalyze the binding of xenobiotic chemicals to humus monomers, as well as to humic substances [25,26]. A large number of substituted phenols and amines were cross-coupled and copolymerized with phenolic intermediates of lignin degradation during the humification process. The action of microbial laccases, tyrosinases, and peroxidases allowed predictions concerning the environmental fate of xenobiotic residues.

Inorganic components of soil, such as clay minerals; oxides of manganese, iron, aluminum, and silicon; and certain primary minerals, also mediate various condensation and polymerization reactions [8,9,15,27,28], including those involved in the detoxification of recalcitrant aromatic pollutants [10,23,29]. Under natural conditions, inorganic catalysts may be less active than enzymes, but as a result of the longer persistence of inorganic catalysts, they appear to be equally efficient. The rate of polymerization depends on the chemical structure of both the catalyst and the organic compound and on the pH of the reaction mixture. Clay minerals have also been used as catalysts in organic chemistry synthesis on a preparative scale for industrial purposes [30,31].

Every reaction pathway proposed for the catalytic polymerization of naturally occurring and xenobiotic phenols proceeded through generation of phenolic radicals or reactive hydroxybenzoquinones and the nucleophilic additions of amino groups to free radicals [26,32]. Although a mechanism involving free radicals is generally assumed, Hamilton [33] suggested that mechanisms involving ions should also be considered. He proposed that the metal ion at the active site of an enzyme served to transfer electrons from the coupling phenols to the oxidant (O_2 or H_2O_2) and that protons were transferred through the solvent. The polymerization reactions were accompanied by decarboxylation or dechlorination of phenolic substrates [26,211].

E. Phototransformation

Photolytic transformation of chemical substances is an important abiotic process in the atmosphere and in water, as well as on various surfaces such as leaves and soil. However, as the effects of sunlight on the soil surface are limited to the upper 1-mm layer, the efficiency of photochemical processes in soil under natural conditions is not easily predicted. It is not known whether photocatalytic transformation of chemicals on the soil surface is a significant mode of transformation. Nevertheless, increasing interest in the photochemical fate of xenobiotics has stimulated recent studies using model systems that presumably simulate natural conditions.

The amount and spectral quality of UV irradiation at the soil surface varies greatly with season, time of day, location, meteorological conditions, and thickness of the ozone layer. The absorption of a photon of appropriate energy by a molecule raises an electron to an orbital of higher energy. The new state of the molecule is referred to as the excited singlet. The singlet molecule may undergo a spin inversion, leading to a lower energy state, which is referred to as the triplet. The molecule in an excited state can also return to its ground state through rapid dissipation of the absorbed energy, and in the process, heat may be generated or radiation emitted in the form of fluorescence. The energy of excitation may also be transferred to another molecule or used in a chemical reaction. If there is an excess of excitation above a certain energy level, the molecule may even dissociate.

Although humic substances have been shown to catalyze the phototransformation of various classes of susceptible compounds, relatively little is known about photoreactions induced by humic substances on soil surfaces. Irradiation of humic substances with visible and UV light might enhance their indigenous free radical character [24]. In turn, humic substance radicals may react with ground-state O_2 to give a variety of highly reactive oxygen radicals or with xenobiotic compounds to generate various photoproducts on soil surfaces. Triplet excited-state humic soil fractions may also participate in the photoproduction of singlet oxygen by a triplet-triplet reaction [34].

Alternative methods for the production of singlet oxygen are also possible. According to Gohre et al. [35], the photogeneration of singlet oxygen is a general characteristic of sunlight-irradiated soil surfaces. In addition to humic fractions, clay and primary mineral fractions may be involved in the photochemical event, as surface defects on mineral structures probably enhance photosensitization. Miller et al. [36] hypothesized that the lifetime of singlet oxygen in the vapor phase is sufficient for penetration into soil via gaseous diffusion, thereby extending the depths of the effects of photolysis.

Little has been published on photolytic reactions that do not involve humic free radicals or generation of singlet oxygen. The extent to which pH, mineral composition, and the type and concentration of xenobiotics of other reactants affect photolysis is unknown. Light-catalyzed reactions might include oxidation, isomerization, or substitution. However, the exact mechanisms are often unclear. According to Wolfe et al. [37], phototransformation products of pesticides are often similar to those resulting from other abiotic processes and biotransformation.

Differences in the yield of light energy combined with environmental conditions probably affect transformation pathways and reaction rates. Miller et al. [36] found that phototransformation of organic compounds on irradiated soil surfaces was slow compared with that in solution, primarily as the result of the light-screening action of soil and the photochemical quenching of electronically excited states on surfaces. As a consequence, photolytic reactions do not follow first-order kinetics; they may be relatively rapid in the first phase, but subsequently become slower.

V. ENZYMES AS CATALYSTS IN SOIL

Vast quantities of organic debris resulting from the breakdown of agricultural crop remains, dead animals, and microorganisms are present in soil. This debris

Table 1 Major Organic Polymers from Plant, Microbial, and Animal Residues in Soil [39]

Macromolecular substrate (origin[a])	Structure	Molecule entering cell[b] (known or suspected)
Cellulose (P, M)	β-(1-4)-D linked glucan	Glucose, cellobiose
Hemicelluloses (P)	β-(1-4)-D linked xylan	Xylose, xylobiose
	Glucuronans	Glucuronic acid
	Galacturonans	Galacturonic acid
Pectin (P, M)	Xyloglucan	Xylose
	Galacturonans	Galacturonic acid
Starch (P, M)	α-(1-4) and α-(1-6) linked glucans	Glucose, maltose
Lignin (P)	Polymers of *p*-hydroxycinnamyl alcohols	Mono-lignols (coniferyl, sinapyl and *p*-coumaryl alcohols), di- and tri-lignols
Chitin, (A, M)	β-(1-4)-linked *N*-acetylglucosamine	*N*-acetylglucosamine, chitobiose
Proteins and peptides (A, M, P)	Polymers of amino acids	Amino acids, low number peptides
Lipids (A, M, P)	Triglycerides, phospholipids	Glycerols, fatty acids
Peptidoglycan (M)	Polymers of *N*-acetylglucosamine and *N*-acetylmuramic acid	*N*-acetylglucosamine, *N*-acetylmuramic acid low number peptides
Teichoic acid (M)	Polymers of polyol phosphates with saccharides and D-alanine	Glycerol, ribitol, mono- and disaccharides, alanine
Microbial exopolysaccharides (M)	Mannans, dextrans, levans, xanthans, pullulan, alginate	Mono- and disaccharides

[a] P = plant; A = animal; M = microbial
[b] Differs for different microorganisms

is the source of raw material for the synthesis of new humus. It decomposes with the release of nutrients essential for living organisms. However, a great many of the organic residues liberated into soil by cellular lysis consists of complex macromolecules that cannot pass through cell membranes to undergo further intracellular degradation. The major nonhumic polymers found in soil are listed in Table 1. Minor amounts of other constituents, such as resins, pigments, terpenes, alkaloids, tannins, and related polyphenols are also present.

The continuous input of dead biomass is supplemented by the influx of waste materials, pesticides, and various other toxins of both natural and anthropogenic origin, which, upon degradation, also serve as sources of carbon, nitrogen, and energy for microbial growth and reproduction. Some natural materials, such as lignins, are recalcitrant as a consequence of their insolubility and molecular size. In addition, degradable, nonaromatic polymers, such as polysaccharides or proteins, may become recalcitrant through formation of chemical or physical linkages with the resistant lignin, humus, and soil minerals [39]. Certain anthropogenic compounds to which microorganisms have not been exposed in the course of evolution may accumulate in soil and become less susceptible to degradation, resulting in an increased threat to the environment.

Extracellular enzymes, derived principally from microorganisms and, to a lesser extent, from plant debris, plant root exudates, and soil fauna, have a crucial role in the decomposition processes occurring in soil. They constitute, for instance, the critical factor for the conversion of complex molecules into forms more readily available to plants and microorganisms. They are also responsible for transformation processes leading to the detoxification of xenobiotics. According to Burns [39], extracellular enzymes are those that are excreted by microorganisms and accumulate in soil independently of current microbial proliferation. In order to accumulate, they must be stabilized against the degradative action of proteolytic enzymes, unfavorable temperature, and other denaturing factors. Certain processes may require the concerted and complex action of more than one extracellular enzyme.

No combination of environmental factors can result in optimal conditions for uniform accumulation of all enzymes. Therefore, specific enzyme activities may develop depending on pH, soil moisture, oxygen concentration, and availability of nutrients to enzyme-producing microorganisms. Sorption of microbial cells on humic substances or clay minerals may diminish the ability of intracellular enzymes to metabolize contaminants. As a result, exoenzymes, which remain active outside the cells, can gain in importance as transformers of xenobiotics.

The initial breakdown of various polysaccharide polymers is effected by extracellular cellulases, hemicellulases, amylases, pectinases, pentosanases, and chitinases. Lignin components are believed to be the primary precursor of the aromatic nuclei in humus molecules. The pattern of lignin transformation is unclear; however, the literature suggests that the breakdown of the lignin structure is the result primarily of enzymes secreted by white rot fungi [40].

The depolymerization of proteins and nucleic acids may involve extracellular proteolytic and nuclealytic enzymes. Peptides and amino acids resulting from proteolysis are readily assimilated by a variety of organisms. Proteolytic products may also be incorporated into soil organic matter. Urease, the most investigated soil enzyme, which catalyzes the breakdown of urea to ammonia, and enzymes (phosphatases, sulfatases, hydrolases) capable of releasing certain chemical groups participate in the overall production of low molecular weight molecules that are able to penetrate plant or microbial cells.

Enzymes in soil also participate in the synthesis of organic substances. Polyphenoloxidases catalyze the oxidation of phenolic compounds that are subsequently incorporated into humic polymers. Other synthetic reactions include the enzymatic oligomerization of aromatic xenobiotics or their cross-coupling with humus constituents [17,25,26]. As a general consequence, both enzyme-mediated degradative and synthetic processes may reduce soil pollution and lead to an improvement in soil fertility [41]. In fact, the use of enzymes is considered to be a means for the clean-up of soil environments. For instance, hydrophobic contaminants, as a result of their insolubility in water, may be adsorbed on soil and become unavailable for microbial degradation. In such a case, enzymatic treatment may be highly suitable to degrade these pollutants after their desorption with organic solvents [42]. At high concentrations, certain toxic xenobiotics, such as chlorinated phenols, are resistant to microbial degradation, but they can be enzymatically detoxified through binding to soil.

A. Location and Categories

The location of an enzyme in the terrestrial system depends on the nature and stage of growth of the originating cells, and the physicochemical properties of the soil. Kiss et al. [43], Skujins [44], and Burns [39,41,45] proposed several categories of enzyme location in soil. In Figure 2, the various categories are divided into four groups. Terms such as extracellular, accumulated, abiontic, and immobilized are generally used for differentiating enzymes within a category. Extracellular enzymes, in the classical sense, are considered to be only those secreted by living organisms during cell growth and division [38]. Glenn [46] extended the term extracellular to include enzymes released by cell lysis and enzymes bound on the cell wall and retained within the periplasmic space. In soil, extracellular enzymes may be defined as all enzymes that catalyze reactions outside of the organism that synthesized them. Extracellular enzymes, although readily degradable, may survive in soil through stabilization as immobilized enzymes, that is, bound on cell constituents and organic and mineral soil components. Both the free and bound enzymes that belong to categories B, C, and D of Figure 2 are considered accumulated [43]. These enzymes are the major contributors to the catalytic transformation of substrates in soil. According to Skujins [44], abiontic enzymes are those secreted by proliferating cells and those in groups B, C, and D that are not contained in dead or viable but nonproliferating cells. The distribution of enzymes in the categories listed in Figure 2 must be considered in a dynamic sense. The fate of enzymes secreted from living cells or released from dead cells into the soil environment may differ from enzyme to enzyme, depending on soil properties and the duration of exposure to soil [41].

Despite difficulties inherent in studying the compartmentalization of enzymes in soil, several researchers have successfully explored this question. Frankenberger and Dick [47] studied the activities of 11 soil enzymes as indicators of microbial growth and concluded that the activities of phosphatase, amidase, and catalase were clearly correlated with both microbial respiration and

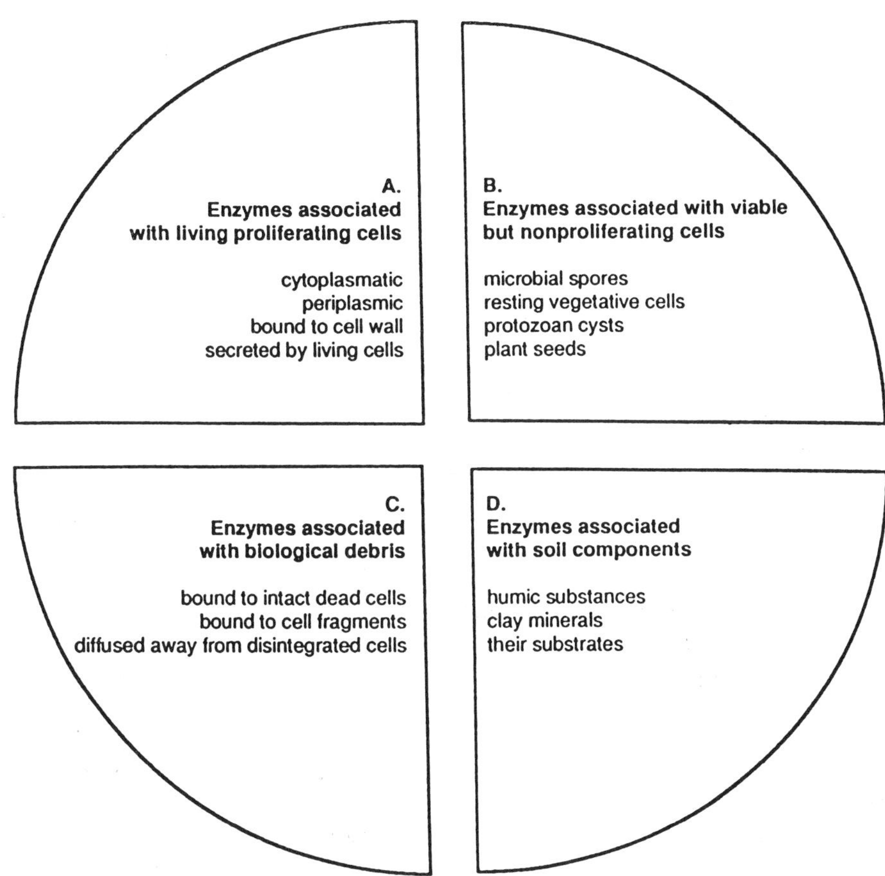

Figure 2 Location and categories of enzyme activity in soil (modified from [39,41,43–45]).

total biomass. Similar relations between microbial biomass and the activity of catalase, dehydrogenase, alkaline phosphatase, amylase, and protease were also reported [48]. Urease activity, determined in total biomass, was highly correlated with some soil properties (i.e., organic carbon, total nitrogen, cation exchange capacity) that affect microbial growth [47]. Regarding the components of cellulase activity in soil, Hope et al. [49] proposed that ß-D-glucosidase was bound to soil colloids, whereas exo- and endoglucanases were either free in the soil solution or attached to the outer surfaces of cellulolytic microorganisms. Mayaudon [50] claimed that enzymes in soil are of fungal origin and that they are bound to carbohydrates derived from bacterial lipopolysaccharides, with the entire structure immobilized on humic surfaces. Correlations between soil microporosity and the activities of urease and phosphatase were also observed [51], and the activity of glucosidase and galactosidase was found to decrease with depth in the soil profile [52].

B. Humus-Enzyme Complexes

The three main methods that have been developed to evaluate interactions between enzymes and humic substances are (1) examination of natural complexes extracted from soil; (2) studies with model complexes synthesized using enzymes from various sources and humic acids extracted from soil; and (3) studies with model complexes synthesized using enzymes and synthetic, humic-like polymers.

An objective of method 1, extraction of natural complexes from soil, is to achieve high yields of organic matter with enzyme activity without the formation of artifacts. Unfortunately, an ideal extractant does not exist. Various mild extractants have been employed (Table 2), some to extract enzymes themselves, and others to isolate humus-enzyme complexes. The recovery of enzyme activity has been generally less than 20% [53–59], although higher yields have been reported. Ladd [60] extracted a soil with borate/TRIS buffer to obtain 50 to 75% of the protease activity, and Maggioni and Cacco [61] produced extracts that retained 37% of acetyl-naphthyl-esterase activity. After a 24-h extraction of three soils with pyrophosphate, Nannipieri et al. [55] recovered 31 to 66% of phosphatase, 5 to 9% of urease, 57 to 100% of casein, and 15 to 68% of benzoylarginine hydrolyzing proteases. Perez-Mateos et al. [62], using a diluted solution of pyrophosphate, extracted catalase with a yield of 31% of the original activity.

The efficiency of extraction depends on the type of enzyme and soil, the extraction time, the soil:extractant ratio, and the chemical nature, concentration, and pH of the extractant solution. The evaluation of extraction efficiency is difficult, and many problems are encountered. For example, it is likely that both free enzymes and those associated with humic fractions are extracted. Enzymes may be partially destroyed, or the humic-enzyme complexes may be modified during the extraction procedure. In certain cases, extraction may release enzyme inhibitors. Intracellular enzymes may be released as a consequence of microbial lysis during the procedure. Enzymes in soil may also be bound tightly on clay colloids or other soil components and, therefore, prove difficult to extract. Finally, extraction yields may be affected by the enzyme assays.

The correlation between the enzyme activity in soil and extracted organic carbon or nitrogen suggested binding of enzymes to humic material [55]. To obtain clear evidence for the existence of persistent and active humus-enzyme complexes, various purification steps have been applied to crude enzyme extracts from soil. However, biochemical procedures, commonly adopted for protein purification were largely ineffective [64]. Fractionation of enzymes by

Table 2 Natural Humus-Enzyme Complexes Extracted from Soil

Enzyme	Extractant	Concentration (M)	Extraction pH	Reference
Oxygenase-decarboxylase	Sodium carbonate	0.2	—	[145]
Decarboxylase	Phosphate/EDTA	0.3	8.0	[170]
β-Glucosidase	Phosphate/EDTA	0.2	8.0	[53]
Invertase	Phosphate/EDTA	0.2	8.0	[53]
Cellulase	Phosphate/EDTA	0.2	8.0	[53]
Diphenoloxidases	Phosphate/EDTA	0.3	8.0	[202–205]
Laccase	Pyrophosphate	0.1	7.1	[54]
Catalase	Pyrophosphate	0.01	7.0–7.3	[62]
Protease	Borate/TRIS	0.1	8.1	[60]
	Phosphate/EDTA	0.3	8.0	[205]
	Phosphate/EDTA	0.2	8.0	[53]
	Pyrophosphate	0.1	7.1	[55,65,67]
Esterase	Sodiumhydroxide	0.2	—	[206,207]
	Phosphate/urea/NaCl/ EDTA/mercaptoethanol	—	7.0	[61,208]
	Phosphate/KCl/EDTA	—	7.0	[209]
Phosphatase	Phosphate/EDTA	0.2	8.0	[53]
	Pyrophosphate	0.1	7.1	[55,65,66]
	Ammonium acetate	1	7.0	[56]
Urease	Phosphate/urea/NaCl/ EDTA/mercaptoethanol	—	7.0	[57]
	Citrate/phosphate/glycine/ NaCl	—	6.3	[58,210]
	Pyrophosphate	0.1	7.0	[55,59,64, 65,151]

salting-out treatments caused coprecipitation of humic substances, and no separation on ion exchange resins was obtained [64]. In contrast, partially purified fractions of enzymatically active humic matter were obtained by gel chromatography [53,54,64–66], ultrafiltration [61,64–66], and isoelectric focusing [61,67]. These results proved that a portion of the enzyme activity in soil was firmly associated with humic substances.

The preparation of model humic-enzyme copolymers appears to be a useful approach for understanding the complex relation between soil enzymes and their polyanionic supports. Humic acids can be extracted from soil [63,68–73], and model phenolic polymers can be synthesized from the structural units of humic substances [74–78] (Table 3). Stable and insoluble enzymatic complexes have been prepared by coprecipitation of humic acids with enzyme using Ca^{2+} [63,68,69,71,72], Fe^{3+} [69,73], or Co^{2+}, Cu^{2+}, Mg^{2+}, or Al^{3+} [69]. Serban and

Nissenbaum [70] succeeded in isolating humic-peroxidase and humic-catalase complexes by gel chromatography. Some inhibitory effects on the activity of catalase (up to 24%) [70] and tyrosinase (up to 14%) [71,72], related to the presence of humic acids, were also observed. Despite this inhibition, the complexes maintained significant activity. K_m and V_{max} values, thermal and proteolytic stability, pH-activity dependence, and activation energy were affected by the immobilization on humic polymers [70,74–79]. The pH determined not only the net charge of the enzyme molecule but also the configuration of the humic acid [77,80]. Calcium ions themselves, which were added to provoke complex formation, could alter the electron structure of the humic acid particles [78].

Experiments using humic-like polymers were initiated 25 years ago by Rowell et al. [74] and continued by other authors [75,81–83] (Table 3). The preparation of synthetic humus was based on the ability of phenols to polymerize, either by auto-oxidation or in the presence of catalysts [75,84,85]. The expected advantage afforded by the preparation of synthetic humic-enzyme complexes was stabilization of the enzyme during its copolymerization with humic constituents [74,75].

However, some phenolic monomers commonly known as building blocks of humic substances may affect the retention of enzyme activity. Sarkar and Burns [75,82], for instance, showed that the incorporated enzymes were less active than the free enzymes. The percentage of retained activity, determined for 11 different phenolic-β-glucosidase complexes, ranged from 14 to 55%, probably as the result of varying degrees of inactivation caused by phenol monomers or oligomers during copolymer formation [75,82]. Grego et al. [83] used three different phenols (resorcinol, pyrogallol, and catechol) in the synthesis of humic-pronase copolymers and demonstrated that only resorcinol was able to produce enzymatically active copolymers, although only 20% of the activity initially applied was retained. This percentage increased to 80% after gel chromatography of the sample, and the authors suggested that the fractionation through Sephadex eliminated inhibitors of pronase associated with the crude complexes.

Table 3 Synthetic Humus-Enzyme Complexes

Enzyme	Source of humic acids	Reference
Invertase	Soil	[68,69,73]
Catalase	Soil	[70]
Peroxidase	Soil	[70]
Cellulase	Soil	[63]
Tyrosinase	Soil	[71,72]
Proteases	Artificial phenolic polymers	[74,83]
β-Glucosidase	Artificial phenolic polymers	[75,81,82]
Cellulase	Artificial phenolic polymers	[82]

On the other hand, the early report by Rowell et al. [74] showed that abiotically formed benzoquinone-trypsin and benzoquinone-pronase polymers lost from 32 to 98% of the applied activity, depending on the nature of the complex and of the substrate assayed. These findings were not surprising, as immobilized enzymes generally show reduced activity, probably as the result of modifications of the tertiary structure that result in restricted access of substrates to the active sites [45].

The inhibitory action of humic acids may be another factor responsible for losses in enzyme activity [70-72,78,86–94]. Various mechanisms are possible, such as (1) complexation by humic acids or the metal ions that are part of the structure of the active site of some enzymes; (2) conformational changes in the enzyme; (3) competition with substrates for the catalytically active site; or (4) binding of the substrate to humic acids. Inhibition was found to be dependent on concentration [71,72,88,94] and molecular weight [89,91,92] of the humic acids, and on the number of acidic groups on the humic acid molecule. Some authors attribute the inhibition only to carboxyl groups [87,88], whereas others implicate phenolic groups [78,91].

Ideas regarding the nature of interactions between humic acids and enzymes still appear to be more speculative than empirically supported. Moreover, differences in experimental procedures have often resulted in conflicting results. Rowell et al. [74], for instance, assumed that covalent as well as hydrogen bonds are important in the complexes, but they excluded the involvement of ionic bonds. In contrast, Ruggiero and Radogna [72] suggested, based on infrared spectroscopic data, that both ionic and hydrogen bonds link the enzyme molecule in humic-tyrosinase-Ca^{2+} complexes. A similar conclusion was made by Sarkar [63] from chromatographic and electrophoretic studies with humic-cellulase-Ca^{2+} complexes. The author also hypothesized the formation of covalent bonds between the enzyme and functional groups of humic acids. Maignan [68] and Serban and Nissenbaum [70] assumed that the enzymes were trapped within the macromolecular net of the humic acids, not excluding that some molecules may be immobilized at the surface by adsorption forces. Finally, Gosewinkel and Broadbent [56], who added electron donors or acceptors to humic-phosphatase complexes isolated from soil, considered the possibility that the bonding mechanism between humic acids and phosphatase was due to an electron donor-acceptor complex, with the enzyme functioning as the electron donor.

C. Clay-Enzyme Complexes

Similar to humic materials, clays have a profound influence on the activity and stability of enzymes in soil. This is principally the result of the small size of clay particles, which en masse, provide a large specific surface area and significant cation exchange capacity. Some clay minerals belong to the 1:1 layer silicates (e.g., nonexpanding kaolinite and halloysite), and others to the 2:1 layer silicates (e.g., expanding smectites and nonexpanding illite). The most frequently studied clay minerals are kaolinite and smectites, because they are responsible for a large portion of the physicochemical properties of soils.

Interactions of enzymes with clay minerals have been reviewed in detail by many authors [39,45,95–100]. Much research has been directed at either modeling natural clay-enzyme complexes or developing inexpensive techniques to immobilize enzymes on supporting matrices for industrial applications. The enzymes studied belonged principally to two classes—hydrolases and oxidoreductases—and their adsorption by clay minerals has been investigated by numerous workers in the past 40 years (Table 4). Generally, in vitro studies using relatively pure clay minerals saturated with a specific cation were performed. However, the practical difficulty of extracting natural clay-enzyme complexes and the need to approximate in situ soil conditions in the artificial systems studied stimulated the use of models that re-create natural environments. In soil,

Table 4 Synthetic Clay-Enzyme Complexes

Enzyme	Clay	Reference
Invertase	Bentonite	[101]
Proteases	Kaolinite, allophane, smectite, vermiculite, illite	[102–109]
Urease	Montmorillonite, organo-smectite	[110–113]
Phosphatase	Kaolinite, illite, montmorillonite	[114–115]
Glucose oxidase	Kaolinite, bentonite, montmorillonite nontronite, illite, allophane	[116–119]
Catalase	Kaolinite, illite, montmorillonite [clean or coated with polymeric oxyhydroxides of Fe(III)], smectite	[106,107,120]
Chitinase	Kaolinite	[121]
Peptidases	Kaolinite, montmorillonite	[122,123]
Arylsulfatase	Kaolinite, montmorillonite	[124,125]
Cellulase	Kaolinite, montmorillonite, palygorskite, bentonite	[126,127]
α-Amylase	Kaolinite, montmorillonite, palygorskite	[126]
Amyloglucosidase	Kaolinite, montmorillonite, palygorskite	[126]
Pectinase	Kaolinte, montmorillonite, palygorskite	[126]
Arginase	Organo-smectite	[113]
Tyrosinase	Kaolinite, smectite, bentonite clay-humus complexes	[128,129]
β-D-glucosidase	Kaolinite, montmorillonite, goethite, talc	[130–132]
Laccase	Kaolinite, bentonite, clay-humus complexes	[129]
Peroxidase	Kaolinite, bentonite, clay-humus complexes	[129]

clay surfaces are constantly renewed by microenvironmental factors and they interact with other inorganic and organic soil components to form more complex associations. Hence, both clean and unpurified clays have been considered as supports to which enzymes could be anchored [112,113,120,129]. Not surprisingly, however, unpurified clays showed surface properties different from those of clean clays, including modified surface:volume ratios, reduced access to the interlayer spaces of expanding clays, and changes in the pH–net surface charge relationship.

The affinity for the surfaces of clay minerals, whether pristine or chemically modified, appeared considerable for enzymes and for proteins in general [95]. A variety of clay-enzyme complexes have been obtained. Enzymes may intercalate the interlayer spaces of expanding smectites, but in many cases, they appear to be adsorbed only on external sites, especially near the edges of clays. Harter and Stotzky [106] were among the first to observe that these complexes can be partially disrupted and the adsorbate released by repeated washings with water. To evaluate whether a complexed protein was really active or available, it was deemed necessary to distinguish clearly between the quantity of protein adsorbed and the quantity tightly complexed or bound by clay minerals [97,98,106]. By measuring the amount of protein remaining in solution after equilibration with clay and after several cycles of washing of the clay-protein complex, adsorption and binding isotherms could be determined. According to the classification system of Giles et al. [133], many clays exhibited the L-type (Langmuir) adsorption isotherm [106,108,120,122,124,126,128]. Among other adsorption patterns, the C-type (constant-partitioning) [106,108,109,128], H-type (high affinity) [108,109], and S-type (surface) isotherms have been distinguished [124]. The similarity among isotherms implies involvement of the same adsorption mechanism.

The adsorption of enzymes on clays is, to varying extents, dependent on the type of clay mineral; the ratio of enzyme to clay; the pH of the clay suspension; the cation exchange capacity and specific surface area of the clay; the oxidation state of the cation saturating the clay; the isoelectric point of the enzyme; and the structure, shape, and molecular weight of the enzyme. The complexity of these relationships has been studied and reviewed by many authors [96–98,106,108, 109,118,130].

Because of its amphoteric nature, an enzyme carries a net electric charge that is positive at pH values below its isoelectric point (IEP) and negative when the pH is raised above the IEP. Conversely, most clay minerals show a permanent and pH-independent negative charge, arising from isomorphic substitutions in the tetrahedral and octahedral layers, although a fraction of the total charge is variable with pH as a result of the dissociation of Si-OH and Al-OH groups at the edges of clays. This fraction, which is negligible for smectite clays, is considerably higher for 1:1 clays such as kaolinite. It would appear that the surface interactions between an enzyme and a clay are restricted to ionic bonds between opposite charges. However, recent research showed that other adsorption mechanisms may also be involved, including van der Waals forces and hydrogen bonds, as well as hydrophobic interactions [130]. The formation of covalent bonds was never reported [96].

Electrostatic forces, as the most energetic, are important when the enzyme is in the cationic form, that is, at a pH below the isoelectric point. The following reasons have been suggested for interactions when the net charge on both the protein and clay is negative: (1) a lower pH at the clay surface than that in the bulk solution; (2) positive charges retained at specific sites on the enzyme or on edge sites of the clay mineral; (3) presence of polyvalent inorganic cations that reduce the electrokinetic potential on both the protein and clay minerals; and (4) a combination of other attractive non-coulombic forces that at high ionic strength may be greater than the repulsive electrostatic forces [97–99,106,130].

Immobilization on the clay surface may considerably change enzymatic activity. In general, adsorbed enzymes exhibit reduced activity compared with that of the initially free enzymes. Morgan and Corke [118] showed that the activity of glucose oxidase when adsorbed on montmorillonite and kaolinite was sharply reduced. Similarly, the activities of clay-adsorbed arylsulphatase [124,125] and bacteriolytic endopeptidase [123] were considerably lower than those of the unadsorbed enzymes. Adsorption of arylsulphatase on kaolinite was confined to the external surfaces, whereas that on montmorillonite also involved the interlayer surfaces. With both clays, arylsulphatase was partially inactivated by adsorption. In studies by Claus and Filip [129], phenoloxidases (laccase and tyrosinase) were strongly bound on homoionic bentonites and less so on kaolinite. Binding on bentonite suppressed all laccase activity and more than 90% of tyrosinase activity, whereas up to 50 and 15% of tyrosinase and laccase activity, respectively, was retained when bound on kaolinite. Enzyme molecules less tightly bound on exterior surfaces (secondary surface adsorption) may retain the activity of free enzymes, as was the case with phosphatase-clay complexes [115].

The reduced activity of clay-enzyme complexes is generally interpreted in terms similar to those used for enzymes immobilized on other solid supports, that is, steric and diffusional restrictions, direct involvement of the active site in binding to the support, and modified conformation of the immobilized enzyme. According to Martinek and Mozhaev [134], the structural differences between free and immobilized enzymes result from linkages between the enzyme and the support. Quiquampoix and coworkers [130–132] demonstrated that the reduced activity of β-D-glucosidase adsorbed on different mineral surfaces under varied pH conditions could not be assigned to microenvironmental effects but to a conformational change induced in the adsorbed enzyme by electrostatic and hydrophobic forces.

When clay surfaces are coated with organic and inorganic polymers, their ability to bind enzymes may be modified. Garwood et al. [119] and Boyd and Mortland [112,113] studied the hydrophobic mechanism of a pH-independent immobilization of glucose oxidase and urease on smectite saturated with hexadecyltrimethylammonium cation ($HDTMA^{+}$). The enzymes became strongly bound and retained, respectively, 60 and 100% of their activity in the free state. Peroxidase and arginase were also strongly adsorbed on HDTMA-smectite, but they lost all their activity [113]. These differences in activity may have depended on the relation of binding sites to active sites on the enzyme molecule. Different kinds of organic coatings may differentially alter enzyme activity. Arginase and urease, for example, were immobilized on smectite saturated with bipyridyl cations without a loss of activity [113]. Studies on the adsorption of β-D-glucosidase on different mineral and organomineral surfaces indicated that binding mechanisms may involve the molecular π orbitals of the aromatic nuclei of certain peptides and the bipyridil surface.

Overall the action of the mineral support of the enzyme was often unfavorable, and the observed losses of enzyme activity resulted in a search for alternative ways to bind enzymes on clays. Ruggiero et al. [136] and Sarkar et al. [137] immobilized various enzymes on different clays by using 3-aminopropyltriethoxysilane plus glutaraldehyde to ensure that the enzymes remained far enough from the clay surface to retain most of their original activity.

D. Kinetics of Enzymatic Reactions in Heterogeneous Systems

Kinetic studies of enzyme-substrate systems in free solution provide only background information. Inasmuch as enzymes in vivo are frequently associated with soil components (Figure 2), they may behave differently from those in free solution. Many properties of natural or synthetic soil-, clay-, and humus-enzyme complexes exhibited features similar to those of immobilized enzymes [138]. Detailed reviews on the kinetic behavior of immobilized enzymes are available elsewhere [134,139–141]. In this section, some of the most important kinetic principles that apply to immobilized enzymes are emphasized, thus simulating the essential features of in vivo function.

Goldstein [140] classified the effects of immobilization on the kinetic behavior of an enzyme into four groups:

1. *Conformational and steric effects*: The immobilization may cause conformational alterations of the enzyme molecule, as well as render the active site less accessible to the substrate.
2. *Partitioning effects*: Because no support is completely inert, its polar or apolar interactions with low molecular weight species may induce differences in the concentrations of substrate, product, effector, and

hydrogen and hydroxyl ions present in the vicinity of the support and in the bulk solution.

3. *Microenvironmental effects*: The immobilized enzyme exists in an environment different from that when free in solution. Therefore, the enzyme-substrate interaction, as well as subsequent steps in the catalytic reaction, may undergo significant modifications.
4. *Diffusional or mass-transfer effects*: The diffusion of substrate to, or of product from, the site of the enzymatic reaction may be hampered by the presence of the support. Such an effect would be particularly pronounced in the case of fast enzymatic reactions. As a consequence of diffusional restrictions, the concentration of substrate, product, or effector in the immediate vicinity of the attached enzyme will be different from that in the bulk solution.

The behavior of an immobilized enzyme is in certain cases predictable. If, for instance, the enzyme is immobilized on a polyelectrolyte support and the partitioning effect is caused only by electrostatic interactions, the substrate concentration, $[S]_m$, in the microenvironment of the enzyme can be represented, according to the Boltzmann distribution, as

$$[S_m] = [S]_0 \exp(-ze\psi/kT) \tag{1}$$

where z is the charge on the substrate; e is the charge on the electron; ψ is the electrostatic potential; k is the Boltzmann constant; T is the absolute temperature; and $[S]_0$ is the concentration of charged substrate in the outer solution. Note that $[S]_m > [S]_0$ when z and ψ are of opposite sign; that is, when the immobilized enzyme and substrate have an opposite charge. Conversely, $[S]_m < [S]_0$ when z and ψ are of the same sign.

In general, the reaction catalyzed by an immobilized enzyme obeys the hyperbolic kinetics of Michaelis-Menten, and the observed initial rate of the reaction, v, depends on substrate concentration in the domain of the immobilized enzyme:

$$v = \frac{k_c\,[E]_0\,[S]_0}{K_m + [S]_m} \tag{2}$$

where $[E]_0$ is the total enzyme concentration; k_c is the catalytic constant; and K_m is the Michaelis constant. Insertion of Equation (1) into Equation (2) results in

$$v = \frac{k_c\,[E]_0[S]_0 \exp(-ze\psi/kT)}{K_m + [S]_0 \exp(-ze\psi/kT)} \tag{3}$$

By dividing the numerator and denominator of Equation (3) by $\exp(-ze\psi/kT)$, the following expression is obtained:

$$v = \frac{k_c [E]_0[S]_0}{K_m/\exp(-ze\psi/kT) + [S]_0} = \frac{k_c [E]_0[S]_0}{K_m' + [S]_0} \tag{4}$$

where the apparent Michaelis constant, K_m', is related to the intrinsic Michaelis constant of the free enzyme, K_m, by the expression

$$K_m' = K_m \exp(ze\psi/kT) \tag{5}$$

Equation (5) clearly shows that $K_m' > K_m$ when the charges of the support and substrate are the same and that $K_m' < K_m$ when the charges are opposite.

The humic polymer and mineral moieties of complexes with enzymes usually carry a net charge; therefore, partitioning effects of charged substrates significantly affect the kinetics of the reactions. At high ionic strengths, K_m values of free and bound enzymes should be similar, provided that diffusion effects are negligible. Goldstein [140] proposed equations relating the apparent Michaelis constant, K_m', to the ionic strength of the medium. Partitioning effects may also arise from hydrophobic interactions between apolar substrates and hydrophobic groups contained in organic or organomineral supports.

According to Johansson and Mosbach [142], if the concentration of substrate is less in the environment of the bound enzyme than in the bulk of the substrate solution, then an increase in the observed K_m can be expected.

When additional effects must be considered, it is difficult to forecast the kinetics of the immobilized enzyme. An enzymatic reaction is brought to an end through various catalytic steps involving, in general, more than one transition state. Even a small perturbation arising from the physical and chemical properties of the support can, in principle, modify the intrinsic kinetic parameters of immobilized enzymes.

Two types of restriction to substrate transport by diffusion can be distinguished [140]: (1) external resistance during diffusion of substrate from the bulk solution to the active site of the bound enzyme across a boundary layer of liquid; and (2) internal resistance during diffusion of substrate within the support. The effects of external and internal diffusion on the activity of an immobilized enzyme were examined using porous membranes of varying thicknesses by Sundaram et al. [143], Thomas et al. [144], and Laidler and Bunting [141]. Sundaram et al. [143] suggested that external diffusional hindrances could be eliminated by increasing the stirring rate of the system. According to Laidler and Bunting [141], the kinetics that apply to enzymes immobilized in thick and thin membranes as the support also apply, with some modifications, to other particle shapes. The validity of conclusions derived from theoretical considerations was experimentally demonstrated in several studies [134,140,141]. The theory indicates that Lineweaver-Burk plots for immobilized enzymes controlled by diffusional effects should be convex to the $1/[S]$ axis. This prediction was confirmed by experimental data, suggesting that kinetic models may be in agreement with the behavior of enzyme systems in vivo [141].

E. Performance of Enzymes in Soil

The soil environment has a great effect on the properties and performance of enzymes. For instance, in contrast to enzymes in fungi and plants, enzymes in soil exhibit limited substrate specificity. Interactions of enzymes with humic polymers and clay minerals, as well as the binding of substrates on soil components, may lead to restrictions in the range of compounds that enzymes in soil may transform [79]. Numerous practical and theoretical considerations make it difficult to study the properties of enzymes in different soils. A major obstacle is the variability of enzymatic-assay conditions used in different laboratories. Tabatabai [146] and Malcolm [147] have offered useful suggestions for a more uniform approach to this problem. The following section summarizes information concerning the effect of soil components and various environmental factors on the behavior of enzymes in soil.

pH Effects

The optimal pH for enzymes in soil is usually higher than that of enzymes from other sources [79,139,140,148]. Based on Equation (1), the partitioning of hydrogen ions between the solution and the immobilized enzyme can be described by

$$(a_{H^+})_m = (a_{H^+})_0 \exp(-ze\psi/kT) \tag{6}$$

where $(a_{H^+})_m$ and $(a_{H^+})_0$ are the activities of hydrogen ion in the vicinity of the enzyme and in the outer solution, respectively. Equation (6) may be expressed in a different form:

$$\Delta pH = pH_m - pH_0 = 0.43\, e\psi/kT \tag{7}$$

where pH_m and pH_0 are the local and outer pH values. Equations (6) and (7) indicate that the concentration of hydrogen ions near a polyanionic enzyme associated with soil colloids ($\psi < 0$) will be higher than that measured in the bulk solution (accordingly, the pH will be lower).

Also, if the enzyme reaction generates or transforms acids, the displacement of the pH optimum can be represented by changes in the value of the dissociation constant (K_a) for the free enzyme. If K_a is the dissociation constant for the ionization of a group, -BH, attached to the active site of the enzyme (E), then

$$E - BH \Leftrightarrow E - B^- + H^+$$

If there is a negatively charged group in the neighborhood of the -BH group, the K_a value will be altered to a K_a' value:

$$E - BH + X^- \Leftrightarrow E - B^- + H^+ + X^-$$

X^- symbolizes the negatively charged microenvironment around the enzyme. The dissociation constant K_a' will be less than K_a because the X^- attracts the dissociated protons; thus pK_a' will be larger than pK_a and, in analogy to Equation (7):

$$\Delta pH = pK_a - pK_a' = 0.43\, e\, \psi/kT \tag{8}$$

However, the influence of pH on the activity of enzymes in soil may be simultaneously modified by other factors, and shifts of the pH optima can be easily misinterpreted. Quiquampoix [130] investigated the effect of electrostatic forces on the alteration of the activity of a β-D-glucosidase adsorbed on different mineral surfaces. The pH versus activity profiles varied with ionic strength and the nature of the anions in the buffer. Frankenberger and Johanson [149] attributed the variation in enzyme activity to diversity in vegetation, microorganisms, and soil fauna, as well as to the adsorption properties of the soil.

In addition, the effects of pH on enzyme activity were usually studied in buffered solutions under conditions of substrate concentration that were supposedly sufficient to saturate the enzyme. Because the assumed experimental parameters might not always be achieved in soil, the results could be conflicting and their relevance to natural soil conditions uncertain.

Resistance to Proteolysis and Thermal Denaturation

The resistance of enzymes in soil to proteolysis and thermal denaturation has been addressed in a number of studies [66,76,142,150,151]. Proteolytic enzymes are very active in hydrolyzing proteins in soil, and the very presence of enzyme activity in soil suggests the existence of protective mechanisms. Protection is usually attributed to the association of enzymes with organomineral complexes that shield the enzymes against proteolytic activity without hindering the diffusion of substrate molecules to the active sites [63,70,75,136,150,152]. However, some studies with model humic-tyrosinase [72] and clay-organic-urease [112] complexes demonstrated enhanced proteolysis of the complexed enzymes relative to that of free enzymes, probably as the result of the coadsorption of the enzymes and proteases on neighboring sites on the support materials.

According to Skujins [44], the temperature needed to inactivate enzymes in soil must be 10 to 20°C higher than that needed to inactivate free enzymes. The thermostable fraction of enzymes in soil was found in organic matter extracts [54,70,75,76,79,151,153], indicating that the organic matrix preserved the tertiary structure of enzyme proteins [70]. The contribution of clay minerals to enzyme stabilization in soil probably lies in the formation of porous clay-humic complexes in which enzymes may be trapped and protected. On the other hand, humic-tyrosinase complexes [72], glucose oxidase adsorbed to clays [118], and urease immobilized on a clay-organic complex [112] were found to be less thermostable than the corresponding free enzymes. A possible explanation for these observations is that increased temperature induced modifications of the supporting surface of the enzyme, resulting in a blockage of the active site. Because of the variety of humus- and clay-enzyme complexes that can be formed, considerable differences in the thermal stability of a given enzyme may occur in different soils [135].

Activation Energy and Temperature Coefficient

The temperature dependence of the reaction rate constant, k, can be described by the Arrhenius equation: $k = A \exp(-E_a/RT)$, where A is the preexponential factor, E_a is the activation energy, R is the universal gas constant, and T is the absolute temperature. Theoretically, the E_a value for a specific enzyme-catalyzed reaction should not vary. In practice, the enzyme activity in each soil depends on many factors, some of which cannot be controlled. A low E_a value indicates that the reaction is the result of enzyme activity and not of abiotic catalysts, and that the catalytic efficiency of the enzymatic complex is higher than that of the free enzyme. The measured activation energies for enzymes catalyzing reactions related to the cycling of carbon, nitrogen, phosphorus, and sulfur in different soils ranged between 30.8 and 57.0 kJ mol^{-1} for carbon [154], between 19.8 and 65.4 kJ mol^{-1} for nitrogen [155–159], between 30.2 and 50.1 kJ mol^{-1} for phosphorus [160–162], and between 42.4 and 46.9 kJ mol^{-1} for sulfur [163], whereas the activation energies for natural humus-laccase and model humus-invertase complexes were 23.8 [54] and 24.5 kJ mol^{-1} [73], respectively. The large range in activation energy values reported for an enzyme was often the result of a lack of control of other factors that affect reaction rate [155].

The temperature coefficient, Q_{10}, representing the increment of reaction rate with a temperature increase of 10°C, was found to be directly correlated with the E_a value. For an enzymatic reaction, the Q_{10} was always less than two [54,73], indicating a reduced demand for the kinetic energy. Values higher than two that were found for enzymes mediating the nitrogen cycle in various soils were probably the result of differences in assay procedures [155,157,158].

Kinetics

The Michaelis constant, K_m, is considered to be the most fundamental parameter in enzyme chemistry. It remains to be determined whether the K_m of enzymes in soil has the same physical meaning as defined by pure enzymology: that is, the measure of substrate affinity for the enzyme molecule. According to many authors [54,151,153,164], the K_m values for enzymes in soil are higher than those for the corresponding enzymes in solution in vitro. This finding can be explained on the basis of conformational, steric, partitioning, and diffusional effects that are typical of heterogeneous molecular environments. The wide range of K_m values determined for the same enzyme in various soils may be the result of differences in assay conditions or in the type and amount of clay minerals and organic matter present. In model experiments, Makboul and Ottow [165] investigated the influence of humic acid and different clay minerals homoionic to calcium on the activities of urease and acid phosphatase. Decreases in enzyme activity were observed with increasing amounts of each clay at different substrate concentrations. In general, maximum inactivation was determined for approximately 0.6 g of each sorbent. The inactivation of urease decreased in the sequence, humic acid > montmorillonite > illite > pyrophillite > halloysite > kaolinite > bentonite. The order for acid phosphatase was pyrophillite > montmorillonite > bentonite > halloysite > kaolinite > illite. The K_m values were quite high in the urease-montmorillonite complex and somewhat lower in complexes with kaolinite and illite. Maximum reaction velocities (V_{max}) were reduced with increasing amounts of kaolinite and illite, but increased with increasing concentrations of montmorillonite [166]. In another study [167], illite inhibited the activity of alkaline phosphatase to a much greater extent than did montmorillonite or kaolinite. However, the V_{max} was diminished by illite and kaolinite and increased in the presence of montmorillonite. The most remarkable observation, however, was the marked decrease of K_m in the presence of illite. Taken together, the K_m and V_{max} values in the presence of illite indicated enhanced enzyme-substrate formation and restricted hydrolysis of the enzyme-substrate complex. The latter effect could result from steric modification of the sorbed alkaline phosphatase.

Dick and Tabatabai [115], studying the effect of clay minerals on root phosphatases, showed a noncompetitive inhibition of acid phosphatase and pyrophosphatase by montmorillonite and illite (reduced V_{max} but an unchanged K_m). Conversely, the inhibition by kaolinite followed a partially competitive model (an unchanged V_{max} value and an increase in the K_m value). The authors suggested the formation of a clay-enzyme complex with a somewhat reduced activity. As in a model for arylsulfatase-clay complexes [125], most of the phosphatase appeared to be only loosely associated with clay; a certain proportion, however, was apparently tightly bound on the clay surface.

Among the enzymes of the carbon-cycle, different degrees of inhibition by various clay minerals were observed [168,169]. The activity of glucose oxidase was markedly reduced by montmorillonite, but only slightly reduced by allophane [116,117]. In a study of the inhibition of glucose oxidase by clay minerals, Morgan and Corke [118,171] considered the effect of the cation exchange capacity of the clay, the pH of the suspending solution, the protein conformation, and the concentration of the coenzyme FAD of the holoenzyme. Complexation of oxidoreductive enzymes, such as laccase or tyrosinase, with humic acid had inhibitory effects (enhanced K_m values) on the transformation of phenolic substrates [54,72]. According to Maignan [73], the kinetic parameters of synthetic humus-Fe^{3+}-invertase complexes depended on the method of preparation.

One of the features of enzymatic catalysis is a decrease in reaction velocity at relatively high substrate concentrations, which is referred to as substrate inhibition. This type of inhibition was observed for pyrophosphatase at pyrosphosphate concentrations above 60 mM [160], for urease activity [172], and for a humus-laccase complex extracted from soil using quinol as the substrate [54]. From the differences between theoretical and experimental reaction rates at each concentration, the percentage of inhibition could be calculated [54].

Deviation from Michaelis–Menten kinetics may also result from the fact that more than one form of enzyme activity may catalyze the same reaction in soil [173,174]. Another reason may be allosteric interaction involving active sites on a multisite enzyme. In studies with L-histidine NH_3-lyase, Frankenberger [157] found a sigmoidal pattern for the curve of velocity as a function of substrate concentration, a deviation from classical Michaelis–Menten kinetics. Because sigmoidicity is considered to be a kinetic characteristic of allosterically regulated enzymes whose activity is closely associated with cell metabolism, Frankenberger concluded that the L-histidine NH_3-lyase activity in soil was probably biontic in nature [157].

Effects of Salinity on Enzyme Activities in Soil

The accumulation of soluble salts in soils as a consequence of pedogenetic factors, frequent irrigation, and intensive fertilization promotes significant changes in soil fertility. The effect of salinity on the activity of various enzymes in soil, such as amidase, urease, acid and alkaline phosphatase, phosphodiesterase, inorganic pyrophosphatase, arylsulphatase, rhodanese, α- and β-glucosidase, α- and β-galactosidase, dehydrogenase and catalase, was investigated by Elivazi and Tabatabai [52] and Frankenberger and Bingham [175]. The authors observed a reduction in the activity of these enzymes in soils treated with salts of sodium, potassium, calcium, and magnesium. The degree of inhibition varied among the enzymes and with the nature and amounts of salts added. The decline in activity was probably associated with a change in the osmotic potential of the soil water phase that led to the desiccation of microbial cells, resulting in release of intracellular enzymes vulnerable to enzymatic proteolysis in soil. Also, specific ionic effects on enzyme structures and microbial growth may have been of importance [175].

Effects of Trace Elements on Enzyme Activities in Soil

The term trace elements refers to elements that occur at very low levels in a given system and that, when present in sufficient concentrations, are toxic to living organisms [176]. Trace elements may enter the soil as impurities in agrochemicals or as components of municipal and industrial wastes. Aerosols, particulate matter, and airborne dusts deposited on soils may also contribute to the trace element load. Trace elements are often heavy metals, and some of them are considered to be among the most harmful pollutants. Their presence in soils may affect enzymatic activities because most trace elements have a significant affinity for sulfur and attack disulfide bonds of enzymes. Carboxylic and amino groups of both protein and substrate molecules may also chemically bind heavy metals.

Much of the research on the effects of trace elements on enzyme activity in soils has been done by Tabatabai and coworkers [52, 176–182]. The activities of 12 enzymes involved in carbon, nitrogen, phosphorus, and sulfur cycling were assayed in different soils before and after addition of about 20 water-soluble metallic salts, some of which are listed in Table 5. The trace elements inhibited enzyme activity at concentrations ranging from 2.5 to 25 μmol g^{-1} of soil, with the amount of inhibition dependent upon the type of trace element and soil used.

It is noteworthy that Ba(II) enhanced the activity of nitrate reductase in a neutral soil, but had no effect in calcareous and acid soils [178]. A similar pattern was observed for B(III), which caused, on average, a 1% inhibition of nitrate reductase (enhancement of activity was observed in a neutral soil, inhibition in an acid soil, and no effect in a calcareous soil).

Effects of Pesticides on Enzyme Activities in Soil

Pesticides used in modern agriculture constitute another factor that may affect enzymes in soil. Both inhibition and activation of enzyme activity by pesticides have been observed in field and in vitro experiments [190].

According to Cervelli et al. [191], pesticides may indirectly affect enzyme activity by altering the life functions of soil organisms. Reviews by Kiss et al. [43], Cervelli et al. [191], and Schäffer [190] provided a first reference for a variety of pesticide-enzyme interactions. Table 6 presents a list of the most significant papers related to this topic. The effects reported differed markedly for different pesticides and depended on the type of enzyme, soil properties, experimental procedure, and rate of pesticide application. The wide range of

Table 5 Effects of Trace Elements on Enzyme Activity in Soils [52,176–182]

		Enzyme						
		Percentage inhibition (average for various soils)						
Element	Oxidation state	α-Glucosidase	β-Glucosidase	α-Galactosidase	β-Galactosidase	Amidase	Nitrate reductase	L-Asparaginase
As	V	19	10	51	38	2	87	5
Mo	VI	15	14	73	37	3	15	4
W		35	15	17	10	2	90	18

Table 6 Effects of Pesticides on Enzyme Activity in Soils

Pesticides	Enzymes	References
Herbicides	Amidase	[177]
	Amylase	[193,194]
	L-Asparaginase	[179]
	Dehydrogenase	[192,195,196]
	1,3-β-Glucanase	[183]
	Invertase	[193,194]
	Laccase	[184]
	Nitrogenase	[196]
	Phosphatase	[185,192,195,196]
	Phytase	[186]
	Urease	[183,187–189,192,195–197]
Insecticides	Amidase	[177]
	Amylase	[193,194,198]
	L-Asparaginase	[179]
	Dehydrogenase	[195,196,198]
	1,3-β-Glucanase	[183]
	Invertase	[193,194,198]
	Nitrogenase	[196,198]
	Phosphatase	[195,196,198,199]
	Phytase	[186]
	Rhodanese	[200]
	Urease	[183,195,196,198,201]
Fungicides	Amidase	[177]
	Amylase	[193,194]
	L-Asparaginase	[179]
	Dehydrogenase	[195,196]
	Invertase	[193,194]
	Nitrogenase	[196]
	Phosphatase	[195,196]
	Rhodanese	[200]
	Urease	[195,196]

soils used and variable assay conditions were often the source of discrepancy [192].

VI. CONCLUDING REMARKS

Transformation reactions in soil result from a combination of chemical and biological factors. Despite the multitude and interdependence of soil processes, model experiments provide some insight into the functioning of enzymes under field conditions. The results indicated that practically all major transformation reactions observed in terrestrial systems are enzyme mediated. In particular, various hydrolases and oxidoreductases appear to be largely involved in the decomposition of dead biomass and systematic replenishment of losses in humus. Furthermore, the enzyme-mediated reactions seem to facilitate assimilation of the raw organic and inorganic material by plants and other organisms. In view of the above findings, enzymatic activity in soil is perceived as being beneficial and deserving stimulation.

The catalytic properties of enzymes may be modified to a great extent by the presence of certain chemicals and by the immobilization of enzyme molecules on humus and clay minerals. The inhibitory effects are usually considered to be a disadvantage, but on the other hand, an unrestricted accumulation of enzyme activity in soil hardly appears to be desirable, since it might affect the stability of the soil system.

Immobilization of enzymes on clay or humus may result in loss of enzyme activity, but once immobilized the enzyme is more resistant to inactivation and degradation under the harsh in situ soil conditions. Enzyme immobilization appears to be the norm in the soil environment. Therefore the mechanism of immobilization and the role of this phenomenon in transformation processes are important issues for current research on enzymes in soil.

REFERENCES

1. Eyring, D. L. 1986. Kinetics of reactions in pure and in mixed systems, p. 83–145. *In* D. L. Sparks (ed.), Soil physical chemistry. CRC Press, Boca Raton, Florida.
2. Bartlett, R. J. 1986. Soil redox behavior, p. 179–207. *In* D. L. Sparks (ed.), Soil physical chemistry. CRC Press, Boca Raton, Florida.
3. Lehmann, R. G., H. H. Cheng, and J. B. Harsh. 1987. Oxidation of phenolic acids by soil iron and manganese oxides. Soil Sci. Soc. Am. J. 51:352–356.
4. Pohlman, A. A., and J. G. McColl. 1989. Organic oxidation and manganese and aluminium mobilization in forest soils. Soil Sci. Soc. Am. J. 53:686–690.
5. McBride, M. B. 1987. Adsorption and oxidation of phenolic compounds by iron and manganese oxides. Soil Sci. Soc. Am. J. 51:1466–1472.
6. Stone, A. T., and J. J. Morgan. 1984. Reduction and dissolution of manganese (III) and manganese (IV) oxides by organics: 2. Survey of the reactivity of organics. Environ. Sci. Technol. 18:617–624.
7. Stone, A. T. 1987. Reductive dissolution of manganese (III/IV) by substituted phenols. Environ. Sci. Technol. 21:979–988.
8. Shindo, H., and P. M. Huang. 1984. Catalytic effects of manganese(IV), iron(III), aluminium, and silicon oxides on the formation of phenolic polymers. Soil Sci. Soc. Am. J. 48:927–934.
9. Wang, M. C., and P. M. Huang. 1989. Pyrogallol transformations as catalyzed by nontronite, bentonite, and kaolinite. Clays Clay Miner. 37:525–531.

10. Mortland, M. M., and L. J. Halloran. 1976. Polymerization of aromatic molecules on smectite. Soil Sci. Soc. Am. J. 40:367–370.
11. Laszlo, P. 1987. Chemical reactions on clays. Science 235:1473–1477.
12. Senesi, N. 1990. Application of electron spin resonance (ESR) spectroscopy in soil chemistry. Adv. Soil Sci. 14:77–130.
13. Steelink, C., and G. Tollin. 1967. Free radicals in soil, p. 147–169. *In* A. D. McLaren and G. H. Peterson (ed.), Soil biochemistry. Marcel Dekker, New York.
14. Dragun, J., and C. S. Helling. 1985. Physicochemical and structural relationships of organic chemicals undergoing soil- and clay-catalyzed free-radical oxidation. Soil Sci. 139:100–111.
15. Wang, T. S. C., S. W. Li, and Y. L. Ferng. 1978. Catalytic polymerization of phenolic compounds by clay minerals. Soil Sci. 126:15–21.
16. Vogel, T. M., C. S. Criddle, and P. L. McCarthy. 1987. Transformation of halogenated aliphatic compounds. Environ. Sci. Technol. 21:722–736.
17. Sjoblad, R. D., and J.-M. Bollag. 1981. Oxidative coupling of aromatic compounds by enzymes from soil microorganisms, p. 113–152. *In* E. A. Paul and J. N. Ladd (ed.), Soil biochemistry, Vol. 5. Marcel Dekker, New York.
18. Robb, D. A. 1984. Tyrosinase, p. 207–240. *In* R. Lontie (ed.), Copper proteins and copper enzymes, Vol. II. CRC Press, Boca Raton, Florida.
19. Suflita, J. M., M. J. Loll, W. C. Snipes, and J.-M. Bollag. 1981. Electron spin resonance study of free radicals generated by a soil extract. Soil Sci. 131:145–150.
20. Choudry, G. G., and G. R. B. Webster. 1985. Protocol guidelines for the investigations of photochemical fate of pesticides in water, air, and soils. Res. Rev. 96:79–136.
21. Larson, R. A., and M. R. Berenbaum. 1988. Environmental phototoxicity. Environ. Sci. Technol. 22:354–360.
22. Pignatello, J.J. 1992. Dark and photoassisted Fe^{3+}-catalyzed degradation of chlorophenoxy herbicides by hydrogen peroxide. Environ. Sci. Technol. 26:944–951.
23. Boyd, S. A., and M. M. Mortland. 1986. Radical formation and polymerization of chlorophenols and chloroanisole on copper(II)-smectite. Environ. Sci. Technol. 20: 1056–1058.
24. Senesi, N., and C. Steelink. 1989. Application of ESR spectroscopy to the study of humic substances, p. 373–408. *In* M. H. B. Hayes, P. MacCarthy, R. L. Malcolm, and R. S. Swift (ed.), Humic substances. II. In search of structure. John Wiley & Sons, Chichester.
25. Bollag, J.-M. 1983. Cross-coupling of humus constituents and xenobiotic substances, p. 127–141. *In* R. F. Christman and E. T. Gjessing (ed.), Aquatic and terrestrial humic materials. Ann Arbor Science, Ann Arbor, Michigan.
26. Bollag, J.-M., and W. B. Bollag. 1990. A model for enzymatic binding of pollutants in the soil. Int. J. Environ. Anal. Chem. 39:147–157.
27. Degens, E. T., J. Matheja, and T. A. Jackson. 1970. Template catalysis: asymmetric polymerization of amino acids on clay minerals. Nature 227:492–493.
28. Wang, T. S. C., P. M. Huang, C.-H. Chou, and J.-H. Chen. 1986. The role of soil minerals in the abiotic polymerization of phenolic compounds and formation of humic substances, p. 251–281. *In* P. M. Huang and M. Schnitzer (ed.), Interactions

of soil minerals with natural organics and microbes. Special Publication No. 17. Soil Science Society of America, Madison, Wisconsin.

29. Boyd, S. A., and M. M. Mortland. 1985. Dioxin radical formation and polymerization on Cu(II)-smectite. Nature 316:532–535.
30. Ballantine, J. A., J. H. Purnell, and J. M. Thomas. 1983. Organic reactions in a clay microenvironment. Clay Min. 18:347–356.
31. Adams, J. M. 1987. Synthetic organic chemistry using pillared, cation-exchanged and acid-treated montmorillonite catalysts. A review. Appl. Clay Sci. 2:309–342.
32. Liu, S.-Y., R. D. Minard, and J.-M. Bollag. 1987. Soil-catalyzed complexation of the pollutant 2,6-diethylaniline with syringic acid. J. Environ. Qual. 16:48–53.
33. Hamilton, G. A. 1969. Mechanisms of two- and four-electron oxidations catalyzed by some metalloenzymes. Adv. Enzym. 32:55–96.
34. Zepp, R. G., P. F. Schlotzhauer, and R. M. Sink. 1985. Photosensitized transformations involving electronic energy transfer in natural waters: role of humic substances. Environ. Sci. Technol. 19:74–81.
35. Gohre, K., R. Scholl, and G. H. Miller. 1986. Singlet oxygen reactions on irradiated soil surfaces. Environ. Sci. Technol. 20:934–938.
36. Miller, G. C.., V. R. Hebert, and W. W. Miller. 1989. Effect of sunlight on organic contaminants at the atmosphere-soil interface, p. 99–110. *In* B. L. Sawhney and K. Brown (ed.), Reactions and movement of organic chemicals in soils. Special Publication No. 22. Soil Science Society of America, Madison, Wisconsin.
37. Wolfe, N. I., U. Mingelgrien, and G. C. Miller. 1990. Abiotic transformation in water, sediments, and soil, p. 103–168. *In* H. H. Cheng (ed.), Pesticides in the soil environment: processes, impacts, and modeling. Soil Science Society of America, Madison, Wisconsin.
38. Pollock, M. R. 1962. Exoenzymes, p. 121–178. *In* I. C. Gunsalus and R. Y. Stanier (ed.), The bacteria, Vol. 4. Academic Press, New York.
39. Burns, R. G. 1983. Extracellular enzyme-substrate interactions in soil, p. 249–298. *In* J. H. Slater, R. Whittenburry, and J. W. T. Wimpemy (ed.), Microbes in their natural environments. Cambridge University Press, Cambridge.
40. Ander, P., and K.-E. Eriksson. 1976. The importance of phenol oxidase activity in lignin degradation by the white-rot fungus *Sporotrichum pulverulentum*. Arch. Microbiol. 109:1–8.
41. Burns, R. G. 1982. Enzyme activity in soil: location and a possible role in microbial ecology. Soil Biol. Biochem. 14:423–427.
42. Efroymson, R. A., and M. Alexander. 1991. Biodegradation by an *Arthrobacter* species of hydrocarbons partitioned into an organic solvent. Appl. Environ. Microbiol. 57:1441–1447.
43. Kiss, S., M. Dragan-Bularda, and D. Radulescu. 1975. Biological significance of enzymes accumulated in soil. Adv. Agron. 27:25–87.
44. Skujins, J. J. 1976. Extracellular enzymes in soil. CRC Crit. Rev. Microbiol. 4:383–421.
45. Burns, R. G. 1986. Interaction of enzymes with soil mineral and organic colloids, p. 429–451. *In* P. M. Huang and M. Schnitzer (ed.), Interactions of soil minerals with natural organics and microbes. Special Publication No. 17. Soil Science Society of America, Madison, Wisconsin.

46. Glenn, A. R. 1976. Production of extracellular proteins by bacteria. Annu. Rev. Microbiol. 30:41–62.
47. Frankenberger, W. T., Jr., and W. A. Dick. 1983. Relationships between enzyme activities and microbial growth and activity indices in soil. Soil Sci. Soc. Am. J. 47: 945–951.
48. Domsch, K. H., T. Beck, J. P. E. Anderson, B. Söderström, D. Perkinson, and G. Trolldenier. 1979. A comparison of methods for soil, microbial population and biomass studies. Z. Pflanzenernaehr. Bodenkd. 142:520–533.
49. Hope, C. F. A., and R. G. Burns. 1987. Activity, origins and location of cellulases in a silt loam soil. Biol. Fertil. Soils 5:164–170.
50. Mayaudon, J. 1986. The role of carbohydrates in the free enzymes in soil, p. 263–309. *In* C. H. Fuchsman (ed.), Peat and water. Elsevier, London.
51. Sequi, P., G. Cercignani, M. De Nobili, and M. Pagliai. 1985. A positive trend among two soil enzyme activities and a range of soil porosity under zero and conventional tillage. Soil Biol. Biochem. 17:255–256.
52. Elivazi, F., and M. A. Tabatabai. 1990. Factors affecting glucosidase and galactosidase activities in soils. Soil Biol. Biochem. 22:891–897.
53. Batistic, L., J. M. Sarkar, and J. Mayaudon. 1980. Extraction, purification and properties of soil hydrolases. Soil Biol. Biochem. 12:59–63.
54. Ruggiero, P., and V. M. Radogna. 1984. Properties of laccase in humus-enzyme complexes. Soil Sci. 138:74–87.
55. Nannipieri, P., B. Ceccanti, S. Cervelli, and E. Matarese. 1980. Extraction of phosphatase, urease, proteases, organic carbon, and nitrogen from soil. Soil Sci. Soc. Am. J. 44:1011–1016.
56. Gosewinkel, U., and F. E. Broadbent. 1986. Decomplexation of phosphatase from extracted soil humic substances with electron donating reagents. Soil Sci. 141:261–267.
57. Burns, R. G., M. H. El-Sayed, and A. D. McLaren. 1972. Extraction of an urease-active organo-complex from soil. Soil Biol. Biochem. 4:107–108.
58. McLaren, A. D., A. H. Pukite, and I. Barshad. 1975. Isolation of humus with enzymatic activity from soil. Soil Sci. 119:178–180.
59. Nannipieri, P., B. Ceccanti, S. Cervelli, and P. Sequi. 1974. Use of 0,1M pyrophosphate to extract urease from a podzol. Soil Biol. Biochem. 6:359–362.
60. Ladd, J. N. 1972. Properties of proteolytic enzymes extracted from soil. Soil Biol. Biochem. 4:227–237.
61. Maggioni, A., and G. Cacco. 1977. Acetyl-naphthyl-esterase activity in humus-enzyme complexes of different molecular size. Soil Sci. 123:122–125.
62. Perez-Mateos, M., S. Gonzales-Carcedo, and M. D. Busto Nuñez. 1988. Extraction of catalase from soil. Soil Sci. Soc. Am. J. 52:408–411.
63. Sarkar, J. M. 1986. Formation of [^{14}C]cellulase-humic complexes and their stability in soil. Soil Biol. Biochem. 18:251–254.
64. Ceccanti, B., P. Nannipieri, S. Cervelli, and P. Sequi. 1978. Fractionation of humus-urease complexes. Soil Biol. Biochem. 10:39–45.
65. Nannipieri, P., B. Ceccanti, D. Bianchi, and M. Bonmati. 1985. Fractionation of hydrolase-humus complexes by gel chromatography. Biol. Fertil. Soils 1:25–29.

66. Nannipieri, P., B. Ceccanti, and D. Bianchi. 1988. Characterization of humus-phosphatase complexes extracted from soil. Soil Biol. Biochem. 20:683–691.
67. Ceccanti, B., M. Bonmati-Pont, and P. Nannipieri. 1989. Microdetermination of protease activity in humic bands of different sizes after analytical isoelectric focusing. Biol. Fert. Soils 7:202–206.
68. Maignan, C. 1982. Activité des complexes acides humiques-invertase: influence du mode de préparation. Soil Biol. Biochem. 14:439–445.
69. Maignan, C. 1983. Activité des complexes acides humiques-invertase: influence des cations floculants. Soil Biol. Biochem. 15:651–659.
70. Serban, A., and A. Nissenbaum. 1986. Humic acid association with peroxidase and catalase. Soil Biol. Biochem. 18:41–44.
71. Ruggiero, P., and V. M. Radogna. 1987. Tyrosinase activity in a humic-enzyme complex. Sci. Total Environ. 62:365–366.
72. Ruggiero, P., and V. M. Radogna. 1988. Humic acids-tyrosinase interactions as a model of soil humic-enzyme complexes. Soil Biol. Biochem. 20:353–359.
73. Maignan, C. 1990. Paramètres cinétiques de complexes humates- Fe^{3+}-invertase. C. R. Acad. Sci. Paris, Series III, 310:571–576.
74. Rowell, M. J., J. N. Ladd, and E. A. Paul. 1973. Enzymically active complexes of proteases and humic acid analogues. Soil Biol. Biochem. 5:699–703.
75. Sarkar, J. M., and R. G. Burns. 1984. Synthesis and properties of β-D-glucosidase-phenolic copolymers as analogues of soil humic-enzyme complexes. Soil Biol. Biochem. 16:619–625.
76. Nannipieri, P., B. Ceccanti, C. Conti, and D. Bianchi. 1982. Hydrolases extracted from soil: their properties and activities. Soil Biol. Biochem. 14:257–263.
77. Gosh, K., and M. Schnitzer. 1980. Macromolecular structures of humic substances. Soil Sci. 129:266–271.
78. Pflug, W., and W. Ziechmann. 1981. Inhibition of malate dehydrogenase by humic acids. Soil Biol. Biochem. 13:293–299.
79. Ladd, J. N., and J. H. A. Butler. 1975. Humus-enzyme systems and synthetic, organic polymer-enzyme analogs, p. 143–194. *In* E. A. Paul and A. D. McLaren (ed.), Soil biochemistry, Vol. 4. Marcel Dekker, New York.
80. Senesi, N., Y. Chen, and M. Schnitzer. 1977. Aggregation-dispersion phenomena in humic substances, p. 143–155. *In* Proceedings of symposium on soil organic matter studies, Braunschweig, Germany, Vol. II. IAEA, Vienna.
81. Sarkar, J. M., and R. G. Burns. 1983. Immobilization of β-D-glucosidase and β-D-glucosidase-polyphenolic complexes. Biotechnol. Lett. 5:619–624.
82. Sarkar, J. M., and R. G. Burns. 1984. Characterization of cellulase in soils and sediments and the evaluation of synthetic humic-cellulase complexes, p. 168–178. *In* G. L. Ferrero, M. P. Ferranti, and H. Naveau (ed.), Anaerobic digestion and carbohydrate hydrolysis of waste. Elsevier, London.
83. Grego, S., A. D'Annibale, M. Luna, L. Badalucco, and P. Nannipieri. 1990. Multiple forms of synthetic pronase-phenolic copolymers. Soil Biol. Biochem. 22:721–724.
84. Ladd, J. N., and J. H. A. Butler. 1966. Comparison of some properties of soil humic acids and synthetic phenolic polymers incorporating amino derivatives. Aust. J. Soil Res. 4:41–54.

85. Brannon, C. A., and L. E. Sommers. 1985. Preparation and characterization of model humic polymers containing organic phosphorus. Soil Biol. Biochem. 17:213–219.
86. Ladd, J. N., and J. H. A. Butler. 1969. Inhibitory effect of soil humic compounds on the proteolytic enzyme pronase. Aust. J. Soil Res. 7:241–251.
87. Butler, M. H. A., and J. N. Ladd. 1969. The effect of methylation of humic acids on their influence on proteolytic enzyme activity. Aust. J. Soil Res. 7:263–268.
88. Ladd, J. N., J. H. A. Butler. 1971. Inhibition by soil humic acids of native and acetylated proteolytic enzymes. Soil Biol. Biochem. 3:157–160.
89. Butler, J. H. A., and J. N. Ladd. 1971. Importance of the molecular weight of humic and fulvic acids in determining their effects on protease activity. Soil Biol. Biochem. 3:249–257.
90. Mato, M. C., E. Fàbregas, and J. Méndez. 1971. Inhibitory effect of soil humic acids on indoleacetic acid-oxidase. Soil Biol. Biochem. 3:285–288.
91. Mato, M. C., M. G. Olmedo, and J. Méndez. 1972. Inhibition of indoleacetic acid-oxidase by soil humic acids fractionated on Sephadex. Soil Biol. Biochem. 4:469–473.
92. Pflug, W. 1979. Über die Hemmung der Aminopeptidase K (aus *Tritirachium album* limber) durch Huminstoffe. Z. Pflanzenernaehr. Bodenkd. 142:290–298.
93. Pflug, W. 1980. Effect of humic acids on the activity of two peroxidases. Z. Pflanzenernaehr. Bodenkd. 143:432–440.
94. Pflug, W. 1981. Effect of humic acids on the activity of soil enzymes. Z. Pflanzenernaehr. Bodenkd. 144:423–425.
95. Boyd, S. A., and M. M. Mortland. 1990. Enzyme interactions with clays and clay-organic matter complexes, p. 1–28. *In* J.-M. Bollag and G. Stotzky (ed.), Soil biochemistry, Vol. 6. Marcel Dekker, New York.
96. Theng, B. K. G. 1979. Formation and properties of clay-polymer complexes. Elsevier, New York.
97. Stotzky, G. 1980. Surface interactions between clay minerals and microbes, viruses and soluble organics, and the probable importance of these interactions to the ecology of microbes in soil, p. 231–247. *In* R. C. W. Berkely, J. M. Linch, J. Melling, P. R. Rutter, and B. Vincent (ed.), Microbial adhesion to surfaces. Ellis Horwood, Chichester.
98. Stotzky, G., and R. G. Burns. 1982. The soil environment: clay-humus-microbe interactions, p. 105–133. *In* R. G. Burns and J. H. Slater (ed.), Experimental microbial ecology. Blackwell Scientific, Oxford.
99. Theng, B. K. G. 1982. Clay-polymer interactions: summary and perspectives. Clays Clay Miner. 30:1–10.
100. Mortland, M. M. 1986. Mechanisms of adsorption of nonhumic organic species by clays, p. 59–76. *In* P. M. Huang and M. Schnitzer (ed.), Interactions of soil minerals with natural organics and microbes. Special Publication No. 17. Soil Science Society of America, Madison, Wisconsin.
101. Fischer, E. H., L. Kohtés, and J. Fellig. 1951. Propriétés de l'invertase putifieé. Helv. Chim. Acta 34:1132–1138.
102. McLaren, A. D. 1954. The adsorption and reactions of enzymes and proteins on kaolinite. Int. J. Phys. Chem. 58:129–137.

103. McLaren, A. D. 1954. The adsorption and reactions of enzymes and proteins on kaolinite. II. The action of chymotrypsin on lysozyme. Soil Sci. Soc. Am. Proc. 18: 170–174.

104. McLaren, A. D., and E. F. Estermann. 1956. The adsorption and reactions of enzymes and proteins on kaolinite. III. The isolation of enzyme-substrate complexes. Arch. Biochem. Biophys. 61:158–173.

105. Aomine S., and Y. Kobayashi. 1964. Effects of allophane on the enzyme activity of a protease. Soil Sci. Plant Nutr. 10:28–32.

106. Harter, R. D., and G. Stotzky. 1971. Formation of clay-protein complexes. Soil Sci. Soc. Am. Proc. 35:383–389.

107. Harter, R. D., and G. Stotzky. 1973. X-ray diffraction, electron microscopy, electrophoretic mobility, and pH of some stable smectite-protein complexes. Soil Sci. Soc. Am. Proc. 37:116–123.

108. Albert, J. T., and R. D. Harter. 1973. Adsorption of lysozyme and ovalbumin by clay: effect of clay suspension pH and clay mineral type. Soil Sci. 115:130–136.

109. Harter, R. D. 1975. Effect of exchange cations and solution ionic strength on formation and stability of smectite-protein complexes. Soil Sci. 120:174–181.

110. Pinck, L. A., and F. E. Allison. 1961. Adsorption and release of urease by and from clay minerals. Soil Sci. 91:183–188.

111. Pinck, L. A. 1962. Adsorption of proteins, enzymes and antibiotics by montmorillonite. Clays Clay Miner. 9:520–529.

112. Boyd, S. A., and M. M. Mortland. 1985. Urease activity on a clay-organic complex. Soil Sci. Soc. Am. J. 49:619–622.

113. Boyd, S. A., and M. M. Mortland. 1985. Manipulating the activity of immobilized enzymes with different organo-smectite complexes. Experientia 41:1564–1566.

114. Ramirez-Martinez, J. R., and A. D. McLaren. 1966. Some factors influencing the determination of phosphatase activity in native soils and in soils sterilized by irradiation. Enzymologia 31:23–38.

115. Dick, W. A., and M. A. Tabatabai. 1987. Kinetics and activities of phosphatase-clay complexes. Soil Sci. 143:5–15.

116. Zvyagintsev, D. G., and L. L. Velikanov. 1968. Influence of soils and clay minerals on the activity of glucose oxidase and invertase. Sov. Soil Sci. 1:789–794.

117. Ross, D. J., and B. A. McNeilly. 1972. Some influences of different soils and clay minerals on the activity of glucose oxidase. Soil Biol. Biochem. 4:9–18.

118. Morgan, H. W., and C. T. Corke. 1976. Adsorption, desorption, and activity of glucose oxidase on selected clay species. Can. J. Microbiol. 22:684–693.

119. Garwood, G. A., M. M. Mortland, and T. J. Pinnavaia. 1983. Immobilization of glucose oxidase on montmorillonite clay: hydrophobic and ionic modes of binding. J. Molec. Catal. 22:153–163.

120. Fusi, P., G. G. Ristori, L. Calamai, and G. Stotzky. 1989. Adsorption and binding of protein on "clean" (homoionic) and "dirty" (coated with Fe oxyhydroxides) montmorillonite, illite and kaolinite. Soil Biol. Biochem. 21:911–920.

121. Skujins, J., A. Pukite, and A. D. McLaren. 1974. Adsorption and activity of chitinase on kaolinite. Soil Biol. Biochem. 6:179–182.

122. Haskå, G. 1975. Influence of clay minerals on sorption of bacteriolytic enzymes. Microb. Ecol. 1:234–245.

123. Haskå, G. 1981. Activity of bacteriolytic enzymes adsorbed to clays. Microb. Ecol. 7:331–341.
124. Simpson, G. H., and J. D. Hughes. 1978. Arylsulphatase-clay interactions. I. Adsorption of arylsulphatase by kaolinite and montmorillonite. Aust. J. Soil Res. 16: 27–33.
125. Hughes, J. D., and G. H. Simpson. 1978. Arylsulphatase-clay interactions. II. The effect of kaolinite and montmorillonite on arylsulphatase activity. Aust. J. Soil Res. 16:35–40.
126. Hamzehi, E., and W. Pflug. 1981. Sorption and binding mechanism of polysaccharide cleaving soil enzymes by clay minerals. Z. Pflanzenernaehr. Bodenk. 144:505–513.
127. Chhonkar, P. K., and Y. S. S. Rao. 1987. Desorption of C_1-, C_x-cellulases adsorbed on Na-, Ca-, and Al-kaolinite and bentonite. J. Indian Soc Soil. Sci. 35:736–738.
128. Ruggiero, P., and V. M. Radogna. 1987. Interactions between soil colloids and enzymes which oxidize phenols and polyphenols. *In* G. Giovannozzi-Sermanni and P. Nannipieri (ed.), Current perspectives in environmental biogeochemistry. C.N.R.–I.P.R.A., Rome.
129. Claus, H., and Z. Filip. 1988. Behaviour of phenoloxidases in the presence of clays and other soil-related adsorbents. Appl. Microbiol. Biotechnol. 28:506–511.
130. Quiquampoix, H. 1987. A stepwise approach to the understanding of extracellular enzyme activity in soil. I. Effect of electrostatic interactions on the conformation of a β-D-glucosidase adsorbed on different mineral surfaces. Biochimie 69:753–763.
131. Quiquampoix, H. 1987. A stepwise approach to the understanding of extracellular enzyme activity in soil. II. Competitive effects on the adsorption of a β-D-glucosidase in mixed mineral or organo-mineral systems. Biochimie 69:765–771.
132. Quiquampoix, H., P. Chassin, and R. G. Ratcliffe. 1989. Enzyme activity and cation exchange as tools for the study of the conformation of proteins adsorbed on mineral surfaces. Progr. Colloid Polym. Sci. 79:59–63.
133. Giles, C. H., T. H. MacEwan, S. N. Nakhwa, and D. Smith. 1960. Studies in adsorption. Part XI. A system of classification of solution adsorption isotherms, and its use in diagnosis of adsorption mechanisms and in measurement of specific surface areas of solids. J. Chem. Soc. 1960:3973–3993.
134. Martinek, K., and V. V. Mozhaev. 1985. Immobilization of enzymes: an approach to fundamental studies in biochemistry. Adv. Enzymol. 57:179–249.
135. O'Toole, P., and M. A. Morgan. 1984. Thermal stabilities of urease enzymes in some Irish soils. Soil Biol. Biochem. 16:471–474.
136. Ruggiero, P., J. M. Sarkar, and J.-M. Bollag. 1989. Detoxification of 2,4-dichlorophenol by a laccase immobilized on soil or clay. Soil Sci. 147:361–370.
137. Sarkar, J. M., A. Leonowicz, and J.-M. Bollag. 1989. Immobilization of enzymes on clays and soils. Soil Biol. Biochem. 21:223–230.
138. McLaren, A. D. 1978. Kinetics and consecutive reactions of soil enzymes, p. 97–116. *In* R. G. Burns (ed.), Soil enzymes. Academic Press, London.
139. McLaren, A. D., and L. Packer. 1970. Some aspects of enzyme reactions in heterogeneous systems. Adv. Enzymol. 33:245–308.

140. Goldstein, L. 1976. Kinetic behavior of immobilized enzyme systems, p. 397–443. *In* K. Mosbach (ed.), Methods in enzymology, Vol. XLIV. Academic Press, Orlando, Florida.
141. Laidler, K. J., and P. S. Bunting. 1980. The kinetics of immobilized enzyme systems, p. 227–248. *In* D. L. Purich (ed.), Methods in enzymology, Vol. 64. Academic Press, New York.
142. Johansson, A.-C., and K. Mosbach. 1974. Acrylic copolymers as matrices for the immobilization of enzymes. II. The effect of a hydrophobic microenvironment on enzyme reactions studies with alcohol dehydrogenase immobilized to different acrylic copolymers. Biochim. Biophys. Acta 370:348–353.
143. Sundaram, P. V., A. Tweedale, and K. J. Laidler. 1970. Kinetic laws for solid-supported enzymes. Can. J. Chem. 48:1498–1504.
144. Thomas, D., G. Broun, and E. Sélégny. 1972. Monoenzymatic model membranes: Diffusion-reaction kinetics and phenomena. Biochimie 54:229–244.
145. Chalvignac, M. A., and J. Mayaudon. 1971. Extraction and study of soil enzymes metabolising tryptophan. Plant and Soil 34:25–31.
146. Tabatabai, M. A. 1982. Soil enzymes, p. 903–947. *In* A. L. Page, R. H. Miller, and D. R. Keeney (ed.), Methods of soil analysis, Part 2. American Society of Agronomy, Madison, Wisconsin.
147. Malcolm, R. E. 1983. Assessment of phosphatase activity in soils. Soil Biol. Biochem. 15:403–408.
148. Goldstein, L., Y. Levin, and E. Katchalski. 1964. A water insoluble polyanionic derivative of trypsin: 2. Effect of the polyelectrolyte carrier on the kinetic behavior of the bound trypsin. Biochemistry 3:1913–1919.
149. Frankenberger, W. T., Jr., and J. B. Johanson. 1982. Effect of pH on enzyme activity in soils. Soil Biol. Biochem. 14:433–437.
150. Burns, R. G., A. H. Pukite, and A. D. McLaren. 1972. Concerning the location and persistence of soil urease. Soil Sci. Soc. Am. Proc. 36:308–311.
151. Nannipieri, P., B. Ceccanti, S. Cervelli, and P. Sequi. 1978. Stability and kinetic properties of humus-urease complexes. Soil Biol. Biochem. 10:143–147.
152. Tul'skaya, Y. M., and D. G. Zvyagintsev. 1981. Immobilization of catalase by soils. Sov. Soil Sci. 13:66–71.
153. Pettit, N. M., L. J. Gregory, R. B. Freedman, and R. G. Burns. 1977. Differential stabilities of soil enzymes. Assay and properties of phosphatase and arylsulphatase. Biochim. Biophys. Acta 485:357–366.
154. Eivazi, F., and M. A. Tabatabai. 1988. Glucosidases and galactosidases in soils. Soil Biol. Biochem. 20:601–606.
155. Moyo, C. C., D. E. Kissel, and M. L. Cabrera. 1989. Temperature effects on soil urease activity. Soil Biol. Biochem. 21:935–938.
156. Frankenberger, W. T., Jr., and M. A. Tabatabai. 1980. Amidase activity in soils: II. Kinetic parameters. Soil Sci. Soc. Am. J. 44:532–536.
157. Frankenberger, W. T., Jr. 1983. Kinetic properties of L-histidine ammonia-lyase activity in soils. Soil Sci. Soc. Am. J. 47:71–74.
158. Tena, M., J. A. Pinilla, and M. Magallanes. 1986. L-Phenylalanine deaminating activity in soil. Soil Biol. Biochem. 18:321–325.

159. Frankenberger, W. T., Jr., and M. A. Tabatabai. 1991. L-Asparaginase activity in soils. Biol. Fertil. Soils 11:6–12.
160. Dick, W. A., and M. A. Tabatabai. 1978. Inorganic pyrophosphatase activity of soils. Soil Biol. Biochem. 10:59–65.
161. Dick, W. A., and M. A. Tabatabai. 1984. Kinetic parameters of phosphatases in soils and organic waste materials. Soil Sci. 137:7–15.
162. Juma, N. G., and M. A. Tabatabai. 1988. Comparison of kinetic and thermodynamic parameters of phosphomonoesterases of soils and of corn and soybean roots. Soil Biol. Biochem. 20:533–539.
163. Perucci, P., and L. Scarponi. 1984. Arylsulphatase activity in soils amended with crop residues: kinetic and thermodynamic parameters. Soil Biol. Biochem. 16:605–608.
164. Cervelli, S., P. Nannipieri, B. Ceccanti, and P. Sequi. 1973. Michaelis constant of soil acid phosphatase. Soil Biol. Biochem. 5:841–845.
165. Makboul, H. E., and J. C. G. Ottow. 1979. Einfluss von zwei- und Dreischichttonmineralen auf die Dehydrogenase-, saure Phosphatase- und Urease-aktivität in Modellversuchen. Z. Pflanzenernaehr. Bodenkd. 142:500–513.
166. Makboul, H. E., and J. C. G. Ottow. 1979. Clay minerals and the Michaelis constant of urease. Soil Biol. Biochem. 11:683–686.
167. Makboul, H. E., and J. C. G. Ottow. 1979. Alkaline phosphatase activity and Michaelis constant in the presence of different clay minerals. Soil Sci. 128:129–135.
168. Pflug, W. 1982. Effect of clay minerals on the activity of polysaccharide cleaving soil enzymes. Z. Pflanzenernaehr. Bodenkd. 145:493–502.
169. Ross, D. J. 1983. Invertase and amylase activities as influenced by clay minerals, soil-clay fractions and topsoils under grassland. Soil Biol. Biochem. 15:287–293.
170. Mayaudon, J., M. El Halfawi, and C. Bellinck. 1973. Decarboxylation des acides aminés aromatiques-1-^{14}C par les extraits de sol. Soil Biol. Biochem. 5:355–367.
171. Morgan, H. W., and C. T. Corke. 1977. Release of flavin adenine dinucleotide on adsorption of the enzyme glucose oxidase to clays. Can. J. Microbiol. 23:1109–1117.
172. Rachhpal-Singh, and P. N. Nye. 1984. The effect of soil pH and high urea concentrations on urease activity in soil. J. Soil Sci. 35:519–527.
173. Nannipieri, P., B. Ceccanti, S. Cervelli, and C. Conti. 1982. Hydrolases extracted from soil: kinetic parameters of several enzymes catalysing the same reaction. Soil Biol. Biochem. 14:429–432.
174. Irving, G. C. J., and D. J. Cosgrave. 1976. The kinetics of soil acid phosphatase. Soil Biol. Biochem. 8:335–340.
175. Frankenberger, W. T., Jr., and F. T. Bingham. 1982. Influence of salinity on soil enzyme activities. Soil Sci. Soc. Am. J. 46:1173–1177.
176. Tabatabai, M. A. 1977. Effect of trace elements on urease activity in soils. Soil Biol. Biochem. 9:9–13.
177. Frankenberger, W. T., Jr., and M. A. Tabatabai. 1981. Amidase activity in soils: IV. Effects of trace elements and pesticides. Soil Sci. Soc. Am. J. 45:1120–1124.
178. Fu, M. H., and M. A. Tabatabai. 1989. Nitrate reductase activity in soils: effects of trace elements. Soil Biol. Biochem. 21:943–946.

179. Frankenberger, W. T., Jr., and M. A. Tabatabai. 1991. Factors affecting L-asparaginase activity in soils. Biol. Fertil. Soils 11:1–5.
180. Juma, N. G., and M. A. Tabatabai. 1977. Effects of trace elements on phosphatase activity in soils. Soil Sci. Soc. Am. J. 41:343–346.
181. Stott, D. E., W. A. Dick, and M. A. Tabatabai. 1985. Inhibition of pyrophosphatase activity in soils by trace elements. Soil Sci. 139:112–117.
182. Al-Khafaji, A. A., and M. A. Tabatabai. 1979. Effects of trace elements on arylsulfatase activity in soils. Soil Sci. 127:129–133.
183. Lethbridge, G., A. T. Bull, and R. G. Burns. 1981. Effects of pesticides on 1,3-β-glucanase and urease activities in soil in the presence and absence of fertilisers, lime and organic materials. Pestic. Sci. 12:147–155.
184. Ruggiero, P., and V. M. Radogna. 1985. Inhibition of soil humus-laccase complexes by some phenoxyacetic and *s*-triazine herbicides. Soil Biol. Biochem. 17:309–312.
185. Perucci, P., L. Scarponi, and M. Monotti. 1988. Interference with soil phosphatase activity by maize herbicidal treatment and incorporation of maize residues. Biol. Fertil. Soils 6:286–291.
186. Cervelli, S., and A. Perna. 1985. Phytase inhibition by insecticides and herbicides. Water Air Soil Pollut. 24:397–403.
187. Cervelli, S., P. Nannipieri, G. Giovannini, and A. Perna. 1976. Relationships between substituted urea herbicides and soil urease activity. Weed Res. 16:365–368.
188. Cervelli, S., P. Nannipieri, G. Giovannini, and A. Perna. 1977. Effect of soil on urease inhibition by substituted urea herbicides. Soil Biol. Biochem. 9:393–396.
189. Hong, J.-U., and S. M. Cho. 1979. The changes of the activity of nitrogen-containing herbicides in soils. Part 1. Effects of nitrogen-containing herbicides on the urease activity in soils. J. Korean Agric. Chem. Soc. 22:217–225.
190. Schäffer, A. 1993. Pesticide effects on enzyme activities in the soil ecosystem, p. 273–340. *In* J.-M. Bollag and G. Stotzky (ed.), Soil biochemistry, Vol. 8. Marcel Dekker, New York.
191. Cervelli, S., P. Nannipieri, and P. Sequi. 1978. Interaction between agrochemicals and soil enzymes, p. 251–293. *In* R. G. Burns (ed.), Soil enzymes. Academic Press, London.
192. Davies, H. A., and M. P. Greaves. 1981. Effects of some herbicides on soil enzyme activities. Weed Res. 21:205–209.
193. Tu, C. M. 1982. Influence of pesticides on activities of invertase, amylase and level of adenosine triphosphate in organic soil. Chemosphere 11:909–914.
194. Tu, C. M. 1988. Effects of selected pesticides on activities of invertase, amylase and microbial respiration in sandy soil. Chemosphere 17:159–163.
195. Tu, C. M. 1981. Effects of some pesticides on enzyme activities in an organic soil. Bull. Environ. Contam. Toxicol. 27:109–114.
196. Tu, C. M. 1981. Effects of pesticides on activities of enzymes and microorganisms in a clay soil. J. Environ. Sci. Health B16:179–191.
197. Tafuri, F., M. Businelli, C. Marucchini, and M. Patumi. 1979. Effetto inibente di alcuni erbicidi sull'attività ureasica. Riv. Agron. 13:392–395.
198. Tu, C. M. 1990. Effect of four experimental insecticides on enzyme activities and levels of adenosine triphosphate in mineral and organic soils. J. Environ. Sci. Health B25:787–800.

199. Cervelli, S., P. Nannipieri, G. Giovannini, and A. Perna. 1978. Alkaline phosphatase inhibition by vinyl phosphate insecticides. Water Air Soil Pollut. 9:315–321.
200. Ray, R. C., and N. Sethunathan. 1989. Effect of rice straw, organic acids, and pesticides (HCH and benomyl) on rhodanese activity of submerged soils. J. Indian Soc. Soil Sci. 37:706–711.
201. Lethbridge, G., and R. G. Burns. 1976. Inhibition of soil urease by organophosphorus insecticides. Soil Biol. Biochem. 8:99–102.
202. Mayaudon, J., M. El Halfawi, and M.-A. Chalvignac. 1973. Propriétés des diphenol oxydases extraites des sols. Soil Biol. Biochem. 5:369–383.
203. Mayaudon, J., and J. M. Sarkar. 1974. Etude des diphenol oxydases extraites d'une litière de forêt. Soil Biol. Biochem. 6:269–274.
204. Mayaudon, J., and J. M. Sarkar. 1974. Chromatographie et purification des diphenol oxidases du sol. Soil Biol. Biochem. 6:275–285.
205. Mayaudon, J., L. Batistic, and J. M. Sarkar. 1975. Propriétés des activités protéolitiques extraites des sols frais. Soil Biol. Biochem. 7:281–286.
206. Getzin, L. W., and I. Rosefield. 1971. Partial purification and properties of a soil enzyme that degrades the insecticide malathion. Biochim. Biophys. Acta 235:442–453.
207. Satanarayana, T., and L. W. Getzin. 1973. Properties of a stable cell-free esterase from soil. Biochemistry 12:1566–1572.
208. Cacco, G., and A. Maggioni. 1976. Multiple forms of acetyl-naphthyl-esterase activity in soil organic matter. Soil Biol. Biochem. 8:321–325.
209. Hayano, K. 1977. Extraction and properties of phosphodiesterase from a forest soil. Soil Biol. Biochem. 9:221–223.
210. Burns, R. G., A. H. Pukite, and A. D. McLaren. 1972. Concerning the location and persistence of soil urease. Soil Sci. Soc. Am. Proc. 36:308–311.
211. Dec, J., and J.-M. Bollag. 1994. Dehalogenation of chlorinated phenols during oxidative coupling. Environ. Sci. Technol. 28:484–490.

4

Influence of Natural and Anthropogenic Factors on Enzyme Activity in Soil

Liliana Gianfreda Università degli Studi di Napoli Federico II, Naples, Italy

Jean-Marc Bollag The Pennsylvania State University, University Park, Pennsylvania

I. INTRODUCTION

Increased attention is being directed to ameliorating the problems of soil pollution—an undesired by-product of modern life. Of special interest is the potential role that bioremediation may have in the decontamination process. Inasmuch as naturally occurring enzymes in soil offer an important means by which this might be accomplished, it is crucial that their behavior is thoroughly understood. An extensive literature in the area of soil enzymology is available, and both general [1–7] and particular [8–15] aspects have been described. However, a broad overview of the field is lacking, and this chapter is an effort to fill this need. The classification, compartmentalization, and physicochemical properties of soil enzymes together with the various factors, both natural and anthropogenic, that influence their activities are discussed. Also included are suggestions about areas in which there is a need for future research.

II. CLASSIFICATION SYSTEMS FOR ENZYME ACTIVITIES IN SOIL

Soil is an extremely heterogeneous environmental system in which different physical phases (e.g., solid, liquid, and gaseous) and numerous biotic (e.g., microorganisms, small animals, enzymes) and abiotic (e.g., clay minerals, humus materials, organomineral aggregates) components are involved in physical, chemical, and biological processes [3,16]. Through coordinated, enzyme-driven, cyclic pathways, enzymes ensure the movement of materials among the biotic and abiotic portions of the soil environment. All biochemical transformations in soil are dependent on, or related to, the presence of enzymes. In early 1946, Quastel [17] suggested that soil may be considered a biological entity, i.e. a "tissue." As stated by Kuprevich and Shcherbakova [7], each type of soil may have its own characteristic pattern of specific enzymes and its own inherent level of enzyme activity, because "the differences in level of enzymatic activity are caused primarily by the fact that every soil type, depending on its origin and developmental conditions is distinct from every other in its content of organic matter, in the composition and activity of living organisms inhabiting it and, consequently, in the intensity of biological processes."

Soil enzyme activities are controlled by four classes of enzymes: hydrolases, lyases, oxidoreductases, and transferases. A partial list of enzyme activities found in soil is shown in Table 1 [18–90]. Some of these (e.g., urease and β-glucosidase) catalyze reactions both inside and outside of the organisms that synthesized them; some (e.g., cellulase and proteinase) can only function outside of the producing cell because of the large size of their target substrate; and others (e.g., dehydrogenase and nitrate reductase), being involved in the central aspects of metabolism, do not function extracellularly. Furthermore, most examples refer to enzyme-like activities detected in soil, rather than to purified enzyme proteins. As reviewed by Tabatabai and Fu [14], only a limited number of enzymes have been extracted from soil and they have only been partially purified. The majority of enzymes extracted from soils (which can be considered free enzymes) may be present in soils strongly associated with carbohydrates [14,19]. These carbohydrate-enzyme complexes—with the carbohydrate portion chemically bound to the peptidic part through the R groups of specific amino acids (e.g., asparagine or glutamine)—are considered to be true glycoenzymes [92]. It has been hypothesized that the carbohydrate chain is not necessary for enzyme function but contributes to stabilization of the active component by enveloping it to form vesicles resistant to proteolytic activity, microbial denaturation, or oxidizing agents [14,91].

A. Classification by Location

The overall enzyme activity of a soil is a composite of various intracellular and extracellular enzyme components produced by microorganisms (e.g., bacteria, fungi) or derived from animal and plant sources (e.g., roots, lysed plant residues, digestive tracts of small animals). The exact origins of these enzymes components, as well as the temporal and spatial relationships of their sources, remain undefined and difficult to identify. In addition, the various enzymes may exist in different states or forms, have nonuniform characteristics, and act under a range of microenvironmental conditions that affect their catalytic behavior. Methodologies for assaying the enzyme activities in soil may contribute to this inconsistent information base. Various research groups have developed their own assays for the measurement of enzyme activity in soil. Different storage conditions, soil treatments, and enzyme assay conditions (e.g., pH, temperature, choice and concentration of the substrate, and composition of the buffer) have been utilized, resulting in conflicting results. Consequently, the interpretation of findings in soil enzymology is typically hindered by the difficulty of comparing data obtained by different methods. Furthermore, it is difficult to relate enzyme activities determined in vitro with those occurring in situ (see Refs. 1, 4, 6, and several chapters of Ref. 3).

In general, enzymes can be classified into two main groups: endo- or intracellular and exo- or extracellular. In living cells, endoenzymes may exist in different cellular compartments, such as the cytoplasm or periplasm, or they may be associated with the cell membrane and/or cell wall. In contrast, according to the usual definition of intracellular and extracellular enzymes [93], extracellular enzymes are produced and secreted by living cells and operate at a distance from the parent cell, either as free enzymes in the liquid phase or as enzymes still associated with the external surface of root epidermal or microbial cell wall (i.e., partially attached to or trapped within a viscous network of homo- and heteropolysaccharides secreted by several cells which still allows the passage of substrate and products [11,12]). In this case, they are often named ectoenzymes.

Components of the enzyme activity of soil were classified first by Kiss et al. [8] and subsequently by Skujins [2]. These classifications are combined and summarized in Figure 1. Kiss et al. [8] used the term "accumulated enzymes," to indicate those "present and active in soil in which no microbial proliferation takes place." They include (1) exoenzymes released from living cells or endoenzymes released from disintegrated cells, acting either free in the soil solution or adsorbed on soil inorganic and organic constituents, and (2) enzymes bound to cell constituents, i.e., enzymes presents in disintegrated (cells fragments), or in dead but not disintegrated cells, or in viable but nonproliferating cells, either acting in the adsorbed state or in suspension.

Skujins [2] used the term "abiontic enzymes" to describe all enzymes present in soils except those present within proliferating cells. Enzymes attached to microbial cellular components, such as nonproliferating cells, intact dead cells, or cellular fragments may be considered abiontic enzymes. Extracellular enzymes, continuously released by proliferating cells during normal growth, and present as enzymes free in the soil liquid phase or bound to soil constituents but not associated with cell components, are also considered abiontic enzymes.

An exhaustive description of the location of enzyme activity in soil was made by Burns [10], who listed ten distinct categories:

1. Enzymes acting within the cytoplasm of proliferating microbial, animal, and plant cells. These enzymes may be considered truly endo-

Table 1 Functional Classification of Enzymatic Activities in Soil [a]

Enzyme EC number	Common name	Biogeochemical cycle	Reaction	Reference
Hydrolases				
3.2.1.1	Amylase	Carbon	Hydrolysis of α-1,4-glucan bonds	18–22
3.2.1.4	Cellulase	Carbon	Hydrolysis of β-1,4-glucan bonds	23–25
3.2.1.14	Chitinase	Carbon		26
3.2.1.22/23	Galactosidase (α or β)	Carbon	α- or β-galactoside ↔ ROH + galactose	27–29
3.2.1.6	1,-3-β-glucanase	Carbon	Hydrolysis of β-1,3-glucan bonds	30–32
3.2.1.20/21	Glucosidase (α or β)	Carbon	α- or β-glucoside ↔ ROH + glucose	27–29
3.2.1.26	Invertase	Carbon	Saccharose ↔ glucose + fructose	18–22
3.1.1.3	Lipase	Carbon	Trygliceride + H_2O ↔ glycerol + fatty acids	32–33
3.2.1.8	Xylanase	Carbon	Hydrolysis of β-1,4-xylan bonds	34–35
3.5.4.4	Adenosine deaminase	Nitrogen	Adenosine ↔ NH_4 + inosine	36
3.5.1.4	Amidase	Nitrogen	monocarboxylacid amide + H_2O ↔ monocarboxyl acid + NH_3	37–40
3.5.1.1.	L-Asparaginase	Nitrogen	Asparagine ↔ aspartic acid + NH_3	41–42
3.5.1.2	L-Glutaminase	Nitrogen	Glutamine ↔ glutamic acid + NH_3	43–44
3.4. groups	Protease	Nitrogen	Protein hydrolysis	45–47
3.5.1.5	Urease	Nitrogen	Urea ↔ 2 NH_3 + CO_2	48–50
3.1.3.1	Acid phosphatase	Phosphorus	Phosphate esters + H_2O ↔ ROH + H_3PO_4	51
3.1.3.2	Alkaline phosphatase	Phosphorus	Phosphate esters + H_2O ↔ ROH + H_3PO_4	51
3.1.4. groups	Phosphodiesterase	Phosphorus		52–54
	Pyrophosphatase	Phosphorus	Pyrophosphate + H_2O ↔ $2PO_4^{3-}$	55–58
3.1.6.1	Sulfatase	Sulfur	Sulfate esters + H_2O ↔ ROH + H_2SO_4	59–61
Lyases				
4.1.1. groups	Aminoacid decarboxylase	Nitrogen	Aspartic acid ↔ alanine Glutamic acid ↔ γ-aminobutyric acid	62–65

Oxidoreductases				
1.10.3.3	Ascorbate oxidase	Carbon	Ascorbic acid ↔ dehydroascorbic acid	66
1.1.3.4	Glucose oxidase	Carbon	Glucose + O_2 ↔ gluconic acid + H_2O	67–68
1.10.3.2	Laccase	Carbon	Phenol + O_2 ↔ Quinone + H_2O	69–70
1.14.18. groups	Phenol oxidase	Carbon	Phenol + 1/2 O_2 ↔ Quinone + H_2O	71–72
1.11.1.7	Peroxidase	Carbon	S + H_2O_2 ↔ oxidized S + H_2O	73–74
1.11.1.6	Catalase	Oxygen	2 H_2O_2 ↔ 2 H_2O + O_2	75
1.1.1.1	Dehydrogenase	Oxygen	XH_2 + A ↔ X + AH2	76–78
1.12.7.1	Hydrogenase	Oxygen	H_2 ↔ H_2O	79–80
1.7.99.4	Nitrate reductase	Nitrogen	NO_3^- ↔ NO_2^-	81
	Nitrogenase	Nitrogen	N_2 ↔ $2NH_3$	82–83
1.7.3.3	Urate oxidase	Nitrogen	Uric acid + O_2 ↔ allantoin, CO_2, other products	84
Transferases				
2.6.1. groups	Transaminase	Nitrogen	R_1R_2-CH-NH_2 + R_3R_4CO ↔ R_1R_2-CO + R_3R_4CH-NH_2	85–86
2.4.1. groups	Transglycosylase	Carbon	n($C_{12}H_{22}O_{11}$) + HOR ↔ H($C_6H_{10}O_5$)nOR + n($C_6H_{12}O_6$)	87–88
2.8.1.1	Rhodanese	Sulfur	$S_2O_3^{2-}$ + CN ↔ SCN^- + SO_3^{2-}	89–90

ROH = Hydroxylated compound; R = Alkyl, aryl, or glycosil group.

S = Reduced organic substrate.

XH_2 = Hydrogen donor organic compound.

A = Hydrogen acceptor.

[a] The Enzymes are Classified according to the International Union of Biochemistry.

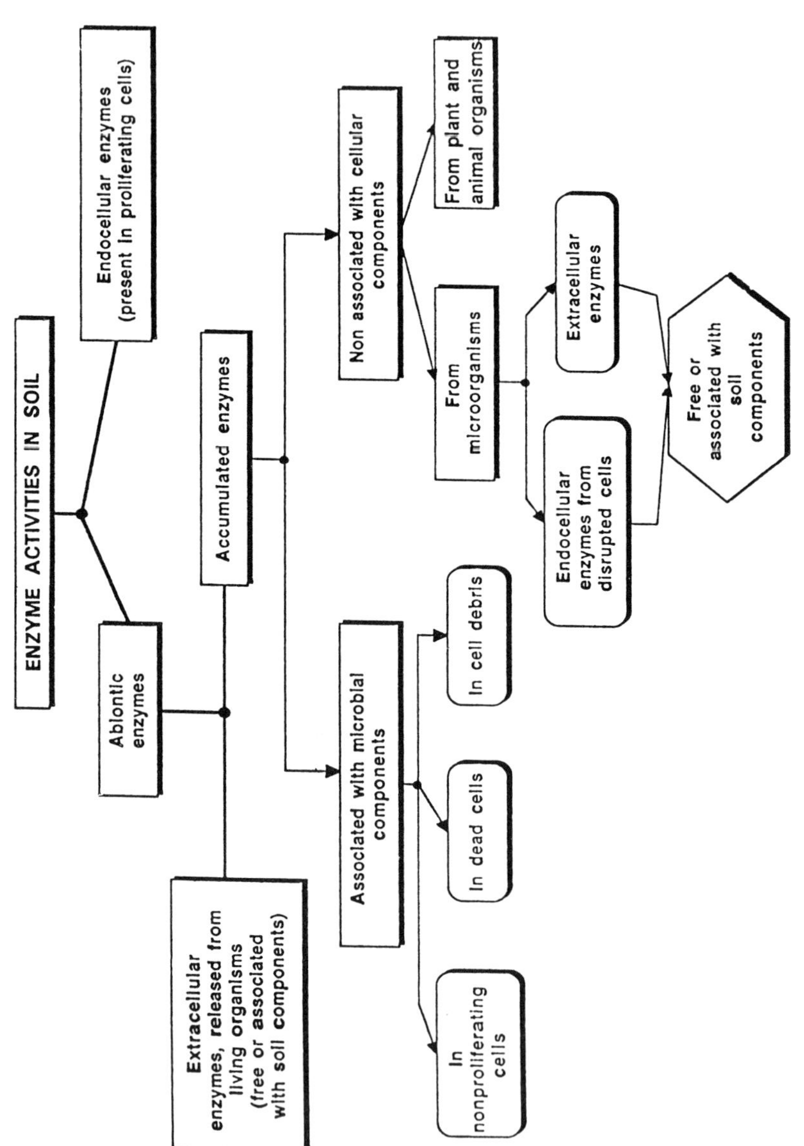

Figure 1 Localization of enzyme activities in soil

cellular enzymes, and several of their activities are strictly related to, and dependent on, the viability of the host cells.

2. Enzymes present in the periplasmatic space of proliferating microorganisms.
3. Enzymes bound to the external surface of viable cells and functioning within the liquid outside of the cell. On the basis of the definition of intra- and extracellular enzymes, these are extracellular enzymes of the ectoenzyme subclass.
4. Enzymes produced by living organisms during normal cell growth, secreted outside the cells, and acting as free enzymes in the aqueous phase of the soil. These can be considered extracellular enzymes and are usually characterized by a low molecular weight.
5. Enzymes within nonproliferating cells, such as spores, cysts, plant seeds, etc.
6. Enzymes bound to whole dead cells or cell debris.
7. Enzymes leaked or released from lysed cells.
8. Enzymes partially associated within soluble or insoluble enzyme-substrate complexes.
9. Enzymes adsorbed on internal and external surfaces of clay minerals.
10. Enzymes associated with humic colloids, as a result of adsorption, entrapment, polymerization, or copolymerization processes taking place during the genesis of humic matter.

Burns' classification [10] includes elements of the classifications of both Kiss et al. [8] and Skujins [2]. In fact, "accumulated enzymes" are those described in categories 5 to 10, whereas "abiontic enzymes" belong to categories 4 to 10. The enzymes grouped in categories 9 and 10 are considered soil-bound enzymes. Present in these enzymatic soil-colloid complexes are both free enzymatic proteins (secreted by living cells or released to the soil from lysed cells) and enzymes associated with dead cells and cell debris. Individual enzymes may belong to more than one classification category and may change from one to another with time. Depending on the type of enzyme, it may be difficult to differentiate among categories or to assign enzyme activities to a single category.

In fact, the most serious problem encountered by soil enzymologists is identification of the soil category or categories giving rise to a certain measured enzyme activity. Although the field of soil enzymology has reached maturity, methods currently available for assaying enzymatic activities in whole soil suspensions are still not sufficiently selective to discriminate and quantify intracellular versus extracellular activities or to distinguish the activity of live, nongrowing cells from that of all other enzymatic components. However, if a microbial inhibitor is incorporated in the reaction mixture or a brief incubation period is used for the activity assay, the activities of enzymes produced by proliferating cells (categories 1–4) might be restricted or eliminated and the measured activity may be assigned more properly to the contribution of "accumulated enzymes." Furthermore, as will be specified later, much experimental evidence supports the assumption that a large proportion of enzymes in soil belong to the last two categories described by Burns [10].

B. Classification by Function

Two groups of enzyme activities appear to predominate in soil: hydrolases and oxidoreductases (Table 1). Some of these will be described briefly.

Hydrolases, such as polysaccharidases and proteinases, that are essentially extracellular enzymes, transform macromolecular substrates into smaller compounds. The depolymerization of carbohydrates and proteins and the subsequent mineralization of their monomeric products have special importance in the biological cycling of both carbon and nitrogen in soils [11].

Polysaccharidases (e.g., amylases, cellulases, xylanases, dextranases) catalyze the first stage of the depolymerization of polysaccharides to oligo- and monosaccharides [18]. These, in turn, can be assimilated by living organisms or undergo further decomposition in soil by the action of hydrolytic enzymes such as α- or β-glucosidases and α- or β-galactosidases. The final hydrolysis products, such as glucose or galactose, of these enzymes constitute important energy sources for soil microorganisms.

Hydrolysis of saccharose by invertase is another significant reaction of carbohydrate metabolism that occurs in soils. The presence of invertase in soil has been correlated with microbial activity in several studies [19–21], and therefore, it was assumed to be a useful "fertility index." However, as stressed recently by Nannipieri [15], it is somewhat "unrealistic to assume that a simple relationship exists between measurements of a single enzyme activity and the fertility status of soil."

Observations that proteins added to soils are hydrolyzed to peptides and amino acids indicate the presence of proteinase and peptidase activities in soil. Limitations in the porosity of cell walls or the absence of suitable peptide transport systems restrict the direct uptake of proteins or large peptides by microbial cells. Thus, the depolymerizing activities of proteinases in soil allow exogenous proteins to be utilized by soil microorganisms [11].

Because of the central role occupied by ammonia in the soil nitrogen cycle, urease, which catalyzes the hydrolysis of urea (an important fertilizer) into ammonia and carbon dioxide, has been one of the most studied enzymatic activities in soil. The origin, nature, properties, and inhibition of this enzyme have drawn the attention of numerous researchers. However, it is still uncertain whether the urease in soil is exclusively of microbial origin or if it derives largely from plants. It is also unknown whether the enzyme is essentially a free extracellular enzyme or if it is associated with soil constituents [48–50]. The various methods used to assay urease may account for these discrepancies.

Sulfur [59] and phosphorus [94] are two nutrients essential for the growth of plants. In the cycles of both these elements, the inorganic and organic forms are closely associated through mineralization immobilization processes mediated by abiotic and biotic activities [59,94,95]. In the mineralization process, organic sulfur and phosphorus fractions, which represent a large amount of sulfur and phosphorus in soils, are transformed into inorganic forms that are usually taken up by plant roots. In soils, the release of inorganic sulfate or phosphate from organic sulfur or phosphorus is effected by the action of sulfatases and phosphatases [51]. These two enzymes have been recognized as acting outside their cells of origin, and, in some cases (e.g., phosphatases), they also exist as ectoenzymes [96].

There has been increasing interest in the prospect of using hydrolytic enzyme activities, detected in soil, for transformation of pesticides and other xenobiotics. Herbicide- and insecticide-hydrolyzing activities, such as those of propanil-acylamidases or malathion- and parathion-esterases, have been studied in various soils [13,97]. However, knowledge concerning pesticide-transforming enzymes present in soil remains sparse.

Assays for dehydrogenase activity in soil have often been used to achieve correlative information on the biological activities of microbial populations in soils (i.e., as an index of total microbial activity) rather than on the enzyme itself [1,76–78]. These enzymes, which in the presence of active cofactors catalyze the oxidation of numerous organic compounds by removal of electrons and hydrogen, are believed to be localized only within intact viable cells; treatment of soils with bacteriostatic or bactericidal compounds destroys their activity. However, no consistent correlation between dehydrogenase activity and the number of microorganisms in soils has been demonstrated [13]. The overall process of dehydrogenation can be summarized as

$$XH_2 + A \leftrightarrow X + AH_2$$

where XH_2 is a hydrogen-donor organic compound and A is a hydrogen acceptor. Several different dehydrogenases are involved in the dehydrogenase activity of soils, and these enzymatic systems probably have an important role in the oxidation processes of soil organic matter.

Furthermore, comparison of changes in dehydrogenase activity with that of other soil enzymes in response to agricultural practices, environmental stresses, and changes in climatic conditions, could be used as a tool for discriminating intracellular from extracellular activities [15]. This approach, however, requires some caution; the use of synthetic substrates, such as 2,3,5-triphenyltetrazolium chloride (TTC) [4] or 2-(*p*-iodophenyl)-3-(*p*-nitrophenyl)-5-phenyl tetrazolium chloride (iodonitrotetrazolium chloride, INT) [98-100] in the dehydrogenase assay could lead to inaccurate estimation of the enzymatic activity, since both substrates are less effective than O_2 [13,15].

The utilization of atmospheric H_2 by moist soils was assumed to be catalyzed by hydrogenases [79,80]. However, these activities are difficult to detect as a result of their high sensitivity to soil treatments such as autoclaving, fumigation with organic compounds, UV irradiation, addition of bacteriostatic or bactericidal agents, or prolonged storage of dried soils [6].

Reduced organic compounds, mostly phenolic, may be oxidized by the action of peroxidase [73,74]. The presence of hydrogen peroxide, which is transformed to water and O_2, is required during this reaction. Hydrogen peroxide, a potential toxic intermediate of respiration, may also be destroyed by catalase, which produces molecular oxygen and water. This decomposition of hydrogen peroxide makes catalase an important enzyme of soil biochemical activity [75]. Both peroxidase and catalase have been extracted from soils and partly purified [73–75].

Peroxidases and, more generally, polyphenol oxidases such as laccases and tyrosinases catalyze the oxidation of a variety of phenols, aromatic amines, and other compounds [69–74]. Although the mechanisms and products of the reactions may be different, these enzymes are involved in several important biochemical transformations in soil. Their oxidative activities, such as the oxidative coupling reaction, have an important role in humus biochemistry (e.g., synthesis of humic substances). These oxidoreductive enzymes also participate in the degradation and/or incorporation of xenobiotic substances (e.g., phenolic or anilinic compounds derived from agricultural and industrial processes) into soil organic matter [9,71,72].

III. DO ENZYMES EXIST THAT ARE UNIQUE TO SOIL?

Establishing whether or not enzymes present in soil are unique to soil is hindered by the complexity of enzyme activities in soil and persisting limitations of the methodologies. However, the extensive characterization of soil and soil extracts found in the literature reveals several interesting properties of enzyme-catalyzed reactions in soil [12,101]. Such properties suggest that a large number of enzymes in soil exist as stable complexes with inorganic and organic colloids which persist in the absence of living organisms. From this perspective, it could be assumed that these enzymatic fractions are "soil enzymes," because their behavior as catalysts is strictly inherent to the habitat (i.e., type of soil) in which they act.

A. Properties of Soil Enzymes

Soil enzymes frequently display special properties. They usually exhibit high stability when subjected to thermal denaturation, attack by proteases, or irradiation [102,103] and a characteristic kinetic behavior [6,11,12].

Several studies have demonstrated that "native" soil enzymes are more stable, especially against proteolytic attack, than enzymes added to soil. For example, urease from jack bean [48] and phosphatase and inorganic pyrophosphatase from corn roots [58] were partially or completely inactivated when added to different soils. This inactivation was the result of inhibition by soil constituents or by the activity of protease. In contrast, urease and phosphatase as well as other enzymes extracted from soils showed considerable stability when subjected to proteolytic enzymes or high temperatures [14 and references therein].

The kinetic behavior of soil enzymes is usually different from that exhibited by the same enzyme in purified form: the Vmax values may be lower and Km constants are higher, indicating lower catalytic efficiency and a reduced substrate affinity. Similarly, different effects of pH or temperature were demonstrated for soil enzymes [6,64,104–107].

These characteristics can, at least in part, be explained by assuming that soil enzymes are immobilized on soil supports, i.e., are "naturally immobilized enzymes," and consequently, work as "heterogeneous catalysts" [108–110]. An immobilized enzyme is a protein physically localized in a certain region of space or converted from a water-soluble, mobile state to a water-insoluble, immobile one. The immobilization process involves binding of enzymes to solid supports by covalent bonds, cooperative adsorptive interactions, entrapment and encapsulation in stable aggregates, or cross-linking or co-cross-linking with bifunctional agents. The most significant changes in enzymes occurring upon immobilization are (1) an increased stability toward physical, chemical, and biological denaturing agents due to minimization or prevention of conformational changes [111], and (2) an altered catalytic behavior caused by the "heterogeneous nature" of catalysis associated with a "static enzyme" [108]. Although it can be assumed that all of the above-described immobilization mechanisms are involved in the formation of soil enzymes, the exact mechanism by which enzymes are immobilized and consequently stabilized in soils remains obscure.

Numerous studies examining the adsorption of proteins on clay minerals and the properties of the resulting clay-protein complexes suggest that the clay fraction has an important role in the stabilization of soil enzymes [16,112,113]. Experimental evidence obtained on the interactions between proteins and organic substances have indicated that soil organic matter is also important to the stabilization process [114].

Various types of interactions between enzymes and soil constituents have been proposed [115]. Cation exchange adsorption mechanisms [113], van der

Waals forces [116], and hydrophobic bonds have been suggested as possible mechanisms whereby enzymes are held to clay surfaces [115]. Intercalation of protein molecules in expandable clays may also occur [12,113,115]. Ion exchange, entrapment within organic networks, and covalent bonding may account for the stable association between enzymes and humic materials [114].

B. Methodological Approaches in the Study of Soil Enzymes

Two methodological approaches have been applied for understanding the relationships between immobilized enzymes in soil and their clay or humic supports. The first approach is based on the isolation, purification, and characterization of active enzyme fractions from soils. About 20 enzymes have been extracted from soils, including oxidoreductases, hydrolases, and a lyase. Various procedures have been employed for the extraction, separation, and purification of these enzymes [14 and references therein], and one of the most important problems encountered is the choice of the extracting solution. An efficient solvent system should (1) provide high enzyme yield without denaturation of the protein structure; (2) remove soil organic matter; and (3) avoid the undesired lysis of soil organisms. Although several biochemical techniques have been used to purify extracted enzymes, few researchers have succeeded in isolating pure enzymes from soil. As previously reported [91], carbohydrate-enzyme complexes were often obtained as the final products of separation procedures. Separation of the predominating carbohydrate portion from the small amount of enzymatic protein in these complexes requires more efficient separation techniques than are currently available.

The second methodological approach involves in vitro preparation of synthetic enzyme complexes by simulating interactions between enzymes and clay minerals, humic substances, or organomineral complexes. The properties and behavior of these complexes may be affected by such parameters as the nature and properties of both support and enzyme (for example, isoelectric point seems to have a great influence on the binding of enzymes to clay colloids), and chemical or physical mechanisms involved in the immobilization process [12,114–117].

In soil, clay mineral surfaces are usually coated with polymeric species or noncrystalline oxides of aluminum, iron, and manganese. "Clean" clays have been utilized more than "dirty" clays for studying clay-protein interactions. Only a few examples are reported in the literature on the adsorption of enzymes to dirty clays and on the properties of the resulting complexes [118–121]. Studies with clay-enzyme complexes showed that the kinetic parameters, pH and temperature activity profiles, and stability of the enzymes generally changed upon enzyme adsorption to clays and that the properties of some of these enzymatic complexes resembled those of the corresponding soil enzymes [12].

The interactions of enzymes with organic matter and the properties of humus-enzyme systems were exhaustively reviewed by Ladd and Butler in 1975 [114]. Numerous synthetic enzyme complexes, formed by utilizing natural or synthetic humic acids have been studied [114]. A later review extends to the more complicated systems obtained by association of enzymes with clay-organo matter complexes [115]. Both reviews provide insights into the stabilization and improvement in kinetic properties that can be achieved by extracellular enzymatic activities in soils upon their interaction with organic or organomineral matrices.

IV. PROBLEMS CONNECTED WITH THE DETERMINATION OF ENZYME ACTIVITIES IN SOIL

Understanding the cause-and-effect relationship between an observed change in soil enzyme activity and the physical, chemical, or biological factors that have produced it requires accurate and reliable methodologies. Buffer pH, temperature, type and concentration of substrate, and treatment of the soil before or during assays may affect enzyme activities (Table 2). Examples of variable results caused by different experimental conditions are described briefly.

A. Soil Handling and Storage

Regardless of whether the overall activity of an enzyme in soil or only the contribution of its extracellular fraction has to be measured, soil samples must be collected and stored prior to analysis. Therefore, the impact of various soil pretreatments on enzyme activity has been investigated to establish suitable procedures for handling and storage of soils before assays. Variables examined have included air drying and storage for different times at different temperatures in the air-dried or moist states. Depending upon the enzyme under investigation, considerable changes in enzyme activity were demonstrated during storage and handling of soils [1,6].

Air drying soils before assaying enzyme activities is a common practice. Air drying may significantly affect enzyme levels, depending on the temperature of drying and subsequent storage [49,51,122-124]. For example, air-drying of field-moist soils at 22 to 24°C caused an increase in the activity of arylsulfatase (average increase of 43%) [60], glucosidase, and galactosidase [27]. In another study, air drying resulted in a 21% average decrease in amidase activity [37]. Similarly, a marked decrease in rhodanese activity (44%) was found when field-moist soils were air dried [89]. A reduction was also detected in the activities of urease and phosphatase after the air drying of a clay-loam soil and subsequent reinoculation with fresh soil. In contrast, dehydrogenase activity in the

Table 2 Experimental Procedures that Can Affect the Determination of Enzyme Activities in Soil.

Soil pretreatment	
Handling	— air dried state
	— moist state
Storage	— room temperature
	— refrigerated
	— frozen
Soil sterilization	
Physical	— autoclaving
	— steam heating
	— dry heating
	— irradiation
Chemical	— bacteriostatic
	— antibiotic
	— plasmolytic
Assay conditions	
Substrate	
Buffer	
pH	
Temperature	
Duration	
Shaking	

same soil was not affected even after treatments such as lyophilization, freezing (–20°C and with dry ice), autoclaving, and oven drying. These results were accounted for by the increase in microbial population detected after 2 to 7 days of incubation with the fresh soil [125]. Palma and Conti [126] studied the effect of four treatments (moist and air dried, stored at room temperature, refrigerated, and frozen) on urease activity in surface samples of Argentinean agricultural soils. By comparing urease activity after each storage treatment with that of a moist-fresh control sample, they concluded that storage at 4°C in sealed plastic bags, which allowed the natural soil moisture content to be maintained, was sufficient to assure urease values similar to those obtained at the time of sampling and prevented increases in enzyme activity resulting from microbial proliferation, which were favored by room temperature storage.

In a comparison between air drying and acetone dehydration on the activities of invertase, amylase, cellulase, xylanase, urease, protease, phosphatase, and sulfatase in nine New Zealand pasture soils, acetone dehydration resulted in smaller losses of activity than did air drying for most of the enzymes [123].

Storage in the field-moist state at 4°C appeared most suitable for retaining the activities of protease, urease, and phosphatase in three pasture soils planted with grass and clover, whereas sulfatase activity was best retained when soil was stored air dried at 4°C [124]. No significant decrease was observed in the activities of urease, invertase, and amylase after 30-day storage of an Oklahoma Mollisol at 4°C [127].

A significant loss in the activity of soil dehydrogenase was noted after 15 and 30 days of storage at both 4°C and room temperature [128]. Over half of the dehydrogenase activity was lost in samples of air-dried soils, and further inactivation occurred when the air-dried soils were stored at 20°C. A freezing treatment (–20 or –10°C) retained the activities of soil dehydrogenase [128], catechol oxidizing [129], rhodanese [90], and arylsulfatase [61]. An increase in amidase activity (9%) was observed after freezing of a field-moist soil sample at –20°C for 3 months [39,40].

The divergent findings so far examined seem to confirm the conclusion drawn by Skujins as early as 1967 [1] that the best method of preservation of soil samples for the assay of enzyme activity depends upon the enzyme under assay and the temperature of drying and storage. To account for the contrasting results obtained with drying and storage of urease from different soil samples, Bremner and Mulvaney [49] hypothesized that the amount and type of plant residue present were important influences. The absence of a significant effect on urease from cultivated soils (used for corn and soybean production) relative to that from soils under pasture could be accounted for by the higher levels of nonhumified plant material contained in the latter [49].

B. Soil Sterilization

Soil sterilization is necessary when only extracellular enzymes are under investigation and microbial contribution to enzyme activity is not desired. It may be achieved by physical treatments (e.g., autoclaving, dry heating at high temperature, and high-energy irradiation) or chemical procedures involving chemical agents such as sterilizants, antiseptics, and bacteriostatics. The advantages and disadvantages of these procedures were reviewed by Skujins [1], Ladd [6], and Kiss et al. [8], and in several chapters of the book *Soil Enzymes* [3]. A perfect sterilizing agent must completely inhibit all microbial activity without lysing cells or affecting the activity of extracellular enzymes. Unfortunately, no such method has yet been found.

Enzymes in soil are usually more resistant to high temperatures than enzymes in pure preparations and solutions. Some relative heat resistance may occur, depending on the level of temperature, period of exposure to that temperature, and method of heating (whether dry or steam).

Several investigations showed that invertase in soil retained high residual activity after various extreme heat treatments [1,6,8]. No changes in enzyme activity were found when invertase of dry soil samples were heated at 100°C for

3 h [130] or at 50°C for 25 days [8]. In addition, invertase retained a residual activity of 47% after dry heating of a chernozem soil for 80 min at 150°C.

No loss of protease activity was detected after dry heating at 180°C for 3 h [130], and high levels of phosphatase, urease, and decarboxylase activities remained in soils after an 8-week exposure to dry heat at 100°C [131].

Autoclaving has proven to be more effective than dry heat for the inactivation of various enzymatic activities in soil. It completely destroyed the activity of invertase in several cultivated soils [130]. The adenosine deaminase activity of several surface soils (forest soils and samples from field soils under Chinese cabbage, tomato, cucumber, and strawberry cultivation) was completely inactivated by autoclaving at 120°C for 15 min. Similarly, a pronounced decrease in the activities of dehydrogenase, urease, and phosphatase was observed when a broad spectrum of soil types (orchard, forest, garden, and grassland) was subjected to ignition or autoclaving [132]. Steam treatment completely destroyed glucosidase and galactosidase activities in various soils [28].

Toluene is the most utilized chemical sterilization agent, although contradictory results from its use have been reported [133]. The activities of urease [49] and glucose oxidase [68] were significantly reduced when toluene was used for sterilization, whereas the activities of arylsulfatase [59] and phosphatase [51] were increased. In searching for an agent that would minimize the contribution of microbial growth without interfering with catalysis, Frankenberger and Johanson [134] evaluated the effects of various plasmolytic and antiseptic agents (toluene, dimethyl sulfoxide (DMSO), ethanol, and Triton X-110) on the activities of intracellular and extracellular enzymes in soil. Comparison was made with the effects of these agents on purified enzyme preparations obtained from plant, animal, or bacterial sources. Toluene, ethanol, and Triton X-100 were effective in preventing microbial proliferation during the assay of amidase in soil. DMSO was an effective plasmolytic agent, but it severely inhibited many of the purified enzyme reactions. Triton X-100 had no effect on the activities of purified enzymes but resulted in low activities when soil was assayed. Ethanol inhibited several purified enzymes; however, it was more suitable than toluene for the assay of catalase and acid and alkaline phosphatases in soil, showing little effect on the activity of the purified enzymes and preventing microbial proliferation. The activities of arylsulfatase and urease in soil were enhanced 1.30- to 1.34-fold by toluene, suggesting that the plasmolytic character of this agent increased the contribution of the intracellular enzymatic fraction of the measured activity [134].

Irradiation of soils by hard X-rays, gamma rays, or an electron beam of sufficient intensity [5 to 10 Mev] may be effective for achieving total microbial inactivation (i.e., loss of the ability to divide and produce new enzyme molecules) but does not result in inactivation of cell enzymes. Most cellular biochemical activities may continue for some time following irradiation, thereby allowing both intracellular and extracellular enzymatic activities to be measured.

Enzyme activities in soil have shown variable sensitivity to irradiation. Exposure to 2 Mrad of gamma irradiation did not affect the activities of urease, invertase, and protease [135], whereas phosphatase activity was reduced to 37% of its initial level by treatment of a clay-loam soil with a 19-Mrad electron beam [136]. After 7.5 Mrad of gamma irradiation of a sandy-loam soil, urease activity remained unchanged and phosphatase and decarboxylase activities were reduced to 70% and 0.5% of their initial levels, respectively [131]. The responses of arylsulfatase, phosphatase, urease, and β-1,3-glucanase activities to gamma irradiation (5 to 50 Mrad) of air-dried and moist samples of a silt-loam soil have also been investigated [137]. The levels of irradiation required to induce a 90% loss in activity ranged from 7 Mrad for urease in the wet soil to 48 Mrad for phosphatase in the dry soil. Higher levels of irradiation were necessary for all four enzymes in dry samples, probably because activated water radicals in wet soil cause denaturation of the enzymes.

Factors such as the inherent structural and chemical characteristics of enzymes, their location within the soil microenvironment, changes in the permeability of cell membranes, and the release and/or production of inhibitors were suggested to have relevance for explaining the differential sensitivities of enzymes to irradiation [131,137].

C. Assay Conditions

As emphasized by Burns [138], a primary obstacle impeding advances in soil enzymology is the need to develop a suitable assay for determination of enzyme activity in soil. Although numerous volumes (currently about 244 volumes of Methods in Enzymology [139]) have been dedicated to experimental techniques for determining enzyme activity, methods for the measurement of enzyme activity in soil are not yet standardized. For example, the variable methodology utilized in assays of soil urease is summarized in Table 3. Attempts to standardize methods for determining the activity of soil enzymes have been made [140,141]. An ordered and detailed description of methods for assaying activities of the principal enzymes in soil was presented by Tabatabai [4].

In devising assays for enzymatic activities in soil, several factors should be carefully considered. Primarily it is important to guarantee that the reactions are mediated by enzymes and not by inorganic catalysts, and that the rates of reactions, whether evaluated by the disappearance of substrates or the formation of products, are not affected by other biological or physicochemical reactions such as microbial utilization of products (especially during long incubation assays) or adsorption of substrates or products on soil particles. The choice of substrate and

Table 3 Experimental Conditions of Urease Assays in Soil

Soil (g)	Urea (mM)	Buffer nature	Buffer (mM)	pH	T (°C)	Reaction time (min)	Analytical methods for NH_3 determination
1	1500	Tris/maleate	500	7.0	20	60	Titrimetric
5	20	Tris/sulfate	50	9.0	37	120	Microdistillation
3	7	Phosphate	67	8.8	37	4	Microdistillation
5	500	K-Citrate	500	6.8	35	2.5	Colorimetric
5	830	Phosphate	Unknown	6.7	37	24	Microdistillation
0.2	14	Water	Unknown	Soil pH	30	20	Colorimetric
2	Unknown	Phosphate	100	7.0	28	72	Titrimetric
1	330	K-Acetate	50	5.5	Unknown	Continuous	Radiometric

Adapted from Ref. 145.

its concentration, presence or absence of a buffer, incubation temperature, duration of the assay, and shaking conditions are all factors that can influence the estimation of enzyme activities in soil.

Because several enzymes in soil have broad substrate specificity, one problem is the choice of a natural or an artificial substrate. For example, phosphatase activity has been estimated by hydrolysis of natural substrates such as β-glycerophosphate, nucleic acids, and phytate, and of artificial substrates such as phenylphosphate, α- and β-naphthyl phosphate, and *p*-nitrophenylphosphate [4,51]. Similarly, *p*-nitrophenyl sulfate or *p*-nitrophenylglycosides have been used for assaying the activities of soil arylsulfatases [59–61] or α- and β-glucosidases and galactosidases [27,28], respectively. Several of these substrates have the advantage of a simple colorimetric estimation of *p*-nitrophenol, which is released as a product of enzyme hydrolysis. Recently, new approaches based on dye-labeled substrates have been introduced as a sensitive assay for some enzymes. Fitzgerald et al. [142] introduced a ^{35}S-labeled tyrosine sulfate that is considered environmentally more relevant and sensitive than *p*-nitrophenyl sulfate for the assay of arylsulfatase activity. Soluble, dye-labeled, and acid-precipitable polysaccharide derivatives that offer a sensitive and reproducible assay of polysaccharide endohydrolases, such as cellulases, xylanases, chitinases, 1,3-glucanases, and amylases, were proposed by Wirth and Wolf [143].

Another consideration is the proper concentration of the substrate for the assay. Many studies have demonstrated that enzymes in soil follow simple Michaelis and Menten kinetics. Therefore, enzyme activities of soil are usually assayed under saturating substrate concentrations to maintain zero-order kinetics. However, substrate concentration in soil may vary considerably, thus affecting the kinetics of the related enzymes. For example, in studies on urease activity in two silt loam soils exposed to a wide range of urea concentrations

(0.01 to 10 M) and adjusted to pH values of 5.5, 6.6, 7.5, 8.5, and 9.5 [144], the results were best described by a kinetic model involving two enzymatic reactions, both following simple Michaelis and Menten kinetics but having very different affinities for urea. Two sets of kinetic parameters (V_{max1} and K_{m1}, and V_{max2} and K_{m2}; with $V_{max1} > V_{max2}$, and $K_{m1} << K_{m2}$) for high-affinity and low-affinity reactions, respectively, were determined. These findings suggested that two enzymatic fractions, differing in their kinetic parameters, contributed to overall urease activity in the soils.

The choice of buffer and pH for assays of enzyme activity in soil is frequently discussed in the literature. Results may be significantly affected, especially in quantitative terms, by whether or not a buffer is used. In an unbuffered assay, pH values can change during the incubation period. Zantua and Bremner [145] compared the activity levels of urease in 16 soils (with and without buffer) and demonstrated that the buffered method gave markedly higher values than the unbuffered method. Enzyme assays are usually carried out at the optimum pH because enzymes are generally most stable at that pH. Enzymes may be irreversibly deactivated under extremely acid or alkaline conditions. However, optimum pH in the laboratory may differ from the actual pH of a natural soil-water suspension and result in a different activity level. Furthermore, the pH and, more importantly, the buffer solution (nature, ionic strength, etc.) should exclude extraction of soluble organic or inorganic components from a soil sample that would interfere with the analytical method. With a proper control this problem can be eliminated.

Temperature is another factor that strongly affects the enzyme activity of soil. The rate of an enzyme-catalyzed reaction increases with rising temperature, but at higher temperatures the rate decreases, probably because of denaturation of the enzyme. To avoid contributions made by soil abiotic catalysts to the rates of reactions under investigation, temperatures selected for soil enzyme assays are usually lower than those yielding maximum activity. However, because of the great stability exhibited by some enzymes in soil, assay temperatures can be higher than those used for assaying enzymes from other sources. A combined influence of temperature and pH on the activity and stability of enzymes in soil may also occur, as demonstrated by Vorobe'yeva and Gviniashvili [146]. When studying the effect of temperature and pH on the activities of protease and invertase in two mountain soils, they observed that both enzymes exhibited a characteristic stabilization in an acidic pH range with rising temperature. At 80°C, protease retained its activity at a pH of 4.9, and invertase still exhibited activity at a pH of 3.2. These results were attributed to a protection effect of soils on the activities of both the enzymes.

When selecting the duration of the activity assay, it is important to consider that the incubation time must be long enough to ensure that the rate of product formation is constant and only a small fraction of the substrate is utilized (about

Table 4 Assay of Enzyme Activities in Soil: Conventional Laboratory Versus Field Conditions

	Conventional Assays	Field Conditions
Substrate	Excess	Limiting
	Homogeneous	Heterogeneous
	Soluble	Insoluble
	Artificial	Natural
Buffer	Present	Absent
pH	Optimal	Variable
Temperature	Constant	Variable
Shaking	Often	Stationary
Diffusional problems	Absent	Present
Fauna and flora	Absent	Present
Reproducibility	High	Low
Activity level	Potential	Actual

Source: Adapted from Ref. 138.

5% or less), in order to ensure that a detectable level of activity is achieved and substrate concentration is saturating throughout the incubation period. However, it should be short enough to avoid deactivation of the enzyme or inhibition by the product. If the activity of "accumulated enzymes" is to be evaluated, the duration of the assay should be chosen so as to avoid the growth of microorganisms and possible contributions from intracellular catalysis or newly produced enzymatic molecules.

Because methodological shortcomings may be potential sources of error that limit the reliability of a given response, it is important to determine the activity of enzymes in soil under carefully controlled and well-established experimental conditions. Furthermore, in interpreting the results of enzyme measurements, it should be remembered that laboratory assay conditions usually ensure optimal rates of catalysis and measure the maximum potential rather than the actual catalytic activity that would occur under natural conditions of substrate supply (Table 4) [15,138].

V. FACTORS AFFECTING ENZYME ACTIVITY IN SOIL

Numerous factors may affect enzyme activity in soil. Natural (pedogenetic, geographical, or physicochemical properties of soils; their content of organic matter, clay, or biomass) and anthropogenic (management; environmental pollution; additives, such as fertilizers, pesticides, salts, heavy metals) factors may influence the production, activity, catalytic behavior, and persistence of soil enzymes.

A. Natural Factors

Seasonal and Geographical Factors

Seasonal Changes. Several studies have examined the influence of the season on soil enzyme levels; however, a direct, defined, and precise correlation between soil enzyme activities and seasonal variation is still lacking. Other factors such as geographical location, climatic changes, or position of the soils may affect enzyme activities. Furthermore, if studies are performed during the same time period, but under different vegetative covers, the response may drastically change.

Rastin et al. [147] investigated the seasonal variation of an acid phosphatase, a phosphodiesterase, and a β-glucosidase in a beech forest soil. Results showed that the maximum activity for all enzymes occurred in spring. In studies on soils at two different altitudes (100 and 1500 ft above sea level) in northeast India, Kshattriya et al. [148] demonstrated that the activities of particular enzymes showed a marked seasonal variation, depending on the altitude from which the samples were collected. The peak of activity was observed in May-June for cellulase and amylase at both altitudes, whereas a second peak was observed for cellulase during September at the higher altitude. Unlike that for cellulase and amylase, invertase activity decreased over the sampling period (1 year).

Seasonal variation in the activities of amylase, invertase, and urease was also observed during 2 years of sampling soils under three sorts of vegetal covers (a sward of *Brachyposium ramosum*, and bushes of *Quercus coccifera* and *Q. ilex*) from the Mediterranean area [149,150]. In another study, invertase and amylase activities in the surface litter under a hard beech (*Nothofagus truncata*) forest were least in summer and greatest in winter; soil depth did not affect enzyme activities seasonally [19]. Urease in surface samples of Argentinean soils showed highest activity during summer under natural pasture, and at the end of summer and beginning of autumn under Eucalyptus forest. The lowest levels of enzyme activity were encountered during the winter under both types of vegetation [126].

Marked season-related activities of dehydrogenase, urease, and phosphatase were demonstrated by Kumar et al. [151] in studies performed on two forest stands in different stages of regeneration and at two altitudes. At both altitudes, the highest dehydrogenase activity was recorded in May, and it correlated with high spring fungal and bacterial populations. After a peak in May, urease activity decreased in July; thereafter, a steady increase in activity was observed up to

September. Phosphatase was also significantly influenced by season as well as by altitude. At the lower altitude, the enzyme activity was highest in May and significantly correlated with fungal population size, while at the higher altitude it showed two peaks, one in March and another in September, and no significant correlation with the microbial population was found.

Complex seasonal trends (i.e., no minimum or maximum but fluctuating values at different sites and at different seasons) were observed in phosphatase activity of soils, measured at low- and high-fertility pasture sites over 12 months [152]. Seasonal fluctuation in urease activity was also demonstrated by Stojanovic [153] in studies performed on six Mississippi soils belonging to four different textural classes and sampled in March, June, September, and December. Most of the June sampled soils exhibited the highest level of urease activity, whereas the lowest values were observed in December. A parallel decline in microbial population was observed.

Geographical location. Systematic studies to assess the correlation between enzyme activity of a soil and its geographical location have not been performed. Usually, higher enzyme levels are expected in soils from temperate environments. However, enzymes such as urease, phosphatase, and sulfatase were shown to be active at significant rates in soils associated with arctic tundra communities [154] characterized by very low temperatures. In contrast, denitrifying enzymes were active in desert soils, demonstrating tolerance of extended periods of desiccation [155].

Investigations of temperature effects on enzymes in different soils might be useful for estimating the potential enzyme activities of soils originating from geographical areas having different temperature ranges. McClaugherty and Linkins [156] showed that the temperature response curves (estimated as Arrhenius plots) for five enzymes—endocellulase (i.e., endo-1,4-β-glucanase), exocellulase (i.e., exo-1,4-β-glucanase), exochitinase, laccase, and peroxidase—of two forest soils from central Massachusetts were linear over the range of ambient temperature. Furthermore, these results were comparable to those of soils from arctic tundra areas. In comparative studies on enzyme activity in soils from pre-Ural forest, forest-steppe, steppe, and adjacent soils, a general regularity was demonstrated: the activity of hydrolytic enzymes, such as invertase, phosphatase, urease, and protease, as well as that of redox enzymes, such as dehydrogenase and catalase, was highest in the steppe zone and decreased in the forest-steppe and forest zones [157].

In Situ Distribution. Much information is available on the relationship between enzyme activity and depth in the soil profile [1-3,6,18,49,51]. Studies on the distribution or spatial variability of some enzyme activities and other factors, such as organic carbon, total nitrogen, humus content, and vegetation have been carried out.

Enzyme activity in soil profile samples usually decreases with depth [6] and is paralleled by a decrease in organic matter [7]. In a study of Iowa soils, Frankenberger and Tabatabai [44] found that L-glutaminase activity decreased ten fold at a sample depth of 105 to 115 cm relative to that observed at 0 to 15 cm. The authors showed a similar distribution when amidase activities were studied in 21 Iowa soils differing in pH, organic carbon, total nitrogen, and texture [39]. A marked decrease in the activities of rhodanese [90], glucosidase, and galactosidase [27] was observed at greater sample depths, and this decrease was associated with a decrease in the content of organic carbon. Similarly, in soils of northern New South Wales, urease activity was higher at the surface than in deeper horizons, and a positive correlation ($r = 0.875$; $P = 0.01$) was found between urease activity and soil organic carbon [158]. By changing the sampling depth of a peatland soil from 0 to 8 m, invertase, amylase, urease, and sulfatase decreased by about 15-, 48-, 85-, and 40-fold, respectively, whereas phosphatase remained virtually unchanged. However, appreciable differences in the variation patterns among the enzymes were observed [159]. Activities of dehydrogenase, protease, β-glucosidase, urease, and, to a lesser extent, alkaline phosphatase, amylase, and invertase of three different soil types under different vegetative covers decreased rapidly with increasing soil depth. The higher the soil acidity, the more rapid the decline [160].

These results seem to support the suggestion of Kuprevich and Shcherbakova [7] that several enzymes in soil are closely associated with soil organic matter. The deeper the horizon, the lower the content of organic matter and the lower the level of enzyme activity.

A systematic study of several enzyme activities are affected by their distribution was conducted by Ross and Speir: dehydrogenase and phenoloxidase [161], invertase and amylase [19–21], and urease, phosphatase, and sulfatase [162–164] activities were determined at different sites in New Zealand soils. Dehydrogenase as well as phenol-oxidizing activities appeared to change less with depth [161]. Furthermore, studies were performed on soil fractions separated by flotation techniques and ultrasonic vibrations employing density differences [22,163]. After dispersion in a dense liquid, Nemagon, and ultrasonic treatments, a "light fraction" and a "soil residue" were obtained and separated [165]. Higher activities of urease, phosphatase, sulfatase, invertase, and amylase were found in the light fractions as compared to unfractionated soils. On a percentage basis in the whole soil, however, most of the enzyme activity occurred in soil residue. The authors suggested that enzymes found in the light fractions—which are predominantly formed by nonhumified or partly humified plant and animal remains [165]—were predominantly derived from microbial synthesis, because these fractions generally showed higher enzyme activities than did dead plant materials and roots from the same sites.

Kinetic and stability studies performed on soil catalase, dehydrogenase, urease, and protease associated with different soil structural fractions [50,166] indicated that catalase and urease showed similar behavior across the soil fractions studied, whereas protease activity varied.

Studies performed on fractions of calciferous alkaline and acidic soils, obtained by a granulometric procedure, showed that phosphatase activity was concentrated in the larger-particle soil fraction (100 to 2000 mm in diameter), containing fresh less-humified organic matter and high levels of plant debris. These results suggest that phosphatase activity was associated with plant material, probably plant roots; and in stable humified fractions, association of phosphorus compounds with humates can occur, protecting them from chemical or biochemical attack [167].

A correlation between the distribution of phosphatase activity and its significance in the phosphorus nutrition of various plants were attempted by studying the levels of these enzymes and the rate of organic phosphorus transformation in the rhizosphere of different soil-plant systems (rhizosphere effect) [168–171]. In stands of 60- to 100-year-old Norway spruce trees, the activity of acid phosphatase was measured in bulk soil, rhizosphere soil, and mycorrhizal rhizoplane soil, and along the soil profile [168]. Acid phosphatase activity increased from the bulk soil toward the rhizosphere and mycorrhizal rhizoplane soil by a factor of 2.5. In the rhizosphere zone acid phosphatase activity was positively correlated ($r = 0.83$, $P = 0.001$) with the length of mycelial hyphae. Concurrently, lower concentrations of readily hydrolyzable organic phosphorus compounds were found in the rhizosphere and rhizoplane soil than in the bulk soil.

A significant correlation between the depletion of organic phosphorus and the presence of phosphatase activity was found in the rhizosphere of 16 crop-soil ecosystems consisting of both leguminous and nonleguminous crops [169]. Higher levels of acid, neutral, and alkaline phosphatases were detected in rhizosphere than in nonrhizosphere soils. Subsequently, Tarafdar and Junk [170] demonstrated that the increase in activities of both alkaline and acid phosphatase, observed at the soil-root interface of four plants (*Brassica oleracea, Allium cepa, Triticum aestivum, and Trifolium alexandrinum*) was associated with a concomitant removal of various phosphorus fractions. Furthermore, the increase in phosphatase activity depended upon plant age, plant species, and soil type.

The localization of phosphatase as well as that of other enzyme activities in situ in soil microorganisms and plant roots, in both rhizosphere and soil aggregates, have been inferred by electron microscopy studies performed with ultracytochemical/histochemical methods, as reviewed by Ladd et al. in Chapter 2 of this book.

Physicochemical Properties

A large number of studies has been devoted to correlating the presence, type, levels, and kinetic parameters (V_{max} and K_m) of enzyme activities with soil characteristics such as pH, moisture, cation exchange capacity, and content of organic carbon, total nitrogen, clay minerals, and soluble salts.

Content of Organic Carbon, Total Nitrogen, and Organic Matter. Enzyme activities of soils often correlate directly with their content of organic carbon and total nitrogen, possibly because organic carbon and total nitrogen reflect the level of organic matter.

Urease activity in soils of three different geographical and climatic areas (tropical soils from India [172], surface soils from Trinidad [173], and Iowa soils [174]) was positively correlated with organic carbon ($r = 0.88$, $P = 0.01$; $r = 0.96$, $P = 0.01$; and $r = 0.72$, $P = 0.001$, respectively) and with total nitrogen content ($r = 0.52$, $P = 0.01$; $r = 0.70$, $P = 0.01$; and $r = 0.71$, $P = 0.001$, respectively). Positive correlations with organic carbon and total nitrogen ($r > 0.7$ and 0.5, respectively; $P = 0.01$) were also demonstrated for amidase [39], L-glutaminase [44], glucosidase and galactosidase [27], phosphatase [175,176], cellulase, chitinase, and xylanase [143], and invertase and amylase [20]. Multiple regression analyses showed that the content of organic matter accounted for most of the variation in enzyme activities [44,61,170,174,175,177].

An attempt to correlate phosphatase activity with organic matter content and its role in the soil phosphorus cycle was suggested by Trasar-Cepeda and Gil-Sotres [176,178]. They studied the depthwise distribution, kinetic properties, and seasonal variation of phosphatase activities in acid soils with high levels of organic matter in Galicia (northwest Spain). These soils contained only acid phosphatase and no other phosphatase. The enzyme activity was largely concentrated in a shallow surface layer, and it diminished rapidly as the depth increased in spite of the still high content of organic matter. Trasar-Cepeda and Gil-Sotres attributed the severe lack of phosphate, from which Galician soils generally suffer, to this evident loss of acid phosphatase. They hypothesized that the loss was probably due to the inclusion of the enzyme in humic material, with a consequent drastic reduction of its catalytic activity. In another study, no significant correlations between various enzyme activities (amylase, cellulase, invertase, dehydrogenase, and urease) and carbon or nitrogen content were found during successive stages of revegetation to climax stands of three vegetation types (prairie grass, oak forest, and oak-pine forest), and it was concluded that the variations of enzyme levels were based on the type of vegetation, and thus the type of organic matter added to the soil [179].

pH. Several studies have been devoted to examining relationships between enzyme activities and soil pH; results have varied with positive, negative, or no correlations being reported [23,39,44,51,58,171,174–176,178,180–184]. Abramyan and Galstyan [180] claimed that the degree of soil acidity or basicity was an important factor in regulating several enzyme activities in soil, including those of invertase, phosphatase, urease, ATPase, dehydrogenase, and catalase.

No significant correlations between pH and the activity or kinetic parameters of urease in soils of different geographical areas have been demonstrated [172,174,181,182]. Similar results were also obtained for amidase [39], L-glutaminase [44], invertase and amylase [183], and cellulase [23]. A positive correlation was found for dehydrogenase activity when pH changed from 3.9 to 10.6 [98]. In contrast, glucosidase and galactosidase [27] were negatively correlated with soil pH; the activities of both enzymes decreased when pH increased from 4.5 to 8.5.

Significant relationships between soil pH and both acid and alkaline phosphatase activities have often been demonstrated [51 and references therein, 58,175,176,184]. Several investigations showed that phosphatase activities in soil often appear to have two pH optima, one acid and the other alkaline. Studies performed on alkaline and acid soils [51 and references therein, 176,178,184, 185] support the suggestion of Dick and Tabatabai [186] that acid phosphatase dominates in acid soils and alkaline phosphatase dominates in alkaline soils, thus indicating that hydrolysis of phosphorus compounds approaches its maximum rate at the pH of the soil. The strict association between the pH of the soil and the maximum activity of phosphatases seems to suggest that the stability and rates of production and release of these enzymes by soil biota are likely related to soil pH.

Moisture and Salinity. Water is not only the major cellular component of soil microorganisms (the origin of a large proportion of soil enzymes) but also the medium in which catalytic processes, especially hydrolytic reactions, occur. Therefore, it is important to the synthesis, production, and activity of intracellular and extracellular enzymes. Salts also influence the catalytic behavior of enzymes by acting directly on their activity (inhibition or activation effects) or by affecting the solubility of the enzymatic protein through a "salting-out effect." Furthermore, both factors may have an impact on the osmotic potential of soils. Either an increase in salt concentration or a decrease in soil water content may lead to hypertonic osmotic pressure, resulting in decreased metabolic activity of soil microbiota or in the osmotic desiccation of microbial cells. Active intracellular enzymes may be released and their activity may be partly or totally reduced because of the deactivating processes (e.g., immobilization on soil colloids, proteolysis) occurring in the soil.

El-Shinnawi and El-Shimi [187] demonstrated that the activity of dehydrogenase was directly associated with an increase in moisture content up to saturation; in contrast, an optimum moisture content of 60% water-holding capacity (w.h.c.) was found for the activity of both urease and phosphatase. When the soil water content decreased to air dryness, phosphatase and sulfatase activity declined exponentially [188]. Although invertase and amylase activities in seven soils were found to be uncorrelated with soil moisture content [183], some differences in the magnitude of the ratio of invertase to amylase activity among groups of soils in tussock grasslands were partly explained by variations in soil moisture [20].

In studies assessing the levels of soil enzyme activity in relation to nitrogen, carbon, phosphorus, and sulfur cycles in saline soils, Frankenberger and Bingham [189] showed that the activities of amidase, urease, acid and alkaline phosphatase, phosphodiesterase, inorganic pyrophosphatase, arylsulfatase, rhodanese, α-glucosidase, α-galactosidase, dehydrogenase, and catalase decreased with increasing salinity. The degree of inhibition varied with the type of enzyme and the nature and amount of salts. Greater inhibition was observed for dehydrogenase than for the hydrolases. Dehydrogenase was also inhibited by high concentrations (>150 mg/ml) of $(NH_4)_2SO_4$ and $NaNO_3$. However, when the activity was measured in the presence of KNO_3, no significant effects were seen [98]. Increasing amounts of NaCl, either by its addition or by the drying of soils, caused a marked decrease in amidase and urease activity. Both enzymes recovered activity when soils were rewetted to the initial level [190].

Inhibition or activation of enzyme activity due to the anions present in salts was demonstrated by Singh and Tabatabai [90]. When rhodanese activity was measured in the presence of 1 mM of $NaNO_2$, NaN_3, Na_2SO_4, NaF, and NaCl, an activation effect was detected, whereas $NaNO_3$, $NaSO_3$, Na_2S, KH_2PO_4, and $NaHCO_3$ were inhibitory.

Inhibition (0 to 70%) exerted by 12 inorganic salts (at 8 mM concentration) on the activities of α-glucosidase and α-galactosidase was greater than that exerted on their beta counterparts (0 to 40%) [27]. No explanation was reported for this difference in response, although it may be accounted for by the molecular, physical, and catalytic characteristics of beta-enzymes.

Clay Minerals. As exhaustively reviewed by Stotzky [16], clay minerals may strongly influence microbial growth, adhesion, and metabolic processes and the ecology of microbes and viruses. This is accomplished through surface interactions between clays and microbiota (direct effects) and through modification of the microbial and viral environment (indirect effects). These influences may ultimately be reflected in the enzymatic activities of soil.

The formation and properties of complexes between clays and enzymatic and nonenzymatic proteins have been extensively studied, and several papers and a review have been dedicated to this topic [12,110,112,113,115,118–120,191,193–197].

Clay minerals may have inhibitory as well as stabilizing effects on enzyme activities in soil. These changes in enzyme properties are an expected consequence of the interactions (adsorption, interlayering) of enzymes with clays. Kiss et al. [191] reviewed more than 120 papers concerning the effects of clay minerals on enzyme activity. Two basic conclusions were reached by the authors:

1. "In general, the clay minerals inhibit the activity of enzymes, which is followed by stabilization and protection of the residual activity."
2. "Clay minerals, through their stabilizing and protecting effect on enzymes, contribute to the accumulation of enzymes in soil, and to the survival of enzyme molecules in the soil microenvironment. But the contribution of clay minerals to enzyme accumulation in soil and probably also in aquatic sediments is achieved in association with humic substance" [191].

The data analyzed by Kiss et al. [191] were usually produced in model experiments; i.e., clay minerals and enzyme proteins were mixed under laboratory conditions and the effects of clays on the behavior of enzymes were determined. The studies reviewed by Kiss et al. [191] demonstrated that enzymes adsorbed by clay minerals frequently changed properties, such as pH optimum, kinetic parameters, activity, and stability. Generally, upon interaction with clays,

1. the apparent optimum pH was shifted by one or two units toward the alkaline region [109,192];
2. the activity decreased, but the extent of deactivation varied widely [112];
3. the kinetic parameters (V_{max} and K_m) differed from those for the corresponding free enzymes, indicating a lower reaction rate as well as a diminished affinity for the substrate [12]; and
4. the stability of enzymes against thermal and proteolytic deactivation usually increased [115].

Further experimental evidence seemed to confirm the conclusions drawn by Kiss et al. [191]. Claus and Filip [193] showed that phenoloxidases were adsorbed and inactivated by clays, but that clays, such as kaolinite, and quartz sand may enhance the production of phenoloxidases by some fungi and actinomycetes [194].

A protective mechanism provided by clay (and humus) colloids was suggested by Lähdesmäki and Piispapen [195], who studied the effects of various physical and chemical treatments on the activity of protease, cellulase, and amylase in a forest soil. Experiments examining the influence of repeated freezing and thawing, heating and drying, changes in pH, and the presence of different concentrations of $ZnSO_4$ and $CuSO_4$ showed that clay colloids actually protected soil enzyme activities from these environmental fluctuations.

Both inhibition and activation were detected by Novakova [196] in studies on the influence of montmorillonite and kaolinite on the activity of soil catalase, α-amylase, and cellulase. The activity of catalase was depressed by both clays, whereas that of cellulase was decreased by montmorillonite, but increased by kaolinite. No clay effects on the activity of amylase were observed. In another paper, Novakova and Sisa [197] demonstrated that soil cellulase activity was stimulated by the addition of kaolinite, but inhibited by montmorillonite. The stimulating effect of kaolinite and the depressing effect of montmorillonite were interpreted by the authors as a stimulation or an inhibition of bacterial growth, respectively, with a consequent influence on production of the enzyme [197].

Studies examining the correlation between soil clay contents and enzyme activities have produced variable results. The activities of ureases [172], L-glutaminase [44], and phosphatases [175,177] were not significantly correlated with the presence of clay. On the other hand, the activities of amidase [39] and protease [45] were positively correlated with the percentage of clay ($r = 0.53$ and 0.69, respectively).

Humus. Humic colloids have an important role in the immobilization, stabilization, and expression of enzyme activity in soil. Humic substances, such as humic and fulvic acids, tannins, and melanins, or humic precursors, such as phenols and quinones, may associate with enzyme proteins through different bonding mechanisms. As a result of immobilization, the catalytic behavior of enzymes can be altered and an enhanced stability to biological attack can be achieved.

Inhibition of enzyme activity can be caused by several humic compounds and related substances. As reviewed by Ladd and Butler [114], tannins, melanins, and humic and fulvic acids differentially inhibit enzyme activity. In a series of papers [198–202], Ladd and Butler showed that the activity of proteolytic enzymes, such as pronase, carboxypeptidase, chymotrypsin, and trypsin, were inhibited up to 50% by soil humic acids, whereas other peptidases (e.g., papain, ficin, subtilopeptidase, and thermolysin) were stimulated. Methylation of humic acid, acetylation of proteolytic enzymes, or interactions of humic acid and enzymes in the presence of inorganic cations markedly reduced inhibition. From these observations, the authors concluded that the mechanisms of inhibition involved primarily carboxyl groups of humic acids.

Inhibition of nonproteolytic enzymes was also observed. Pflug [203] demonstrated that the activities of two peroxidases were inhibited by synthetic and natural humic acids by competitive interactions between humic acids and the donor substrate (e.g., guaicol or NADH) for the enzyme. Binding of H_2O_2 to any of the enzymes was not influenced. Inhibition of malate dehydrogenase [204], indoleacetic acid oxidase [205], and oxidoreductases (peroxidase, tyrosinase, and laccase) [206] was also detected. A direct inhibitory effect of soil organic constituents on the activities of acid phosphatase and inorganic pyrophosphatase was observed by Dick et al. [58] and Gianfreda and Bollag [207]. Gianfreda et al. [121,208,209, and unpublished results] showed that tannic acid, a humic acid precursor, inhibited the activities of jack bean urease more than invertase from *Saccharomyces cerevisiae* and acid phosphatase from potato. At 0.1 mM tannic acid, the urease activity was reduced by 73%, whereas that of invertase and acid phosphatase decreased by 10 and 25%, respectively. The glycoprotein nature of both invertase and phosphatase probably offered a higher resistance to complexation between tannic acid and enzymatic molecules, or permitted the complexation of substantial amounts of tannic acid even without noticeable effect on the residual activity of the enzymes. Furthermore, it could be hypothesized that Ni^{2+} ions, present on the urease active center, enhanced the inhibition of urease activity by tannic acid [208].

Data demonstrating correlations between enzyme activities in soil and humus content are not sufficient to infer an obvious influence of humic substances on the formation and stability of catalytically active humus-enzyme complexes. A better understanding may be obtained by isolating humus-enzyme derivatives from soil or preparing them synthetically. Humic-enzyme derivatives extracted from soil show two main characteristics: (1) a strong resistance to storage, high temperature, and proteolytic attack; and (2) changed kinetic properties (kinetic parameters, pH and temperature profiles) [12,114,210–213].

The preparation of synthetic humic-enzyme complexes have required two basic procedures:

1. Natural humic acids have been extracted from soil and their interactions with enzymes studied in vitro [214–219].
2. Synthetic humus-enzyme complexes have been prepared by the binding of phenolic compounds with enzyme proteins and subsequent polymerization [12,121,208,209,220–222]. The enzyme complexes usually exhibited stability and kinetic properties different from those of free enzymes. In addition, the synthetic humic moiety displayed characteristics similar to those of extracted humic acids [221]. Furthermore, useful information on bonding mechanisms involved in the immobilization of soil enzymes on soil organic matter could be acquired.

As summarized by Müller-Wegener [223], there were three fundamental interactions of humic substances with enzymes:

1. In soil, a direct interaction between enzymes and humic substances occurred. Several physical and chemical mechanisms, including adsorption, steric effects, and hydrogen or amidic bonds were probably

involved. A direct action of humic substances on the active sites of enzymes, rather than a modification of the active site due to conformational changes induced by humic substances, could be assumed. This assumption seems justified by the inhibitory effects, previously reported, that humic substances have shown on the activities of several enzymes.

2. As a result of their biochemical heterogeneity and complexity, humic substances may have acted as analog substrates and interfered with the equilibrium of the enzyme reaction.
3. It is possible that humic substances, with their cation exchange properties, interacted with cations involved as cofactors in enzymatic catalysis or in stabilization of the active enzyme structure.

Abramyan [224] claimed that the immobilization and behavior of enzymes in soil were regulated by natural and anthropogenic factors acting on the level of organic matter. The author studied the activities of invertase, phosphatase, urease, amidase, arylsulfatase, ATPase, dehydrogenase, and catalase in 10 different soil types. Factors studied included the degree and nature of soil acidity and basicity, temperature, humidity, and fertilizer addition. Abramyan concluded that activity levels of most enzymes in the various soils were regulated by the contents of organic compounds of nitrogen, phosphorus, and sulfur, which in turn were regulated by the content of available forms (organic and inorganic) of nitrogen, phosphorus, and sulfur. This conclusion was based on positive correlations ($r > 0.90$; $P = 0.01$) between the amount of organo-nitrogen, -phosphorus, and -sulfur compounds and the activities of related enzymes (urease, amidase, phosphatase, and arylsulfatase). The organo compounds would have increased the enzyme activity through their beneficial effects on microbial activity. The consequently higher enzyme activity would have led to increased transformation of the organo compounds to their inorganic forms, which, in turn, would have inhibited (feedback effect) the activity of enzymes. This continuous interplay of the induction-repression mechanism regulated the activity levels of enzymes.

Biomass. Although enzymes from plant or animal origins may be important contributors to the activities of enzymes in soil, enzymes of microbial origin are considered most important. The overall microbial activity can be considered an appropriate index of soil fertility.

Several studies have examined correlations between the enzyme activities in soil and the number of microorganisms in soil or the level of soil respiration. Stevenson [76] and Casida et al. [77] considered dehydrogenase activity a reliable index of microbial activity in soil, although no relationships between levels of dehydrogenase activity and numbers of microorganisms were found. Negative or no significant correlations were demonstrated between numbers of aerobic bacteria and the activities of invertase and amylase [183], or between invertase and the number of fungi and bacteria [148]. When the activities of 11 soil enzymes (alkaline and acid phosphatases, amidase, α-glucosidase, dehydrogenase, phosphodiesterase, arylsulfatase, invertase, α-galactosidase, urease, and catalase) were correlated with microbial respiration, biomass, and viable plate counts in surface samples of 10 diverse soils, a variety of results was obtained. Phosphodiesterase and α-galactosidase activities were significantly related to microbial numbers obtained on some selective culture media, whereas phosphatase, amidase, and catalase were highly correlated with microbial respiration and total biomass in soils [225]. The fractionation of an arable brown soil into organic and mineral soil particles of different sizes demonstrated that the larger organic particles accumulated most of the soil enzyme activity (β-glucosidase and protease) and showed the highest counts of bacteria, including actinomycetes and fungi per gram of soil [226].

Several authors have attempted to determine the relative contributions of enzyme activity from bacterial and fungal sources. Using various biochemical criteria (CO_2 evolution, activities of urease, phosphatase, and protease, content of ATP, content and nature of amino acids and amino sugars), Nanniperi et al. [227] concluded that urease and phosphatase activities were significantly correlated with bacterial but not fungal biomass. The results obtained with antibiotic (cycloheximide and chloroamphenicol) treatment of seven Japanese soils suggested that the activity of phospholipase C was produced predominantly by bacteria and, in some soils, to a lesser degree by fungi. After treatment, the number of bacteria as well as the activity of the enzyme generally increased, whereas fungal number decreased. Only in some soils was the opposite effect observed [228]. Rhee et al. [24] exploited the glycoprotein nature of fungal carboxymethylcellulase (CMCase) in their studies. By using a separation technique based on the varying affinities of glycoproteins for Concavalin A, the authors demonstrated that the fungal enzyme contributed six times the activity of the bacterial enzyme, thus revealing the dominant role of fungi in the decomposition of cellulosic materials in soil ecosystems. Similarly, in studies on the properties of β-glucosidase derived from various fungal and bacterial isolates, it was demonstrated that fungi (mainly mucoraceous) were probably the primary source of β-glucosidase activity in soil [229].

A positive correlation between dehydrogenase activity and fungal population numbers was found in soil samples from sites at two different altitudes. A positive and significant correlation of both dehydrogenase and urease activity with bacteria numbers was demonstrated but only at lower altitude [151]. However, these correlations were strongly affected by seasonal changes and vegetative cover.

Soil management influences the biomass content of a soil, thus affecting enzyme activity. If soil microorganisms are considered the primary sources of enzymes in soil, every management factor influencing microbial population will indirectly affect enzyme activity. Macura [230] demonstrated that the enrichment of soils with various energy sources (mono-, di- and polysaccharides; amino acids; proteins) influenced the mechanisms controlling the production and activity of catabolic enzymes. Induction and catabolite repression occurred in the soil biological population, which reflected on the production of enzymes. A direct inhibition by catabolites on enzyme activities in soil was also observed. A direct relationship between microbial biomass and the activities of dehydrogenase, phosphatase, arylsulfatase, and amidase was demonstrated in a silty clay-loam soil [231].

B. Anthropogenic Factors

In addition to natural factors, significant changes in enzyme activities in soil may arise through anthropogenic impacts. The use of vegetative cover, soil amendments and management, and the application of fertilizers and pesticides are all agricultural practices that are designed to affect soil productivity. However, enzyme activities in soil may be adversely affected by some of these interferences, e.g., by pollution with heavy metals or xenobiotics, or by acidification caused by acid rain. Moreover, anthropogenic and natural events may interact synergistically, with the result that it is sometimes difficult to distinguish which of these two influences may have caused the changes detected in an enzyme activity.

Human interventions can be grouped into two categories: those resulting in soil pollution, and those devoted to improve the productivity of soil.

Environmental Pollutants

Acid Precipitation. In the last decade, the acidity of precipitation has continuously increased as the resulting of increasing levels of CO_2, sulfur, and nitrogen gases released into the atmosphere. Transformation of such gases to the corresponding acids have raised H^+ concentrations in the atmosphere, and consequently, the amount of H^+ deposited on the soil. When considering the effects of acid rain on soils, attention must be given to the ability of a soil to maintain its present pH, i.e., its buffering capacity. Clays, organic matter, oxides of aluminum and iron, and calcium and magnesium carbonates (limestone) are the components responsible for pH buffering in most soils. Therefore, it is probable that acidic precipitation will impact more on sandy, low organic matter soils than on those higher in clays, organic matter, and carbonates.

Because soil microbial and enzyme processes may be affected by the acidity of the medium, several studies have been devoted to investigating the effects of acidity on biological processes in soils [232–236]. Reactions catalyzed by enzyme activities in soil may be directly or indirectly influenced by changes in pH. Variations of pH may alter the ionization and solubility of enzymes, substrates, and cofactors. Furthermore, pH fluctuations may significantly affect the stability of active protein structures or increase the solubility of metals (such as nickel, copper and zinc) that often behave as enzyme inhibitors.

Studies examining the effect of acid precipitation on enzyme activities in soil were performed using soil exposed to simulated acidic conditions. Short-term (92 days) exposure of soil columns to acid rain at pH 3.0 showed no alterations in the activities of dehydrogenase, urease, phosphatase, or protease. Long-term (690 days) exposure caused changes in these activities, although responses varied with soil type [237]. Similar studies performed on field soils showed no significant reduction of dehydrogenase, phosphatase, or urease activities, while a detectable decrease in protease activity was found [238].

A 1-year exposure of a soil to polluted air containing SO_2 decreased the pH of the soil from 4.2 to 3.7. However, no significant changes in the activity of arylsulfatase, cellulase, dehydrogenase, acid phosphatase, rhodanese, and urease were found [239]. Moreover, experiments performed on a forest soil (pH 6.4) exposed to simulated rain (pH ranging from 2 to 5.6) showed that in the surface layer receiving the pH input, urease activity was unaffected, whereas arylsulfatase activity increased and phosphatase activity decreased [240].

The activities of dehydrogenase, acid phosphatase, and arylsulfatase in the rhizosphere of loblolly pine were studied in experiments that combined the influence of simulated acid rain and ozone (O_3) [241]. Seedlings were exposed to various O_3 concentrations and pH values ranging from 3.3 to 5.3 during 11 weeks. Dehydrogenase and acid phosphatase were inhibited by increasing acidity; O_3 and pH influences were independent of each other. Variations in arylsulfatase activity depended on particular combinations of pH and O_3 concentration. In agricultural and peat soils, the activity of arylsulfatase was significantly reduced by high concentrations of sulfur ions (i.e., HSO_3^-, SO_3^{2-}, SO_4^{2-}), resulting from deposition of acid precipitation [240,242,243].

Heavy metals. Soil pollution by metals, particularly by heavy metals, constitutes a persistent environmental problem. Metals may be released to the environment through municipal and industrial wastes or deposited on soil as impurities in fertilizers or combustion products of fuel oils. The presence of these elements in soil may influence biochemical processes by affecting both microbial proliferation and enzyme activity [244].

Heavy metals may inhibit enzyme activity through the masking of catalytically active groups, denaturing effects on protein conformation, or competition with metal ions involved in the formation of enzyme-substrate complexes, and many investigations have been devoted to assessing the influence of metals on the activity of enzymes in soil. Juma and Tabatabai [245] studied the effect of 20 trace elements on the activity of acid and alkaline phosphatase in 10 surface soils. Soils were selected to obtain a range in physicochemical properties and phosphatase activity. The relative effectiveness of the trace elements in inhibiting phosphatase activity depended on soil and phosphatase type. Hg(II), As(V), W(VI), and Mo(VI) were the most effective inhibitors of acid phosphatase, whereas Ag(I), Cd(II), V(IV), and As(V) caused the greatest inhibition of alkaline phosphatase (average inhibition >50%), although all trace elements that inhibited acid phosphatase activity also inhibited alkaline phosphatase. Furthermore, the authors demonstrated an additional inhibition by phosphate ions, although no further effects were detected in the presence of other anions such as NO_2^-, NO_3^-, Cl^-, and SO_4^{2-}.

The effect of trace elements on the activities of other enzymes, including α- and β-glucosidase [27], α- and β-galactosidase [27], arylsulfatase [246], nitrate reductase [247], rhodanese [90], L-glutaminase [44], L-asparaginase [41], and L-amidase [38], was also studied in several soils. In general, the effects varied considerably among the elements, enzymes, and soils. The inhibition of α-glucosidase ranged between 4 and 9% for Mo(VI) and V(IV) and 66% for Ag(I) and Cd(II); under similar conditions β-glucosidase activity was much less reduced [27]. The inhibition of α-galactosidase was 10 and 13% after the addition of Hg(II) and Sn(II), respectively, and 73% after the addition of Mo(VI) [27]. On average, more than 50% inhibition of arylsulfatase and rhodanese activities was also caused by several elements (e.g., Ag(I), Hg(II), B(III), V(IV), and Mo(VI); and Mn(II), Sn(II), Ni(II), Cr(III), B(III), Al(III), As(III), Se(IV), V(IV), Ti(IV), As(V), W(VI), and Mo(VI), respectively), with a positive correlation between the amount of trace element added per gram of soil and the extent of inhibition [190,246]. Greater inhibition (>75%) of the activity of nitrate reductase was seen with the addition of Ag (I), Cd(II), Hg(II), Se(IV), As(V), and W(VI). A strong correlation between the degree of effectiveness and the acidity or alkalinity of the soil was noted [247].

The effects of trace elements (e.g., Ag(I), Cu(I), Ba(II), Hg(II), Cu(II), Cd(II), Co(II), Fe(II), Mn(II), Ni(II), Pb (II), Sn(II), Zn(II), Al(III), B(III), As(III), Cr(III), Fe(III), V(IV), Se(IV), Ti(IV), As(V), W(VI), Mo(VI)) on the activity of L-glutaminase in soil was comparable to those detected for L-asparaginase and amidase [38,41,44]. The responses of three amidohydrolases were similar in that Hg(II) was the most effective inhibitor of their activities; this implied that thiol groups were probably involved in their active sites. Ag(I), Hg(II), Sn(II), Cr(III), Ti(IV), and W(VI) showed an average inhibition of >25%, whereas Ba(II), Co(II), Fe(II), Mn(II), and all others caused an inhibition lower than 10%.

In a study conducted on a spruce more polluted with copper and zinc from a brass foundry in Sweden, highly significant negative regression coefficients were found for the relationship between log(Cu + Zn) and phosphatase and urease activities, whereas no relationship with β-glucosidase activity was observed [248]. Kinetic studies performed on soil catalase activity demonstrated that cadmium increased and silver decreased K_m and V_{max} values, whereas lead had no significant effect [249].

A dependence on the form of added metal was demonstrated in studies examining the influence of vanadium on acid phosphatase activity in an organic Swedish soil [250]. At a concentration of added vanadium of 100 mg/kg of soil (dry weight), comparable to that occurring in urban and suburban areas largely as a result of fuel oil combustion, enzyme activity decreased in the order $Na_3VO_4 > NaVO_3 > VOSO_4 > V_2O_5$.

To quantify the effect of hazardous chemicals on soil microbiological activity, Babich and Stotzky [251] introduced the concept of the ecological dose (EcD), defined as the concentration of chemical that decreases a tested microbial activity in soil to some percentage lower than 100%. Subsequently, in studies devoted to establishing acceptable limits for heavy metal contamination of soil, Doelman and Haanstra [252,253] defined the 50% effective ecological dose (ED_{50}) as the concentration of a chemical that reduces a given microbial enzyme activity to 50% of its initial value. The authors studied the short- and long-term effects of copper, cadmium chromium and zinc on phosphatase activity of five diverse soil types, and results were presented graphically as logistic dose response curves. In clay soils, the ED_{50} values were comparable to each other for all the metals (approximately 45 m mol/kg of dry soil). In sandy loam and silty loam soils, metal toxicity varied between 12 and 88 m mol/kg of dry soil. A similar study previously conducted by Doelman and Haanstra [253] examined the inhibitory effects of cadmium, chromium, lead, copper nickel and zinc on urease activity in five different soils during two different periods. The ED_{50} value decreased significantly only for copper in the sandy soil, whereas it tended to increase over an 18-month period for cadmium, copper, and zinc. With nickel and lead, however, the toxicity stabilized in all except sand and clay soils.

Activity levels of numerous enzymes (C_1-cellulase, C_x-cellulase, cellobiase, xylanase, chitinase, lichenase, inulase, pectinase, amylase, invertase, lactase, maltase, lipase, protease, and acid phosphatase) were studied in organic soils of differing copper content [33]. Representative pairs of muck and peat (sapric and hemic Histosol) samples, collected from a mildly cupriferous virgin bog under a normal mixed forest flora in Canada, were exposed to varying amounts of copper for different incubation times. Within each pair of samples, which were air dried following a 6-week incubation, enzyme activities were lower in those containing higher copper concentrations. An average decrease of 70% was found. Assays of invertase, amylase, phosphatase, and protease activities in moist soil samples incubated at 21°C for 368 days indicated that the inhibitory effect of copper on hydrolytic enzyme activities also applied to microbially active soils.

Several metal ions (Ba^{2+}, Ca^{2+}, Co^{2+}, Mg^{2+}, Mn^{2+}, Ni^{2+}, and Zn^{2+}) activated pyrophosphatase activity in three soils after removal of soluble metal salts and exchangeable metal ions by leaching with 1 N NH_4OAc (pH 8) and addition of metals at various concentrations [57]. The highest and lowest activations were shown by Ca^{2+} and Mn^{2+}, which increased pyrophosphatase activity in 10 soils by 47 and 15%, respectively. Similar activation effects were shown by Zn^{2+} on rhodanese activity [90].

A stimulating effect on enzyme activity was demonstrated in soils treated with the sulfate, sulfite, and carbonate salts of zinc, whereas an inhibitory effect was caused by the salts of copper. A dose of 7500 ppm, administered at one time or distributed over a 7-week period, was used. The overall enzyme activity in soil was determined by the kind and number of microorganisms inhabiting the soil [254].

Some authors have proposed enzyme activity as a bioindicator of soil contamination with heavy metals. Krasnova [255] suggested that urease and invertase be used for this purpose because they were the enzymes most sensitive to heavy metal pollution of a soddy podzolic soil. In another study, dehydrogenase activity was utilized as a method to estimate microbial activity in metal-contaminated soils [256]. However, care is required in the use of enzyme activity for evaluating the degree of soil contamination by heavy metals. In fact, the lower activity of dehydrogenase observed in copper-contaminated versus uncontaminated soils appeared to be an artefact: it was caused by a reaction between copper and triphenylformazan (TPF) (the transformation product from the conversion of triphenyltetrazolium chloride) which decreased the absorbance of TPF [257].

Agricultural Influences

Enzyme Substrates. Rates of enzyme reactions are limited by the amounts of both enzymes and substrates when other environmental factors are constant. In addition, the availability of substrates in soil may affect the induction of specific enzymes by particular organisms or enhance microbial growth, thus affecting the level of overall enzyme activity.

Hayano [25] demonstrated that addition of cellulose induced cellulase activity in a soil sample from a tomato field. Microbial amylase production in three soils (leached chernozem, brown forest, and alluvial) depended on the nature of added substrate (starch, glycogen, and dextrin) and the soil type [258]. The efficacy of amylase induction followed the decreasing order: starch > dextrin > glycogen in the leached chernozem soil, and starch > glycogen > dextrin in the other two soils. The addition of NH_4NO_3 had no significant influence on the production of amylase. This indicated that efficient nitrogen sources for the amylase-producing microorganisms preexisted in the soils.

From studies performed on unamended and substrate- or enzyme-amended soils, Tateno [259] concluded that enzyme reactions in natural soils were limited by substrate supply and not by the amount of enzymes. Tateno measured the activities of cellulase and protease in soil samples under three conditions: amended with carboxymethylcellulose (CMC), amended with a casein solution, or not amended; the amended soils were tested both with and without added cellulase and protease. In the CMC- and casein-amended soils, both enzyme activities were much higher than in unamended or enzyme-amended soils. In contrast, neither the potential CMC-cellulase activity nor the total degree of CMC mineralization significantly differed in a soil continuously supplied with CMC. The addition of NH_4^+ affected the maximum CMC-cellulase activity and the degree of substrate mineralization [260].

The addition of starch did not influence the significant correlation found between glycoside hydrolase activities in a New Zealand topsoil and physicochemical parameters such as mean annual rainfall, soil moisture, and organic carbon content [261]. The activity of phosphatases, produced by several plants and microorganisms, was investigated in soils treated with organic phosphorus compounds, such as sodium glycerophosphate, lecithin, and phytin [262]. In general, phosphatase activity (acid and alkaline) and microbial population increased in all the phosphorus treatments.

Fertilizers. The influence of fertilizers (inorganic and organic) on soil enzyme activity depends on soil type, enzyme type, and application time. In long-term experiments, effects may be caused by changes in soil characteristics such as plant yields, moisture content, and concentration and availability of organic and inorganic nutrients.

A complex study of the effects of mineral and organic fertilizers on soil fauna [number of protozoa (testate amoebae, ciliates); number of small metazoa (rotifers, nematodes)] and soil enzymes (catalase, cellulase) was conducted in a high-altitude reforestation trial [263]. Mineral (NPK) and organic (dried bacterial and fungal biomass) fertilizers were applied alone and in combination with magnesite. Catalase activity increased 52 to 84% after application of the organic fertilizer. Cellulolytic activity showed no significant fertilizer effects, although the activity was usually lower in the fertilized treatments than in the controls. No correlations between enzyme activities and abundance of protozoa or nematodes were obtained. In another study involving long-term exposure (9 years), organomineral fertilization (farmyard manure + NPK) proved more efficient than manure or mineral fertilization alone in increasing the activity of invertase and phosphatase in a technogenic soil located in the southern zone of the capus iron strip mine in Romania [264].

The influence of phosphate or phosphorus-fertilizer addition on soil enzymes was investigated. Haynes and Swift [265] studied the effects of adding lime and/or phosphorus to an acidic phosphorus-deficient soil on microbial and enzyme activities, levels of biomass, and extractable nitrogen, sulfur, and phosphorus. Additions of lime and phosphorus stimulated mineralization of carbon, nitrogen and sulfur; a concomitant rise in soil pH was measured. Additions of lime generally increased protease and sulfatase activities but decreased phosphatase activity. Additions of phosphate decreased the activities of all three enzymes, but increased the accumulation of mineral nitrogen and SO_4^{2-}. This latter result suggested that protease and sulfatase activities were not reliable indicators of the relative amounts of mineral nitrogen and SO_4^{2-} accumulated in the soil during incubation. A decrease in phosphatase activity and extractable phosphorus fractions was also measured after a 39-day incubation of an acidic soil rich in organic matter with different doses of lime. The higher the dose of lime, the greater the decrease in phosphatase activity [266]. The effect of lime on phosphatase activity is complex. It increases the enzyme activity because of a beneficial effect on microbial activity. On the other hand, it increases the amount of soluble inorganic phosphate, which represses enzyme activity [51].

When phosphatase and dehydrogenase activities were determined in acidic soils sampled under spruce (*Picea abies* L.) subjected to acid deposition before and after liming, liming increased dehydrogenase activity and available phosphorus but no phosphatase activity [267].

Enhancing effects on the activity of nitrogenase (a typical intracellular enzyme) in soil were observed when soils were inoculated with *Azotobacter chroococcum* [268] or earthworms (*Lumbricidae*) [269]. Inoculation of agricultural soils with *A. chroococcum* led to increases in the nitrogenase activity of maize roots, suggesting that the nitrogenase activity in these roots was associated with the presence of appreciable densities of *Azotobacter* [268]. Higher levels of nitrogenase activity found in soils amended with *Lumbricus rubellus* indicated that nitrogenase activity in soil was significantly increased by the burrowing and feeding activity of these worms [269].

Nitrogen fertilizers affect the activity of several enzymes. Their effects are often influenced by the presence of carbon sources. Ladd and Paul [270] measured several biological variables during a period of immobilization of nitrate-^{15}N and mineralization of organic nitrogen in a sandy-loam soil including (a) the concentration of an added carbon source (glucose-^{14}C); (2) evolution of $^{14}CO_2$; (3) bacterial populations; (4) distribution and concentration of newly synthesized, acid-soluble amino acid ^{15}N; and (5) distribution and activities of several oxidative and hydrolytic enzyme systems. All of these were measured in various soil fractions (e.g., extractable proteins, extractable amino acids and peptides, particulate material containing microbial cells, cell debris, and microbial metabolites mainly bound to soil colloids) obtained by a relatively mild fractionation procedure. Enzyme activities generally increased with a concomitant increase of viable bacterial populations. However, changes in enzyme activities were not associated with changes in protein-^{15}N concentrations of the different fractions.

Stimulating effects on dehydrogenase activity in a soil cultivated with potatoes were measured only when nitrogen-fertilizers were supplemented with the carbon source, sucrose [271]. There were no consistent effects from adding nitrogen, either alone or with straw. Negative effects of different forms of nitrogen on urease production were detected in soils amended with organic carbon (glucose) [272]. The addition of NH_4^+ or NO_3^- to carbon amended soils repressed urease production, although microbial activity, as measured by CO_2 evolution, was stimulated. The addition of inhibitors of nitrogen transformations (inorganic nitrogen assimilation and NO_3^- reduction to NH_4^+) by microorganisms relieved the repression of urease production in carbon-amended soils. These observations indicated that microbial production of urease in these soils was probably repressed by products formed by microbial assimilation of NH_4^+ and NO_3^-.

The activities of glutamine synthetase and glutamate dehydrogenase, nitrogen uptake, and corn growth were related to ammonium concentration when soils were exposed to NH_4^+ concentrations ranging from 0 to 4500 mg N/kg of soil [273]. High levels of nitrogen induced very high NH_4^+ concentrations in the soil solution and limited plant growth. Glutamine synthetase activity peaked at NH_4^+ concentrations of 3 mM and then decreased; glutamate dehydrogenase activity increased up to 1.3 mM NH_4^+ and then remained constant.

The addition of NH_4NO_3 with glucose increased the activity of soil acid phosphatase sixfold [274]. Glucose addition also increased rhodanese [90] and protease [275] activity. Preincubation of six soils for 24 and 48 h with 0.1% glucose gave rise to average rhodanese increases of 9 and 23%, respectively [90]. Increases of soluble protease (the soluble fraction of newly produced protease separated from soil by a sterilized Millipore filter) and total protease activities in soils amended with glucose coincided with increases in ATP content, total counts of bacteria, growth of fungi, and CO_2 evolution. Similar effects were also observed with agar amendments [275].

A grassland soil amended with casamino acids increased alkaline phosphatase activity after 2 days of incubation [276]. A decrease to the original level was detected after 24 days. When fungi populations were suppressed by addition of antibiotics, no increase in alkaline phosphatase activity was recorded in soils supplemented with casamino acids, suggesting that fungi were responsible for the increase.

Several nitrogen, phosphorus, potassium and sulfur fertilizers (e.g., NH_4NO_3, $(NH_4)_2CO_3$, $(NH_4)_2SO_4$, KNO_3, $CaHPO_4$, $NH_4H_2PO_4$, $(NH_4)_2HPO_4$, KCl, K_2SO_4, $CaSO_4$, and $MgSO_4$) as well as urea had no effect on the activity of urease in Iowa soils when applied at levels equivalent to 500 ppm (soil basis) of nitrogen, phosphorus, potassium, or sulfur [49].

Organic Amendments. Organic amendments (plant materials, animal residues, various waste products, sewage sludge) to soils are decomposed by microbial transformations, causing a release of essential nutrients such as nitrogen, phosphorus, and sulfur. Microbial and enzyme activity in soil may be promoted and soil physical characteristics improved by the incorporation of these organic materials. Stimulation of enzyme activities in soil is usually greater than that induced by inorganic fertilizers. However, organic matter may also contain pollutants (e.g., heavy metals, toxic organic compounds) that, if present in inhibitory concentrations, may decrease enzyme activities in soil.

The activity of arylsulfatase in two soils (a sandy clay loam and a clay loam) from Perugia, Italy, was modified by the addition of various crop residues [277]. Kinetic parameters (V_{max} and K_m) changed significantly, and the changes depended on soil properties and on the type of crop residues incorporated. The V_{max} values in the amended clay loam soil were always higher than those for the unamended soil. A similar result was obtained with the sandy clay loam soil, but only with tobacco and sunflower amendments. These findings appear to confirm the hypothesis that arylsulfatase in soil is mostly of microbial origin. An increase of V_{max}, which is related to the concentration of enzyme, could indicate enhanced production of enzymatic molecules by arylsulfatase-producing microorganisms stimulated by the addition of organic matter.

A general increase in amylase, dehydrogenase, and arylsulfatase activities, but not in phosphomonoesterase activity, was measured in a clay loam soil amended with seven different crop residues (tobacco, sunflower, maize, wheat straw, capsicum, sorghum, and tomato) [278]. The addition of tobacco and sunflower residues caused an increase in most enzyme activities, whereas the addition of tomato residues caused an increase only in amylase and phosphodiesterase activities. A subsequent study on the activities of phosphomonoesterase (acid and alkaline) and phosphodiesterase in a soil treated with the same crop residues showed that the majority of the residues stimulated phosphodiesterase activity (by 28 to 48%) and inhibited acid phosphatase activity (by 6 to 32%) [279]. Slight increases in alkaline phosphatase and phosphotriesterase activities (by 2 to 20%, and by 6 to 13%, respectively) were found. Variations in enzyme activities following the application of different types of crop residues were attributed to stimulatory or inhibitory effects on the growth of the enzyme-producing microbes.

Several studies have demonstrated that the influence of crop residues and manures on enzyme activity in soil may also be affected by other factors, including soil aeration, availability of nutrients (mainly nitrogen compounds), and the presence of particular microorganisms. McCarty and Bremner [280] showed that under aerobic conditions microbial production of urease was markedly stimulated by plant residues and other readily decomposable organic materials, whereas no appreciable effects were detected under anaerobic conditions, although hydrolysis of urea by soil urease was not inhibited. Aerobic or anaerobic conditions used in processing manures before application affected their influence on the activities of dehydrogenase and nitrogenase in clay loam and sandy soils [281]. Fermented manures (maize stalks and cow dung anaerobically processed) usually produced higher dehydrogenase and lower nitrogenase activities in both soils than did aerobically processed manures.

The effect of a 55-year-old straw and nitrogen fertilization treatment on the activity of acid and alkaline phosphatase, arylsulfatase, β-glucosidase, urease, and amidase was determined in a winter wheat (*Triticum aestivum* L.) fallow system on semiarid soils of the Pacific Northwest [282]. Several treatments were examined: straw, straw with fall or spring burn, straw plus 45 or 90 kg N/ha, straw burned in spring plus 45 or 90 kg N/ha, and straw plus 2.24 T/ha of straw-manure. All enzyme activities were significantly ($P < 0.001$) affected (increases or decreases) by residue management. The highest activities were observed in the manure-treated soil (increases from 36 to 190%), whereas the lowest activities (alkaline phosphatase, amidase, and urease) occurred in soils treated with straw plus 90 kg N/ha.

The effect of introduced populations of diazotrophs (*Azospirillum brasilense* Sp7 and *Cellulomonas* sp. CS1-17) on the activity of nitrogenase (C_2H_2 reduction assay) was studied in 20 different soils from the New South Wales wheat belt amended with sterile straw [283]. In the absence of inoculation, all soils showed increased nitrogenase activity following incorporation of straw, indicating that indigenous diazotrophs were widely distributed, and cellulolytic organisms were present, capable of supplying the indigenous diazotrophs with energy from the breakdown of straw. No significant effect of coinoculation with *A. brasilense* and *Cellulomonas* was observed on nitrogenase activity.

Sludge application may constitute a useful practice for improving soil fertility; however, because sludge generally contains heavy metals, it is particularly important to study the effect of its application on the biochemical activity of soil in the absence and presence of vegetation. In fact, as previously reported, several enzyme activities may be affected at rhizosphere level by the presence of plant roots and associated microorganisms. In a greenhouse experiment, Reddy et al. [284] studied the activity of dehydrogenase, urease, and phosphatase in an Enon sandy loam soil amended with different quantities of sewage sludge before and after planting with ransom soybean (*Glycine max.* L. Merr.). Forty days after planting, soil (0 to 5 cm) and roots were removed; the soil was then separated from the roots and referred to as rhizosphere soil. Soil from plant-free experiments was also collected and referred to as nonrhizosphere soil. High sludge levels inhibited dehydrogenase and phosphatase activities, but the inhibitory effects were markedly diminished by the presence of plants. In contrast, sludge amendments enhanced urease activity in rhizosphere soils, probably as a result of the higher microbial proliferation and activity in the root zone.

The inhibitory effect of sewage sludge on soil dehydrogenase activity was later confirmed by measuring the enzyme activity in soil-sludge mixtures after various incubation times and with different sludge levels [285]: the greater the amount of sludge applied and the longer the incubation time, the lower the dehydrogenase activity.

High levels of biomass, nitrogen mineralization, and enzyme activities involved in nitrogen, phosphorus and carbon cycling (urease, alkaline phosphatase, and xylanase) were measured after long-term application of cattle slurry to grassland soils [286]. The cattle slurry improved enzyme production by microbial biomass. No significant effects on enzyme activities were found in a soil treated with clear, black, and white polyethylene mulches during growth of a strawberry crop [287], although enzyme activities showed temporal fluctuations. Protease and phosphatase activities initially increased, declined sharply in early spring, and then rose again, whereas sulfatase activity initially declined and then remained relatively constant during the rest of the study period (1 year). No significant differences in concentrations of biomass nitrogen, sulfur and phosphorus among treatments were detected, but seasonal fluctuations were observed. The absence of discernible relationships between seasonal fluctuations in enzyme activities and levels of biomass nitrogen, sulfur, and phosphorus, or extractable mineral nitrogen, sulfur, and phosphorus reflected the complexity of interactions among soil microbial and biochemical properties and nutrient availability.

Temporal fluctuations in the activities of dehydrogenase, amylase, phosphatase, urease, protease, deaminase, catalase, and arylsulfatase were examined in a clay loam soil amended with municipal solid-waste compost at various application levels over 1 and 3 years of incubation [288,289]. All activities increased after addition of the compost, and were greater at higher compost doses. Lower activities were observed following successive applications. The increases in enzyme activity were attributed to an increase in microorganisms, resulting either from the growth stimulating effects of the easily biodegradable organic matter in the compost or from the addition of exogenous microorganisms. The successive decreases were attributed to exhaustion of biodegradable organic matter or to the toxicity of undesirable materials introduced into the soil with the compost.

A field study examining the effect of repeated additions of several organic residues on the activity and persistence of enzyme activities in soil was recently conducted by Martens et al. [290]. The activities of 10 soil enzymes involved in the cycling of carbon, nitrogen, phosphorus, and sulfur were assayed in a coarse loamy soil amended with poultry manure, sewage sludge, barley straw (*Hordeum vulgare*), or fresh alfalfa (*Medicago sativa*) over a 31-month period. The enzyme activity in amended soil increased by an average of two- to fourfold when compared with unamended soil. Any additions after 1 year of organic amendments failed to sustain high enzyme activity. Straw most effectively enhanced the activity of all the enzymes except urease. Sewage sludge was not inhibitory. The enhanced enzyme activity was thought to reflect an increase in protective sites within the soil resulting from the elevated humus content. Two hypotheses were advanced to account for these results: (1) with repeated additions of organic residues the activities of enzymes were inhibited by a negative feedback mechanism, controlled by the supply of energy; and (2) the synthesis and release of enzymes by microorganisms were no longer needed because the fraction of immobilized enzymes increased.

Vegetation Cover. Enzyme activity in soil may be particularly affected by the presence and nature of the plant cover. Plants may behave as sources of enzymes or contribute to favorable conditions for microbial synthesis of enzymes.

Several studies have provided indirect evidence for the persistence of plant enzymes in soils. The activities of some carbohydrases (amylase, cellulase, and invertase) decreased, while those of dehydrogenase and urease increased, during two successional stages of fields revegetated with tall grass prairie, post oak forest (*Q. stellata, Q. marilandica*), and oak-pine (*Quercus, Pinus*) forest [179,291]. The type of vegetation, and thus the type of organic matter entering the soil during succession, was considered the chief determinant of the variability in activity of the enzymes under study.

Restoration of invertase, amylase, cellulase, xylanase, urease, phosphatase, and sulfatase activities was detected in a silt loam soil after the removal of the topsoil and reestablishment of a grass-clover pasture [292]. A rapid recovery of dehydrogenase activity occurred in soil cores that were fumigated with methyl bromide and then returned to their original pasture and forest sites [293]. No differences were observed between the two forest sites, whereas a consistently lower level of enzyme activity was found in the fumigated than in the untreated samples. The presence of vegetation was also critical for the recovery of microbial or enzymatic activities in mine soils varying in age and vegetation [182].

The urease, sulfatase, and protease activities of moistened, unplanted soils gradually decreased over 5 months, whereas increases were observed in soils planted with ryegrass [294]. In addition, higher arylsulfatase activities were found in the rhizosphere of climax vegetation in coastal sands than in samples from areas lacking vegetation cover [295].

The activity of enzymes in soil is influenced by the type of vegetation. For example, phosphomonoesterase and sulfatase activities differed with respect to vegetative zone: the greatest amount of enzyme activity occurred in arctic soils in which farb-grass vegetation dominated, and the lowest level was associated with soil from the fell-field zone [296]. Furthermore, the levels of both activities were generally higher than those found in Virginia temperate soils, dominated by single physiognomic vegetation, whether expressed on a dry weight or volume basis. A marked decrease in amylase activity was effected by the presence of grass and two types of crops (*Zea mays* L. and *Paspalum scrobiculatum* L.) according to the order: grass > maize > *Paspalum*; no significant differences were found in the activities of urease, protease, invertase, and cellulase in the three different botanical zones [297]. The presence of three *Azolla* fern species (*A. pinnata, A. filiculoides,* and *A. microphylla*), as well as *Lemna* and water hyacinth, increased the activity of amylase, invertase, dehydrogenase, and urease in soil from a flooded rice field [298]. The nature of vegetation (legumes, grass, pasture) markedly influenced the activities of starch- and sucrose-hydrolyzing enzymes [299] and urease [126,300].

In 16 diverse crop-soil ecosystems involving both leguminous and nonleguminous crops, the legume rhizosphere showed higher levels of acid, neutral, and alkaline phosphatase than did the rhizosphere of nonleguminous crops [301]. The rhizosphere of all the crops had higher levels of phosphatases than did the nonrhizosphere soils [170]. Soils from a permanent pasture and an alfalfa field showed significantly higher arylsulfatase activities than soils from cultivated wheat fields [59]. The levels of polyphenoloxidase and invertase activities increased in a soddy podzolic soil under potatoes and flax when subjected to crop rotation during long-term fertilization experiments [302].

Management. Various management practices are commonly used for increasing crop production, and there is increasing concern about the effect of these practices on enzyme activity in soils. In fact, enzyme activities in soil are presumed to be useful for discriminating among soil management treatments, probably because they are related to microbial biomass, which is sensitive to such treatments. However, this relationship may fail for repressible or inducible enzymes [15].

Soils receiving no-tillage treatments usually have exhibited higher enzyme activity levels than have tilled soils [303–305]. Long-term conventional tillage decreased the activities of phosphatase, dehydrogenase, urease, and protease in both topsoils and subsoils. However, the levels of enzyme activities were greater in subsoil than in topsoil samples [303–304]; tillage practices would have had a greater effect in the near-surface layers because of their influence on the quantity and quality of soil organic matter [305]. Dick [306] found significantly higher activities of acid phosphatase, arylsulfatase, invertase, amidase, and urease in no-tillage soils than in soils with conventional tillage. These effects were very conspicuous in the 0- to 7.5-cm profile soils, and were significantly affected by crop rotation, with the highest activities detected for the corn-oats-alfalfa rotation and the lowest for the corn-soybean rotation. In soil profiles (7.5 to 30 cm), the effect of tillage on enzyme activities varied with the soil and/or the enzyme studied. Tillage effects on enzyme activities in soil were generally correlated with microbial populations and the contents of organic carbon, total nitrogen, and water. Of the total variation in enzyme activities observed in no-till plots, Dick [306] calculated that 78 to 92% could be accounted for by the concentration of organic carbon.

No significant effects of tillage treatments on the activities of dehydrogenase, phosphatase, and urease were observed in soils planted with spring barley over a 4-year period [307]. Comparisons between the A_p horizon of agricultural soils and the A horizon of black spruce forest soils revealed greater enzyme activities in the forest soils [307]. Changes in aerobic and anaerobic microbial populations explained the differences in enzyme activity between tilled and no-till plots. Compared with soils subjected to conventional tillage practices, no-till soils usually present a less oxidative environment [6].

Waterlogging a soil often results in chemical and microbiological processes that influence nutrient cycling and the accumulation of toxins. As a result of waterlogging, a retardation in the gaseous exchange between soil and air, changes in microbial population and a decrease of soil redox potential (Eh) with the formation of a reduced environment may arise. All these variations may reflect on enzyme activities in soil; in particular, changes in redox potential should do so markedly because the reduced metal ions produced could act as enzyme inhibitors or activators of enzymes. Relationships between changes in enzyme activities and variation in Eh of soils were examined in soils after waterlogging. Pulford and Tabatabai [308] studied the activities of eight enzymes involved in carbon, nitrogen, phosphorus, and sulfur cycling in 10 Iowa soils before and after waterlogging for 7 days at 22°C. The percentage changes ranged from −59 to +192% for phosphatases (acid and alkaline phosphatase, phosphodiesterase, and pyrophosphatase); from −66 to −19% (average = −50%) for urease; from −31 to −5% (average = −21%) for arylsulfatase; from −43 to +43% for β-glucosidase; and up to 71% for amidase. Acid and alkaline phosphatase, urease, and arylsulfatase activities were significantly positively correlated, whereas the activities of phosphodiesterase and amidase were significantly, but negatively, correlated with Eh_7 (redox potential of soils normalized to pH 7). A significant reduction of arylsulfatase activity was also observed in waterlogged arctic soils [296].

Flooding of soil leads to changes in the microbial population; primarily increases in anaerobic microorganisms. Consequently, enhancement or inhibition of enzyme activity in soil may result. Dehydrogenase activity increased several-fold in three soils following flooding, whereas invertase activity decreased considerably [309]. However, the addition of rice straw enhanced the activity of invertase as well as that of dehydrogenase under both flooded and nonflooded conditions [309]. Higher rhodanese activities were measured in both flooded and nonflooded rhizosphere than in either flooded or nonflooded nonrhizosphere soils [310]. The activities of dehydrogenase and urease in soils planted with rice were measured under three different agricultural practices prevailing in hill regions (hill slope, terrace, and valley agriculture) [311]. The effects of these practices were examined for two cropping seasons. The valleys and terraces were kept flooded and the hill slopes were cultivated following dryland practices. The highest activities of both enzymes were measured in the valley soils, followed by terrace and hill slope sites.

Disturbances caused by vehicular traffic resulted in significant decreases in phosphomonoesterase, phosphodiesterase, and arylsulfatase activities in tundra soil ecosystems that were approaching water saturation, whereas no significant effects on enzyme activities were measured when vehicles disturbed soils from drier, well-drained sites [185,312]. Compacted skid trails in forest soils showed lower activity levels of dehydrogenase, phosphatase, arylsulfatase, and amidase than control soils [231]. Decreases in the activities of arylsulfatase and acid phosphatase in different aggregate size classes of an orthic brown chernozemic soil after 69 years of cultivation have also been demonstrated, and the declines in enzyme activities were directly related to decreases in microbial biomass and content of organic carbon [313].

Urease inhibitors. The effectiveness of urea-based fertilizers is limited by the rapid hydrolysis of urea to NH_3 and CO_2 by urease in soil. As a consequence, several problems in the use of urea as a fertilizer may occur: a rise in pH with severe losses of NH_3 through volatilization; accumulation of NO_2^- resulting from the inhibition of NO_2^--oxidizing microorganisms by the high concentration of NH_4^+ and high pH; and damage to germinating seedlings.

Several investigations have been devoted to finding compounds capable of controlling the hydrolysis rate of urea in soils, i.e., by inhibition of urease activity [49,314]. Numerous compounds capable of inhibiting soil urease have been found, and a variety of mechanisms responsible for the inhibition have been hypothesized. Cervelli et al. [315–317] tested substituted urea herbicides (fenuron, monuron, diuron, linuron, siduron, and neburon) as urease inhibitors. The inhibition of urea hydrolysis in three soils, which differed markedly in physicochemical properties, ranged from 9 to 39% for all the compounds, and a linear relationship between Hammet sigma values and log K_i (inhibition constant) was obtained. Kinetic analysis suggested the formation of a complex between herbicides and enzymes and a mixed-type inhibition mechanism. In this type of inhibition both V_{max} and K_m are modified by the inhibitor, and thus the inhibition takes on both competitive and noncompetitive characteristics. Studies performed on jack bean urease in the presence of the same herbicides confirmed this hypothesis [315] and suggested a possible effect of the soil matrix on formation of the herbicide-enzyme complex [317]. In fact, soil colloids may adsorb the herbicides, resulting in a decrease of their inhibitory effect [9].

An inhibition mechanism involving a thiol-disulfide exchange reaction between a disulfide and one or more of the urease sulphydryl groups was consistent with results obtained in studies on the inhibition of jack bean urease by heterocyclic sulfur compounds [318]. When compared with other known urease inhibitors in soil (e.g., quinones and related phenols, urea derivatives), only two heterocyclic mercaptans (1,3,4-thiadiazole-2,5-dithiol and 5-amino-1,3,4-thiadiazone-2-thiol) inhibited soil urease to a detectable degree (27 to 46% inhibition).

A simple and rapid method of evaluating the ability of various compounds (acetohydroxamic acid, sodium *p*-chloromercuribenzoate, hydroquinone, *p*-benzoquinone, and catechol) to inhibit urease activity in soil was proposed by Douglas and Bremner [319]. The method involved determination of the effect of the test compound on the amount of urea hydrolyzed during incubation of soil with urea and toluene added (the latter being added to reduce the possible proliferation of microorganisms during the activity assay). Experimental conditions (e.g., substrate concentration, incubation time, amount of toluene used, reaction volume, and incubation temperature) were selected to give a precise assay of urease activity. The method led to reproducible results that were not confounded by soil constituents and could be modified to study the inactivation rate of urease inhibitors in soils. A different approach for evaluating the effectiveness of urease inhibitors was used by Praveen-Kumar et al. [320], taking into consideration two factors: (1) urea inhibitors are only effective when both urea and inhibitors are present at the same site, and (2) the efficiency of the inhibitors can be reduced by their failure to move along the urea. The mobility and patterns of movement of urea and nine urease inhibitors (hydroquinone, catechol, phenylmercuric acetate, and several derivatives of quinones) in acid, alkaline, and saline soils were studied by soil thin-layer chromatography. Results showed that all inhibitors were less mobile than urea, which moved with the water front, and that the mobility and movement patterns of urease inhibitors were controlled by the functional groups and their positions in the molecules. Mobility increased with the introduction of a methyl group on the aromatic ring or a shift in the position of hydroxyl groups from *ortho* to *para* positions. Praveen-Kumar et al. [321] also studied the effect of the same compounds on thermodynamic parameters [activation energy (E_a), entropy (ΔS), and free energy (ΔF) of activation] of urea hydrolysis in the three soils. All tested inhibitors influenced the thermodynamic parameters, and similar changes were caused by structurally related compounds. E_a and ΔS generally increased in the presence of inhibitors. The increase in E_a suggested that the complex formed upon interaction of the inhibitor with the enzyme affected the bonding of substrate to the enzyme and the subsequent transformation of the product. The increase in ΔS was explained by a gradual unfolding of the enzyme molecule caused by the presence of the inhibitor.

The efficiency of urease inhibitors may vary markedly depending on soil type and cultural and management conditions. When phenylphosphorodiamidate

(PPD) and N-(*n*-butyl) thiophosphoric triamide (NBPT) were tested in moist and flooded soils in laboratory, greenhouse, and rice field experiments, the relative effectiveness of the two inhibitors varied considerably with moisture content, soil type, and presence or absence of light [322]. In general, while both inhibitors were ineffective in moist soils, PPD was a more effective inhibitor than NBPT in flooded soils. Moreover, inhibition appeared to be greater when the flooded soils were maintained under dark as compared to light conditions. Apparently, photosynthetic algae, stimulated by light, were affecting urea metabolism. Enhanced growth of algae stimulated the disappearance of urea, possibly because they contain other enzymes, such as urea carboxylase, which catalyze the decomposition of urea. Furthermore, NBPT increased the depth of penetration, immobilization, and retention of urea nitrogen in a flooded soil. In the greenhouse experiments, 35 days after ^{15}N-labeled urea application, not only was more nitrogen (about twofold) retained in the soil treated with NBPT than in the soil receiving urea alone, but also appreciable nitrogen had diffused to the 20- to 45-mm layer. An increased net immobilization of added labeled nitrogen into soil organic matter was also found. These effects, however, were not observed in rice field experiments and were not translated into increased nitrogen recovery in the plant-soil system or into grain yield because of a possible short-lived effect of inhibitors on urea hydrolysis. These results indicated that urease inhibitors certainly will not impede nitrogen loss from flooded rice fields [322]. The presence of PPD did not eliminate increases in the concentration and subsequent evolution of NH_3 in a waterlogged soil in the presence of rice, thus suggesting that degradation and/or inactivation of PPD may have occurred, thereby eliminating the inhibition of urease [323].

Laboratory experiments conducted in an Aquic Udifluvent soil in the presence of PPD, NBPT, and hydroquinone showed that the addition of urease inhibitors increased urea nitrogen immobilization by 5 to 30%, with NBPT having the stronger effect. Increasing the soil organic carbon content decreased the effect significantly [324]. In a greenhouse experiment, when catechol, hydroquinone, phenylmercuric acetate, and PPD were applied to an alkaline soil in which volatilization of NH_3 was the major route of nitrogen loss, there were no adverse effects on the germination of wheat, except in the case of hydroquinone. At harvest, hydroquinone and PPD increased grain yield by 20 and 25% and nitrogen uptake by 7.4 and 13.8%, respectively [325].

In studies comparing the effect of NBPT and PPD on urea hydrolysis by urease in seven Iowa soils and by urease preparations obtained from microbial (*Bacillus pasteurii*) and plant (jack bean) origins, PPD was less effective than NBPT in inhibiting the activity of urease in soil. In contrast, PPD was much more effective in inhibiting urea hydrolysis by plant and microbial urease [326]. Further studies [327] showed that NBPT, one of the most effective urease inhibitors among the phosphoroamides [328], was rapidly decomposed in soil to a compound more effective in inhibiting urease activity. After isolation by reverse-phase HPLC, mass spectral analysis revealed the active compound to be N-(*n*-butyl)phosphoric triamide, the oxon analog of NBPT [326,327]. Phenol, formed as a product of the decomposition of PPD in soil incubated at temperatures above 25°C, showed inhibitory effects equivalent to those observed with PPD [329]. Similarly, studies on the decomposition of urea phosphate in soils demonstrated that the phosphoric acid released from this compound acted as an inhibitor of urea hydrolysis by soil urease, and reduced losses of urea nitrogen as NH_3 [330].

The application of phenyl mercuric acetate, hydroquinone, and an alcohol extract of the kernel of neem (*Azadiracta indica*) to a sandy clay loam soil resulted, after an initial inhibitory effect, in increases in urease activity [331]. This was probably the result of breakdown or biochemical degradation of the inhibitors to simpler compounds that, acting as readily available sources of carbon favored microbial activity and thus the production of new extracellular microbial urease [331]. A decrease in inhibitory activity was also observed, over time, for the phenoxy (e.g., 2-phenoxy, 2,4-diphenoxy, and 2,4,6-triphenoxy) derivatives of the phosphazene compound, 2,2,4,4,6,6-hexaaminocyclotriphosphazatriene [332]. Inhibition after a 16-h incubation (immediate inhibition) decreased with an increase in the number of phenoxy substitutions, e.g., 95 and 22% for the monophenoxy and triphenoxy derivatives, respectively; the compound, phenyl phosphorodiamidate, by comparison, inhibited 100%. In contrast, the sustained inhibition (i.e., more than 4 days) tended to increase with an increase in the number of phenoxy substitutions.

Pesticides. Although pesticides may have a beneficial impact on agricultural productivity, there is considerable concern about the potential environmental hazard of these agrochemicals. Pesticides reach the soil primarily by direct application and secondarily as a result of nonagricultural activity (accidental spillage, industrial and domestic wastes, etc.). Once in soil, they usually influence soil biochemical and microbial processes.

Because of the important role of soil enzymes in nutrient cycling and soil fertility, numerous investigations have examined the side effects of pesticides on enzyme activities in soil. These studies have been reviewed by Cervelli et al. [9], by Wainwright [333], and by Schäffer [334]. In general, the effects of pesticides on the enzyme activities in soil appear to depend on several factors such as the chemical nature and dose of the pesticide, the type of enzyme and soil, and the type of experiment (field or laboratory). At normal field application levels, pesticides usually did not exhibit serious and appreciable effects (neither stimulation nor inhibition) on the activity of enzymes in soil [334] or on the microbial activity of soil [333]. However, at elevated concentrations, pesticides either increased or inhibited soil enzyme activity. These effects may have been the result of direct interactions between pesticide and enzyme molecules or of indirect interactions resulting from changes in the number or activity of microorganisms, which, in turn, resulted in altered levels of intra- and extracellular enzymes.

Contrasting and contradictory effects of the same pesticide on the same enzyme have often been reported [9,334 and the references therein]. A pesticide that inhibited an enzyme activity in one soil may have had no influence or may have increased the same activity in another soil. These results may depend on several factors: (1) the presence in soil of various active enzyme fractions (see Section II.A) that may change over time and, possibly, respond in different ways to the presence of additives; (2) complete decomposition or transformation of the applied pesticides to less toxic by-products; and (3) adsorption of pesticides by soil colloids resulting in a decrease in their concentration and, possibly, in their effect on enzyme activity.

Recently a new experimental approach for elucidating the nature of the interactions between pesticides and the catalytic behavior of enzymes in soil was proposed by Gianfreda et al. [335–337], who studied the effects of pesticides on model enzyme systems (free enzymes and enzymes immobilized on inorganic, organic, and organomineral soil colloids) to stimulate the extracellular enzyme components of soils.

The effects of pesticides on selected enzyme activities in soils with different physicochemical properties and from different sites were also considered. The results varied depending on the physical state of the enzyme, i.e., whether free in solution, immobilized on inorganic, organic, and organomineral supports in vitro, or present in soil. For example, glyphosate and paraquat enhanced the urease activity of soils, whereas no significant effects on the activity of jack bean urease, free in solution or adsorbed on montmorillonite, were found [336]. In contrast, the activity of urease associated with organic and organomineral matrices was enhanced by both pesticides. Thus, the increase of urease activity in soils was explained by assuming that in soil such enzyme associations predominated and the activity of both free and intracellular enzymes was probably not important under conditions of the assay. Conflicting results were obtained with two additional enzymes: invertase and acid phosphatase. Both pesticides markedly increased the activity of free invertase, whereas no effects on acid phosphatase were observed [335]. A general decrease was found for invertase immobilized on organic and organomineral matrices, and contrasting results were detected for invertase in soils. Both pesticides significantly enhanced the invertase activity of a loam soil (>30%), but inhibited this activity in a sandy loam soil by 33%, and left the activities of all other soils relatively unaltered [337].

VI. CONCLUSIONS

The activity of enzymes in soil is the result not only of the integration of complex processes of synthesis, persistence, stabilization, regulation, and catalytic behavior but also of location (intracellular or extracellular), inside microaggregates, or present on the surface of aggregates. All these processes may be dynamically interrelated and influenced by changes in the physical, chemical, and biological composition of soil. Furthermore, the detected activity reflects the actual enzyme activity of a soil enzyme only if a reliable methodology (e.g., satisfactory soil storage and treatment, optimum assay conditions) is adopted. As summarized in Figure 2, several factors may influence one or more of the processes implicated in the maintenance and manifestation of enzyme activity in soil. Ecological parameters (e.g., seasonal changes, geographic location, or hydrothermal soil regimes) usually affect enzyme activity levels by influencing both the production of enzymes by plants, microorganisms, and other soil biota and their persistence under natural conditions. Agricultural activities (e.g., addition of fertilizers, organic amendments, and pesticides; other management practices) may affect the chemical composition and structural characteristics of soil, which in turn will influence (1) the species composition and abundance of soil microorganisms and/or their metabolic activity, (2) the enhancement or suppression of enzyme production, and (3) the overall activity of an enzyme in soil.

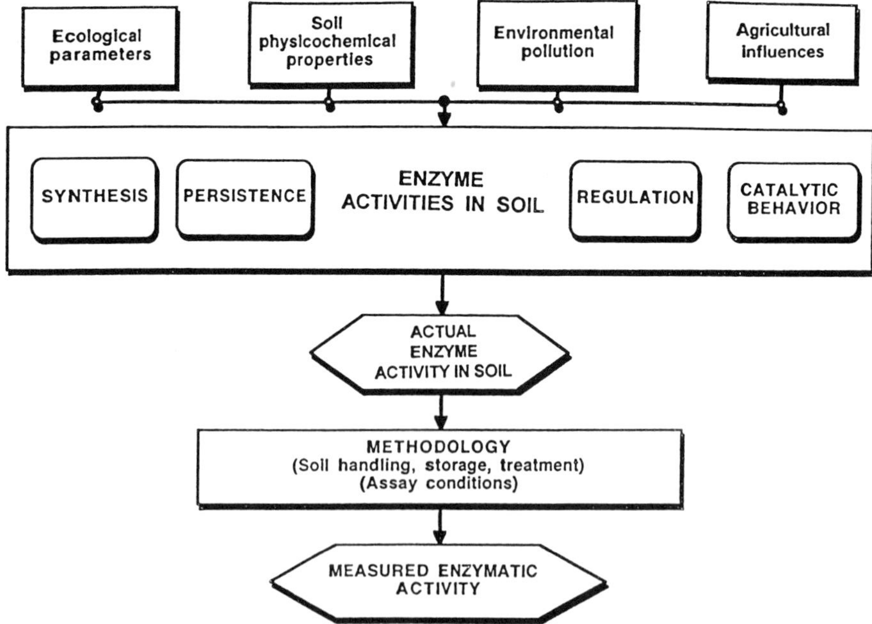

Figure 2 Factors influencing enzyme activities in soil.

The physical and chemical properties of a soil are involved in the immobilization and stabilization processes for most extracellular enzymes. High contents of clay or humus colloids are usually associated with stable but less active enzymes. Therefore, factors that influence or change the physical and chemical properties of soil will probably affect the immobilization, stability, and catalytic behavior of enzymes in soil.

The wide variability and frequent inconsistency of results reviewed in this chapter demonstrate that it is difficult to generalize cause-and-effect relationships between influencing factors and enzyme activities in soil. The biology, physics, and inorganic and organic chemistry of soil are highly complicated, and despite an extensive research effort, considerable scientific debate remains as to the specific mechanisms and phenomena involved, and the situations in which such factors will have a significant impact on the enzyme activity of soil.

In our opinion, much attention by soil enzymologists should be devoted to performing studies in the field of immobilized enzymes for biotechnological purposes. For example, it has been shown that, when operating in the presence of organic solvents, enzymes have much higher stability relative to that prevailing under free-solution conditions [111 and the references therein, 338–341]. These unusual properties of enzymes in organic solvents have been explained by assuming that only a small amount of water (a monolayer covering the protein) is essential to maintain the native catalytically active conformation. If the organic solvent is sufficiently less hydrophilic to prevent distortion or removal of the essential water layer from the protein surface, the enzyme molecules remain "frozen" in their active conformation and retain unaltered their catalytic activity. This picture is analagous in several ways to the molecular environment of enzymes in soil. The localization of enzymes in humus aggregates, or inside biological membranes or particular subcellular structures with physical properties and chemical composition much different from those of aqueous solutions, could account for most of the stability exhibited by enzyme activities in soil.

In conclusion, the statement that "It is difficult to imagine a more complicated environment for microbial activity than that of soil" [11] can also be extended to enzyme activity in soil. In this chapter an attempt has been made to present a general description of this broad and complex topic, and only a portion of the existing literature in this field could be mentioned. New advances in soil enzymology will be possible when methodological problems are solved (e.g., development of better methods for assaying the enzyme activity of soils). Further studies are needed to elucidate the role of enzymatic activities in the transformation of nutrients and xenobiotic substances; soil fertility and productivity; and the complex process of humus formation. A better understanding of the relationship between the activity, kinetic properties, and stability of enzymes in soil and the characteristics of immobilizing soil colloids is also required.

REFERENCES

1. Skujins, J. J. 1967. Enzymes in soils, p. 371–414. *In* A. D. McLaren and G. H. Peterson (eds.), Soil biochemistry, Vol.1. Marcel Dekker, New York.
2. Skujins, J. J. 1987. History of abiontic soil enzymes research, p. 1–49. *In* R. G. Burns (ed.), Soil enzymes. Academic Press, London.
3. Burns, R. G. (ed.). 1978. Soil enzymes. Academic Press, London.
4. Tabatabai, M. A. 1982. Soil enzymes, p. 903–947. *In* A. L. Page, R. H. Miller, and D. R. Keeney (eds.), Methods of soil analysis, Pt. 2. American Society of Agronomy, Madison, WI.
5. Chhonkar, P. K. 1985. Soil enzymes: activity and agricultural significance, p. 25–37. Proc. Soil Biol. Symp., Hisar.
6. Ladd, J. N. 1985. Soil enzymes, p. 175–221. *In* D. Vaugham and R. E. Malcolm (eds.), Soil organic matter and biological activity. Martinus Nijhoff/Dr. W. Junk Publishers, Dordrecht, Netherlands.
7. Kuprevich, V. F., and T. A. Shcherbakova. 1971. Comparative enzymatic activity in diverse types of soil, p. 167–201. *In* A. D. McLaren and J. Skujins (eds.), Soil biochemistry, Vol. 2. Marcel Dekker, New York.
8. Kiss, S., M. Dragan-Bularda, and D. Radulescu. 1975. Biological significance of enzymes accumulated in soil. Adv. Agron. 27:25–87.
9. Cervelli, S., P. Nannipieri, and P. Sequi. 1978. Interactions between agrochemicals and soil enzymes, p. 251–293. *In* R. G. Burns (ed.), Soil enzymes. Academic Press, London.
10. Burns, R. G. 1982. Enzyme activity in soil: Location and a possible role in microbial ecology. Soil. Biol. Biochem. 14:423–427.
11. Burns, R. G. 1983. Extracellular enzyme-substrate interactions in soil, p. 249–298. *In* J. H. Slater, R. Whittenburry, and J. W. T. Wimpemy (eds.), Microbes in their natural environments. Cambridge University Press, Cambridge.
12. Burns, R. G. 1986. Interaction of enzymes with soil mineral and organic colloids, p. 429–451. *In* P. M. Huang and M. Schnitzer (eds.), Interactions of soil minerals with natural organics and microbes. Special Publication No. 17. Soil Science Society of America, Madison, WI.
13. Nannipieri, P., S. Grego, and B. Ceccanti. 1990. Ecological significance of the biological activity in soil, p. 293–355. *In* J.-M. Bollag and G. Stotzky (eds.), Soil biochemistry, Vol. 6. Marcel Dekker, New York.
14. Tabatabai, M. A., and M. Fu. 1992. Extraction of enzymes from soils, p. 197–227. *In* G. Stozky and J.-M. Bollag (eds.), Soil biochemistry, Vol. 7. Marcel Dekker, New York.
15. Nannipieri, P. 1994. The potential use of soil enzymes as indicators of productivity, sustainability and pollution, p. 238–244. *In* C. E. Pankhurst, B. M. Doube, V. V. S. R. Gupta, and P. R. Grace (eds.), Soil biota: Management in sustainable farming systems. CSIRO, East Melbourne, Victoria, Australia.
16. Stotzky, G. 1986. Influence of soil mineral colloids on metabolic processes, growth, adhesion, and ecology and microbes and viruses, p. 305–428. *In* P. M. Huang and M. Schnitzer (eds.), Interactions of soil minerals with natural organics and microbes. Special Publication No. 17. Soil Science Society of America, Madison, WI.

17. Quastel, J.H. 1946. Soil metabolism. Royal Institute of Chemistry of Great Britain and Ireland, London.
18. Kiss, S., M. Dragan-Bularda, and D. Radulescu. 1978. Soil polysaccharidases: Activity and agricultural importance, p. 117–147. *In* R. G. Burns (ed.), Soil enzymes. Academic Press, London.
19. Ross, D. J., L. Hutt, and H. S. Roberts. 1973. Biochemical activities in a soil profile under hard beech forest. 1. Invertase and amylase activities and relationships with other properties. New Zeal. J. Sci. 16:209–224.
20. Ross, D. J. 1975. Studies on a climosequence of soils in tussock grasslands. 5. Invertase and amylase activities of topsoils and their relationships with other properties. New Zeal. J. Sci. 18:511–518.
21. Ross, D. J. 1975. Studies on a climosequence of soils in tussock grasslands. 6. Invertase and amylase activities of tussock plant materials and of soil. New Zeal. J. Sci. 18:519–526.
22. Ross, D. J. 1975. Studies on a climosequence of soils in tussock grasslands. 7. Distribution of invertase and amylase activities in soil fractions. New Zeal. J. Sci. 18:527–534.
23. Chhonkar, P. K., and Y. S. S. Rao. 1982. A kinetic study of cellulases in soils. J. Indian Soc. Soil Sci. 30:398–400.
24. Rhee, Y. H., Y. C. Hah, and S. W. Hong. 1987. Relative contributions of fungi and bacteria to soil carboxymethylcellulase activity. Soil Biol. Biochem. 19:479–481.
25. Hayano, K. 1986. Cellulase complex in a tomato field soil: Induction, localization and sole properties. Soil Biol. Biochem. 2:215–219.
26. Ueno, H., K. Miyashita, Y. Sawada, and Y. Oba. 1991. Assay of chitinase and *N*-acetylglucosaminidase activity in forest soils with 4-methylumbelliferyl derivatives. Zeitschr. Pflanzenk. Bodenk. 154:171–176.
27. Eivazi, F, and M. A. Tabatabai. 1990. Factors affecting glucosidase and galactosidase activities in soils. Soil Biol. Biochem. 22:891–898.
28. Eivazi, F, and M. A. Tabatabai. 1988. Glucosidases and galactosidases in soils. Soil Biol. Biochem. 20:601–606.
29. Rysavy, P., and J. Macura. 1972. The formation of β-galactosidase in soil. Folia Microbiol. 17:375–380.
30. Lethbridge, G. A., A. T. Bull, and R. G. Burns 1978. Assay and properties of 1,3-glucanase in soil. Soil Biol. Biochem. 10:389–391.
31. Lethbridge, G. A., A. T. Bull, and R. G. Burns. 1978. 1,3-glucanase activity in soil and its response to soil amendment. Rev. Ecol. Biol. Sol. 17:479–489.
32. Pancholy, S. K., and J. Q. Lynd. 1972. Quantitative fluorescence analysis of soil lipase activity. Soil Biol. Biochem. 4:257–259.
33. Mathur, S. P., J. I. MacDougall, and M. McGrath. 1980. Levels of activities of some carbohydrases, protease, lipase and phosphatase in organic soils of differing copper content. Soil Sci. 129:376–385.
34. Wong, K. K. Y., L. U. L. Tan, and J. N. Saddler. 1988. Multiplicity of 1,4-xylanase in microorganisms: Functions and applications. Microbiol. Rev. 52:305–317.
35. Sorensen, H. 1955. Xylanase in the soil and the rumen. Nature 176:74.
36. Sato, F., H. Omura, and K. Hayano. 1986. Adenosine deaminase activity in soils. Soil Sci. Plant Nutr. 32:107–112.

37. Frankenberger, W. T., and M. A. Tabatabai. 1980. Amidase activity in soils. I. Method of assay. Soil Sci. Soc. Am. J. 44:282–287.
38. Frankenberger, W. T., and M. A. Tabatabai. 1981. Amidase activity in soils IV. Effects of trace elements and pesticides. Soil Sci. Soc. Am. J. 45:1120–1124.
39. Frankenberger, W. T., and M. A. Tabatabai. 1981. Amidase activity in soils. III. Stability and distribution. Soil Sci. Soc. Am. J. 45:333–338.
40. Frankenberger, W. T., and M. A. Tabatabai. 1980. Amidase activity in soils. Kinetic parameters. Soil Sci. Soc. Am. J. 44:532–536.
41. Frankenberger, W. T., and M. A. Tabatabai. 1991. Factors affecting L-asparaginase activity in soils. Biol. Fertil. Soils 11:1–5.
42. Frankenberger, W. T., and M. A. Tabatabai. 1991. L-Asparaginase activity of soils. Biol. Fert. Soils 11:6–12.
43. Frankenberger, W. T., and M. A. Tabatabai. 1991. L-Glutaminase activity of soils. Soil Biol. Biochem. 23:869–875.
44. Frankenberger, W. T., and M. A. Tabatabai. 1991. Factors affecting L-glutaminase activity of soils. Soil Biol. Biochem. 23:875–879.
45. Ladd, J. N., and J. H. A. Butler. 1972. Short-term assays of soil proteolytic enzyme activities using proteins and dipeptide derivatives as substrates. Soil Biol. Biochem. 4:19–30.
46. Ladd, J. N. 1972. Properties of proteolytic enzymes extracted from soil. Soil Biol. Biochem. 4:227–237.
47. Mayaudon, J., L. Batistic, and J. M. Sarkar. 1975. Propriétés des activités protéolytiques extraited des sols frais. Soil Biol. Biochem. 7:281–286.
48. Pettit, N. M., A. R. J. Smith, R. B. Freedman, and R. Burns. 1976. Soil urease: Activity, stability and kinetic properties. Soil Biol. Biochem. 8:479–484.
49. Bremner, J. M., and R. L. Mulvaney. 1978. Urease activity in soils, p. 149-196. *In* R. G. Burns (ed.), Soil enzymes. Academic Press, London.
50. Perez Mateos, M., and S. Gonzales-Carcedo. 1985. Distribution of urease activity in soil structure units 2. State and behavior of the enzyme. Ann. Edaf. Agrobiol. 43:699–710.
51. Speir, T. W., and D. J. Ross. 1978. Soil phosphatase and sulphatase, p. 197–250. *In* R. G. Burns (ed.), Soil enzymes. Academic Press, London.
52. Hayano, K. 1977. Extraction and properties of phosphodiesterase from a forest soil. Soil Biol. Biochem. 9:349–351.
53. Browman, M. G., and M. A. Tabatabai. 1978. Phosphodiesterase activity of soils. Soil Sci. Soc. Am. J. 42:284–290.
54. Hayano, K. 1987. Characterization of a phosphodiesterase component in a forest soil extract. Biol. Fertil. Soils 3:159–164.
55. Dick, W. A., and M. A. Tabatabai. 1978. Inorganic pyrophosphatase activity of soils. Soil Biol. Biochem. 10:59–65.
56. Tabatabai, M. A., and W. A. Dick. 1979. Distribution and stability of pyrophosphatase in soils. Soil Biol. Biochem. 11:655–659.
57. Dick, W. A., and M. A. Tabatabai. 1983. Activation of soil pyrophosphatase by metal ions. Soil Biol. Biochem. 15:359–363.
58. Dick, W. A., N. G. Juma, and M. A. Tabatabai. 1983. Effects of soils on acid phosphatase and inorganic pyrophosphatase of corn roots. Soil Sci. 136:19–25.

59. Germida, J. J., M. Wainwright, and V. V. S. R. Gupta. 1992. Biochemistry of sulfur cycling in soil, p. 1–53. *In* G. Stotzky and J.-M. Bollag (eds.), Soil biochemistry, Vol. 7. Marcel Dekker, New York.
60. Tabatabai, M. A., and J. M. Bremner. 1970. Arylsulfatase activity of soil. Soil Sci. Soc. Am. Proc. 34:225–229.
61. Tabatabai, M. A., and J. M. Bremner. 1970. Factors affecting arylsulfatase activity of soil. Soil Sci. Soc. Am. Proc. 34:427–429.
62. Drobnik, J. 1956. Degradation of asparagine by the enzyme complex of soils. Cesk. Mikrobiol 1:47–51.
63. Chalvignac, M. A., and J. Mayaudon. 1971. Extraction and study of soil enzymes metabolising tryptophane. Plant Soil 34:25–31.
64. Frankenberger, W. T. 1983. Kinetic properties of L-histidine ammonia-lyase activity in soils. Soil Sci. Soc. Am J. 47:71–80.
65. Mayaudon, J., M. El Halfawi, and C. Bellinck. 1973. Decarboxylation des acides amines aromatiques-1-^{14}C par les extraits de sol. Soil Biol. Biochem. 5:355–367.
66. Galstyan, A. S., and L. G. Marukyan. 1973. Determination of ascorbate oxidase activity in the soil. Dokl. Akad. Nauk. Arm. 57:100–102.
67. Ross, D. J. 1974. Glucose oxidase activity in soil and its possible interference in assays of cellulase activity. Soil Biol. Biochem. 6:303–306.
68. Ross, D. J. 1968. Some observations on the oxidations of glucose by enzymes in soil in the presence of toluene. Plant Soil 28:1–9.
69. Mayaudon, J., and J. M. Sarkar. 1975. Laccases de *Polyporus versicolor* dans le sol et la litière. Soil Biol. Biochem. 7:31–34.
70. Leonowicz, A., and J.-M. Bollag. 1987. Laccases in soil and the feasibility of their extraction. Soil Biol. Biochem. 19:237–242.
71. Sjoblad, R. D., and J.-M. Bollag. 1981. Oxidative coupling of aromatic compounds by enzymes from soil microorganisms, p. 113–152. *In* E. A. Paul and J. N. Ladd, (eds.), Soil biochemistry, Vol 5. Marcel Dekker, New York.
72. Bollag, J.-M., and M. J. Loll. 1983. Incorporation of xenobiotics into soil humus. Experientia 39:1221–1231.
73. Bartha, R., and L. Bordeleau. 1969. Cell-free peroxidases in soil. Soil Biol. Biochem. 1:139–143.
74. Bollag, J.-M., C.-M. Chen, J. M. Sarkar, and M. L. Loll. 1987. Extraction and purification of a peroxidase from soil. Soil Biol. Biochem. 19:61–67.
75. Perez Mateos, M., S. Gonzales-Carcedo, and M. D. Busto Nunez. 1988. Extraction of catalase from soil. Soil Sci. Soc. Am. J. 52:408–411.
76. Stevenson, I. L. 1959. Dehydrogenase activity in soils. Can. J. Microbiol. 5:229–235.
77. Casida , L. E., D. A. Klein, and T. Santoro. 1964. Soil dehydrogenase activity. Soil Sci. 98:371–376.
78. Casida, L. E. 1977. Microbial metabolic activity in soil as measured by dehydrogenase determinations. Appl. Environ. Microbiol. 34:630–636.
79. Conrad, R., and W. Seiler. 1981. Decomposition of atmospheric hydrogen by soil microorganisms and soil enzymes. Soil Biol. Biochem 13:43–49.
80. Schuler, S., and R. Conrad. 1991. Hydrogen oxidation activities in soil as influenced by pH, temperature, moisture and season. Biol. Fertil. Soils 12:127–130.

81. Abdelmagid, H. M., and M. A. Tabatabai. 1987. Nitrate reductase activity of soils. Soil Biol. Biochem. 19:421–427.
82. Gajendiran, N., and A. Mahadevan. 1989. Nitrogenase activity associated with leaf, root, and rhizosphere soils in an Indian tropical forest. Biol. Fertil. Soils 8:71–74.
83. Tann, C. C., and J. Skujins. 1985. Soil nitrogenase assay by ^{14}C-acetylene reduction: Comparison with the carbon monoxide inhibition method. Soil Biol. Biochem. 17:109–112.
84. Martin-Smith, M. 1963. Uricolytic enzymes in soil. Nature 197:361–362.
85. Hoffman, G. 1959. Investigations on the synthetic effect of enzymes in soil. Z. Pflanzenernähr. Düng. Bodenkd. 85:193–201.
86. Hoffman, G. 1963. Synthetic effects of soil enzymes. Recent Progress in Microbiology, p. 230–234. Symposium 8th International Congress on Microbiology, Montreal.
87. Kiss, S., and M. Dragan-Bularda. 1972. Persistence of levansucrase in soil. Stud. Univ. Babes-Bolay Ser. Biol. 2:139–144.
88. Dragan-Bularda, M., and S. Kiss. 1972. Occurrence of dextransucrase in soil, p. 119–128. Symposium on Soil Biology, National Society of Soil Scientists, Bucharest.
89. Tabatabai, M. A., and B. B. Singh. 1976. Rhodanese activity in soils. Soil Sci. Soc. Am. J. 40:381–385.
90. Singh, B. B., and M. A. Tabatabai. 1978. Factors affecting rhodanese activity in soils. Soil Sci. 125:337–342.
91. Mayaudon, J. 1986. The role of carbohydrates in the free enzymes in soil, p. 263–309. *In* C. H. Fuchsman (ed.), Peat and water. Elsevier, New York.
92. Pazur, J. H., Y. Tominaga, L. S. Forsberg, and D. L. Simpson. 1980. Glycoenzymes: An unusual type of glycoprotein structure for a glucoamylase. Carbohydrate Res. 84:103–116.
93. Pollok, M. R. 1962. Exoenzymes, p. 121–178. *In* I. C. Gunsalus and R. Y. Stanier (eds.), The bacteria, Vol. 4. Academic Press, New York.
94. Halstead, R. L., and R. B. McKercher. 1975. Biochemistry and cycling of phosphorus, p. 31–63. *In* E. A. Paul and A. D. McLaren (eds.), Soil biochemistry, Vol. 4. Marcel Dekker, New York.
95. Cosgrove, D. J. 1967. Metabolism of organic phosphates in soil, p. 216–228. *In* A. D. McLaren and G. H. Peterson (eds.), Soil biochemistry, Vol. 1. Marcel Dekker, New York.
96. Dankelaker, B., and H. Marshner. 1992. In vivo demonstration of acid phosphatase activity in the rhizosphere of soil-grown plants. Plant Soil 144:199–205.
97. Burns, R. G., and J. A. Edwards. 1980. Pesticide breakdown by soil enzymes. Pestic. Sci. 11:506–512.
98. Trevors, J. T. 1984. Effect of substrate concentration, inorganic nitrogen, O_2 concentration, temperature and pH on dehydrogenase activity in soil. Plant Soil 77:285–293.
99. Trevors, J. T. 1984. Dehydrogenase activity in soil. A comparison between the INT and TTC assay. Soil Biol. Biochem. 16:673–674.

100. von Mersi, W., and F. Schinner. 1991. An improved and accurate method for determining the dehydrogenase activity in soils with iodonitrotetrazolium chloride. Biol. Fertil. Soils 11:216–220.
101. Stotzky, G., and R. G. Burns. 1982. The soil environment: Clay-humus-microbe interactions. p. 105–133. *In* R. G. Burns and J. H. Slater (eds.), Experimental microbial ecology. Blackwell Scientific, Oxford, England.
102. Skujins, J. J. 1973. Persistence of enzymes in soil, p. 395–408. *In* E. Ingerson (ed.), Proceedings of Symposium on Hydrogeochemistry and Biogeochemistry. Clarke Co., Washington.
103. Galstyan, A. S. 1982. Soil enzyme stability. Pochvovednie 2:108–110.
104. Beri, V., K. P. Goswami, and S. S. Brar. 1978. Urease activity and its Michaelis constant for soil systems. Plant Soil 49:105–116.
105. Tarafdar, J. C., and P. K. Chhonkar. 1978. Thermal sensitivity and Michaelis constants of acid, neutral and alkaline phosphatases in different soils and root exudates. Z. Pflanzenernähr. Bodenkd. 141:753–759.
106. Pal, S., and P. K. Chhonkar. 1979. Thermal sensitivity and kinetic properties of soil urease. J. Indian Soc. Soil Sci. 27:43–47.
107. Paulson, K. N., and L. T. Kurtz. 1970. Michaelis constant of soil urease. Soil Sci. Soc. Amer. Proc. 34:70–72.
108. McLaren, A. D. 1978. Kinetics and consecutive reaction of soil enzymes, p. 97–116. *In* R. G. Burns (ed.), Soil enzymes. Academic press, London.
109. McLaren, A. D., and L. Packer. 1970. Some aspects of enzyme reaction in heterogenous system. Adv. Enzymol. 33:245–308.
110. McLaren, A. D. 1975. Soil as a system of humus and clay immobilized enzymes. Chemica Scripta 8:97–99.
111. Gianfreda, L., and M. R. Scarfi. 1991. Enzyme stabilization: state of the art. Molec. Cellul. Biochem. 100:97–128.
112. Theng, B. K. G. 1979. Formation and properties of clay-polymer complexes. Elsevier, New York.
113. Harter, R. D., and G. Stotzky. 1971. Formation of clay-protein complexes. Soil Sci. Soc. Amer. Proc. 35:383–398.
114. Ladd, J. N., and J. H. Butler. 1975. Humus-enzyme systems and synthetic, organic polymer-enzyme analogs, p. 143–194. *In* E. A. Paul and A. D. McLaren (eds.), Soil biochemistry, Vol. 4. Marcel Dekker, New York.
115. Boyd, S. A., and M. M. Mortland. 1990. Enzyme interaction with clays and clay-organo matter complexes, p. 1–28. *In* J.-M. Bollag and G. Stotzky (eds.), Soil biochemistry, Vol. 6. Marcel Dekker, New York.
116. Hamzehi, E., and W. Pflug. 1981. Sorption and binding mechanisms of polysaccharide cleaving soil enzymes by clay minerals. Z. Pflanzenernähr. Bodenkd. 144:505–513.
117. Stotzky, G. 1980. Surface interactions between clay minerals and microbes, viruses, and soluble organics, and the probable importance of these interactions to the ecology of microbes in soil, p. 231–249. *In* R. C. W. Berkeley, J. M. Lynch, J. Melling, P. R. Rutter, and B. Vincents (eds.), Microbial adhesion to surfaces. Ellis Horwood Limited, Chichester, England.

118. Gianfreda, L., M. A. Rao, and A. Violante. 1991. Invertase (β-fructosidase): Effects of montmorillonite, Al-hydroxide and $Al(OH)_x$-montmorillonite complex on activity and kinetic properties. Soil Biol. Biochem. 23:581–587.
119. Gianfreda, L., M. A. Rao, and A. Violante. 1992. Adsorption, activity and kinetic properties of urease on montmorillonite, aluminium hydroxide and $Al(OH)_x$-montmorillonite complexes. Soil Biol. Biochem. 24:51–58.
120. Fusi, P., G. G. Ristori, L. Calamai, and G. Stotzky. 1989. Adsorption and binding of protein on "clean" (homoionic) and "dirty" (coated with Fe oxyhydroxides) montmorillonite, illite and kaolinite. Soil Biol. Biochem. 21:911–920.
121. Gianfreda, L., M. A. Rao, and A. Violante. 1993. Interactions of invertase with tannic acid, OH-Al-species and/or montmorillonite. Soil Biol. Biochem. 25:671–677.
122. Golimbert, V. E. 1982. Methods for studying soil enzyme activity. Pochvovedenie 6:127–130.
123. Speir, T. W., and D. J. Ross. 1981. A comparison of the effects of air drying and acetone dehydration on soil enzyme activities. Soil Biol. Biochem. 13:225–230.
124. Speir, T. W., and D. J. Ross. 1975. Effects of storage on the activities of protease, urease, phosphatase and sulphatase in three soils under pasture. New Zeal. J. Sci. 18:231–237.
125. Tu, C. M. 1982. Physical treatment and re-inoculation of soil effects on microorganisms and enzyme activities. Soil Biol. Biochem. 14:57–62.
126. Palma, R. M., and M. E. Conti. 1990. Urease activity in Argentine soils: Field studies and influence of sample treatment. Soil Biol. Biochem. 22:105–108.
127. Pancholy, S. K., and E. L. Rice. 1972. Effect of storage conditions on activities of urease, invertase, amylase and dehydrogenase in soil. Soil Sci. Soc. Amer. Proc. 36:536–537.
128. Ross, D. J. 1970. Effects of storage on dehydrogenase activities of soils. Soil Biol. Biochem. 2:55–61.
129. Ross, D. J., and B. A. McNeilly. 1973. Biochemical activities in a soil profile under hard beech forest. 3. Some factors influencing the activities of polyphenol-oxidising enzymes. New Zeal. J. Sci. 16:241–257.
130. Zvyagintsev, D. G., E. A. Vorob'eva, and E. B. Gviniashvili. 1987. Stabilization of proteases and invertase in soils. Vestn. Mosk. Univ. Ser. XVII Pochvoved. 2:59–63.
131. Shih, K. L., and K. A. Souza. 1978. Degradation of biochemical activity in soil sterilized by dry heat and gamma radiation. Origins Life 9:51–64.
132. Tiwari, S. C., B. K. Tiwari, and R. R. Mishra. 1988. Enzyme activities in soils: Effects of leaching, ignition, autoclaving and fumigation. Soil Biol. Biochem. 20:583–585.
133. Kaplan, D. L., and R. Hartenstein. 1979. Problems with toluene and the determination of extracellular enzyme activity in soils. Soil Biol. Biochem. 11:335–338.
134. Frankenberger, W. T., and J. B. Johanson. 1986. Use of plasmolytic agents and antiseptics in soil enzyme assays. Soil Biol. Biochem. 18:209–213.
135. Voets, J. P., M. Dedeken, and E. Bessems. 1965. The behaviour of some amino acids in gamma irradiated soils. Naturwissenschaften 52:476.
136. Skujins, J. J., L. Braal, and A. D. McLaren. 1962. Characterization of phosphatase in a terrestrial soil sterilized with an electron beam. Enzymol. 25:125–133.

137. Burns, R. G., L. J. Gregory, G. Lethbridge, and N. M. Pettit. 1978. The effect of γ-irradiation on soil enzyme stability. Experientia 34:301–302.

138. Burns, R. G. 1978. Enzyme activity in soil: Some theoretical and practical considerations, p. 295–340. *In* R. G. Burns (ed.), Soil enzymes. Academic Press, London.

139. Methods in enzymology, Vols. 1–244. 1955–1994. Academic Press, New York.

140. Galstyan, A. S. 1978. Standardization of methods for determining the activity of soil enzymes. Pochvovedenie 2:107–114.

141. Roberge, M. R. 1978. Appendix-Methodology of soil enzyme measurement and extraction, p. 341–370. *In* R. G. Burns (ed.), Soil enzymes. Academic Press, London.

142. Fitzgerald, J. W., M. E. Watwood, and F. A. Rose. 1985. Forest floor and soil arylsulfatase hydrolysis of tyrosine sulphate, an environmentally relevant substrate for the enzyme. Soil Biol. Biochem. 17:885–887.

143. Wirth, S. J., and G. A. Wolf. 1992. Micro-plate colorimetric assay for endo-acting cellulase, xylanase, chitinase, 1,3-glucanase and amylase extracted from forest soil horizons. Soil Biol. Biochem. 24:511–519.

144. Cabrera, M. L., D. E. Kissel, and B. R. Bock. 1991. Urea hydrolysis in soil: Effect of urea concentration and soil pH. Soil Biol. Biochem. 23:1121–1124.

145. Zantua, M. I., and J. M. Bremner. 1975. Comparison of methods of assaying urease activity in soils. Soil Biol. Biochem. 7:291–295.

146. Vorobe'yeva Y. A., and E. B. Gviniashvili. 1992. Stabilization of protease and invertase in the soil as a function of temperature and pH. Pochvovedenie 2:82–90.

147. Rastin, N., K. Rosenplaenter, and A. Hüttermann. 1988. Seasonal variation of enzyme activity and their dependence on certain soil factors in a beech forest soil. Soil Biol. Biochem. 20:637–642.

148. Kshattriya, S., G. D. Sharma, and R. R. Mishra. 1992. Enzyme activities related to litter decomposition in forest of different age and altitude in North East India. Soil Biol. Biochem. 24:265–270.

149. Cortez, J, P. Lossant, and G. Billes. 1972. L'activité biologique des sols dans les écosystèmes méditerranées. III. Activités enzymatiques. Rev. Ecol. Biol. Sol. 9:1–19.

150. Cortez, J., G. Billes, and P. Lossant. 1975. Étude comparative de l'activité biologique des sols sous peuplements arbustifs et herbacés de la ganigue méditerranéenne. II. Activités enzymatiques. Rev. Ecol. Biol. Sol. 12:141–156.

151. Kumar, J. D., G. D. Sharma, and R. R. Mishra. 1992. Soil microbial population number and enzyme activities in relation to altitude and forest degradation. Soil Biol. Biochem. 24:761–767.

152. Speir, T. W., and J. C. Cowling. 1991. Phosphatase activities of pasture plants and soils: Relationship with plant productivity and soil P fertility indices. Biol. Fertil. Soils 12:189–194.

153. Stojanovic, B. J. 1959. Hydrolysis of urea in soils as affected by season and by added urease. Soil Sci. 89:251–255.

154. Neal, J. L. 1990. Phosphatase enzyme activity at subzero temperatures in arctic tundra soils. Soil Biol. Biochem. 22:883–884.

155. Peterjohn, W. T. 1991. Denitrification: Enzyme content and activity in desert soils. Soil Biol. Biochem. 23:845–855.

156. McClaugherty, C. A., and A. E. Linkins. 1990. Temperature responses of enzymes in two forest soils. Soil Biol. Biochem. 22:29–33.
157. Khaziev, F. K., and I. K. Khanirov. 1983. Physico-geographical factors and soil enzyme activity. Pochvovedenie 11:57–65.
158. Myers, M. G., and J. W. McGarity. 1968. The urease activity in profiles of five great soil groups from Northern New South Wales. Plant Soil 28:25–37.
159. Speir, T. W., and D. J. Ross. 1990. Temporal stability of enzymes in a peatland soil profile. Soil Biol. Biochem. 22:1003–1005.
160. Dutzler-Franz, G. 1977. Enzyme activities of different soil types under the influence of some chemical and physical soil properties. J. Plant Nutr. Soil Sci. 140:329–350.
161. Ross, D. J. 1973. Biochemical activities in a soil profile under hard beech forest. 3. Some factors influencing oxygen uptake and dehydrogenase activities. New Zeal. J. Sci. 16:225–240.
162. Speir, T. W. 1976. Studies on a climosequence of soils in tussock grasslands. 8. Urease, phosphatase, and sulphatase activities of tussock plant materials and of soil. New Zeal. J. Sci. 19:383–387.
163. Speir, T. W. 1977. Studies on a climosequence of soils in tussock grasslands. 10. Distribution of urease, phosphatase, and sulphatase activities in soil fractions. New Zeal. J. Sci. 20:151–157.
164. Speir, T. W. 1977. Studies on a climosequence of soils in tussock grasslands. 11. Urease, phosphatase, and sulphatase activities of top soils and their relationships with other properties including plant available sulphur. New Zeal. J. Sci. 20:159–166.
165. Ford, G. W., D. J. Greenland, and J. M. Oades. 1969. Separation of light fraction from soils by ultrasonic dispersion in halogenated hydrocarbons containing a surfactant. J. Soil Sci. 20:291–296.
166. Perez Mateos, M., and S. Gonzales-Carcedo. 1987. Effect of fractionation on the enzymatic state and behavior of enzyme activities in different structural soil units. Biol. Fertil. Soils 4:151–154.
167. Rojo, M. J., S. Gonzales-Carcedo, and M. P. Mateos. 1990. Distribution and characterization of phosphatase and organic phosphorus in soil fractions. Soil Biol. Biochem. 22:169–174.
168. Haussling, M., and H. Marschner. 1989. Organic and inorganic soil phosphate and acid phosphatase activity in the rhizosphere of 80-year-old Norway spruce (*Picea abies* L. Karst.) trees. Biol. Fertil. Soils 8:128–133.
169. Chhonkar P. K., and J. C. Tarafdar. 1981. Characteristics and location of phosphatases in soil-plant system. J. Indian Soc. Soil. Sci. 29:215–219.
170. Tarafdar, J. C., and A. Junk. 1987. Phosphatase activity in the rhizosphere and its relation to the depletion of soil organic phosphorus. Biol. Fertil. Soils 3:199–204.
171. Fox, T. R., and N. B. Comerford. 1992. Rhizosphere phosphatase activity and phosphatase hydrolyzable organic phosphorus in two forest spodosols. Soil Biol. Biochem. 24:579–583.
172. Pal, S., and P. K. Chhonkar. 1981. Urease activity in relation to soil characteristics. Pedobiologia 21:152–158.
173. Dalal, R. C. 1975. Urease activity in some Trinidad soils. Soil Biol. Biochem. 7:5–8.

174. Zantua, M. I., L. C. Dumenil, and J. M. Bremner. 1977. Relationships between soil urease activity and other soil properties. Soil Sci. Soc. Am. J. 41:350–352.
175. Chhonkar, P. K., and J. C. Tarafdar. 1984. Accumulation of phosphatases in soils. J. Indian Soc. Soil. Sci. 32:266–272.
176. Trasar-Cepeda, M. C., and F. Gil-Sotres. 1987. Phosphatase activity in acid high organic matter soils from Galicia (NW Spain). Soil Biol. Biochem. 19:281–287.
177. De Prado, R., M. Tena, and J.-A. Pinilla. 1982. Relationship between soil phosphatase activity and organic matter content. Agronomie 2:539–544.
178. Trasar-Cepeda, M. C., and F. Gil-Sotres. 1988. Kinetics of acid phosphatase activity in various soils of Galicia (NW Spain). Soil Biol. Biochem. 20:275–280.
179. Pancholy, S. K., and E. L. Rice. 1973. Soil enzymes in relation to oil field succession: Amylase, cellulase, invertase, dehydrogenase, and urease. Soil Sci. Soc. Amer. Proc. 37:47–50.
180. Abramyan, S. A., and A. S. Galstyan. 1981. Acidic basic regulation of soil enzyme action. Pochvovedenie 5:39–45.
181. Dash, M. C., P. C. Mishia, R. K. Mohanty, and N. Bhatt. 1981. Effects of specific conductance and temperature on urease activity in some Indian soils. Soil Biol. Biochem. 13:73–74.
182. Stroo, H. F., and E. M. Jencks. 1982. Enzyme activity and respiration in minesoils. Soil Sci. Soc. Am. J. 46:548–553.
183. Ross, D. J. 1973. Some enzyme and respiratory activities of tropical soils from New Hebrides. Soil Biol. Biochem. 5:559–567.
184. Juma, N. J., and M. A. Tabatabai. 1978. Distribution of phosphomonoesterases in soils. Soil Sci. 126:101–108.
185. Herbein, S. A., and J. L. Neal. 1990. Phosphatase activity in arctic tundra soils disturbed by vehicles. Soil Biol. Biochem. 22:853–858.
186. Dick, W. A., and M. A. Tabatabai. 1984. Kinetic parameters of phosphatases in soils and organic waste materials. Soil Sci. 137:7–15.
187. El-Shinnawi, M. M., and S. A. El-Shimi. 1981. Enzyme activities of soil in relation to moisture content and salinity. Zbl. Bakt. II. Abt. 136:471–477.
188. West, A. W., G. P. Sparling, T. W. Speir, and J. M. Wood. 1988. Dynamics of microbial carbon nitrogen-flush and ATP and enzyme activities of gradually dried soils from a climosequence. Aust. J. Soil Res. 26:519–530.
189. Frankenberger, W. T., and F. T. Bingham. 1982. Influence of salinity on soil enzyme activities. Soil Sci. Soc. Amer. J. 46:1173–1177.
190. Gomah, A. H. M., S. I. Al Nahidh, and H. A. Amer. 1990. Amidase and urease activity in soil as affected by sludge salinity and wetting and drying cycles. J. Plant Nutr. Soil Sci. 153:215–218.
191. Kiss, S., M. Dragan-Bularda, and D. Pasca. 1986. Activity and stability of enzyme molecules following their contact with clay mineral surfaces. Stud. Univ. Babes-Boyai Biol. 2:3–29.
192. Katchalski, E., I. Silman, and R. Goldman. 1971. Effect of the microenvironment on the mode of action of immobilized enzymes. Adv. Enzymol. 34:445–537.
193. Claus, H., and Z. Filip. 1988. Behavior of phenoloxidases in the presence of clays and other soil-related adsorbents. Appl. Microbiol. Biotechnol. 28:506–511.

194. Claus, H., and Z. Filip. 1990. Effects of clays and other solids on the activity of phenoloxidases produced by some fungi and actinomycetes. Soil Biol. Biochem. 22:483–488.
195. Lähdesmäki, P., and R. Piispapen. 1992. Soil enzymology: Role of protective colloid system in the preservation of exoenzyme activities in soil. Soil Biol. Biochem. 24:1173–1177.
196. Novakova, J. 1988. Effect of clay minerals on the activity of soil enzymes. Folia Microbiol. 32:504–505.
197. Novakova, J., and R. Sisa. 1984. Effect of clays on the cellulolytic activity of soil. Zentralbl. Mikrobiol. 139:505–510.
198. Ladd, J. N., and J. H. A. Butler. 1969. Inhibitory effect of soil humic compounds on the proteolytic enzyme pronase. Aust. J. Soil. Res. 7:241–251.
199. Ladd, J. N., and J. H. A. Butler. 1969. Inhibition and stimulation of proteolytic enzyme activities by soil humic acids. Aust. J. Soil. Res. 7:253–261.
200. Butler, J. H. A., and J. N. Ladd. 1969. The effect of methylation of humic acids on their influence on proteolytic enzyme activity. Aust. J. Soil. Res. 7:263–268.
201. Ladd, J. N., and J. H. A. Butler. 1970. The effect of inorganic cations on the inhibition and stimulation of protease activity by soil humic acids. Soil Biol. Biochem. 2:33–40.
202. Ladd, J. N., and J. H. A. Butler. 1971. Inhibition by soil humic acids of native and acetylated proteolytic enzymes. Soil Biol. Biochem. 3:157–160.
203. Pflug, W. 1980. Effect of humic acids on the activity of two peroxidases. Z. Pflanzenernähr. Bodenkd. 143:432–440.
204. Pflug, W., and W. Ziechmann. 1981. Inhibition of malate dehydrogenase by humic acids. Soil Biol. Biochem. 13:293–299.
205. Mato, M. C., and J. Mendez. 1970. Inhibition of indoleacetic acid-oxidase by sodium humate. Geoderma 3:255–258.
206. Sarkar, J. M., and J.-M. Bollag. 1987. Inhibitory effect of humic and fulvic acids on oxidoreductases as measured by the coupling of 2,4-dichlorophenol to humic substances. Sci. Tot. Environ. 62:367–377.
207. Gianfreda, L., and J.-M. Bollag. 1994. Effect of soil on the behavior of immobilized enzymes. Soil Sci. Soc. Am. J. 58:1672–1681.
208. Gianfreda, L., M. A. Rao, and A. Violante. 1995. Formation and activity of urease-tannate complexes as affected by different species of Al, Fe and Mn. Soil Sci. Soc. Am. J. 59:805–810.
209. Gianfreda, L., A. De Cristofaro, M. A. Rao, and A. Violante. 1995. Kinetic behavior of synthetic organo- and organo-mineral-urease complexes. Soil Sci. Soc. Am. J. 59: 811–815.
210. Burns, R. G., M. H. El-Sayed, and A. D. McLaren. 1972. Extraction of an urease-active organo complex from soil. Soil Biol. Biochem. 4:107–108.
211. Ceccanti, B., P. Nannipieri, S. Cervelli, and P. Sequi. 1978. Fractionation of humus-urease complexes. Soil Biol. Biochem. 10:39–45.
212. Nannipieri, P., B. Ceccanti, S. Cervelli, and P. Sequi. 1978. Stability and kinetic properties of humus-urease complexes. Soil Biol. Biochem. 10:143–147.
213. Nannipieri, P., B. Ceccanti, and D. Bianchi. 1988. Characterization of humus-phosphatase complexes extracted from soil. Soil Biol. Biochem. 20:683–691.

214. Maignan, C. 1982. Activité des complexes acides humiques-invertase: Influence du mode de préparation. Soil Biol. Biochem. 14:439–445.

215. Maignan, C. 1983. Activité des complexes acides humiques-invertase: Influence des cations flocculants. Soil Biol. Biochem. 15:651–659.

216. Maignan, C. 1990. Paramètres cinétiques de complexes humates-Fe^{3+}-invertase. C.R. Acad. Sci. Paris t.310, Série III:571–576.

217. Serban, A., and A. Nissenbaum. 1986. Humic acid association with peroxidase and catalase. Soil Biol. Biochem. 18:41–44.

218. Sarkar, J. M. 1986. Formation of [^{14}C]cellulase-humic complexes and their stability in soil. Soil Biol. Biochem. 18:251–254.

219. Ruggiero, P., and V. M. Radogna. 1988. Humic acids-tyrosinase interactions as a model of soil humic-enzyme complexes. Soil Biol. Biochem. 20:353–359.

220. Rowell, M. J., J. N. Ladd, and E. A. Paul. 1973. Enzymically active complexes of proteases and humic acid analogues. Soil Biol Biochem. 5:699–703.

221. Sarkar, J. M., and R. G. Burns. 1984. Synthesis and properties of β-D-glucosidase-phenolic copolymers as analogues of soil humic-enzyme complexes. Soil Biol. Biochem. 16:619–625.

222. Grego, S., A. D'Annibale, M. Luna, L. Badalucco, and P. Nannipieri. 1990. Multiple forms of synthetic pronase-phenolic copolymers. Soil Biol. Biochem. 22:721–724.

223. Müller-Wegener, U. 1988. Interaction of humic substances with biota, p. 170–192. *In* F. H. Frimmel and R. F. Christman (eds.), Humic substances and their role in the environment. John Wiley & Sons, New York.

224. Abramyan, S. A. 1993. Variation of enzyme activity of soil under the influence of natural and anthropogenic factors. Eurasian Soil Sci. 25:57–73.

225. Frankenberger, W. T., and W. A. Dick. 1983. Relation between enzyme activities and microbial growth and activity indices in soil. Soil Sci. Soc. Am. J. 47:945–951.

226. Kanazawa, S., and Z. Filip. 1986. Distribution of microorganisms total biomass and enzyme activities in different particles of brown soil. Microb. Ecol. 12:205–216.

227. Nannipieri, P., F. Pedrazzini, P. G. Arcara, and C. Piovanelli. 1979. Changes in amino acids, enzyme activities, and biomasses during soil microbial growth. Soil Sci. 127:26–34.

228. Kuroshima, T., and K. Hayano. 1982. Phospholipase C activity in soil. Soil Sci. Plant. Nutr. 28:535–542.

229. Hayano, K., and K. Tubaki. 1985. Origin and properties of β-glucosidase activity of tomato-field soil. Soil Biol. Biochem. 17:553–557.

230. Macura, J. 1979. Regulation of enzyme formation and activity in microbial populations in soil. Izv. Akad. Nauk SSSR Ser. Biol. 1:81–87.

231. Dick, R. P., D. D. Myrold, and E. A. Kerle. 1988. Microbial biomass and soil enzyme activities in compacted and rehabilitated skid trail soils. Soil Sci. Soc. Am. J. 52:512–516.

232. Reuss, J. O., and D. W. Johnson. 1986. Acid deposition and the acidification of soils and water. Springer-Verlag, New York.

233. Alexander, M. 1980. Effects of acidity on microorganisms and microbial processes in soil, p. 363–374. *In* T. C. Hutchinson and M. Havas (eds.), Effects of acid precipitation on terrestrial ecosystems. Plenum, New York.

234. Tabatabai, M. A. 1985. Effect of acid rain on soils. CRC Crit. Rev. Environ. Contr. 15:65–110.

235. Babich, H., D. L. Davis, and G. Stotzky. 1980. Acid precipitation: Part I. Cause and effects. Environment 22:6–13.

236. Bewley, R. J. F., and G. Stotzky. 1983. Simulated acid rain (H_2SO_4) and microbial activity in soil. Soil Biol. Biochem. 15:425–429.

237. Bitton, G., and R. A. Boylan. 1985. Effect of acid precipitation on soil microbial activity. 1. Soil core studies. J. Environ. Qual. 14:66–69.

238. Bitton, G., B. G. Volk, D. A. Graetz, J. M. Bossart, R. A. Boylan, and G. E. Byers. 1985. Effect of acid precipitation on soil microbial activity. II. Field studies. J. Environ. Quality 14:69–71.

239. Wainwright, M. 1979. Microbial S-oxidation in soils exposed to heavy atmospheric pollution. Soil Biol. Biochem. 11:95–98.

240. Killham, K., M. K. Firestone, and J. G. McColl. 1983. Acid rain and soil microbial activity: Effects and their mechanisms. J. Environ. Qual. 12:133–137.

241. Reddy, G. B, R. A. Reinert, and G. Eason. 1991. Enzymatic changes in the rhizosphere of loblolly pine exposed to ozone and acid rain. Soil Biol. Biochem. 23:1115–1120.

242. Press, M. C., J. Henderson, and J. A. Lee. 1985. Arylsulfatase activity in peat in relation to acidic deposition. Soil Biol. Biochem. 17:99–103.

243. Jarvis, B. W., G. E. Lang, and R. K. Wieder. 1987. Arylsulfatase activity in peat exposed to acid precipitation. Soil Biol. Biochem. 19:107–109.

244. Babich, H., and G. Stotzky. 1985. Heavy metal toxicity to microbe-mediated ecologic processes: a review and potential application to regulatory policies. Envir. Res. 36:111–137.

245. Juma, N. G., and M. A. Tabatabai. 1977. Effects of trace elements on phosphatase activity in soils. Soil Sci. Soc. Am. J. 41:343–346.

246. Al-Khafaji, A. A., and M. A. Tabatabai. 1979. Effects of trace elements on arylsulfatase in soils. Soil Sci. 127:129–133.

247. Fu, M. H., and M. A. Tabatabai. 1989. Nitrate reductase activity in soils: effects of trace elements. Soil Biol. Biochem. 21:943–946.

248. Tyler, G. 1974. Heavy metal pollution and soil enzyme activity. Plant and Soil 41:303–311.

249. Perez Mateos, M., and S. Gonzales-Carcedo. 1987. Effect of silver, cadmium and lead on soil enzyme activity. Rev. Ecol. Biol. Sol. 24:11–18.

250. Tyler, G. 1976. Influence of vanadium on soil phosphatase activity. J. Environ. Qual. 5:216–217.

251. Babich, H., and G. Stotzky. 1983. Developing standards for environmental toxicants: the need to consider abiotic environmental factors and microbe mediated ecologic processes. Environ. Health Persp. 49:315–323.

252. Doelman, P., and L. Haanstra. 1989. Short and long-term effects of heavy metals on phosphatase activity in soils: An ecological dose-response model approach. Biol. Fertil. Soils 8:235–241.

253. Doelman, P., and L. Haanstra. 1986. Short and long-term effects of heavy metals on urease activity in soils. Biol. Fertil. Soils 2:213–218.

254. Badura, L., J. Pacha, and U. Sliwa. 1980. Effect of zinc and copper on soil enzyme activity. Acta Biol. Katowice 375:128–142.

255. Krasnova, N. M. 1983. Enzyme activity as a bio-indicator of soil contamination with heavy metals. Dokl. Vses. Ordena Lenina Akad. KH. Nauk Im. V I Lenina 7: 41–43.

256. Simon, L., E. Bergerova, and V. Ciganekova. 1984. Effect of some heavy metals on soil microflora. Folia Microbiologica 29:409–412.

257. Chander, K., and P. C. Brookes. 1991. Is the dehydrogenase assay invalid as a method to estimate microbial activity in copper-contaminated soils? Soil Biol. Biochem. 23:909–915.

258. Dragan-Bularda, M., S. Kiss, and D. Radulescu. 1979. Influence of enzyme substrates on microbial amylase production in soil. Stud. Univ. Babes-Bolyai Biol. 24:64–70.

259. Tateno, M. 1988. Limitations of available substrates for the expression of cellulase and protease activities in soil. Soil Biol. Biochem. 20:117–118.

260. Maskova, H. P., and F. Kung. 1988. Microbial decomposition of carboxymethyl cellulose continuously added to the soil. Folia Microbiol. 33:474–481.

261. Ross, D. J., and H. S. Roberts. 1970. Enzyme activities and oxygen uptakes of soils under pasture in temperature and rainfall sequences. J. Soil Sci. 21:368–381.

262. Tarafdar, J. C., and N. Claassen. 1988. Organic phosphorus compounds as a phosphorus source for higher plants through the activity of phosphatases produced by plant roots and microorganisms. Biol. Fertil. Soils 5:308–312.

263. Aescht, E., and W. Foissner. 1992. Effects of mineral and organic fertilizers on the microfauna in a high-altitude reafforestation trial. Biol Fertil. Soils 13:17–24.

264. Dragan-Bularda, M. G. Blaga, S. Kiss, D. Pasca, V. Gherasim, and R. Vulvan. 1987. Effect of long-term fertilization on enzyme activities in a technogenic soil resulting from the recultivation of iron strip mine soils. Stud. Univ. Babes-Bolyai Biol. 32:47–52.

265. Haynes, R. J. and R. S. Swift. 1988. Effect of lime and phosphate addition on changes in enzyme activities, microbial biomass and levels of extractable nitrogen, sulphur and phosphorus in an acid soil. Biol. Fertil. Soils 6:153–158.

266. Trasar-Cepeda, M. C. and T. Barballas. 1991. Liming and the phosphatase activity and mineralization of phosphorus in an andic soil. Soil Biol. Biochem. 23:209–215.

267. Badalucco, L., S. Grego, S. Dell'Orco, and P. Nannipieri. 1992. Effect of liming on some chemical, biochemical, and microbiological properties of acid soils under spruce (*Picea abies* L.). Biol. Fertil. Soils 14:76–83.

268. Martinez-Toledo, M. V., J. Gonzales-Lopez, T. de la Rubia, J. Moreno, and A. Ramos-Cormenzana. 1988. Effect of inoculation with *Azotobacter chroococcum* on nitrogenase activity of *Zea mays* roots grown in agricultural soils under aseptic and non-sterile conditions. Biol. Fertil. Soils 6:170–173.

269. Simek, M., and V. Pizl. 1989. The effect of earthworms (*Lumbridicidae*) on nitrogenase activity in soil. Biol. Fertil. Soils 7:370–373.

270. Ladd, J. N., and E. A. Paul. 1973. Changes in enzymic activity and distribution of acid-soluble amino acid-nitrogen in soil during nitrogen immobilization and mineralization. Soil Biol. Biochem. 5:825–840.

271. Ritz, K., B. S. Griffiths, and R. E. Wheatley. 1992. Soil microbial biomass and activity under a potato crop fertilized with N with and without C. Biol. Fertil. Soils 12:265–271.

272. McCarty, G. W., D. R. Shogren, and J. M. Bremner. 1992. Regulation of urease production in soil by microbial assimilation of nitrogen. Biol. Fertil. Soils 12:261–264.

273. Anghinoni, I., J. R. Magalhaes, and S. A. Narber. 1988. Enzyme activity nitrogen uptake and corn growth as affected by ammonium concentration in soil solution. J. Plant Nutr. 11:131–144.

274. Spiers, G. A., and W. B. McGill. 1979. Effects of phosphorus addition and energy supply on acid phosphatase production and activity in soils. Soil Biol. Biochem. 11:3–8.

275. Asmar, F., F. Eiland, and N. E. Nielsen. 1992. Interrelationship between extracellular enzyme activity, ATP content, total counts of bacteria and carbon dioxide evolution. Biol. Fertil. Soils 14:288–292.

276. Nakas, J. P., W. D. Gould, and D. A. Klein. 1987. Origin and expression of phosphatase activity in a semi-arid grassland soil. Soil Biol. Biochem. 19:13–18.

277. Perucci, P., and L. Scarponi. 1983. Effect of crop residue addition on arylsulphatase activity in soils. Plant Soil 73:323–326.

278. Perucci, P., L. Scarponi, and M. Businelli. 1984. Enzyme activities in a clay-loam soil amended with various crop residues. Plant Soil 81:345–352.

279. Perucci, P., and L. Scarponi. 1985. Effect of different treatments with crop residues on soil phosphatase activity. Biol. Fertil. Soils 1:111–115.

280. McCarty, G. W., and J. M. Bremner. 1991. Production of urease by microbial activity in soils under aerobic and anaerobic conditions. Biol Fertil. Soils 11:228–230.

281. El-Shinnawi, M. M., El-Shimi S. A., and M. A. Badawi. 1988. Enzyme activities in manured soils. Biol. Wastes 24:283–296.

282. Dick, R. P., P. E. Rasmussen, and E. A. Kerle. 1988. Influence of long-term residue management on soil enzyme activities in relation to soil chemical properties of a wheat-fallow system. Biol Fertil. Soils 6:159–164.

283. Halsall, D. M., and A. H. Gibson. 1991. Nitrogenase activity (C_2H_2 reduction) in straw-amended wheat belt soils in response to diazotroph inoculation. Soil Biol. Biochem. 23:987–998.

284. Reddy, G. B., A. Faza, and R. Bennett. 1987. Activity of enzymes in rhizosphere and non-rhizosphere soils amended with sludge. Soil Biol. Biochem. 19:203–206.

285. Reddy, G. B., and A. Faza. 1989. Dehydrogenase activity in sludge amended soil. Soil Biol. Biochem. 21:327.

286. Kandeler, E., and G. Eder. 1993. Effect of cattle slurry in grassland on microbial biomass and on activities of various enzymes. Biol. Fertil. Soils 16:249–254.

287. Haynes, R. J. 1987. The use of polyethylene mulches to change soil microclimate as revealed by enzyme activity and biomass nitrogen, sulfur and phosphorus. Biol. Fertil. Soils 5:235–240.

288. Perucci, P. 1990. Effect of the addition of municipal solid-waste compost on microbial biomass and enzyme activities in soil. Biol. Fertil. Soils 10:221–226.

289. Perucci, P. 1992. Enzyme activity and microbial biomass in a field soil amended with municipal refuse. Biol. Fertil. Soils 14:54–60.

290. Martens, D. A., J. B. Johanson, and W. T. Frankenberger. 1992. Production and persistence of soil enzymes with repeated addition of organic residues. Soil Sci. 153:53–61.

291. Pancholy, S. K., and E. L. Rice. 1973. Carbohydrases in soil as affected by successional stages of revegetation. Soil Sci. Soc. Am. Proc. 37:227–229.

292. Ross, D. J., T. W. Speir, K. R. Tate A. Cairns, K. F. Meyrick, and E. A. Pansier. 1982. Restoration of pasture after topsoil removal: effect on soil carbon and nitrogen mineralization, microbial biomass and enzyme activities. Soil Biol. Biochem. 14:575–581.

293. Yeates, G. W., S. S. Bamforth, D. J. Ross, K. R. Tate, and G. P. Sparling. 1991. Recolonization of methyl bromide sterilized soils under four different field conditions. Biol. Fertil. Soils 11:181–189.

294. Speir, T. W., E. A. Pansier, and A. Cairns. 1980. A comparison of sulphatase, urease and protease activities in planted and in fallow soils. Soil Biol. Biochem. 12:281–291.

295. Skiba, K., and M. Wainwright. 1983. Assay and properties of some sulphur enzymes in coastal sands. Plant Soil 70:125–132.

296. Neal, J. L. 1982. Abiontic enzymes in arctic soils: Influence of predominant vegetation upon phosphomonoesterase and sulphatase activity. Commun. Soil Sci. Plant Anal. 13:863–878.

297. Mahanty, R. K., and S. Padhan. 1992. Comparative studies on soil enzyme activities under two types of crops and adjacent grassland vegetation. Trop. Ecol. 33:205–213.

298. Subraman, S., and S. Kannayan. 1989. The effect of azolla ferns on the enzyme activity of soil under rice. Mikrobiologiya (Russian) 58:118–121.

299. Ross, D. J. 1966. A survey of activities of enzymes hydrolysing sucrose and starch in soils under pasture. J. Soil Sci. 17:1–15.

300. Reynolds, C. M., D. C. Wolf, and J. A. Armbruster. 1985. Factors related to urea hydrolysis in soils. Soil Sci. Soc. Am. J. 49:104–108.

301. Tarafdar, J. C., and P. K. Chhonkar. 1978. Status of phosphatases in the root-soil interface of leguminous and non-leguminous crops. Z. Pflanzenernähr. Bodenkd. 141:347–351.

302. Lykov, A. M., A. F. Safonov, T. Zakuan, and M. A. Zolotarev. 1987. Enzyme activity of soddy-podzolic soil under potatoes and flax in continuous cultivation of the same crop and in crop rotation during long-term fertilization. Izv. Tiniryazevskoi Sel'skohozyaistvennoi Akad. 1:25–32.

303. Doran, J. W. 1980. Soil microbial and biochemical changes associated with reduced tillage. Soil Sci. Soc. Am. J. 44:765–771.

304. Klein, T. M., and J. S. Koths. 1980. Urease, protease and acid phosphatase in soil continuously cropped to corn by conventional or no-tillage methods. Soil Biol. Biochem. 12:293–294.

305. Angers, D. A., B. Bissonnette, A. Legere, and N. Samson. 1993. Microbial and biochemical changes induced by rotation and tillage in a soil under barley production. Can. J. Soil Sci. 73:39–50.

306. Dick, W. A. 1984. Influence of long-term tillage and crop rotation combinations on soil enzyme activities. Soil Sci. Soc. Am. J. 48:569–574.

307. Cochran, V. L., L. F. Elliott, and C. E. Lewis. 1989. Soil microbial biomass and enzyme activity in subarctic agricultural and forest soils. Biol. Fertil. Soils 7:283–288.

308. Pulford, I. D., and M. A. Tabatabai. 1988. Effect of waterlogging on enzyme activities in soils. Soil Biol. Biochem. 20:215–219.

309. Chendrayan, K., T. K. Adhya, and N. Sethunathan. 1980. Dehydrogenase and invertase activities of flooded soils. Soil Biol. Biochem. 12:271–273.

310. Ray, R. C., N. Behera, and N. Sethunthan. 1985. Rhodanese activity of flooded and no-flooded soils. Soil Biol. Biochem. 17:159–162.

311. Tiwari, M.B., B. K. Tiwari, and R. R. Mishra. 1989. Enzyme activity and carbon dioxide evolution from upland and wetland rice soils under three agricultural practices in hilly regions. Biol. Fertil. Soils 7:359–364.

312. Neal, J. L., and S. A. Herbein. 1983. Abiotic enzymes in arctic soils: Changes in sulphatase activity following vehicle disturbance. Plant Soil 70:423–427.

313. Gupta, V. V. S. R., and J. J. Germida. 1988. Distribution of microbial biomass and its activity in different soil aggregate size classes as affected by cultivation. Soil Biol. Biochem. 20:777–786.

314. Mulvaney, R. L., and J. M. Bremner. 1981. Control of urea transformation in soils. p. 153–196. *In* E. A. Paul and J. N. Ladd (eds.), Soil biochemistry, Vol. 5. Marcel Dekker, New York.

315. Cervelli, S., P. Nannipieri, G. Giovannini, and A. Perna. 1975. Jack bean urease inhibition by substituted ureas. Pest. Biochem. Physiol. 5:221–225.

316. Cervelli, S., P. Nannipieri, G. Giovannini, and A. Perna. 1976. Relationships between substituted urea herbicides and soil urease activity. Weed Res. 16:365–368.

317. Cervelli S., P. Nannipieri, G. Giovannini, and A. Perna. 1977. Effect of soil on urease inhibition by substituted urea herbicides. Soil Biol. Biochem. 9:393–396.

318. Gould, W. D., F. D. Cook, and J. A. Bulat. 1978. Inhibition of urease activity by heterocyclic sulfur compounds. Soil Sci. Soc. Am. J. 42:66–72.

319. Douglas, L. A., and J. M. Bremner. 1971. A rapid method of evaluating different compounds as inhibitors of urease activity in soils. Soil Biol. Biochem. 3:309–315.

320. Praveen-Kumar, P. K. Chhonkar, and N. P. Agnihotri. 1987. Mobility of urease inhibitors: Application of soil thin-layer chromatography. Soil Biol. Biochem. 19:687–688.

321. Praveen-Kumar, P. K. Chhonkar, and U. Burman. 1990. Effect of urease inhibitors on energy barriers of urease activity of soils. J. Indian Soc. Soil Sci. 38:34–39.

322. Cai, G. X., J. R. Freney, W. A. Muirhead, J. R. Simpson, D. L. Chen, and A. C. F. Trevitt. 1989. The evaluation of urease inhibitors to improve the efficiency of urea as a N-source for flooded rice. Soil Biol. Biochem. 21:137–145.

323. Pedrazzini, F., R. Tarsitano, and P. Nannipieri. 1987. The effect of phenyl phosphorodiamidate on urease activity and ammonia volatilization in flooded rice. Biol. Fertil. Soils 3:183–188.

324. Zhengping, W., O. van Cleemput, L. Liantie, and L. Baert. 1991. Effect of organic matter and urease inhibitors on urea hydrolysis and immobilization of urea nitrogen in an alkaline soil. Biol. Fertil. Soils 11:101–104.

325. Rao, D. L. N., and S. K. Ghai. 1986. Urease inhibitors: Effect on wheat growth in an alkali soil. Soil Biol. Biochem. 18:255–258.

326. McCarty, G. W., J. M. Bremner, and H. S. Chai. 1989. Effect of N-(*n*-butyl) thiophosphoric triamide on hydrolysis of urea by plant, microbial and soil urease. Biol. Fertil. Soils 8:123–127.
327. Creason, G. L., M. R. Schmitt, E. A. Douglass, and L. L. Hendrickson. 1990. Urease inhibitory activity associated with N-(*n*-butyl) thiophosphoric triamide is due to formation of its oxon analog. Soil Biol. Biochem. 22:209–211.
328. Chai, H. S., and J. M. Bremner. 1987. Evaluation of some phosphoroamide as soil urease inhibitors. Biol. Fertil. Soils 3:189–194.
329. Hendrickson, L. L., and M.-J. O'Connor. 1987. Urease inhibition by decomposition products of phenylphosphorodiamidate. Soil Biol. Biochem. 19:595–597.
330. Bremner, J. M., and L. A. Douglas. 1971. Decomposition of urea phosphate in soils. Soil Sci. Soc. Am. Proc. 35:575–578.
331. Suryanarayan-Reddy, M., and P. K. Chhonkar. 1991. Urease activity in soil and flood water as influenced by regulatory chemicals and oxygen stress. J. Indian Soc. Soil Sci. 39:84–88.
332. Savant, N. K., A. F. James, G. E. Peters, and R. Medina. 1988. Evaluation of phenoxyaminocyclotrienes as sustained-action soil urease inhibitors. J. Agric. Food Chem. 36:390–392.
333. Wainwright, M. 1978. A review of the effects of pesticides on microbial activity in soils. J. Soil Sci. 29:287–298.
334. Schäffer, A. 1994. Pesticide effects on enzyme activities in the soil ecosystem. p. 273–340. *In* J.-M. Bollag and G. Stotzky (eds.), Soil biochemistry, Vol. 8. Marcel Dekker, New York.
335. Gianfreda L., F. Sannino, M. T. Filazzola, and A. Violante. 1993. Influence of pesticides on the activity and kinetics of invertase, urease and acid phosphatase enzymes. Pestic. Sci. 39:237–244.
336. Gianfreda L., F. Sannino, N. Ortega, and P. Nannipieri. 1994. Activity of free and immobilized urease in soil: Effects of pesticides. Soil Biol. Biochem. 26:777–784.
337. Gianfreda L., F. Sannino, and A. Violante. 1995. Pesticide effects on the activity of free, immobilized and soil invertase. Soil Biol. Biochem. 27:1201–1208.
338. Butler, L. G. 1979. Enzymes in non-aqueous solvents. Enzyme Microb. Technol. 1:253–259.
339. Laane, C. 1987. Medium-engineering for bio-organic synthesis. Biocatalysis 1:17–22.
340. Zaks, A., and A. M. Klibanov. 1984. Enzymatic catalysis in organic media at 100°C. Science 224:1249–1251.
341. Zaks, A., and A. M. Klibanov. 1988. Enzymatic catalysis in nonaqueous solvents. J. Biol. Chem. 263:3194–3201.

5

Microbial Processes in Boreal Forest Soils as Affected by Forest Management Practices and Atmospheric Stress

Pertti J. Martikainen National Public Health Institute, Kuopio, Finland

I. INTRODUCTION

The total area of boreal forests is 9.5 to 12×10^{12} m^2 [1]. In some countries in the boreal latitudes (50° to 70° N), the forests cover over 50% of the total area. Therefore, forest industry has a great importance for many countries in the boreal region. A great part of the boreal forests is under silvicultural management because of a high demand for wood. During the last two decades acid deposition caused by the use of fossil fuels has become a stress factor for forests. Many forested areas are under heavy atmospheric loads of sulfur and nitrogen. Some forests are also stressed by atmospheric heavy metals. Because precipitation exceeds evapotranspiration in northern latitudes, the soils in boreal forests are generally podzolized. Most of the podzolic soils of the boreal forests contain minerals from the weathering of granitic bedrocks. These naturally acidic soils have a low content of base cations and are susceptible to further acidification at the present level of acid deposition [2].

The forests have a large influence on the environmental characteristics of the boreal region. The accumulation of carbon in northern forest ecosystems is climatically important. It has been generally accepted that the largest increase in temperature is occurring in high latitudes [3], which may influence the biological processes in boreal forests. The increase in temperature associated with possible changes in the composition of tree species affects the chemical and biological characteristics of soil. The possible changes in soil moisture [4] can have an even greater influence on the function of the northern ecosystems, especially peatlands, than the increase in temperature.

In the past, forest management was done mainly to increase the growth of tree stands. Nowadays, maintaining the vitality of forest ecosystems and retaining biotic diversity are becoming more and more important. Because of atmospheric acid loads, liming and addition of nutrients are sometimes needed to counteract soil acidification and nutrient depletion. Changes in microbial processes after forest management are important to know to reduce unwanted reactions in nutrient cycling. Microbial processes in soils are associated with nutrient leaching, and they produce and consume radiatively active gases (CO_2, CH_4, N_2O). Environmental management and atmospheric pollution may change these microbial processes, thus affecting atmospheric chemistry.

In this chapter, the effects of forest management, anthropogenic air pollution, and climatic change on microbial processes and microbial communities in boreal forest soils (mainly coniferous) are discussed. Some data from temperate forests are also included for comparisons of microbiological processes or to complete lacking data from boreal forests. The topic is very broad. Therefore, many questions, although important in the biochemistry of forest soil, are not considered (e.g., mycorrhizae, heavy metals) or are discussed only briefly (e.g., relationships between soil fauna and microorganisms).

II. MINERAL FOREST SOILS

A. Organic Matter Mineralization

Nitrogen Addition

Forest soils receive extra nitrogen via atmospheric nitrogen and nitrogen fertilization. The effects of nitrogen on the biology of forest soils can be evaluated from fertilization experiments. Conclusions about nitrogen effects using data from sites receiving atmospheric acid loads are generally difficult to interpret because of the many chemical components in acid precipitation.

The addition of nitrogen fertilizers generally increases the growth of tree stand in unpolluted boreal coniferous forests, indicating that tree growth is nitrogen limited and that changes in mineralization of organic matter will affect the growth of forest vegetation. The amounts of nitrogen used for fertilization are known to change the extent of microbial decomposition in boreal forest soils. Urea, a common organic nitrogen compound used in forest fertilization, has an immediate effect in increasing the heterotrophic microbial activity and numbers [5–14]. The effect of inorganic nitrogen compounds on heterotrophic microbes differs from that of urea, as the addition of ammonium nitrate and ammonium sulfate can immediately decrease the activity of heterotrophic microbes [7,9,11, 15–20].

Table 1 Changes in Respiration, ATP Content, Microbial Biomass, and Respiratory Quotient in the Uppermost Soil Profile of a Swedish *Pinus sylvestris* Pine Site After Single Amendments of Ammonium Nitrate or Urea. Measurements Were Done 11 Years After the Treatments (Modified Data from Nohrstedt et al. [25])

		NH_4NO_3–N (kg/ha)		Urea–N (kg/ha)	
Horizon	Control	150	600	150	600
A_{00}					
Respiration[a]	187 (100)[e]	187 (100)	143 (77)	172 (92)	165 (88)
ATP[b]	42 (100)	35 (83)	26 (63)	34 (81)	30 (71)
Biomass C[c]	24 (100)	23 (95)	19 (79)	20 (86)	18 (76)
qCO[d]	2.15 (100)	2.25 (105)	2.07 (96)	2.31 (107)	2.50 (116)
A_{01}					
Respiration	94 (100)	77 (82)	68 (73)	89 (95)	65 (69)
ATP	29 (100)	23 (77)	20 (69)	28 (97)	22 (74)
Biomass C	16 (100)	14 (84)	12 (75)	15 (91)	11 (69)
qCO_2	1.59 (100)	1.56 (98)	1.54 (97)	1.65 (104)	1.59 (100)

[a] $\mu g\ CO_2 * g\ soil\ C^{-1}\ h^{-1}$
[b] $\mu g * g\ soil\ C^{-1}$
[c] $\mu g * g\ soil\ C^{-1}$
[d] Respiratory quotient, calculated from the respiratory activity and biomass values (respiration/biomass) expressed by Nohrstedt et la. [25]
[e] Percentage changes in parentheses, control = 100%.

The long-term effects of nitrogen additions to forest soil differ from their short-term effects. Although urea generally increases immediately the activity of heterotrophic microbes, the biomass of heterotrophic microbes and their activity may then decrease below the level before fertilization. Not only urea but also mineral nitrogen fertilizers have had the same effect [11,14,21–27] (Table 1). The effect of a single nitrogen treatment (150 kg N/ha) on decreasing the activity of heterotrophic microbes has been shown to last at least 14 years [23,24]. There is some evidence, from both in vitro and in situ measurements, that the reduction in soil CO_2 production is more evident in fertilized coniferous forest soils of low fertility than in more fertile soils where no decrease in CO_2 production may occur after fertilization [23,24].

The decrease in decomposition after nitrogen addition is supported by observations that organic matter accumulates in nitrogen-fertilized coniferous forest soils [25,28–31]). In the organic layer, the increase can be 26 to 48% [25,27,30].

The reasons for the decreased microbial activity in forest soils caused by nitrogen addition are not well defined. There are several possibilities:

1. Soil pH. Urea is rapidly hydrolyzed in forest soils and results in increases in soil pH as well as in the amount of soluble organic carbon [12]. These are the probable reasons for the increase in the activity of heterotrophic microbes after the addition of urea [7,11,12]. Inorganic nitrogen compounds initially decrease soil pH [20,32], which may be, in addition to possible osmotic shock, a reason for the decrease in microbial activity after addition of inorganic nitrogen compounds. However, long-term nitrogen fertilization experiments have shown only negligible differences in pH between nitrogen-fertilized and unfertilized forest soils [25,27,30], showing that decreases in pH after nitrogen amendments are probably not the reason for reduced microbial activity.
2. Changes in microbial community and microbial physiology. As mentioned above, there is a decrease in microbial biomass after the addition of nitrogen. There is evidence that in coniferous forest soil, fungi and bacteria make equal contributions to the total soil respiratory activity [33]. Hence, changes in the activity of fungal or bacterial communities after nitrogen addition would cause a reduction in soil CO_2 production. Fertilization with ammonium nitrate or urea (150 to 600 kg N/ha) is known to cause long-lasting (up to 13 years) changes in the structure of the microfungal community. The changes in species composition were more evident with high nitrogen addition, especially of ammonium nitrate [34]. There are contradictory results about the effects of nitrogen fertilization on the amount of microfungi (i.e, number, biomass, hyphal length); both increases [6,27,34] and decreases [14,22] have been reported. Not only microfungi are affected by nitrogen additions; nitrogen fertilization also decreased the number of species of larger fungi [35].

   The ratio of microbial CO_2 production to microbial biomass [metabolic quotient (qCO_2)] varies in forest soils [36]. Different bacteria and fungi have different maintenance energy requirements, and the qCO_2 of a soil reflects the maintenance energy requirements of the microbial community in the soil [37]. As mentioned above, the microfungal community is known to change after nitrogen amendments. Therefore, the qCO_2 could change. Furthermore, it has been suggested that qCO_2 can be used as an index of microbial stress in forest soil, with the qCO_2 being higher in stressed soil [36]. There are few data from fertilization experiments that have studied changes in the qCO_2. A single nitrogen treatment at a high dose slightly increased the qCO_2 in the upper soil horizon of a pine forest [25] (Table 1), whereas a slight decrease in the qCO_2 in the humus layer of a pine site resulting from urea fertilization (200 kg N/ha) has been reported [38]. Repeated fertilization with various nitrogen compounds (first treatment with ammonium sulfate, then with urea, and later with ammonium nitrate; seven treatments together, total 530 to 950 kg N/ha) of spruce forests caused no changes in the qCO_2 [27].

3. Reduction in the availability of easily decomposable organic molecules. The activity of heterotrophic microbes in forest soil is energy limited. If fertilization reduces the availability of organic energy sources, the microbial activity should decrease. Root exudates are important as energy sources for soil microbes. Reduced fine root production and ectomycorrhizal infection after fertilization could be associated with a decrease in microbial activity in forest soil (K. Arnebrant, Ph.D. thesis, Lund University, 1991). These changes in root system could reduce carbon allocation to the soil (i.e., reduction in the amount of root exudates) and partly explain a reduced microbial activity.
4. Reduction in the decomposition of recalcitrant compounds, Nitrogen fertilization increases the nitrogen content of litter, and the high nitrogen content may decrease the final phase of litter decomposition [39–41]. The content of lignin in decomposing litter material generally increases with time, thereby lowering the subsequent rate of decomposition. The faster the initial loss of the mass of litter, the faster is the increase in the lignin content [42]. High nutrient supply increases the first phase of litter decomposition, when cellulose and hemicelluloses are decomposing [43], and thus stimulates the formation of lignin-rich, poorly decomposable organic matter. Nitrogen fertilization has been shown to have a long-term effect on increasing the decomposition rate of pure cellulose in the soil of a nutrient-poor pine site, although the overall microbial activity decreased [24]. This agreed with the generalization presented by Fog [44] that the long-term effects of nitrogen addition on cellulose decomposition are positive, although they are negative on the overall decomposition of soil organic matter. Berg et al. [42] have shown that the decrease in the decomposition rate of litter with time was higher in the nitrogen-fertilized soil, where the initial decomposition rate was higher than in the unfertilized soil (Figure 1). During the last phase of decomposition, the decomposition rate in the fertilized soil may fall below that of the unfertilized soil. It has been suggested that ammonium and amino compounds could react with organic matter to form recalcitrant compounds [44,45]. Ammonium may also have a direct inhibiting effect on the decomposition of lignin by repressing the synthesis of lignolytic enzymes in fungi [46,47].

It is important to note that there are also reports indicating that nitrogen has no long-term effect on decreasing microbial activity and litter decomposition [18,23,48]. Furthermore, the long-term effects can also depend on the type of nitrogen compound added. For example, in

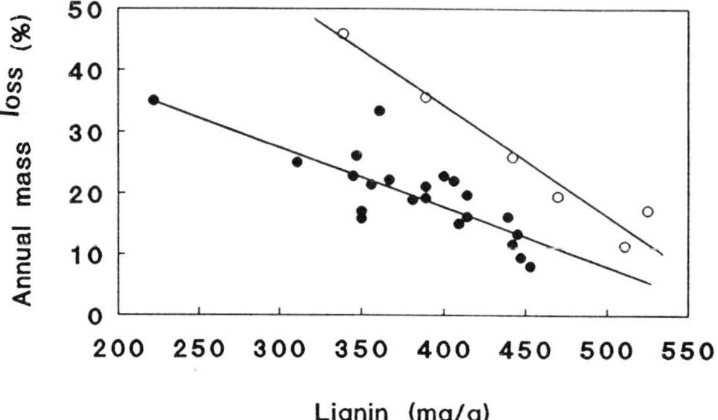

Figure 1 Relationships between loss of mass of Scots pine needle litter and lignin concentration of litter; (•) control soil, (○) soil fertilized with ammonium nitrate and irrigated (data from Berg et al. [42]).

contrast to urea, a slow-releasing urea derivative, ureaformaldehyde, did not decrease soil respiration [23,24,50], indicating again the importance of the initial effect of the added compound (e.g., changes in soil pH and ammonium content). Long-term changes in the structure of soil microbial and faunal communities may also be a reason for the different effects of various nitrogen compounds on decomposition. For example, ureaformaldehyde does not change the biomass and species composition of larger fungi as much as urea or ammonium nitrate [35].

Soil Acidification

The effects of atmospheric acid loads on microbial activities in forest soil has been studied in short-term laboratory experiments, using experimental field acidification or pollution gradients around industrial sources. The last approach should provide the most realistic picture about the long-term effects of acid precipitation on microbial processes and microbial communities in soil. However, using sites with natural atmospheric acid loads, it is difficult to separate the effects of H^+ ions from the effects of other constituents in natural acid precipitation (e.g., nitrogen). In addition to H^+ ions, gaseous SO_2 and NO_2, as a result of their solubility products, affect microbial activities [50,51].

In laboratory experiments, the H^+ load (H_2SO_4 addition) has reduced soil respiration [52–56]. Low doses of acid often have only negligible effects on respiration. It has been suggested that in soils with a pH of 4.0 to 4.5, the acid treatment has to decrease the soil pH by at least 0.3 units to reduce soil respiration by 10% or more [56].

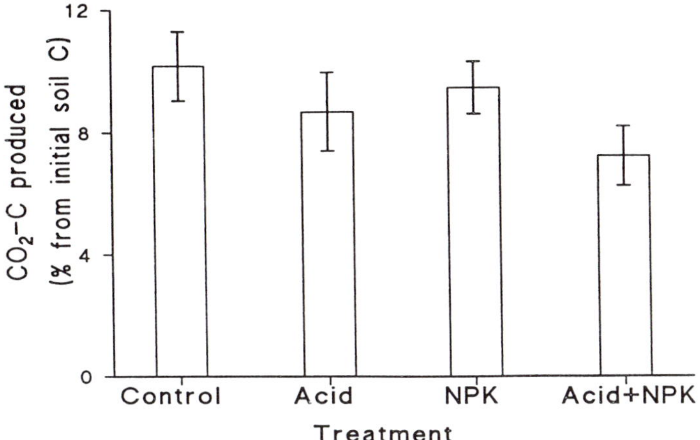

Figure 2 Amount of CO_2-C evolved (%C of initial soil C) during a 9-week incubation of soil samples (20°C, 60% water holding capacity) from untreated, acid-treated, NPK-fertilized, and NPK-fertilized + acid-treated plots of a Scots pine forest site (data from Popovic [58]).

Reductions in respiration in forest soils has been found in field acidification experiments [57–59]. As in laboratory experiments, high doses of acid are also needed in field experiments to reduce soil respiration. Bååth et al. [57] reported that water with a pH of 2 reduced soil respiration by 50%, whereas water at pH 3 had only a minor effect on respiration. However, in long-term field experiments, Persson and Wiren [59] found reduced CO_2 evolution 11 years after the last acid treatment and when the pH and base saturation of the soil were close to the control level. This Norway spruce site has received acidity, as H_2SO_4 or sulfur powder, corresponding to three to four times the annual atmospheric acid deposition. Long-term field acidification studies revealed that combinations of acid treatment with nitrogen addition reduced soil respiration more than acid treatment alone [58] (Figure 2). These results show the difficulties in identifying reasons for decreased microbial activities caused by anthropogenic air pollution.

The effects of anthropogenic emissions of SO_2 and NO_x on microbial activity in boreal forest soils have been studied using pollution gradients. Fritze et al. [60] found no decrease in respiration in Scots pine forest soils around an oil refinery emitting 24×10^6 kg H_2SO_4/yr and 4.4×10^6 kg NO_x/yr. Nohrstedt [61] and Visser and Parkinson [62] also observed no decrease in soil respiration around the sources of sulfur pollution. SO_2 emissions of 9.5 to 23×10^6 kg/yr together with NO_x emissions of 3.1 to 18.7×10^6 kg/yr from an oil refinery or a city region reduced respiration in some soils under Scots pine (Vanhala et al., Water Air Soil Pollut., in press). Similarly, SO_2 emissions of 6.3×10^6 kg/yr and

NO_x emissions of 7.0×10^6 kg/yr from an industrialized city decreased respiration in Scots pine forest soils [63]. The effect of pollution on forest soil microbes depends on the type of forest site. In a large-scale field survey (Vanhala et al., Water Air Soil Pollut., in press), harmful effects were observed in the most nutrient-poor Scots pine forest sites (Calluna forest types), not in the more fertile sites (Vaccinium and Myrtillus forest types). The coarse-textured, dry soil typical of the Calluna forest site obviously was more sensitive to acidic deposition. Furthermore, reduction in respiration was evident only in the humus layer and not deeper in the mineral soil. Soil mineralogy appears to be important in the acid tolerance of microbes because the reduction in respiration was eliminated by adding montmorillonite to acid-stressed soil [64].

Acidification can change the rates of litter decomposition. However, experimental or natural pollution conditions can highly influence decomposition. In a field experiment Bååth et al. [65] found reductions in the decomposition of both root and pine needle litter as a result of heavy additions of H_2SO_4 (150 kg/ha/yr over 6 years). Berg [66] reported that H_2SO_4 (32 or 96 kg S/ha) decreased the decomposition of needle litter on pine sites but not on birch and spruce sites. In another experiment on a spruce site, Berg [67] found that additions of H_2SO_4 or sulfur powder (64 kg S/ha, twice the annual natural sulfur deposition) caused a 20 to 34% reduction in the decomposition of Norway spruce needle litter. Neuvonen et al. [68] reported that high artificial levels of acidification with H_2SO_4 (15 times ambient level) and HNO_3 (11 times ambient level) reduced the decomposition rate of both pine needles and birch leaves. Since high acidity (pH 2 or lower) generally decreases the decomposition of both coniferous needle litter and various leaf litters, lower acidity (pH 2 to pH 3) can increase the decomposition of litters in the early phase; later when litters consist of less degradable substances, the decomposition decreases [69,70]. Decrease in litter decomposition has been observed also around of pollution sources. In a Canadian study, SO_2 emissions of 6.5 to 58×10^6 kg/yr decreased decomposition of coniferous needle litter [71]. However, emissions of the same level (24×10^6 kg H_2SO_4/yr and 4.4×10^6 kg NO_x/yr) did not decrease the decomposition of pine needle litter around a Finnish oil refinery.

Artificial acidification has been reported to have contradictory effects on carbon and nitrogen mineralization in boreal forest soils. Acidification has decreased carbon mineralization, measured as CO_2 production, whereas there has been no change or an increase in net mineralization of nitrogen [56,58,72]. Lohm et al. [72] have shown that although net mineralization of nitrogen was unaffected, acidification decreased turnover rate of nitrogen, indicating changes in the functioning of microbial communities in acid-stressed soil.

There is evidence that it is available carbon that limits the activity of heterotrophic microbes in acidified soil. Bewley and Stotzky [64] found that acidification had a minor effect on CO_2 evolution in soils amended with glucose until the soil pH was lowered below 2.9. Benner et al. [73] suggested that utilization of substrates that require extracellular enzymatic activities is the most sensitive to acidification because extracellular enzyme activities occur outside the buffered intracellular environment. Thus, the degradation rate of recalcitrant macromolecules would decrease. The addition of easily available substrates, e.g., glucose, to acid-stressed soil would then compensate for the lowered natural energy supply and eliminate the reduction in respiration [64].

It is interesting that respiration and nitrogen mineralization can have similar opposite responses to additions of inorganic nitrogen compounds as they do to acidification [20,74]. Nömmik [74] pointed out that the carbon:nitrogen ratio of organic material decomposing in soil treated with inorganic compounds (e.g., phosphorus or nitrogen compounds) may be lower than in untreated soil, which could explain the opposite effects of inorganic compounds on net carbon and nitrogen mineralization. It could be that extra mineral nitrogen in soils treated with acid or inorganic compounds is derived from dead microbes. Osmotic stress in soil amended with inorganic compounds or acids could lyse microbial cells. Hence, high amounts of ammonium are produced in mineralization of nitrogen-rich cell materials (proteins). However, Persson et al. [56] estimated that this can explain only part of the extra mineral nitrogen liberated after acidification.

Acidification experiments have shown changes in the size and composition of microbial populations in forest soil that reflect changes in respiration and litter decomposition. Artificial acidification has decreased the metabolically active bacterial and fungal biomasses [56,59,65]. There also are changes in the composition of microfungal species [75]. Acidification has lowered the counts of heterotrophic bacteria that utilize starch, protein, pectin, and xylan [76] and reduced cellulase activity [77,78]. Similarly, Thomson et al. [79] found in laboratory experiments that acidification changed the physiological characteristics of bacteria in forest soil. These results support the hypothesis, presented above, that the sensitivity of extracellular enzymes to acidification could be associated with reductions in heterotrophic microbial activity in acid-stressed environments.

Changes in Soils Receiving Alkaline Substances

Liming is used to counteract the acidification of forest soil. Wood or peat ash (waste material from electric power stations) have also been tested in acid soils. Some forest sites also receive atmospheric alkaline deposition.

Experiments in boreal forests have shown some problems after liming, which may be associated with the changes in the biology of forest soil. Liming has, in some cases, reduced tree growth [80,81], and it generally causes an accumulation of soil organic matter that can result in a 10 to 50% increase in the organic layer over 20 years [82]. These results indicate disturbances in nutrient cycling and nutrition of trees.

Liming generally has a long-term effect on increasing microbial activities in forest soils, as measured by CO_2 production [74,83–86], head output [83,87], or ATP [83]. However, Persson et al. [84] found that liming had a long-lasting positive effect on CO_2 production only in forest soils rich in nitrogen. In soils low in nitrogen, liming had no long-term effect on the CO_2 production.

Both increases [84,85] and decreases [74,84,88] in net mineralization of nitrogen resulting from liming of forest soils have been reported. Gower et al. [89] showed by the ^{15}N technique that liming increased nitrogen immobilization in sulfur-polluted forest soils. Persson et al. [84] found that liming decreased nitrogen release in the litter layer, whereas it increased nitrogen release in other soil layers if they were rich in nitrogen. However, if the deeper layers had a low nitrogen content, liming had no effect on nitrogen release. There are observations from northwest Europe that liming decreases net nitrogen mineralization, especially in soils of old coniferous stands [80,88]. Nihlgård and Popovic [80] have pointed out that decreases in growth after liming are evident in old stands.

Changes in the C:N ratio of soil may indicate whether there are long-term increases or decreases in the decomposition of organic matter. Decreases in the ratio indicate higher decomposition activity. Both decreases [89] and increases [82,87] in the ratio have been reported.

Some increases in microbial biomass in limed forest soils have been reported [86,90,91]. An increase in the bacterial biomass seems to be primarily responsible for the increase in the total microbial biomass [87]. Fungal biomass has been found to be unaffected [65,91] or decreases [56,87]. For example, a high dose of wood ash has decreased fungal biomass in forest soil (Bååth et al., Soil Biol. Biochem., in press). There are results showing that the relative amount of bacterial biomass to fungal biomass increases in forest soil after liming [87] and wood ash treatment (Bååth et al., Soil Biol. Biochem., in press). The composition of microfungal species has been found to change in some liming experiments, but in some others there have been no changes [92]. Bååth and Arnebrant [92] pointed out that liming seems to affect the microfungal community similar to urea and clear-cutting but opposite to artificial acidification or ammonium nitrate fertilization. Alkaline deposition is known to decrease the number of microfungal species. However, despite of an overall reduction, some species may become more common [93]. Recently, Frostegård et al. [91a] found changes in the bacterial community in limed soil: gram-negative bacteria became more abundant, whereas there was a reduction in gram-positive bacteria.

Alkaline deposition also changes the structure of the bacterial community [94] and, similar to liming, the occurrence of actinomycetes [91a,95]. There is evidence that the acid-tolerant bacterial community diminishes in soils receiving alkaline deposition [94]. The changes in microbial communities and/or soil chemical conditions may be associated with the general increase in qCO_2 in boreal forest soils receiving alkaline substances [26,27,60,86,96]. The increase in the CO_2 production in boreal forest soils treated with lime or other alkaline substances is opposite to the general decrease in CO_2 production in nitrogen- or acid-treated soils (see above). The increase in CO_2 production, as well as in the qCO_2, indicates that heterotrophic microbes in soils receiving alkaline substances are not as energy-limited as microbes in acid-stressed soils. The increase in the qCO_2 may indicate the importance of fast-growing r strategists that use simple substrates. The microbiota thus resembles that found in ecosystems of early succession phases [97,98]. The microbial processes in boreal soils with relatively low atmospheric pollution seem to differ from those in soil of damaged temperate forests where liming has lowered the qCO_2 [99]. As discussed above, the qCO_2 generally increases in limed boreal coniferous forest soil.

It should be pointed out that both nitrogen fertilization (see above) and liming have increased the organic matter content in boreal forest soils. There are still questions about how organic matter production by trees and other vegetation [82] and microbial mineralization/immobilization processes are connected to this phenomenon.

Clear-Cutting and Fire

Clear-cutting has many effects on the physical, chemical, and microbiological properties of the soil. After clear-cutting, there are changes in soil microclimate and moisture. Soil moisture, even more than temperature, influences the fluctuation of bacterial biomass in forest soils [100]. The effects of clear-cutting on the bacterial community are reported to last from 8 to 13 years [101,102]. Soon after clear-cutting there is a large reduction in aboveground and belowground litter production, which decreases the amount of substrates for microbial growth [103]. On the other hand, there is the decomposing root biomass in soil, and if whole-tree harvest has not been applied, there is considerable wood residue left after the removal of the stand.

Bacterial biomass has been shown to increase soon after clear-cutting, but to decrease after 2 years, especially if slash is removed [104]. After clear-cutting, not only the bacterial biomass but also the total microbial biomass, fungal biomass, and soil respiration decrease (Table 2). Fungal biomass decreases more than bacterial biomass, thereby reducing the fungal:bacterial biomass ratio (Bååth et al., Soil Biol. Biochem., in press). According to phospholipid fatty acids (PLFA) analysis of humus there are changes not only in the number of microbes but also in the structure of the microbial community in forest soil after clear-cutting (Bååth et al., Soil Biol. Biochem., in press).

Development of vegetation after clear-cutting greatly affects microbial processes in soil. When CO_2 evolution is measured in situ, it has been reported that 3 to 4 years after cutting the cut site produced more CO_2 than the control

Table 2 Effect of Clear-Cutting and Burning on Some Chemical and Microbial Characteristics of Organic Soil Horizon (F/H Layers) of a Mixed Forest of Norway Spruce and Scots Pine (Modified From Data of Pietikänen and Fritze, Soil Biol. Biochem., in press)

Variable	Control	Clear-Cut[a]	Clear-cut + burned[a]
pH	3.93 (0.04)[b]	4.09 (0.08)	5.85 (0.22)
Org-C (mg/g oven-dry soil)	490 (5.4)	490 (11)	450 (25
Org-N (mg/g oven dry-soil)	11.1 (0.5)	11.5 (0.4)	12.4 (0.2)
Ca[c] (μg/g oven-dry soil)	2700 (160)	3500 (230)	8800 (130)
K[c] (μg/g oven-dry soil)	1200 (58)	570 (21)	490 (63)
Microbial–C (μg/g oven-dry soil)			
Fumigation-extraction (FE)	7300 (400)	5300 (220)	2400 (220)
Substrate-induced respiration (SIR)	11000(720)	8600 (520)	5100 (340)
Ergosterol (μg/g over-dry soil)[d]	250 (15)	170 (3.4)	74 (10)
Respiration (μg/CO_2/g oven-dry soil/h)			
Field-moist samples	31 (2.2)	22 (2.1)	14 (2.2)
Samples adjusted to 60% of the water holding capacity of soil (WHC)	33 (2.5)	26 (1.0)	19 (1.6)
qCO_2 (μg CO_2–C/μg biomass–C × 10^{-3}/h	2.29 (0.11)	2.43 (0.08)	3.99 (0.29)

[a] Clear-cut in 1990, burned in 1992, and sampled in 1993
[b] Standard error in parenthesis
[c] Extractable
[d] Reflects the occurrence of fungi

site. Root respiration is also included in the CO_2 measurements in situ [105]. The change in the qCO_2 after clear-cutting is similar to that in limed soils, i.e., the qCO_2 increases (Table 2). Here, the theory about changes in qCO_2 with successional development fits well.

Not only carbon mineralization but also nitrogen mineralization/immobilization is affected by clear-cutting. For the nitrogen cycle, the changes in litter production and especially in the amount and quality of root-derived organic matter has an important role [106]. Net nitrogen mineralization can increase [107] on deciduous sites, whereas coniferous sites may not respond [108]. In general, the increase in mineralization after disturbance is of short duration [107]. When plant debris is decomposed after clear-cutting, mineral nitrogen is also immobilized. Microbial nitrogen immobilization is considered to be even more important than nitrogen retention by the developing vegetation in preventing nitrogen losses after clear-cutting [109].

Prescribed burning can be applied after clear-cutting, which further modifies the soil conditions. In boreal forest ecosystems, podzolization causes immobilization of carbon and nitrogen, as well as other nutrients, in the organic layer. Soil pH is low in podzolized soil because of the organic acids produced during decomposition of plant debris. Plant uptake of cations is also associated with natural acidification. In boreal podzolized soils, the content of basic cations is low, further favoring soil acidification. Fire-natural or prescribed burning-liberates basic cations from soil organic matter, thereby decreasing the acidity.

Prescribed burning has been found to enhance the effects caused by clear-cutting, i.e., there is a decrease in soil respiration [110–112] and total microbial and fungal biomass, and an increase in the qCO_2 [96,112] (Table 2). The ratio of fungal to bacterial biomass has been found to decrease in burnt forest soil and to be lower in clear-cut soil and burned soil than in only clear-cut soil (Bååth et al., Soil Biol. Biochem., in press). However, the fungal species typical of the early phase of decomposition of herbaceous litter have increased [111]. Fritze et al. [113] found that soil respiration, total microbial biomass, and fungal biomass have a minimum during the first 2 years after burning, then they increase reaching a plateau after 12 years. The ratio of microbial biomass carbon to total soil carbon is reduced by burning [96,113], but it increases during the subsequent 12 years to the ratio generally found in boreal forest soils [113]. These results suggest that microbial changes appear to occur during the first 10 years after clear-cutting and burning.

B. Nitrification

There is still uncertainty as to which microbial processes are responsible for the formation of nitrite and nitrate in acid forest soil. There is strong evidence that if ammonium availability and soil pH are increased, nitrate is produced mainly by chemolithotrophic nitrifiers [114]. However, there are results indicating that in some undisturbed acidic coniferous forest soils nitrate is not produced by chemolithotrophic nitrification, and heterotrophic nitrifiers are suggested to be responsible for nitrification [114–117]. The physiology of heterotrophic nitrifiers differs from that of chemolithotrophic nitrifiers. Heterotrophic nitrifiers include various heterotrophic bacterial and fungal species that are unable to use the energy liberated in the oxidation of nitrogen compounds [118]. Therefore, disturbances in forest ecosystems would affect heterotrophic and chemolithotrophic nitrification differently. One reason why heterotrophic nitrification may be important in forest soils is that chemolithotrophic nitrification requires higher pH levels than those that prevail in acid forest soils. However, recent studies have shown that chemolithotrophic nitrification occurs in acid temperate [119–122] and boreal forest soils [123,124]. Nitrifiers in various layers of the boreal coniferous soil profile have different requirements for pH. Nitrifiers in the litter layer have been shown to be more sensitive to low pH than those living in deeper layers [124].

Nitrogen Addition, Acid Deposition, and Liming

Availability of ammonium is an important factor affecting nitrification in forest soil [125]. Although it is known that acid-tolerant chemolithotrophic nitrifiers exist in forest soil, nitrification is generally stimulated more by the addition of urea, which increases soil pH, than by the addition of mineral nitrogen compounds [21,123,126,127]. The stimulatory effect of urea may be associated with the acid-sensitive chemolithotrophic nitrifiers [128]. After its enzymatic hydrolysis, urea increases both soil pH and the availability of ammonium. In acid soils, chemolithotrophic ammonium oxidizers exist that are capable of urea hydrolysis [129]. The addition of ammonium salts can even inhibit nitrification in coniferous forest soils [13,16,20]. However, a high atmospheric load of ammonium stimulates chemolithotrophic nitrification in boreal forest soils [124], similar to the stimulation in temperate forest soils [119,120,122,130,131]. The stimulation of nitrification by ammonium in situ depends on the amount of nitrogen added. A single fertilization treatment with urea does not always increase net nitrification activity in boreal coniferous forest soils [38]. In contrast to urea, the addition of a slow-releasing urea derivative—ureaformaldehyde—does not stimulate nitrification in boreal coniferous forest soils [123].

Liming generally increases nitrification activity in forest soils [16,53,80,88, 132–137] (Priha and Smolander, Soil Biol. Biochem., in press). Nitrification in soil is associated with the mineralization and immobilization of nitrogen. As discussed above, liming can increase immobilization of nitrogen, which is a probable reason why liming sometimes does not increase nitrification in boreal forest soil. However, there are results that liming can stimulate nitrification, even though liming can decrease net mineralization of nitrogen [136]. As noted by Johnson [138], nitrifiers may compete for ammonium better than generally assumed.

Clear-Cutting and Prescribed Fire

The stimulatory effect of clear-cutting on nitrification has been documented in several forest ecosystems, and the increased availability of ammonium is considered to be an important reason for this [132,139–142]. Prescribed fire has also been found to increase nitrification in forest soil [110] (Pietikåinen and Fritze, Soil Biol. Biochem., in press). However, there are also reports that fire had no effect on nitrification [96].

C. Production of Nitrogen Gases

Denitrification and nitrification are the best-known processes during which nitrous oxide (N_2O) and nitric oxide (NO) are formed [118,143]. In denitrification, N_2O can be the main gas produced [118]. Forest management practices and atmospheric pollution may enhance these processes, which are important for the nutrient balances of forest ecosystems and also in atmospheric chemistry. Unfortunately, most of the information available about microbial processes that produce nitrogen gases is derived from studies in temperate forest soils. Therefore, some publications concerning temperate forest soils are cited here to describe the factors that affect microbial reactions that produce N_2O and NO in acid forest soil.

The availability of nitrogen in undisturbed forest ecosystems is low. Therefore, all environmental changes that enhance the availability of ammonium or nitrate should favor the production of nitrogen gases. Nitrogen fertilization, atmospheric nitrogen load, and clear-cutting all increase the availability of inorganic nitrogen in forest soil. Other factors, such as acidity, both natural and anthropogenic, also affect the microbiological processes responsible for the production of nitrogen gases in forest soils.

Acidity and Liming

Denitrifying populations are known to adapt to low soil pH [144]. Denitrifiers occur in acid boreal forest soils, although the low soil pH limits their activity [145]. The denitrifying potential in boreal coniferous forest soils increases with increasing pH. In podzolized forest soils, the highest denitrification potential has been reported to occur in the B horizon, where the pH was higher than in the A or O horizons [145,146]. Treatments with wood ash or liming, which increased soil pH, enhanced denitrification in boreal forest soils [147] (Priha and Smolander, Soil Biol. Biochem., in press) as well as in temperate forest soils [148]. There are also results that liming, although increasing soil pH, did not stimulate the total production of N_2 and N_2O during denitrification [85,149]. Liming of temperate forest soils has even decreased the evolution of N_2O in situ [150,151]. It may well be that the total denitrification ($N_2 + N_2O$) is not reduced but that the ratio of N_2O to N_2 decreases with increasing pH [149,152–154]. The production of NO in denitrification is also favored by low pH [155,156], which has also been observed in forest soils [157]. In addition to low pH, low temperature is known to increase the $N_2O:N_2$ ratio in denitrification in forest soils [158].

The effect of the atmospheric load of acid on the capacity of forest soil to produce nitrogen gases is complex. It adds nitrate and ammonium to soil, but on the other hand may retard microbiological processes in soil. Treatments with ammonium and nitrate had no effect on or increased the emissions of N_2O and NO in a coniferous forest in Germany [151]. In this forest soil, denitrification was responsible for 0 to 54% of the N_2O and 12 to 42% of the NO emitted. There are also other observations that denitrification is not the source of N_2O in acid forest soil [159]. As discussed above, there are acid-tolerant and acid-sensitive chemolithotrophic nitrifiers in forest soils. Chemolithotrophic ammonium oxidation in soil is known to produce both N_2O and NO [143]. There is evidence that in both boreal [124,160] and temperate [161] coniferous forest soils chemolithotrophic nitrification can produce N_2O, and that acidity favors the production of N_2O associated with the activity of acid-tolerant nitrifiers. Acidity may affect N_2O production associated with nitrification differently in various layers of the podzolized soil profile. The litter layer may be the most sensitive to acidification. On the other hand, liming would enhance N_2O production most in the litter layer where acid-sensitive nitrifiers occur [124]. In nitrogen-stressed forest soil, denitrification can also be very active in the litter layer [162]. In addition to denitrification and chemolithotrophic nitrification, some other reactions that are poorly characterized, such as heterotrophic nitrification [151] or those of nitrate-respiring bacteria, fungi, and yeasts [159], have been suggested to participate in N_2O production in acid forest soils. The microbial community able to produce nitrogen gases in forest soil seems to be very complex, and the total effect of pH on this community is unknown. There is evidence that adaptation to fluctuating pH conditions by nitrate reducers in forest soil is the result of the adaptation of the resident bacteria and not of the development of new bacterial populations [154].

Increased Availability of Inorganic Nitrogen—Fertilization, Nitrogen Deposition, and Clear-Cutting

Various nitrogen fertilizers have different effects on denitrification in boreal coniferous forest soil. Generally the denitrification potential is higher in soil samples from urea-treated soil than in samples from ammonium nitrate–fertilized soil [146,147,158]. Ammonium nitrate at higher concentration inhibits denitrification [147,163]. There are several probable reasons why urea favors denitrification in acid forest soils. First, urea increases nitrification (see above), thus providing a long-term increase in available soil nitrate. Availability of nitrate is generally considered to be an important factor in regulating denitrification in forest soils [159,164–168]. There may also be higher denitrifying populations in soil receiving urea. Second, urea increases soil pH. Third, the availability of organic substrates may also be higher in soil fertilized with urea (see above). Inhibition of denitrification after nitrate addition may be the result of the same reasons, e.g., osmotic stress, that cause an inhibition of respiration after the addition of inorganic nitrogen compounds (Section II.A).

Long-term and high deposition of atmospheric nitrogen can cause nitrogen saturation of forested ecosystems, which then have a high availability of nitrate [169]. Heavy deposition of nitrogen has increased the N_2O emission from temperate forests. However, there is great variation in denitrification rates in forests receiving heavy nitrogen deposition, indicating that in addition to nitrate availability, other factors, such as soil aeration, are important in regulating the denitrification activity in forest soils [170]. Nitrogen saturation of boreal forest ecosystems is presently not common. However, this may change in the near future since nitrogen deposition is exceeding the critical load in many northern

Table 3 Annual Emissions of N_2O in Some Forest Sites

Forest (location)	N_2O–N/ha/yr	Reference
Douglas fir		
(Oregon, USA)		
Virgin	<0.01	166
Clear-cut	0.02–6.1	
Douglas fir		
(Rocky Mountains, USA)		
Virgin	0.03–0.09	223
Clear-cut	0.65–0.67	
White Spruce		
(Alberta, Canada)	0.01–0.03	168
Spruce[a]		
(Southern Sweden)		
Virgin	0.01–1	224
Clear-cut	0.01–2	
Beech[a]		
(Southern Sweden)	0.01–2	224
Alder[a]		
(Southern Sweden)	0.2–4	224
Pine[b]		
(Western Finland)		
Nitrogen deposition		
(kg/ha/yr)		
4.5	0.06	P. J. Martikainen, K. Lång,
10	0.10	H. Nykänen, and A. Ferm,
33	0.26	unpublished data

[a] Nitrogen deposition in southern Sweden can be more than 20 kg/ha/yr [171]. Emissions measures with the acetylene blockage technique.

[b] Measured with a closed-chamber technique (May-October 1993).

areas [171]. Ammonium deposition will increase not only the nitrification activity [124] but also the emission of N_2O from boreal coniferous forests (Table 3).

Clear-cutting causes a transitory increase in nitrogen mineralization and nitrification, (see above) and thus may enhance the production of nitrogen gases. Only a few studies have been conducted on denitrification in boreal forest soils after clear-cutting. Denitrification in well-drained coniferous forest soils after clear-cutting has ranged from 0.01 to 2 kg N_2O-N/ha/yr (Table 2). Even if boreal coniferous forests that are heavily fertilized with nitrogen are clear-cut, the emissions of N_2O have been within this range (Martikainen et al., unpublished data).

D. Methane Oxidation

Well-drained soils, such as forest soils, are sinks for atmospheric methane because of their capacity to oxidize methane. Evidence is accumulating that environmental stress can inhibit the activity of the methanotrophic bacteria responsible for methane oxidation. Nitrogen additions inhibit methane uptake in agricultural as well as in forest soils [172,173]. Pure culture studies with methanotrophic bacteria suggest that ammonia inhibits the functioning of methane monooxygenase [174,175], which is the key enzyme in the oxidation of methane. However, even if enhanced ammonium concentrations in soils are reduced with time, inhibition continues [172]. It has been suggested that nitrite produced from ammonium oxidation of methanotrophic bacteria could be responsible for the long-term inhibition of oxidation of methane after amendments with ammonium [176]. There is no estimate of how much nitrogen deposition has decreased the oxidation of methane by boreal forest soils. Nitrogen deposition has decreased the oxidation of methane by temperate forests and grasslands by 30 to 70% [177]. The effect of inorganic nitrogen on the oxidation of methane emphasizes the important relationship between the processes involved in the nitrogen and methane cycles. Mineralization and immobilization of nitrogen could thus influence the capacity of soil to oxidize methane.

The application of lime to counteract soil acidification has decreased methane oxidation in temperate forest soil [177]. The highest rates of methane oxidation have been found in deeper horizons of forest soils [178,179]. This may protect methane oxidizers from the effects of liming or nitrogen loads that are supposed to have the greatest effects in the uppermost soil layers.

E. Interactions Between Soil Animals, Microbes, and Organic Matter Mineralization

Recent research has shown the great importance of soil microfauna, mesofauna, and macrofauna in enhancing the decomposition of organic matter and nutrient dynamics in forest soil [180–182] (H. Setälä, Ph.D. thesis, University of Jyvåskylå, 1990). This positive effect of soil fauna on nutrient release enhances tree growth [183]. Disturbances in the composition of the soil fauna may reduce nutrient turnover in soil.

Nitrogen

The effect of nitrogen on soil fauna has been studied in fertilization experiments [184]. Fertilization with urea increased the numbers of nematodes, mainly bacterial feeders, shortly after addition [185]. However, this effect, which was transitory [184,185], was probably the result of urea increasing the bacterial population in forest soil (see above). Ammonium nitrate, which has a negative effect on microbial activities (see above), also has lowered the number of nematodes in forest soil [184,185].

Nitrogen fertilizers reduced the biomass of enchytraeid worms for several years after addition [184,186]. This negative effect might be due to changes in microbial populations and the quality of soil organic matter [186].

The effect of nitrogen on microarthropods has been found to be dose-dependent. Nitrogen at 200 kg/ha had a negligible effect on mites (*Acari*) and *Collembola*, whereas higher amounts of nitrogen reduced their numbers [180,187]. This decrease may reflect changes in the microbial community that forms a part of their diet. As discussed above, nitrogen additions can change the structure of microbial communities and reduce the amount of microbial biomass in forest soils.

Acidification and Liming

Acidification experiments have shown that the numbers of bacterial-feeding nematodes are reduced by H_2SO_4, whereas the numbers of root/fungal-feeding nematodes were not reduced [188]. This difference may reflect the differential effect of acid on bacterial and fungal populations. The negative effect on artificial acidification on nematodes has been transitory, and after some years their number may increase to the initial level [188].

Shortly after their addition, lime and ash have been found to decrease the number of nematodes, but subsequently, they can increase the number [184,185]. Other studies have indicated that liming has no long-term effects on nematodes [188]. This variability in observations may reflect differences in how liming affects microbial populations and their activity in various boreal forest soils. As discussed above, liming has no long-term positive effect on microbes in some soils.

Liming and the addition of wood ash have been shown to have a long-term reductive effect on the occurrence of enchytraeids. The negative effects of these treatments differ from that of nitrogen; lime and ash seem to have a continuous, long-term, decreasing influence, whereas nitrogen reduced the number of enchytraeids immediately after addition [184,186]. Liming and ash favor the growth of bacteria, and the ratio of fungal to bacterial biomass decreases (see above). This shift may be a reason for the long-term reduction in the occurrence of enchytraeids [184,186].

The effects of lime and wood ash on microarthropods have been contradictory, as they have both increased and decreased their numbers [184,187].

III. FORESTS ON PEATLANDS

Nutrient dynamics, microbiology, and enzyme activities in peat have been discussed (e.g., by Kuprevich and Shcherbakova [189], Given and Dickinson

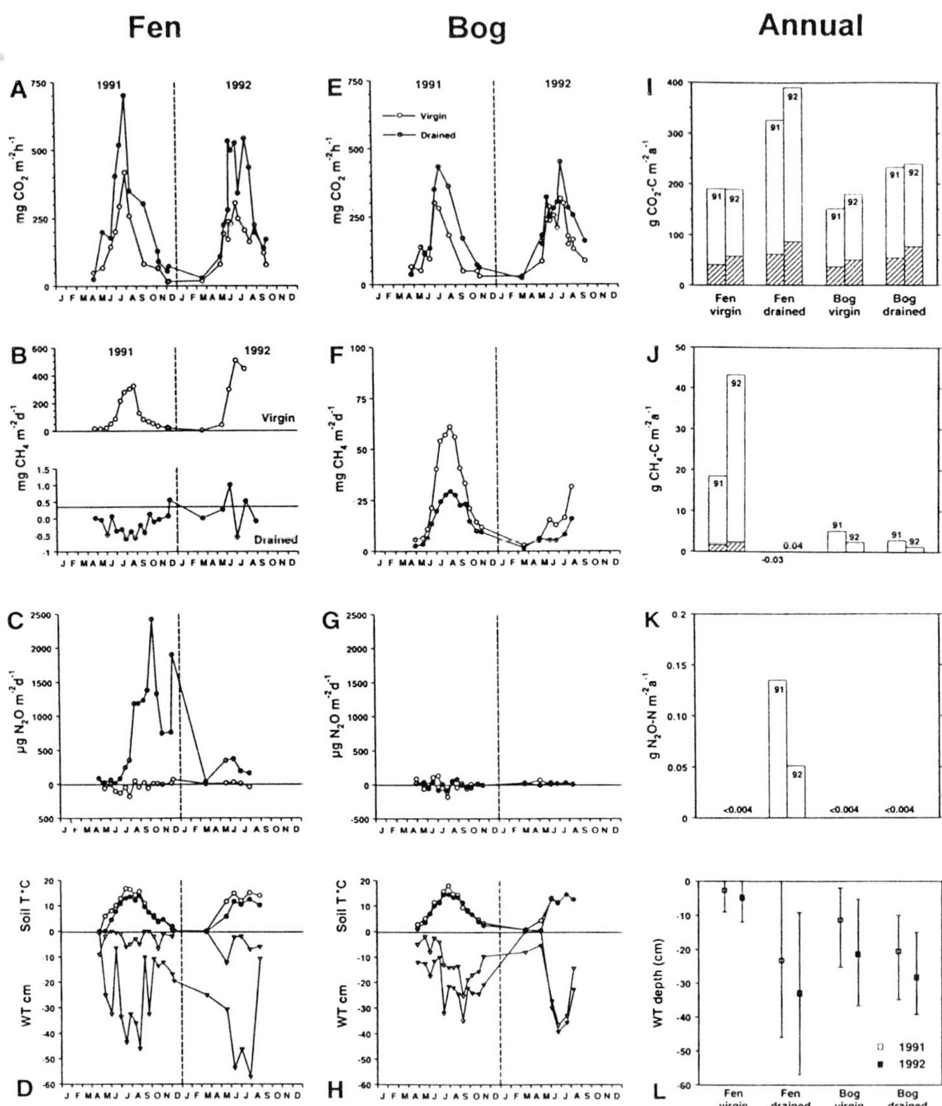

Figure 3 Fluxes of CO_2, CH_4, and N_2O in a fen and a bog in central Finland. The mires have a virgin site and a site drained in 1961 for forestry. Measurements were done in 1991 and 1992. Fluxes of CO_2 were measured with a dynamic chamber method using a portable infrared analyzer, and fluxes of CH_4 and N_2O were measured with a static-chamber technique and the gases were analyzed by gas chromatography. Fluxes of CO_2 (a, e), CH_4, and N_2O (c, g), soil temperature (at 10 cm), and depth of ground water table (WT) (d, h). Open symbols denote virgin areas and solid symbols denote drained areas. Annual (a^{-1}) gas fluxes are given for the years 1991 and 1991 (i, j, k). Hatched parts of

[190], Verhoeven [191], and Richardson et al. [192]). Recently, the impact of the northern peatlands on climate has received much attention [193–195]. Only the effects of the drawdown of the water table and the addition of nutrients on decomposition activity and microbiological processes in peat that are responsible for the fluxes of CO_2, CH_4, N_2O, and NO will be discussed here. Tree growth in virgin peatlands in generally limited by high soil moisture. Therefore, draining is a common practice in peatland forestry. The increased aeration in the uppermost peat profile after draining greatly changes the microbial processes in peat. Nutrient additions further modify the microbial processes in drained peat soil.

A. Decomposition and Enzyme Activities

As a result of the high water table (low availability of oxygen), organic matter decomposition in virgin peat soils is low, and organic matter is accumulated as peat. The increase in aeration after draining results in mineralization of the accumulated organic matter, which can be seen as an increase in CO_2 production [196–198] (Figure 3), especially when peatland with a high natural water table is drained [199]. Nitrogen mineralization [200,201], as well as the release of sulfate, magnesium, and iron, also [202] increase after drainage of peatlands. Tree growth may also increase greatly after drainage, and therefore part of the CO_2 produced is derived from the mineralization of fresh litter material (root litter, root exudates, aboveground litter). The enhanced decomposition activity after drainage is also demonstrated by an increase in the activities of dehydrogenase [49,198], acid phosphatase, and pyrophosphatase [49].

It may be difficult to estimate the effects of fertilization on decomposition activity in drained forest soil because the activity after draining changes without any nutrient addition. There is a peak in CO_2 production shortly after draining [199]. As discussed in Section II, the short-term and long-term effects of fertilizers on the decomposition of organic matter in forest soils may differ. In drained forested peatlands, urea fertilization can have a transitory increasing effect on CO_2 production. Urea additions can also stimulate dehydrogenase activity in the peat profile: in some peat soils this stimulation can persist for a long period. However, negative effects can also occur, and there are differences in the effects in various peat layers [49]. The effect of urea on the activity of acid phosphatase depends on the peatland type, and both decreases and increases in activity have been found [49]. Urea addition generally increases pyrophosphatase activity [49].

Phosphate and potassium, added as salts or as apatite and biotite, respectively, have increased CO_2 production [199] and dehydrogenase activity [49] in

---

the bars show winter fluxes for CO_2 and CH_4 (i,j). The mean WT for the study years are shown by bars; minimum and maximum are included (l) data from P. J. Martikainen, H. Nykänen, J. Alm, and J. Silvola, Plant Soil, in press .

peat. Apatite and biotite, which also slightly increased soil pH, had a longer effect than the salt forms on these activities. However, the salts also inhibited CO_2 production [199]. Phosphorus and potassium fertilizers have decreased acid phosphatase activities in peat soils [49], similar to what has been found in minerals soils [203], and their effects on pyrophosphatase activity can be positive or negative depending on the peatland type [49].

The addition of wood ash generally has a long-term effect on increasing the decomposition activity in peat [199,204–205]. The increase in pH after addition of ash may explain its effect on decomposition activity, as liming also increased decomposition activity in peat [206,207]. Treatment with wood ash decreased phosphatase activities in peat, possibly as a result of the calcium in ash [49].

B. Methane Oxidation

Virgin peatlands are generally sources for CH_4. The lowering of the water table greatly reduces emissions of CH_4 [208] (Figure 3). Emissions of CH_4 are lowered because of the oxidation of CH_4 oxidation in the uppermost aerobic peat profile. The oxidation of CH_4 can be so high that a drained forested peatland can even be a sink for atmospheric methane (Figure 3), similar to forests on mineral soils.

Nitrogen additions inhibit CH_4 oxidation and the uptake of CH_4 in a drained peat profile. However, the relationship between NH_4^+ concentrations and inhibition of CH_4 oxidation in peat has not been definitively established. There is evidence that oxidation of CH_4 in nitrogen-rich peat can tolerate higher concentrations of NH_4^+ than the oxidation of CH_4 in mineral forest soils [209]. Not only NH_4^+ compounds, but also other inorganic compounds, may contribute to the inhibitory effect [209].

C. Nitrification and Production of Nitrous and Nitric Oxides

The high water table in the soil profile of virgin peatlands limits mineralization of nitrogen and nitrification. Nitrification in virgin peats is low [200,210–212], whereas the denitrification potential can be high [145,213]. However, the lack of nitrification in virgin peat is probably the most important reason for the negligible emissions of N_2O [201,214] and NO (K. Lång, personal communication) from virgin peats in situ. Virgin peatland can even exhibit some uptake of N_2O [213,214].

After drawdown of the water table, nitrification may increase in peat. Nitrification is more active in peat of nutrient-rich fens than in peat of nutrient-poor bogs [215,216]. In some peat soils, liming and inoculation with nitrifying bacteria may be needed to start the process [2007,211]. As in mineral forest soils (Section II.C), nitrifiers in drained peat can have different pH requirements: there are nitrifiers that have a higher activity at pH 4 than at pH 6, as well as

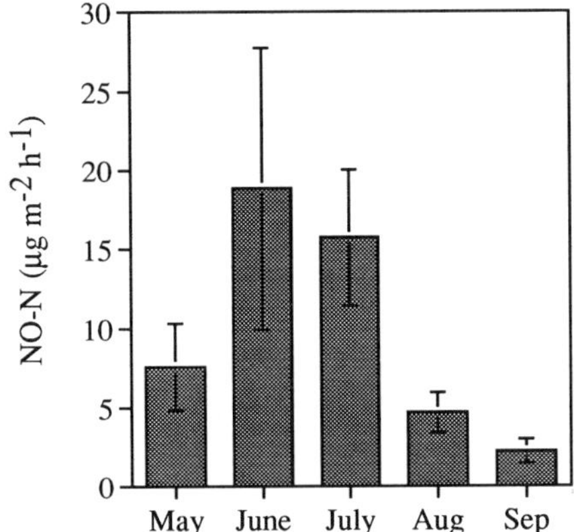

Figure 4 Emission of nitric oxide in a fen in eastern Finland that was drained for forestry in 1940. Measurements were done in 1993 (May–September) in situ with a dynamic-chamber method using an chemiluminescence analyzer. A corresponding virgin fen showed no NO emission (data from K. Lång, unpublished).

nitrifiers that have the reverse pH requirements [217]. Heterotrophic nitrification may be of some importance in peat because nitrification is not always inhibited by acetylene [217]. The draining of boreal peatlands for forestry [214] has increased N_2O emissions in nutrient-rich peatlands (minerotrophic fens) but not in nutrient-poor peatlands (ombrotrophic bogs) (Figure 3). Drainage also increases NO emissions in minerotrophic peatlands (Figure 4). The increase in the emissions of N_2O and NO after draining is probably a result of the increase in nitrification, which provides nitrate for denitrification.

IV. ENVIRONMENTAL AND CLIMATIC IMPACTS OF THE MICROBIAL PROCESSES IN DISTURBED BOREAL FOREST SOILS

A. Forests on Mineral Soils

Undisturbed boreal forest ecosystems are important sinks for atmospheric CO_2 and CH_4. Nutrient cycles in undisturbed boreal forests are closed, i.e., the leaching of nutrients, including nitrate, is negligible. Moreover, the emissions of the nitrogenous gases, N_2O and NO, are low.

Presently, boreal forests can generally retain the nitrogen from acid precipitation or nitrogen fertilizers, as both the tree stand and soil are sinks for nitrogen. The addition of alkaline substances does not enhance decomposition processes, which could then cause reductions in soil organic matter. In contrast, liming, similar to additions of nitrogen, increase the amount of soil organic matter (see Section II). However, as mentioned above, some boreal forests are becoming nitrogen saturated. Nitrogen saturation causes an increase in nutrient leaching and emission of N_2O and NO. However, the nitrogen input must be very high before the emissions of nitrogen oxides in boreal coniferous soils are highly stimulated, as experiments with heavy nitrogen additions have shown. If the emissions of nitrogen oxides from boreal forests increased, this large area would have a significant impact on the global budget of N_2O and NO. The same is true for the capacity of forest soil to consume atmospheric CH_4. As discussed in Section II, addition of nitrogen decreases the oxidation of CH_4 in forest soil. It has been estimated that the present nitrogen load has already decreased the oxidation of CH_4 in temperate grasslands and forests by 30 to 70%, and that decrease might be associated with the increase in the concentration of atmospheric CH_4 [177].

It is possible that the forest vegetation in northern latitudes will change significantly with changing climate and that the boreal coniferous forests may be replaced by temperate deciduous forests. Such a change in vegetation would affect the amount and quality of carbon received by the soils. The possible increase in carbon input, increase in temperature, and decrease in soil moisture would affect microbial processes and organic matter decomposition in forest soils [218]. The decrease in soil carbon and nitrogen storage would, in turn, have an impact on atmospheric chemistry. For example, nutrient turnover rates would increase, thereby causing a higher risk for emissions on nitrogenous gases and leaching of nutrients. However, the loss of carbon and nutrients can be transitory as a result of biomass production and nutrient uptake of the developing vegetation [219–221].

B. Forested Peatlands

Presently the northern peatlands are responsible for over 30% of the total atmospheric load of CH_4 [195]. However, the northern peatlands contain 20 to 30% of the total organic carbon and nitrogen in the world's soils [193], and the accumulation of carbon in the formation of peat counteracts continuously the increase in atmospheric CO_2 concentration. The amount of nitrogen stored in peat is great, and drained peatlands are important sources of N_2O and NO. The emission rates of N_2O from forested nutrient-rich fens (see above) can be higher than those from coniferous forest soils, but they are much lower than the emissions from peatlands drained for agriculture [214]. Long-term experiments with drained forested peatlands can be used to evaluate the fluxes of trace gases if northern peatlands dry with climatic change [3,4]. Recently it has been estimated that the possible increase in N_2O from northern peatlands induced by a changing climate would be not greater than 1% of the present global N_2O emission, despite the large amount of nitrogen stored in northern peatlands [214]. As discussed above, fluxes of CH_4 will decrease with drying of the peatlands.

To evaluate the climatic impact of the possible drying of northern peatlands, it will be necessary to know not only the changes in the fluxes of CH_4 and N_2O but also the long-term changes in their stored carbon. The results from drained and forested mires have shown that although the decomposition rate of peat is accelerated after drying, the changes in the amount of carbon stored may be low because of the increase in carbon input from the developing vegetation [222]. However, drawdown of the water table affects the stored carbon in various peatlands differently, and further study is needed before it will be possible to evaluate the complete climatic significance of the changes in the fluxes of greenhouse gases and stored carbon in different peatland types after drying.

ACKNOWLEDGMENTS

I thank Dr. Hannu Fritze, Dr. Aino Smolander, and Dr. Rauni Ohtonen for providing manuscripts submitted/in press; Kristiina Lång for providing NO data and useful suggestions for the manuscript; Dr. Leena Korhonen for revising the manuscript; and Jukka Alm and Hannu Nykånen for help in preparing the figures. An earlier version of this manuscript was improved by the excellent suggestions of G. Stotzky.

REFERENCES

1. Bouwman, A. F. 1990. Global distribution of the major soils and land cover types, p. 33–59. *In* A. F. Bouwman (ed.), Soils and the greenhouse effect. John Wiley & Sons, Chichester.
2. Sverdrup, H. U., and P. G. Warfvinge. 1988. Assessment of critical loads of acid deposition on forest soils, p. 81–129. *In* J. Nilsson and P. Grennbelt (eds.), Critical loads for sulphur and nitrogen. Report from a workshop held at Skokloster, Sweden, 19–24 March 1988. Nord. Miljörapp, Gotab, Stockholm.
3. Mitchell, J. F. B. 1989. The greenhouse effect and climate change. Rev. Geophys. 24:115–139.
4. Manabe, S., and R. T. Wetherald. 1986. Reduction in summer soil wetness induced by an increase in atmospheric carbon dioxide. Science 232:626–628.
5. Roberge, M. R., and R. Knowles. 1967. The ureolytic microflora in a black spruce (*Picea mariana* Mill) humus. Soil Sci. Soc. Am. Proc. 31:76–79.
6. Schalin, I. 1967. On the effect of nitrogen fertilization on the bacteria and microfungi in humus layer. Silva Fennica 3.1:1–12.
7. Salonius, P. O. 1972. Microbiological response to fertilizer treatments in organic forest soils. Soil Sci. 114:12–19.

8. Salonius, P. O., and M. K. Mahendrappa. 1975. Microbial respiration and exchangeable ammonium in podzol organic horizon materials treated with urea. Can. J. For. Res. 5:731–734.
9. Roberge, M. R. 1976. Respiration rates for determining the effects of urea on the soil-surface organic horizon of a black spruce stand. Can. J. Microbiol. 22:1328–1335.
10. Kelly, J. M., and G. S. Henderson. 1978. Effects on nitrogen and phosphorus additions on deciduous litter decomposition. Soil Sci. Soc. Am. J. 42:972–976.
11. Foster, N. W., E. G. Beauchamp, and C. T. Corke. 1980. Microbial activity in a *Pinus banksiana* Lamb. forest floor amended with nitrogen and carbon. Can. J. Soil Sci. 60:199–209.
12. Foster, N. W., E. G. Beauchamp, and C.T. Corke. 1985. Reactions of ^{15}N-labelled urea with Jack pine forest-floor materials. Soil Biol. Biochem. 17:699–703.
13. Mai, H., H.-J. Fiedler, and F. Leube, 1980. The effect of urea and calcium ammonium nitrate application on the microflora and N-conversion in spruce raw humus. Zbl. Bakt. II. Abt. 135:563–574.
14. Söderström B., E. Bååth, and B. Lundgren. 1983. Decrease in soil microbial activity and biomasses owing to nitrogen amendments. Can. J. Microbiol. 29:1500–1506.
15. Zöttl, H. 1960. Dynamik der Stickstoffmineralisation im organischen Waldbodenmaterial. II. Einfluss des Stickstoffgehaltes auf die Mineralstickstoffnachlieferung. Plant Soil 13:183–206.
16. Viro, P. J. 1962. Factorial experiments on forest humus decomposition. Soil Sci. 95: 24–30.
17. Nömmik, H., and B. Popovic. 1971. Recovery and vertical distribution of ^{15}N labelled fertilizer nitrogen in forest soil. Stud. For. Suec. 92:1–20.
18. Gill, R. S., and D. P. Lavander. 1983. Litter decomposition in coastal hemlock stands: impact of nitrogen fertilizers on decay rates. Can. J. For. Res. 13:116–121.
19. Hendrickson, O. Q. 1985. Variation in the C:N ratio of substrate mineralized during forest humus decomposition. Soil Biol. Biochem. 17:435–440.
20. Martikainen, P. J. 1985. Nitrification in forest soil of different pH as affected by urea, ammonium sulphate and potassium sulphate. Soil Biol. Biochem. 17:363–367.
21. Popovic, B. 1977. Effect of ammonium nitrate and urea fertilizers on nitrogen mineralization, especially nitrification in forest soil. Royal College of Forestry, Stockholm, Departments of Forest Ecology and Forest Soils. Research Notes 30:1–33.
22. Bååth, E., B. Lundgren, and B. Söderström. 1981. Effects of nitrogen fertilization on the activity and biomass of fungi and bacteria in a podzolic soil. Zentralbl. Bakteriol. Mikrobiol. Hyg. I. Abt. Orig. C 2:90–98.
23. Martikainen, P. J., T. Aarnio, V.-M. Taavitsainen, L. Päivinen, and K. Salonen. 1989. Mineralization of carbon and nitrogen in soil samples taken from three fertilized pine stands: long-term effects. Plant Soil 114:99–106.
24. Martikainen, P. J., T. Aarnio, and K. Yrjälä. 1990. Long-term effects of nitrogen addition on mineralization of carbon and nitrogen in forest soil, p. 121–126. *In* O. Brandon and R. F. Huttl (eds.), Nitrogen saturation in forest ecosystems. Proceedings of a workshop held in Aberdeen, United Kingdom, 21–23 September 1988. Kluwer Academic, Dordrecht.

25. Nohrstedt, H. O., K. Arnebrant, E. Bååth, and B. Söderström. 1989. Changes in carbon content, respiration rate, ATP content, and microbial biomass in nitrogen-fertilized pine forest soils in Sweden. Can. J. For. Res. 19:323–328.
26. Aarnio, T., K. Korhonen, A. Palojärvi, and P. J. Martikainen. 1992. Nutrient movements and microbial activity in acid forest soil treated with slow- and fast-releasing fertilizers, p. 810–811. *In* A. Teller, P. Mathy, and J. N. F. Jeffers (eds.), Response of forest ecosystems to environmental changes. Proceedings of the First European Symposium on Terrestrial Ecosystems: Forests and Woodland, Florence, Italy, 20–24 May 1991. Elsevier, London.
27. Smolander, A., A. Kurka, V. Kitunen, and E. Mälkönen. 1994. Microbial biomass C and N, and respiratory activity in soil of repeatedly limed, and N- and P-fertilized Norway spruce stands. Soil Biol. Biochem. 26:957–962.
28. van Cleve, K., and T. A. Moore. 1978. Cumulative effects of nitrogen, phosphorus, and potassium fertilizer additions on soil respiration, pH, and organic matter content. Soil Sci. Soc. Am. J. 42:121–124.
29. Nohrstedt, H.-O. 1990. Effects of repeated nitrogen fertilization with different doses on soil properties in a *Pinus sylvestris stand.* Scan. J. For. Res. 5:3–15.
30. Mälkönen, E. 1990. Estimation of nitrogen saturation on the basis of long-term fertilization experiments. Plant Soil 128:75–82.
31. Mälkönen, E., and M. Kukkola. 1991. Effects of long-term fertilization on the biomass production and nutrient status of Scots pine stands. Fer. Res. 27:113–127.
32. Nömmik, H., and G. Wiklander. 1983. Acidifying and basic effects of nitrogen fertilizers on forest soil. Naturvårdsverket rapport SNV pm 1657.
33. Persson, T., E. Bååth, M. Clarholm, H. Lundkvist, B. E. Söderström, and B. Sohlenius. 1980. Trophic structure, biomass dynamics and carbon metabolism of soil organisms in a Scots pine forest, p. 419–459. *In* T. Persson (ed.), Structure and function of Northern coniferous forests—An ecosystems study. Ecol. Bull. (Stockholm) 32.
34. Arnebrant, K., E. Bååth, and B. Söderström. 1990. Changes in microfungal community structure after fertilization of Scots pine forest soil with ammonium nitrate and urea. Soil Biol. Biochem. 22:309–312.
35. Ohenoja, E. 1988. Effect of forest management procedures on fungal fruit body production in Finland. Acta Bot. Fennica 136:81–84.
36. Anderson, T.-H., and K. H. Domsch. 1993. The metabolic quotient for CO_2 (qCO_2) as a specific activity parameters to assess the effects of environmental conditions such as pH, on the microbial biomass of forest soils. Soil Biol. Biochem. 25:393–395.
37. Anderson, T.-H., and K. H. Domsch. 1985. Maintenance carbon requirements of actively-metabolizig microbial populations under *in situ* conditions. Soil Biol. Biochem. 17:197–203.
38. Aarnio, T., and P. J. Martikainen. 1992. Nitrification in forest soil after refertilization with urea or urea and dicyandiamide. Soil Biol. Biochem. 24:951–954.
39. Berg, B. 1984. Decomposition of root litter and some factors regulating the process: long-term root litter decomposition in a Scots pine forest. Soil Biol. Biochem. 16: 609–617.

40. Berg, B. 1986. Nutrient release from litter and humus in coniferous forest soils—a minireview. Scand. J. For. Res. 1:359–369.

41. McClaugherty, C., and B. Berg. 1987. Cellulose, lignin and nitrogen concentrations as rate regulating factors in late stages of forest litter decomposition. Pedobiologia 30:101–112.

42. Berg, B., C. McClaugherty, and M.-B. Johansson. 1993. Litter mass-loss rates in late stages of decomposition at some climatically and nutritionally different pine sites. Long-term decomposition in a Scots pine fores. VIII. Can. J. Bot. 71:680–692.

43. Berg, B., and H. Staaf. 1980. Decomposing rate and chemical change of Scots pine needle litter. II. Influence of chemical composition. Ecol. Bull. 32:373–390.

44. Fog, K. 1988. The effect of added nitrogen on the rate of decomposition of organic matter. Biol. Rev. 63:433–462.

45. Berg, B., and G. Ekbohm. 1991. Litter mass-loss rates and decomposition patterns in some needle and leaf litter types. Long-term decomposition in a Scots pine forest. VII. Can. J. Bot. 69:1449–1456.

46. Keyser, P., K. T. Kirk, and J. G. Zeikus. 1978. Lignolytic enzyme of *Phanerochaete chrysosporium*: synthesized in the absence of lignin in response to nitrogen starvation. J. Bacteriol. 135:790–797.

47. Kirk, K., E. Schultz, W. J. Gonnors, L. F. Lorbaz, and J. Zeikus. 1978. Influence on culture parameters on lignin metabolism by *Phanerochaete chrysosporium*. Arch. Microbiol. 117:227–285.

48. Prescott, C. E., M. A. McDonald, S. P. Gessel, and J. P. Kimmins. 1993. Long-term effects of sewage sludge turnover in litter in a coastal Douglas fir forest. For. Ecol. Manage. 59:149–164.

49. Martikainen, P. J., R. Ohtonen, J. Silvola, and A. Vuorinen. 1994. Microbiology, p. 40–79. *In* P. Martikainen (ed.), Effects of fertilization on forest ecosystems. Biological Research Reports No. 38, University of Jyväskylä, Finland.

50. Babich, H., and G. Stotzky. 1980. Environmental factors that influence the toxicity of heavy metal and gaseous pollutants to microorganisms. Crit. Rev. Microbiol. 8: 99–145.

51. Ohtonen, R., H. Väre, A. M. Markola, A. Ohtonen, U. Ahonen-Jonnarth, and O. Tarvainen. 1993. A review of forest soil biology under the influence of gaseous pollutants and CO_2. Aquilo Ser. Bot. 32:41–54.

52. Tamm, C. O. 1976. Acid precipitation: biological effects in soil and forest vegetation. Ambio 5:235–238.

53. Francis, A. J. 1982. Effects on acidic precipitation and acidity on soil microbial processes. Water Air Soil Pollut. 18:375–394.

54. Bryant, R. D., E. A. Gordy, and E. J. Laishley. 1979. Effect on soil acidification on the soil microflora. Water Air Soil Pollut. 18:437–445.

55. Klein, T. M., N. J. Novick, J. P. Kreitinger, and M. Alexander. 1984. Simultaneous inhibition of carbon and nitrogen mineralization in forest soil by simulated acid precipitation. Bull. Environ. Contam. Toxicol. 32:698–703.

56. Persson, T., H. Lundkvist, A. Wiren, R. Hyvönen, and B. Wessen. 1989. Effects of acidification and liming on carbon and nitrogen mineralization and soil organisms in mor humus. Water Air Soil Pollut. 45:77–96.

57. Bååth, E., B. Lundgren, and B. Söderström. 1979. Effects on artificial acid rain on microbial activity and biomass. Bull. Environ. Contam. Toxicol. 23:737–740.

58. Popovic, B. 1984. Mineralization of carbon and nitrogen in humus from field acidification studies. For. Ecol. Manage. 8:81–93.

59. Persson, T., and A. Wiren. 1989. Microbial activity in forest soils in relation to acid/base and carbon/nitrogen status. Medd. Nor. Inst. Skogforsk. 42:83–94.

60. Fritze, H., O. Kiikkilä, J. Pasanen, and J. Pietikäinen. 1992. Reaction of forest soil microflora to environmental stress along a moderate pollution gradient next to an oil refinery. Plant Soil 140:175–182.

61. Nohrstedt, H. -O. 1985. Studies of forest floor biological activities in an area previously damaged by sulphur dioxide emissions. Water Air Soil Pollut. 25:301–311.

62. Visser, S., and D. Parkinson. 1989. Microbial respiration and biomass in soil of a lodgepole pine stand acidified with elemental sulphur. Can. J. For. Res. 19:955–961.

63. Ohtonen, R. 1994. Accumulation of organic matter along a pollution gradient: application of Odum's theory of ecosystems energetics. Microb. Ecol. 27:43–55.

64. Bewley, R., and G. Stotzky. 1983. Simulated acid rain (H_2SO_4) and microbial activity in soil. Soil Biol. Biochem. 15:425–429.

65. Bååth, E., B. Berg. U. Lohm, B. Lundgren, H. Lundkvist, T. Rosswall, B. Söderström, and A. Wiren. 1980. Effects on experimental acidification and liming on soil organisms and decomposition in a Scots pine forest. Pedobiologia 20:85–100.

66. Berg, B. 1986. The influence of experimental acidification on nutrient release and decomposition rates of needle and root litter in the forest floor. For. Ecol. Manage. 15:195–213.

67. Berg, B. 1986. The influence of experimental acidification on needle litter decomposition in a *Picea abies* L. forest. Scand. J. For. Res. 1:317–322.

68. Neuvonen, S., J. Suomela, E. Haukioja, M. Lindgren, and K. Ruohomäki. 1990. Ecological effects of simulated acid rain in a subartic area with low ambient sulphur deposition, p. 477–493. *In* P. Kauppi, P. Anttila, and K. Kenttämies (eds.), Acidification in Finland. Springer-Verlag, Heidelberg.

69. Hovland, J., G. Abrahamsen, and G. Ogner, 1980. Effects of artificial acid rain on decomposition of spruce needles and on mobilization and leaching of elements. Plant Soil 56:365–378.

70. Hågvar, S., and B. R. Kjondal. 1981. Decomposition of birch leaves: dry weight loss, chemical changes, and effects of artificial acid rain. Pedobiology 22:232–245.

71. Prescott, C. E., and D. Parkinson. 1985. Effect of sulphur pollution on rates of litter decomposition in a pine forest. Can. J. Bot. 63:1436–1443.

72. Lohm, U., J. Larsson, and H. Nömmik. 1984. Acidification and liming of coniferous forest soil: Long-term effects on turnover rates of carbon and nitrogen during an incubation experiment. Soil Biol. Biochem. 16:343–346.

73. Benner, R., D. L. Lewis, and R. E. Hodson. 1989. Biogeochemical cycling of organic matter in acidic environments: are microbial degradative processes adapted to low pH?, p. 33–45. *In* S. S. Rao (ed.), Acid stress and aquatic microbial interactions. CRC Press, Boca Raton, Fla.

74. Nömmik, H. 1978. Mineralization of carbon and nitrogen in forest humus as influenced by additions of phosphate and lime. Acta Agric. Scand. 28:221–230.

75. Bååth, E., B. Lundgren, and B. Söderström. 1984. Fungal populations in podzolic soil experimentally acidified to simulate acid rain. Microb. Ecol. 10:197–203.
76. Kytöviita, M.-M., H. Fritze, and S. Neuvonen. 1990. The effects of acidic irrigation on soil microorganisms at Kevo, northern Finland. Environ. Pollut. 66:21–31.
77. Bewley, R. J. F., and D. Parkinson. 1986. Sensitivity of certain soil microbial processes to acid deposition. Pedobiology 29:73–84.
78. Ohtonen, R., P. Lähdesmäki, and A. M. Markola. 1994. Cellulase activity in forest humus along an industrial pollution gradient in Oulu, northern Finland. Soil Biol. Biochem. 26:97–101.
79. Thompson, J. P., I. L. Blackwood, and T. D. Davies. 1987. Soil bacterial changes upon snowmelt: laboratory studies of the effect of early and late meltwater fractions. FEMS Microb. Ecol. 45:269–274.
80. Nihlgård, B., and B. Popovic. 1984. Effects of different liming agencies in forests. A literature review. Naturvårdsverket Report 1851 (in Swedish; English summary).
81. Derome, J., M. Kukkola, and E. Mälkönen. 1986. Forest liming on mineral soils. Results of Finnish Experiments. Report No. 3084. National Swedish Environmental Protection Board.
82. Derome, J. 1990/1991. Effects of forest liming on the nutrient status of podzolic soils in Finland. Water Air Soil Pollut. 54:337–350.
83. Zelles, L., I. Scheunert, and K. Kreutzer. 1987. Bioactivity in limed soil of a spruce forest. Biol. Fertil. Soils 3:211–216.
84. Persson, T., A. Wiren, and S. Anderson. 1990/1991. Effects of liming on carbon and nitrogen mineralization in coniferous forests. Water Air Soil Pollut. 54:351–364.
85. Yavitt, J. B., and R. M. Newton. 1990/1991. Liming effects on some chemical and biological parameters of soil (spodosols and histosols) in a hardwood forest watershed. Water Air Soil Pollut. 54:529–544.
86. Priha, O., and A. Smolander. 1994. Fumigation-extraction and substrate-induced respiration derived microbial biomass C and N in limed soil of Norway spruce stands. Biol. Fertil. Soils 17:301–308.
87. Zelles, L., K. Stepper, and A. Zsolnay. 1990. The effect of lime on microbial activity in spruce (*Picea abies* L.) forests. Biol. Fertil. Soils 9:78–82.
88. De Boer, W., M. P. J. Hundscheid, J. M. T. Schotman, S. R. Troelstra, and H. J. Laanbroek. 1993. *In situ* net N transformations in pine, fir, and oak stands of different ages on acid sandy soil, 3 years after liming. Biol. Fertil. Soils 15:120–126.
89. Gower, D. A., M. Nyborg, and N. G. Juma. 1991. Nitrogen and sulphur dynamics is limed, elemental sulphur-polluted forest soils. Soil Biol. Biochem. 23:145–150.
90. von Lutzov, M., L. Zelles, I. Scheunert, and J. C. G. Ottow. 1992. Seasonal effects of liming, irrigation, and acid precipitation on microbial biomass N in spruce (*Picea abies* L.) forest soil. Biol. Fertil. Soils 13:130–134.
91. Smolander, A., and E. Mälkönen. 1994. Microbial biomass C and N in limed soil of Norway spruce stands. Soil Biol. Biochem. 26:503–509.
91a. Frostegård, A., E. Bååth, and A. Tunlid. 1993. Shifts in the structure of soil microbial communities in limed forests as revealed by phospholipid fatty acid analysis. Soil Biol. Biochem. 25:723–730.
92. Bååth, E., and K. Arnebrant. 1993. Microfungi in coniferous forest soils treated with lime or wood ash. Biol. Fertil. Soils 15:91–95.

93. Fritze, H. and E. Bååth. 1993. Microfungal species composition and fungal biomass in a coniferous forest soil polluted by alkaline deposition. Microb. Ecol. 25:83–92.
94. Bååth, E., A. Frostegård, and H. Fritze. 1992. Soil bacterial biomass, activity, phospholipid fatty acid pattern, and pH tolerance in an area polluted with alkaline dust deposition. Appl. Environ. Microbiol. 58:4026–4031.
95. Palmgren, K. 1994. Microbiological changes in forest soil following soil preparation and liming. Folia Forestalia 603:1–27 (in Finnish; English summary).
96. Fritze, H., A. Smolander, T. Levula, and E. Mälkönen. 1994. Wood-ash fertilization and fire treatment in a Scots pine forest stand: Effects on the organic layer, microbial biomass, and microbial activity. Biol. Fertil. Soils 17:57–63.
97. Insam, H., and K. H. Domsch. 1988. Relationship between soil organic carbon and microbial biomass on chronosequences of reclamation sites. Microb. Ecol. 15:177–188.
98. Insam, H., and K. Haselwandter. 1989. Metabolic quotient of the soil microflora in relation to plant succession. Oecologia 79:174–178.
99. Kratz, W., A. Brose, and G. Weigman. 1991. The influence of lime application in damaged pine forest ecosystems in Berlin (FRG), p. 464–471. *In* O. Ravera (ed.), Terrestrial and aquatic ecosystems: Perturbation and recovery. Ellis Horwood Limited, Chichester.
100. Lundgren, B., and B. Söderström. 1983. Bacterial numbers in a pine forest soil in relation to environmental factors. Soil Biol. Biochem. 15:625–630.
101. Niemelä, S., and V. Sundman, 1977. Effects of clear-cutting on the decomposition of bacterial populations of northern spruce forest soil. Can. J. Microbiol. 23:131–138.
102. Sundman, V., V. Huhta, and S. Niemelä. 1978. Biological changes in northern spruce forest soil after clear-cutting. Soil Biol. Biochem. 10:393–397.
103. Hendrickson, O. Q., and J. B. Robinson. 1984. Effects of roots and litter on mineralization processes in forest soil. Plant Soil 80:391–405.
104. Lundgren, B. 1982. Bacteria in a pine forest soil as affected by clear-cutting. Soil Biol. Biochem. 14:537–542.
105. Gordon, A., R. E. Schlentner, and K. van Cleve. 1987. Seasonal pattern of soil respiration and CO_2 evolution following harvesting in the white spruce forest of interior Alaska. Can. J. For. Res. 17:304–310.
106. Gosz, J. R., and F. M. Fisher. 1984. Influence of clear-cutting on selected microbial processes in forest soils, p. 523–530. *In* M. J. Klug and C. A. Reddy (eds.), Current Perspectives in microbial ecology. American Society for Microbiology, Washington, D.C.
107. Attiwill, P. M., and M. A. Adams. 1993. Nutrient cycling in forests. New Phytol. 124:561–582.
108. van Cleve, K., J. Yarie, R. Ericson, and C. T. Dyrness. 1993. Nitrogen mineralization and nitrification in successional ecosystems on the Tana River floodplain, interior Alaska. Can. J. For. Res. 23:970–978.
109. Vitousek, P. M., and P. A. Matson. 1984. Mechanisms of nitrogen retention in forest ecosystems: A field experiment. Science 225:51–52.
110. White, C. S. 1986. Effects of prescribed fire on rates of decomposition and nitrogen mineralization in a ponderosa pine ecosystem. Biol. Fertil. Soil. 2:87–95.

111. Bisset, J., and D. Parkinson. 1980. Long-term effects of the fire on the composition and activity of the soil microflora of a subalpine, coniferous forest. Can. J. Bot. 58: 1704–1721.

112. Pietikäinen, J., and H. Fritze. 1993. Microbial biomass and activity in the humus layer following burning: Short-term effects of two different fires. Can. J. For. Res. 23:1275–1285.

113. Fritze, H., T. Pennanen, and J. Pietikäinen. 1993. Recovery of soil microbial biomass and activity from prescribed burning. Can. J. For Res. 23:1286–1290.

114. Killham, K. 1990. Nitrification in coniferous forest soils. Plant Soil 128:31–44.

115. Schimel, J. P., M. K. Firestone, and K. S. Killham. 1984. Identification of heterotrophic nitrifiers in a Sierran forest soil. Appl. Environ. Microbiol. 48:802–806.

116. Duggin, J. A., G. K. Vogt, and F. H. Bormann. 1991. Autotrophic and heterotrophic nitrification in response to clear-cutting northern hardwood forest. Soil Biol. Biochem. 23:779–787.

117. Klingernsmith, K. M., and K. van Cleve. 1993. Patterns of nitrogen mineralization and nitrification in floodplain successional soils along the Tanana River, interior Alaska. Can. J. For. Res. 23:964–969.

118. Focht, D. D,. and W. Verstraete. 1977. Biochemical ecology of nitrification and denitrification. Adv. Microb. Ecol. 1:135–214.

119. van Breemen, N., J. Mulder, and J. J. M. van Grinsven. 1987. Impacts of acid atmospheric deposition on woodland soils in the Netherlands. II. Nitrogen transformations. Soil Sci. Soc. Am. J. 51:1634–1640.

120. Stams, A. J. M., E. M. Flameling, and E. C. L. Marnette. 1990. The importance of autotrophic versus heterotrophic oxidation of atmospheric ammonium in forest ecosystems with acid soil. FEMS Microbiol. Ecol. 74:337–344.

121. Stams, A. J. M., H. W. G. Booltink, I. J. Lutke-Schipholt, B. Beemsterboer, J. R. W. Woittiez, and N. van Breemen. 1991. A field study on the fate of ^{15}N-ammonium to demonstrate nitrification of atmospheric ammonium in an acid forest soil. Biogeochemistry 13:241–255.

122. Tietema, A., W. De Boer, L. Riemer, and J. M. Verstraten. 1992. Nitrate production in nitrogen-saturated acid forest soils: vertical distribution and characteristics. Soil Biol. Biochem. 24:235–240.

123. Martikainen, P. J. 1984. Nitrification in two coniferous forest soils after different fertilization treatments. Soil Biol. Biochem. 16:577–582.

124. Martikainen, P. J., M. Lehtonen, K. Lång, W. De Boer, and A. Ferm. 1993. Nitrification and nitrous oxide production potentials in aerobic soil samples from the soil profile of a Finnish coniferous site receiving high ammonium deposition. FEMS Microbiol. Ecol. 13:113–122.

125. Robertson, G. P. 1982. Nitrification in forested ecosystems. Phil. Trans. R. Soc. Lond. B. 246:445–457.

126. Overrein, L. N. 1971. Isotope studies on nitrogen in forest soil. Rep. Norweg. For. Res. Inst. 114:261–280.

127. Heilman, P. 1974. Effect of urea fertilization on nitrification in forest soils of the Washington coastal area. Soil Sci. Soc. Am. Proc. 38:664–667.

128. De Boer, W., A. Tietema, P. J. A. Klein Gunnewiek, and H. J. Laanbroek. 1992. The chemolithotrophic ammonium-oxidizing community in a nitrogen saturated acid forest soil in relation to pH-dependent nitrifying activity. Soil Biol. Biochem. 24: 229–234.

129. De Boer, W., H. Duyts, and H. J. Laanbroek. 1989. Urea stimulated autotrophic nitrification in suspensions of a fertilized, acid heath soil. Soil Biol. Biochem. 21: 349–354.

130. Tietema, A., and J. M. Verstraten. 1991. Nitrogen cycling in an acid forest ecosystem in the Netherlands under increased atmospheric nitrogen input. The nitrogen budget and the effect of nitrogen transformations on the proton budget. Biogeochemistry 15:21–46.

131. Tietema, A., L. Riemer, J. M. Verstraten, M. P. van der Maas, A. J. van Wijk, and I. van Voorhuyzen. 1993. Nitrogen cycling in acid forest soils subject to increased atmospheric nitrogen input. For. Ecol. Manage. 57:29–44.

132. Hesselman, H. 1926. Studie över barrskogens humustäcke, dess egenskaper och beroende av skogsvården. Medd. Skogsf. Anst. 22:169–552.

133. Chase, F. E., and G. Baker. 1954. A comparison of microbial activity in an Ontario forest soil under pine, hemlock and maple cover. Can. J. Microbiol. 1:45–54.

134. Corke, C. T. 1958. Nitrogen transformations in Ontario forest podzols, p. 116–121. *In* T. D. Stevens and R. L. Cook (eds.), Proceedings of the First North American Forest Soil Conference, Michigan State University Agricultural Experimental Station, East Lansing.

135. Zöttl, H. 1960. Dynamik der Stickstoffmineralization in organischen Waldbodenmaterial. III. pH-Wert und Mineralstickstoff-Nachlieferung. Plant Soil 13:208–223.

136. Popovic, B. 1967. Kvävemobiliseringsförsök med humusprov från gödlingsförsöksytor i skogsbestånd på fast mark. Royal College of Forestry, Stockholm, Departments of Forest Ecology and Forest Soils. Research Notes 6:1–13.

137. Papen, H., R. von Berg, B. Hellmann, and H. Rennenberg. 1991. Einfluss von saurer Beregnung und Kalkung auf chemolithotrophe und heterotrophe Nitrifikation in Böden des Högwaldes. Forstwiss. Forsch. 39:111–116.

138. Johnson, D. W. 1992. Nitrogen retention in forest soils. J. Environ. Qual. 21:1–12.

139. Smith, W. H., F. H. Bormann, and G.E. Likens. 1968. Response of chemoautotrophic nitrifiers to forest cutting. Soil Sci. 106:471–473.

140. Tamm, C. O., H. Holmen, B. Popovic, and G. Wiklander. 1974. Leaching of plant nutrients from soils as a consequence of forest operations. Ambio 3:211–221.

141. Vitousek, P. M., J. R. Gosz, C. C. Grier, J. M. Melillo, W. A. Reiners, and R. L. Todd. 1979. Nitrate losses from disturbed ecosystems. Science 204:469–474.

142. Prescott, C. E., J. P. Corbin, and D. Parkinson. 1992. Immobilization and availability of N and P in the forest floors of fertilized Rocky Mountain coniferous forests. Plant Soil 143:1–10.

143. Davidson, E. A. 1991. Fluxes of nitrous oxide and nitric oxide from terrestrial ecosystems, p. 219–235. *In* J. F. Rogers and W. R. Whitman (eds.), Microbial production and consumption of greenhouse gases: Methane, nitrogen oxides, and halomethanes. American Society for Microbiology, Washington, D.C.

144. Parkin, T. B., A. J. Sexstone, and J. Tiedje. 1985. Adaptation of denitrifying populations to low soil pH. Appl. Environ. Microbiol. 49:1053–1056.

145. Müller, M., V. Sundman, and J. Skujins. 1980. Denitrification in low pH spodosols and peats determined with the acetylene inhibition method. Appl. Environ. Microbiol. 40:235–239.

146. Pluth, D. J., and H. Nömmik. 1981. Potential denitrification affected by nitrogen source of a previous fertilization of an acid forest soil from central Sweden. Acta Agric. 31:235–241.

147. Uomala, P., P. Martikainen, L. Sköld, and K. Kari. 1982. Effect of fertilization on forest soil denitrification potential, p. 329–336. *In* The Second National Symposium on Biological Nitrogen Fixation, Helsinki, Finland, 8–10 June 1982. Finnish National Fund for Research and Development.

148. Nodar, R., M. J. Acea, and T. Carballas. 1992. Microbiological response to $Ca(OH)_2$ treatments in a forest soil. FEMS Microb. Ecol. 86:213–219.

149. Willison, T. W., and J. M. Anderson. 1991. Denitrification potentials, controls and spatial patterns in a Norway spruce plantation. For. Ecol. Manage. 44:69–76.

150. Brumme, R., and F. Beese. 1992. Effects of liming and nitrogen fertilization on emissions of CO_2 and N_2O from a temperature forest. J. Geophys. Res. 97 D12: 12851–12858.

151. Papen, H., B. Hellman, H. Papke, and H. Rennenberg. 1993. Emission of N-oxides from acid irrigated and limed soils of a coniferous forest in Bavaria, p. 245–260. *In* R. S. Oremland (ed.), Biochemistry of global change: Radiatively active trace gases. Chapman & Hall, New York.

152. Firestone, M., R. B. Firestone, and J. M. Tiedje. 1980. Nitrous oxide from soil denitrification: Factors controlling its biological production. Science 208:749–751.

153. Weier, K. L., and J. W. Gilliam. 1986. Effect of acidity on denitrification and nitrous oxide evolution from Atlantic coastal plain soils. Soil Sci. Soc. Am. J. 50: 1202–1205.

154. Nägele, W., and R. Conrad. 1990. Influence of soil pH on the nitrate-reducing populations and their potential to reduce nitrate to NO and N_2O. FEMS Microbiol. Ecol. 74:49–58.

155. Wijler, J., and C. C. Delwiche. 1954. Investigation of denitrifying processes in soils. Plant Soil 5:155–169.

156. Remde, A., F. Slemr, and R. Conrad. 1989. Microbial production and uptake of nitric oxide and soil. FEMS Microbiol. Ecol. 62:221–230.

157. Nägele, W., and R. Conrad. 1990. Influence of pH on the release of NO and N_2O from fertilized and unfertilized soil. Biol. Fertil. Soils 10:139–144.

158. Melin, J., and H. Nömmik. 1983. Denitrification measurements in intact soil cores. Acta Agric. Scand. 33:145–151.

159. Robertson, G. P., and J. Tiedje. 1984. Denitrification and nitrous oxide production in successional and old-growth Michigan forests. Soil Sci. Soc. Am. J. 48:383–389.

160. Martikainen, P. J. 1985. Nitrous oxide emission associated with autotrophic ammonium oxidation in acid coniferous forest soil. Appl. Environ. Microbiol. 50:1519–1525.

161. Martikainen, P. J., and W. De Boer. 1993. Nitrous oxide production and nitrification in acidic soil from a Dutch coniferous forest. Soil Biol. Biochem. 25:343–347.

162. Tietema, A., W. Bouten, and P. E. Wartenbergh. 1991. Nitrous oxide dynamics in an oak-beech forest ecosystem in the Netherlands. For. Ecol. Manage. 44:53–61.

163. Henrich, M., and K. Haselwandter. 1991. Denitrifying potential and enzyme activity in a Norway spruce forest. For. Ecol. Manage. 44:63–68.

164. Groffman, P. M., and J. Tiedje. 1989. Denitrification in north temperate forest soils: spatial and temporal patterns at the landscape and seasonal scales. Soil Biol. Biochem. 21:613–620.

165. Hulm, S. C., and K. Killham. 1988. Gaseous nitrogen loss from soil under Sitka spruce following the application of fertilizer ^{15}N urea. J. Soil Sci. 39:417–424.

166. Vermes, J.-F., and D. D. Myrold. 1992. Denitrification in forest soils of Oregon. Can. J. For. Res. 22:504–512.

167. Klingensmith, K. M., and K. van Cleve. 1993. Denitrification and nitrogen fixation in floodplain successful soils along the Tanana River, interior Alaska. Can. J. For. Res. 23:956–963.

168. Blew, R. D., and d. Parkinson. 1993. Nitrification and denitrification in a white spruce forest in southwest Alberta, Canada. Can. J. For. Res. 23:1715–1719.

169. Aber, J. D., K. J. Nadelhoffer, P. Steudler, and J. M. Melillo. 1989. Nitrogen saturation in northern forest ecosystems. BioScience 39:378–386.

170. Gundersen, P. 1991. Nitrogen deposition and the forest nitrogen cycle: role of denitrification. For. Ecol. Manage. 44:15–28.

171. Rosen, K., P. Gundersen, L. Tegnhammer, M. Johansson, and T. Frogner. 1992. Nitrogen enrichment of nordic forest ecosystems: the concept of critical load. Ambio 21:364–368.

172. Nesbit, S. P., and G. A. Breitenbeck. 1992. A laboratory study of factors influencing methane uptake by soils. Agric. Ecosyst. Environ. 41:39–54.

173. Steudler, P. A., R. D. Bowden, J. M. Melillo, and J. D. Aber. 1989. Influence of nitrogen fertilization on methane uptake in temperate forest soils. Nature 341:314–316.

174. O'Neill, J. G., and J. F. Wilkinson. 1977. Oxidation of ammonia by methane-oxidizing bacteria and the effects of ammonia on methane oxidation. J. Gen. Microbiol. 100:407–412.

175. Carlsen, H. N., L. Joergensen, and H. Degn. 1991. Inhibition by ammonia of methane utilization in *Methylococcus capsulatus* (*Bath*). Appl. Microbiol. Biotechnol. 35:124–127.

176. King, G. M., and S. Schnell. 1994. Effect of increasing atmospheric methane concentration on ammonium inhibition of soil methane consumption. Nature 370: 282–284.

177. Ojima, D. S., D. W. Valentine, A. R. Mosier, W. J. Parton, and D. S. Schimel. 1993. Effects of land use change on methane oxidation in temperature forest and grasslands soils. Chemosphere 26:675–685.

178. Whalen, S. C., W. Reeburgh, and V. A. Barber. 1992. Oxidation of methane in boreal forest soils: a comparison of seven measures. Biogeochemistry 16:181–211.

179. Adamsen, A. P. S., and G.M. King 1993. Methane consumption in temperate ad subartic forest soils: rates, vertical zonation, and response to water and nitrogen. Appl. Environ. Microbiol. 59:485–490.

180. Persson, T. 1983. Influence of soil animals on nitrogen mineralization in a northern Scots pine forest, p. 117–126. *In* P. Lebrun, H. M. Andre, C. de Mets, C. Gregoire-Wibo, and G. Wauthy (eds.), New trends in soil biology. Proceedings of the VIII

International Colloquium on Soil Zoology, August 9–September 2, 1982, Louvain-la-Neuve, Belgium.

181. Clarholm, M. 1985. Interactions of bacteria, protozoa and plants and leading to mineralization of soil nitrogen. Soil Biol. Biochem. 17:181–187.
182. Verhoef, H. A., and L. Brussaard. 1990. Decomposition and nitrogen mineralization in natural and agroecosystems: the contribution of soil animals. Biogeochemistry 11: 175–211.
183. Setälä, H., and V. Huhta. 1991. Soil fauna increase *Betula Pendula* growth: laboratory experiments with coniferous forest floor. Ecology 72:665–671.
184. Huhta, V. 1994. Soil fauna, p. 79–97. *In* P. Martikainen (ed.), Effects of fertilization on forest ecosystems. Biological Research Report No. 38, University of Jyväskylä, Finland.
185. Hyvönen, R., and V. Huhta. 1989. Effects of lime, ash and nitrogen fertilizers on nematode populations in Scots pine forest soil. Pedobiologia 33:129–143.
186. Huhta, V. 1984. Response of *Cognettia sphagnetorum* (Enchytraeidae) to manipulation of pH and nutrient status in coniferous forest soil. Pedobiologia 27:245–260.
187. Koskenniemi, A., and V. Huhta. 1986. Effects of fertilization and manipulation of pH on mite (Acari) populations of coniferous forest soil. Rev. Ecol. Biol. Sol. 23: 271–286.
188. Hyvönen, R., and T. Persson. 1990. Effects of acidification and liming on feeding groups of nematodes in coniferous forest soil. Biol. Fertil. Soils 9:205–210.
189. Kuprevich, V. F., and T. A. Shcherbakova. 1971. Comparative enzymatic activity in diverse types of soil, p. 167–201. *In* A. D. McLaren and J. Skujins (eds.), Soil biochemistry, Vol. 2. Marcel Dekker, New York.
190. Given, P. H., and C. H. Dickinson. 1975. Biochemistry and microbiology of peats, p. 123–212. *In* E. A. Paul and A. D. McLaren (eds.), Soil biochemistry, Vol 3. Marcel Dekker, New York.
191. Verhoeven, J. T. A. 1986. Nutrient dynamics in minerotrophic peat mires. Aquat. Bot. 25:117–137.
192. Richardson, C. J., D. L. Tilton, J. A. Kadlec, J. P. M. Chamie, and W. A. Wentz. 1978. Nutrient dynamics of northern wetland ecosystems. *In* R. E. God, D. F. Whigham, and R. L. Simpson (eds.), Freshwater wetlands, ecological processes and management potential. Academic Press, New York.
193. Gorham, E. 1991. Northern peatlands: role in the carbon cycle and probable responses to climatic warming. Ecol. Appl. 1:182–195.
194. Tyler, S. C. 1991. The global methane budget, p. 7–38. *In* J. E. Rogers and W. B. Whitman (eds.), Microbial production and consumption of greenhouse gases: methane, nitrogen oxides, and halomethanes. American Society for Microbiology, Washington, D.C.
195. Bartlett, K. B., and R. C. Harriss. 1993. Review and assessment of methane emission from wetlands. Chemosphere 26:261–320.
196. Silvola, J. 1986. Carbon dioxide dynamics in mires reclaimed for forestry. Ann. Bot. Fennica 23:59–67.
197. Moore, T. R., and R. Knowles. 1989. The influence of water table levels on methane and carbon dioxide emissions from peatland soils. Can. J. Soil Sci. 69:33–38.

198. Kim, J., and S.B. Verma. 1992. Soil surface CO_2 flux in Minnesota peatland. Biogeochemistry 18:37–51.

199. Silvola, J., J. Välijoki, and H. Aaltonen. 1965. Effect of draining and fertilization on soil respiration at three ameliorated peatland sites. Acta For. Fenn. 191:1–32.

200. Williams, B. L., and R. E. Wheatley. 1988. Nitrogen mineralization and water-table height in oligotrophic deep peat. Biol. Fertil. Soils 6:141–147.

201. Freeman, C., M. A. Lock, and B. Reynolds. 1983. Climatic change and the release of immobilized nutrients from Welsh riparian wetland soils. Ecol. Engin. 2:367–373.

202. Raskova, N. V. 1984. Activity of oxidoreductases stable and labile to drying in virgin and cultivated soddy-podzolic soils. Vestn. Mosk. Univ. Ser XVII, Pochovoved 01:30–36.

203. Pang, P. C., and H. Kowalenko. 1986. Phosphatase activity in forest soil. Soil Biol. Biochem. 18:35–40.

204. Karsisto, M. 1979. Effect of forest improvement measures on activity of organic matter decomposing microorganisms in forested peatland. II. Effect of ash fertilization. Suo 30:81–91.

205. Weber, A., M. Karsisto, R. Leppänen, V. Sundman, and J. Skujins. 1985. Microbial activities in a histosols: Effects of wood ash and NPK fertilizers. Soil Biol. Biochem. 17:291–296.

206. Kaunisto, S., and M. Norlamo. 1976. On nitrogen mobilization in peat. I. Effect of liming and rotatation in different incubation temperatures. Comm. Inst. For. Fenn. 88.2:1–27.

207. Ivarson, K. C. 1977. Changes in decomposition rate, microbial population and carbohydrate content of an acid peat bog after liming and reclamation. Can. J. Soil Sci. 57:129–137.

208. Roulet, N., T. Moore, J. Bubier, and P. Lafleur. 1992. Northern fens: methane flux and climate change. Tellus 44B:100–105.

209. Crill, P. M, P. J. Martikainen, H. Nykänen, and J. Silvola. 1994. Temperature and N fertilization effects on methane oxidation in a drained peatland soil. Soil Biol. Biochem. 26:1331–1339.

210. Rosswall, T., and U. Granhall. 1980. Nitrogen cycling in a subartic mire. Ecol. Bull. (Stockholm) 30:209–234.

211. Rangeley, A., and R. Knowles. 1988. Nitrogen transformation in a Scottish peat soil under laboratory conditions. Soil Biol. Biochem. 20:385–391.

212. Koerselman, W., H. de Caluwe, and W. M. Kieskamp. 1989. Denitrification and nitrogen fixation in two quaking fens in the Vechtplassen, The Netherlands Biogeochemistry 8:153–165.

213. Hemond, H. F. 1983. The nitrogen budget of Thoreau bog. Ecology 64:99–109.

214. Martikainen, P. J., H. Nykänen, P. Crill, and J. Silvola. 1993. Effect of a lowered water table on nitrous oxide fluxes from northern peatlands. Nature 366:51–53.

215. Waughman, G. J. 1980. Chemical aspects of the ecology of some south German peatlands. J. Ecol. 68:1025–1046.

216. Proctor, M. C. F. 1993. Changes in the chemical composition of mire-water samples during storage. J. Veget. Sci. 4:661–666.

217. Lång, K., M. Lehtonen, and P. J. Martikainen. 1994. Nitrification potentials at different pH values in peat samples from various layers of a drained mire. Geomicrobiol. J. 11:141–147.
218. Anderson, J. M. 1991. The effects of climate change on decomposition processes in grasslands and coniferous forests. Ecol. Appl. 1:326–347.
219. Bonan, G. B., and K. van Cleve. 1992. Soil temperature, nitrogen mineralization, and carbon source-sink relationships in boreal forests. Can. J. For. Res. 22:629–639.
220. Post, W. M., and J. Pastor. 1992. Aspects of the interactions between vegetation and soil under global change. Water Air Soil Pollut. 64:345–363.
221. Oechel, W. C., and G. L. Vourlitis. 1994. The effects of climate change on land-atmospheric feedbacks in arctic tundra regions. Trends Ecol. Evol. 9:324–329.
222. Laine, J., H. Vasander, and A. Puhalainen. 1992/1993. A method to estimate the effect of forest drainage on the carbon store of a mire. Suo (Mires and Peat) 43:227–230.
223. Matson, P. A., S. T. Gower, C. Volkmann, C. Billow, and C. C. Grier. 1992. Soil nitrogen cycling and nitrous oxide flux in a Rocky Mountain Douglas-fir forest: effects of fertilization, irrigation and carbon addition. Biogeochemistry 18:101–117.
224. Klemedtsson, L., and B. H. Svensson. 1988. Effects of acid deposition on denitrification and N_2O emission from forest soils, p. 343–362. *In* J. Nilsson and P. Grennfelt (eds.), Critical loads for sulphur and nitrogen. Reports from a workshop held at Skokloster, Sweden, 19–24 March 1988. NORD miljorapport 1988:15, Nordic Council of Ministers, Copenhagen.

6

Enumeration and Expression of Bacterial Counts in the Rhizosphere

Jay Scott Angle, Joel V. Gagliardi, and Maria S. McIntosh
University of Maryland, College Park, Maryland

Morris A. Levin University of Maryland Biotechnology Institute, Baltimore, Maryland

I. INTRODUCTION

The rhizosphere is that portion of the soil that is influenced by the root [1], and this soil exhibits enhanced microbial numbers and activity in comparison with soil not influenced by the plant [2]. A variety of materials are released from roots including sugars, amino acids, and proteins. These materials are used by rhizosphere organisms as sources of energy and other nutrients [1]. In addition, compounds are released from roots that exhibit hormonal activity on nearby microbial populations. The area of soil influenced by root exudates varies with plant species and soil type. Enhanced activity is greatest immediately adjacent to the root (often called the rhizoplane) and decreases with distance from the root surface. Most estimates suggest that the rhizosphere extends a few millimeters from the root [3].

Microorganisms that respond by increased growth in the rhizosphere have been called rhizosphere competent [4]. Rhizosphere microorganisms are not simply chance inhabitants affected by a passing root but are microorganisms that specifically and preferentially respond to compounds in the rhizosphere [5]. Dahm [6] has shown that rhizosphere microorganisms have different requirements for amino acids, vitamins, and growth factors as compared with nonrhizosphere microorganisms. Many rhizosphere microorganisms have also been shown to respond, via rapid or dense growth, to the presence of exudates in growth media [7].

The zone comprising the rhizosphere is easy to describe—it is the zone of soil influenced by the plant root. However, from a mechanistic point of view, definition of this zone is much more difficult. Giddens and Todd [8] noted that "precise method(s) of determination of microbial numbers in the root zone do not exist." Part of this problem is related to the fact that there is no discrete boundary between the rhizosphere and soil not influenced by the root. Further, the zone influenced by a root may vary within the area of a single plant root system and even within a single root. Consequently, many mechanistic definitions of the rhizosphere have been used in the past. Unfortunately, since many definitions have been used, it is difficult to compare rhizosphere populations between different studies. This chapter examines various methods that have been used to define the rhizosphere and will describe a standardized method that should allow for the comparison of results between studies.

The rhizosphere is influenced in a variety of ways by the root. As root cells die and are sloughed off, microorganisms rapidly degrade cellular components. More important, the root excretes a variety of organic compounds that affect microbial numbers and diversity. Amino acids and sugars are exuded in especially high quantities and the release of these compounds have been reviewed [9,10]. Whipps and Lynch [11] reported that plants can lose up to 80% of their entire photosynthate supply via the loss of root exudates, although losses of approximately 25% are more common. In addition to the biological effects of the plant on the rhizosphere, many soil chemical and physical factors such as water content and pH are affected by the plant [12].

The complexity and abundance of nutritional components within the rhizosphere supports microbial populations that are typically higher than in the associated bulk soil [13,14]. Numerous reports [15–17] have shown that microbial populations are 5 to 20 times higher in the rhizosphere than in the bulk soil, although the differences observed can be as large as 100 times or more [18]. Further, rhizosphere inhabitants are generally larger than identical species in the bulk soil [19] and have thicker capsules than nonrhizosphere counterparts [20]. Many rhizosphere inhabitants are gram negative [18], although some gram positive, nonspore-forming, rod-shaped bacteria are also found [2].

The importance of rhizosphere organisms has only recently been recognized. Rhizosphere organisms, because of their high numbers, provide a buffer against rapid changes in antagonistic organisms that could allow plant pathogens to become established within the rhizosphere. A stable rhizosphere community is believed to protect plants against a variety of pathogens [3]. The ability of rhizosphere microorganisms to inhibit the growth of pathogens is currently being exploited by inoculating plants with these organisms [21–23]. Rhizosphere microorganisms are also involved in the nutrition of their hosts. For example, many rhizosphere bacteria can solubilize iron [24] that would otherwise be unavailable to the plant. Rhizosphere microorganisms may also be used as vectors for recombinant genes, including those that affect biocontrol activity, nodulation of legumes, and bioremediation of toxic organics [25]. Several recent studies have examined the influence of rhizosphere microorganisms on the degradation of industrial organics that contaminate soil [2,26,27]. Past studies have examined the influence of the rhizosphere on stimulating the degradation of pesticides [28,29].

II. ENUMERATION

A. Background

Microbial populations in the rhizosphere are often enumerated for two reasons: to examine inoculum survival and efficacy, and to describe the population within this region. Efficacy is predicated upon successful establishment within the rhizosphere. Densities of 10^6 to 10^9 colony forming units (cfu) per gram in rhizosphere soil are frequently necessary to observe an effect of the microbial inocula. When microbial populations fail to become established at this level, few if any effects are observed. Survival of microorganisms in soil and risk assessment of recombinant microbes also requires that microbial density in the rhizosphere be quantified. Understanding the ability of a microorganism to persist in the environment is essential when assessing the risk of a potential release of a recombinant microorganism into soil. Genetically engineered microbes (GEMs) that survive for exceptionally long periods, possibly as a permanent member of the indigenous microbial community in the rhizosphere, should probably not be released into the environment [30]. Competition between a GEM and indigenous soil microbes, especially those similar to the parent of the GEM, is also an essential measurement, requiring that microbial counts be determined in the rhizosphere.

Enumeration of rhizosphere microorganisms and the subsequent expression of rhizosphere counts is confounded by a lack of methodological standardization. Although methods are available for each step of the analytical process, the plethora of these methods often makes it difficult to compare data between studies. This chapter focuses on methods for the extraction of microorganisms from the rhizosphere and on indices used to express counts within this zone. Procedures for detection in rhizosphere extracts have previously been reviewed in depth [31,32]. Traditional cultural techniques (e.g., plating on artificial media [33] and most probable number techniques [34] have most often been used to assay viable populations. When it is necessary to enumerate the population of microorganisms that have been inoculated into soil, it is possible to genetically mark the introduced strain and recover them on selective and differential media [5]. Molecular (polymerase chain reaction, gene probes) and traditional approaches have also been used for quantification, including viable but nonculturable organisms [13]. Direct counting [35], assays of biomass [36], and the enumeration of gene sequences [37] have also been employed for the enumeration of rhizosphere organisms.

Critical to the enumeration of rhizosphere microorganisms is the definition used to define the physical extent of the rhizosphere. The definition of the rhizosphere will, to some extent, determine the method of isolation [5]. Katznelson [38] mechanistically defined the rhizosphere as that zone around the root where the microbial population was higher than in the rest of the soil. The R/S ratio (rhizosphere numbers/bulk soil numbers) was used as an index of this area. However, the R/S ratio is affected by the amount of soil adhering to the root, which varies greatly with soil type.

Timonin [39] was one of the first who attempted to standardize the expression of rhizosphere counts by shaking soil from the root and analyzing the root-adhering soil matrix. Microbial counts were expressed on the basis of cfu/g dry soil. Rovira and Davey [40] noted that most studies failed to differentiate between rhizosphere and rhizoplane organisms.

No single definition exists to describe this critical zone of the root-soil interface. Often the terms *rhizosphere* and *rhizoplane* are used interchangeably. Some studies have defined the rhizosphere as that area immediately next to, or in contact with, the root; to some extent only including what is more frequently called the rhizosphere. Other studies have assayed roots and a large volume of soil, often extending centimeters into the bulk soil. In the most limited definition of the rhizosphere, as much of the soil as possible is removed from the root before rhizosphere inhabitants are enumerated. Jiang and Sato [41] and Tedla and Stanghellini [42] first washed roots before analysis. This effectively removes much of the visibly adhering soil. Christiansen-Weniger and van Veen [43] scraped as much soil as possible from the root before the root was assayed. These procedures assess only those microorganisms that are intimately associated with the root. Organisms that may not be in contact with the root, yet are still influenced by root exudates, are not enumerated. Through successive washings of roots, Harley and Waid [44] showed that most microorganisms (in this case, fungi) were easily removed from roots and that analysis of only the most tightly adhering soil yielded significantly higher microbial counts.

The other extreme occurs when exudates are considered to affect the soil population far from the root, although this concept may enumerate microorganisms that are not considered true rhizosphere inhabitants. Shaw et al. [45] assayed soil that was only loosely associated with the root. Ten to 20 g of soil and root were collected and assayed, and the observed bacteria were defined as rhizosphere inhabitants. The problem with this approach is that soil less affected by roots may obscure larger effects occurring in the rhizosphere.

Few studies have compared microbial populations in the rhizosphere and bulk soil; there is obviously a need for this information. Richaume et al. [46]

noted that the number of recombinant *Pseudomonas* sp. inoculated into soil was highest in the rhizosphere and significantly lower in the bulk soil. These authors also questioned the extent to which rhizosphere inhabitants are found in or escape into bulk soil.

To examine this question, an experiment was conducted in which *P. aureofaciens* 3732 RNL-11 was inoculated onto seed and the population was assayed over time. *P. aureofaciens* is a recombinant bacterium containing the *lac* ZY insert. This insert allows the organism to metabolize X-Gal (5-bromo-4-chloro-3-idolyl-β-D-galactopyranoside), an analog of lactose that causes colonies to appear blue. The bacterium is also intrinsically resistant to nalidixic acid and rifamycin SV. A single seed of wheat (*Triticum arvense*) was sown into intact soil cores (5 cm in diameter, 15 cm long) from an Adelphia sandy loam soil and 1 ml of a 10^9 CFU/ml inoculum was added to each seed after germination. At periodic intervals, plants were removed, the roots excised, and soil was removed from the roots by gentle shaking. The population of *P. aureofaciens* was analyzed in both the rhizosphere and the bulk soil, as well as organisms tightly associated with the root.

During the first 28 days of growth, total numbers were highest in the bulk soil, and the total number in the rhizosphere was lowest (Figure 1). However, based on concentration, the root supported the greatest number of organisms and the bulk soil population was lowest. Obviously, conclusions derived from these data would be vastly different depending on whether populations were expressed as total number or concentration. This study also shows the importance of examining those portions of the soil in which the organisms of interest are most likely to reside. Many organisms are tightly bound to the root; it is therefore necessary to examine the root and adhering soil. If the organism is not closely associated with the root, then the bulk soil and possibly loosely held rhizosphere soil should be enumerated.

Once the area of the rhizosphere has been mechanistically defined, it is necessary to isolate the rhizosphere so that organisms can be extracted. Many variations exist within the range of available extraction methods. If the rhizosphere is defined as only that soil that adheres very tightly to the root, preliminary methods must first be employed to remove most of the soil from the root. Christiansen-Weniger and van Veen [43] and Kremer et al. [47] first washed roots before shaking in a dilution buffer. As noted, these procedures may fail to assess fully the influence of root exudates on the soil microbial population. If, however, the goal of the study is to isolate organisms that are intimately associated with the root, then this is an acceptable procedure. Most researchers attempt to isolate and enumerate microorganisms in soil that adheres to the root after gentle shaking. Obviously the term "loosely adhering soil" is ambiguous, as both the plant species and soil type affect the amount of soil adhering to roots. Coarse-textured roots retain lesser amounts of soil than do finer-textured roots.

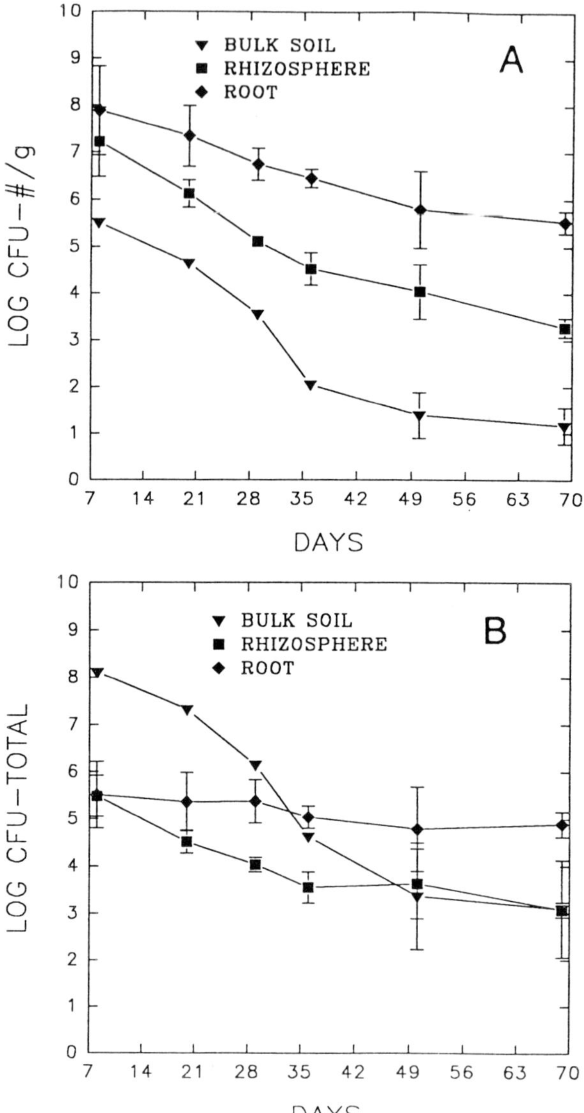

Figure 1 Colony forming units (CFUs) of *Pseudomonas aurofaciens* 3732RN-L11 in rhizosphere soil, bulk soil, and associated with roots of wheat: (A) on the basis of CFUs per gram of soil or root; (B) total in soil or on root. Bars represent the standard error of the mean.

The advantage of this definition, however, is that the soil adhering to roots must be in close contact with the root, typically through small lateral roots and root hairs.

After isolation of that portion of the soil-root system defined as the rhizosphere, the sample must next be treated to release rhizosphere microorganisms. Many techniques are available for release, and many of these techniques will affect the efficiency of release. If roots alone or roots plus very tightly adhering soil are analyzed, then it is usually necessary to macerate the materials to bring the organisms into suspension. This procedure has been used by Seong et al. [48] and Berge et al. [49] and probably recovers most rhizoplane organisms.

When roots and adhering soil are assayed, it is necessary to release organisms from both the root and soil. Most studies have followed, or adapted, the method of Zuberer [33]. The shaking procedure and duration of shaking affect the efficiency of release. Shaking with solid particles (e.g., glass beads) and shaking for longer times often enhances the efficiency of release [30]. Other procedures to facilitate release, such as blending [50,51] and sonication [52,53], have been examined, but the results obtained with these techniques are variable [30]. Efficiency of release and recovery is also affected by such factors as the diluent employed and whether the soil is allowed to settle from suspension [30].

Several other factors have been discussed when examining populations within the rhizosphere. Shafer [54] noted that rhizosphere microorganisms are not evenly distributed over all roots. Procedures that fail to collect most lateral roots may exhibit skewed data if the microorganism(s) being enumerated preferentially colonize younger roots. Alternatively, microorganisms such as rhizobia tend to be found in higher concentrations along the tap root, and if lateral roots are not collected, the rhizosphere population of rhizobia could appear artificially high. Sugimoto et al. [55] chopped roots into 1-cm segments and analyzed only a fraction of the total root system. Inbar and Chet [56] recognized the uneven distribution of organisms in the rhizosphere and suggested that the plant root system be sectioned and each section be analyzed separately. A further confounding factor is the difficulty in analyzing the entire root system, especially for field-grown crops. Kloepper et al. [57] showed that variability was increased when only part of the root was analyzed. When only a fraction of the root can be analyzed, efforts should be made to enumerate a representative portion of the root.

B. Recommendations for Standardization

A modification of the procedures described by Wollum [58] and Zuberer [33] is recommended as the standard for the isolation and processing of rhizosphere samples.

1. Root collection—if possible, the entire root system should be collected. If this is not possible, then it is important to collect a representative portion of the root system. At the point of collection, large clumps of soil, exclusive of roots, should be discarded. The roots and adhering soil should be placed in a plastic bag, kept on ice, and returned to the laboratory for immediate analysis.
2. Soil preparation—Carefully shake roots, using an inoculation needle to remove large clumps of soil attached to the root. If the root system plus adhering soil weighs less than 2.5 g (fresh weight), place into 99 ml of an appropriate diluent. If the sample weighs between 2.5 and 5.0 g, place into a bottle containing 95 ml of an appropriate diluent. When the root weighs more than 5 g, the entire root system can be diluted in a larger volume of diluent (and in a larger bottle) or only part of the root system can be analyzed. If only a portion of the root is analyzed, then it is important to vertically section the root system to ensure that all root types are equally represented.
3. Release of microorganisms—Shaking and maceration should begin immediately after the root system has been placed in the diluent. The most effective method of release for each organism and soil must often be determined experimentally. It is suggested that the sample be shaken for 10 min at 180 rpm. At this point, many researchers prefer to allow soil particles to settle from the solution, usually for approximately 10 min. After shaking and settling, subsequent dilutions in buffered water should be made as needed and the appropriate dilutions plated. A 1:10 dilution scheme of Zuberer [33], should be followed exactly. As will be discussed later, inoculum efficacy should be expressed as CFUs per gram dry soil; consequently, the root system should be removed from the bottle, the remaining soil filtered from solution, dried at 110°C for 24 h, and weighed.

III. EXPRESSION OF RESULTS

A. Background

A review of the literature shows that there are at least 11 methods for calculating and expressing microbial numbers in the rhizosphere (Table 1). The methods usually focus on either the soil or root. Root or soil is diluted and the culturable population is determined by plating on selective media [33] or by MPN procedures [73]. Expression of population counts when evaluating microbial efficacy or inoculum efficiency, whatever the method of enumeration, generally requires that the population be divided by the weight of soil or root from which the

Table 1 Varied Methods Used for Expressing Numbers of Microorganisms in the Rhizosphere

Method	Citation
CFU/g dry soil	Angle et al. [30]
	Li and MacRae [59]
	Richaume et al. [46]
	Seong et al. [48]
	Shafer [54]
	Zuberer [33]
CFU/g wet soil	Jiang and Sato [41]
CFU/g soil	Shaw et al. [45]
	Tedla and Stanghellini [42]
	Young et al. [60]
CFU/g fresh root	DeFreitas and Germida [62]
	Dleep-Kumar and Dube [61]
CFU/g dry root	Fredrickson and Elliott [63]
	Howie and Echandi [64]
CFU/g root	Kluepfel and Tonkyn [65]
	Kremer et al. [47]
	Sugimoto et al. [55]
CFU/g dry soil + dry root	Barraquio et al. [50]
	Berge et al. [49]
	Thompson et al. [66]
CFU/g fresh root + wet soil	Smit et al. [68]
	Walter et al. [67]
CFU/cm root	Kloepper et al. [69,70]
CFU/plant	Christiansen-Weniger and van Veen [43]
	Hartel and Haines [71]
CFU/g dry soil + CFU/g fresh root	Kolb and Martin [72]

population originated. The number of CFUs is typically divided by either the dry or wet weight of the root or the weight of the soil adhering to the root, and the data are normalized to CFU per gram of dry root or soil.

Expression of CFUs by dividing by the fresh weight of the root presents several problems. First, the weight of the root varies depending on the moisture content of soil. Roots are generally about 90% water; however, in extremely dry soil, this value may be 70% or less (unpublished data). If, for example, the entire diluent was found to contain 1×10^6 CFUs and the fresh weight of the root was 1.0 g, the population density would be expressed as 1×10^6 CFU/g of fresh root. If the system of a root was isolated from a soil with a lower moisture content, the same root volume might weigh 0.7 g. The calculated population would therefore be 1.43×10^6 CFU/g of fresh root, a reported difference of almost 50%. Correcting CFU per gram of fresh root to CFU per gram of dry root (by determining the moisture content of the root) avoids this problem. Expression of counts as CFU per gram of dry root, however, requires an additional step, consumes additional time, and increases the potential for measurement or mathematical errors.

A potentially significant problem occurs when CFUs are divided by low root weights. When an observed population is enumerated on young plants having small root systems, this computation can make the population appear artificially high. This is especially true when dividing CFUs by the dry weight of the root system. The weight of a dry root system is about 10% of the fresh weight. For example, the fresh weight of a soybean root only a few days old may be 0.25 g; the dry root weight would then be 0.025 g. If the total number of CFUs in the diluent is 1×10^4, the results would be expressed as 4×10^5 CFU/g of dry root. This is a 40-fold higher concentration of microorganisms than actually present in the rhizosphere. The size of this error decreases as the root weight approaches 1 g. Total populations in the rhizosphere are subsequently underestimated when the root weight divisor exceeds one. Fortunately, root weights, especially for dry weight of plants grown in pots, do not significantly exceed this value.

As also noted in Table 1, many studies have expressed rhizosphere counts on a per gram of soil basis. Using this technique, soil adhering to the root is dislodged and the microbial count within this fraction is determined. Counts may, as with roots, be expressed on a fresh (wet) or dry basis. When counts are divided by the oven dry weight of soil, moisture is eliminated as a variable, allowing counts to be compared between soils of varying moisture content. Calculation and expression of counts by dividing the CFUs by the wet weight of soil introduces uncertainty into the final values. As the moisture content of soils can vary widely, from as low as 2% to as much as 50% on a weight basis, the CFU per gram of wet soil can vary within a similar range. The following is given as an example when the same soil has a moisture content of 2% and 50%. A gram of "wet" soil at 2% moisture contains 0.9804 g of over-dry soil. A "wet" gram of soil at 50% moisture contains 0.6667 g dry soil. When the total number of CFUs in the diluent is 1×10^4, the CFU per gram of dry soil is calculated as 1.02×10^5 and 1.50×10^5 for the soil containing 2% and 50% moisture, respectively.

A further problem associated with the expression of rhizosphere counts as CFU per gram of soil is that methods of detaching soil from the root are extremely variable. As noted by Angle et al. [30], not all methods for extraction and enumeration are equally effective for any given soil. Liddell and Parke [74] have also noted that soil moisture content affects the adhesion of soil to roots. Adhesion of moist soil to roots is stronger than that of dry soil. Thus, a greater volume of bulk soil is analyzed in wet soils, resulting in a lower calculated rhizosphere population. The researcher must determine experimentally the optimum method for detachment or select a standard method for continuous use within a laboratory.

Several researchers have presented the data by dividing the number of CFUs by the weight of the root and detached soil. Various combinations of fresh and dry root weights and wet and dry soil weights have been used (Table 1). Ideally, it is desirable to use the method of Kolb and Martin [72] where soil and roots are analyzed separately. The number of CFUs is expressed on a fresh soil (from around the roots) basis plus the dry root weight. Values are combined to determine the total number within the rhizosphere. A limiting factor associated with this method is that twice the amount of work is required for a single assay. This method is, therefore, unlikely to be adopted as a standard method.

Two other methods of expression are shown in Table 1. Kloepper et al. [69,70] have expressed the number of CFUs per square centimeter of root surface area. This method is useful for determining the rate of root colonization, and it is especially useful for monitoring colonization on a developing root. This approach has the advantage that it accounts for differences in root architecture, both within a root and over time. This approach has been recommended by Kloepper and Beauchamp [5] because colonization is often directly related to the number and size of root hairs. Although methods are available for the determination of the surface area of the root [75,76], these techniques are often complicated and cumbersome. Thus, this approach of expressing the number of microorganisms in the rhizosphere will have limited use until procedures are more automated or simplified.

Christiansen-Wenigen and van Veen [43] and Hartel and Haines [71] have used an extremely simple method for expressing rhizosphere counts. The total number of CFUs are expressed on a per plant basis. Although this method has little use for the study of colonization and efficacy, it is exceptionally well suited for risk assessment. The CFUs per plant represents the total number of organisms associated with the roots of the plant. Another advantage of this method is that it requires no weighing of roots or soil.

To examine closely the various methods of expression presented above, an experiment was conducted that addressed the effects of the method of expression of CFUs in the rhizosphere. The previously described genetically modified bacterium. *P. aureofaciens* 3732 RN–L11, was grown to mid-log phase and 1 ml of the suspension was added to the surface of microcosms sown with a single wheat seed. At periodic intervals, plants were gently separated from the bulk soil and the shoot removed from the root. The rhizosphere population of *P. aureofaciens* was enumerated as previously noted. After microbial analysis, the root system was weighed, then dried at 80°C for 24 h and weighed for dry weight. Rhizosphere soil was filtered from suspension on Whatman No. 4 paper, dried at 110°C and weighed. The results were calculated and expressed as CFUs per plant per gram of fresh root weight, per gram of dry root, and per gram of oven-dry rhizosphere soil.

The CFU per plant represents the total number of CFUs within the rhizosphere (Figure 1). The lowest number (7×10^6 CFU/plant) was obtained by this method of calculation during the first 28 days. All other methods, at least until 28 days, resulted in greater numbers of calculated CFUs, which overestimated the "true" population. Populations can never exceed the total number associated with the entire root system. In particular, when CFUs per gram of dry root were used, the observed population was 5×10^8 CFU. Obviously, the number of CFUs in soil did not vary by nearly two log units; the method of calculation was solely responsible for this difference. This difference between methods of expression continued until day 35. At 35 days, the weight of the roots (dry and wet) and soil associated with the roots began to increase rapidly, while the population of *P. aureofaciens* was decreasing. Throughout the remainder of the study, the CFUs per plant were no longer lower than when compared to other divisors. After 35 days, extremely small roots no longer skewed the expressed number of CFUs. By the end of the study, the values were highest when CFUs were expressed on a per plant basis.

If it is desirable to examine the concentration of microbes in the rhizosphere (such as when determining efficacy), it is necessary to divide the CFUs by either the weight of the soil or root. As seen in Figure 2, the CFUs per gram of rhizosphere soil closely approximated the total number per plant, beginning after 35 days. From 35 to 133 days the population expressed as CFUs per plant and CFUs per gram of rhizosphere soil were often identical. Up to 21 days, no method of expression approached the number observed when expressed as CFUs per plant.

B. Recommendations for Standardization

It is concluded from a review of past studies and current work that there is no single method for enumeration and expression of microbial counts in the rhizosphere that will address all reasons for enumeration. When a GEM or other microbial inoculant is added to soil, its potential to survive within the rhizosphere is of paramount interest. Risk assessment requires that the total number of organisms within the rhizosphere be enumerated. The concentration of GEMs in the rhizosphere soil is less important, since the rhizosphere is such a difficult area to define. For this reason, expression of rhizosphere counts as CFUs per plant is the recommended method. When it is important to know the concentration of an organism in the rhizosphere, such as when examining inoculum efficacy or when comparing studies, the population should be expressed as CFUs per gram of dry rhizosphere soil. Although this method of expression is imperfect during the early growth of a plant, when the weight of the soil adhering to the root is small, later measurements are very accurate.

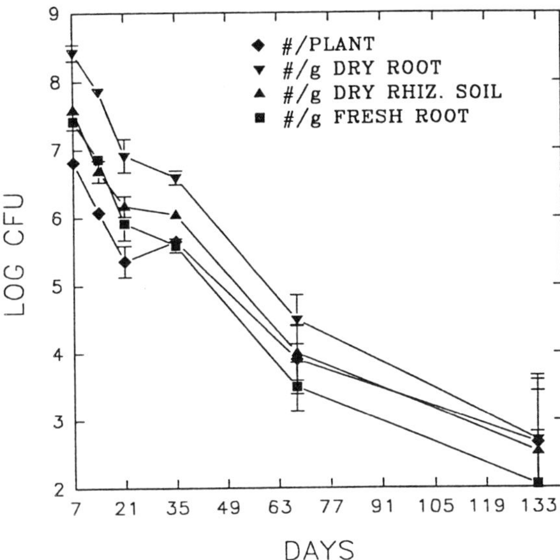

Figure 2 Colony forming units (CFUs) of *Pseudomonas aureofaciens* 3732RN-L11 in rhizosphere soil and associated with roots of wheat. Colony forming units are divided by commonly used divisors. Bars represent the standard error of the mean.

IV. STATISTICAL APPROACHES

Once the number of CFUs has been determined within the rhizosphere and the method of expression established, statistical treatment of the data can further complicate interpretation. It is often desirable to determine whether rhizosphere populations are different among treatments or between treatments and a control group. An analysis of variance (ANOVA) followed by a mean separation procedure such as a least significant difference (LSD) can be used to determine statistically significant differences among means. For a valid ANOVA, the data should be normally distributed and the variances of the treatments homogeneous. However, bacterial populations, especially within the rhizosphere, usually do not meet these conditions. Studies have found bacterial populations to be log normally distributed [30,77,78] and have recommended that data on individual samples be log transformed before analysis. A log transformation will also correct for the heterogeneity of variance that is a related problem associated with rhizosphere data. Means and variances of bacterial populations are usually correlated, and, if the means vary, the variances will be heterogeneous.

The log transformation is particularly important when analyzing populations over time, as the variances may change dramatically over time. In cases where the number of CFUs of the treatment is below the detection limit, a transformation of log10(CFU + 1) should be used. The constant of one will result in a zero rather than a missing number when the log is taken. If missing numbers rather than zeros are used, means will be overestimated. Although the actual bacterial count may be greater than zero, zero is the best estimate of the value. A log transformation has an additional purpose when analyzing populations for survival. Because the change in population often follows an exponential decay model, a log transformation will linearize the data. After transformation, a linear model can be used to determine the rate of decline of the population. An alternative to data transformation when determining survival is to use nonlinear regression to fit an exponential decay model. However, this approach lacks the simplicity of linear regression analysis and makes direct testing of hypotheses difficult.

V. CONCLUSIONS

It is suggested that the rhizosphere be defined as that portion of the soil that remains attached to the root after gentle shaking. The root with attached soil should be added to a buffer and shaken as described. After determination of the number of CFUs in the extract, results should be expressed as CFUs per plant for risk assessment purposes and as CFUs per gram of dry rhizosphere soil for examination of efficacy and inoculum efficiency. No method for the enumeration and expression of rhizosphere populations is perfect or useful for all purposes. With an understanding of the limitations of each method of expression, valid results can be obtained and compared with data from other studies with relative confidence.

REFERENCES

1. Hiltner, L. 1904. Über neuere Erfahrungen und Probleme auf dem Gebiet der Bodenbakteriologie der Gründüngung und Brache. Arb Dtsch Landwirtsch Gesellsch 98:59–78.
2. Aprill, W., and R. C. Sims. 1990. Evaluation of the use of prairie grasses for stimulating polycyclic aromatic hydrocarbon treatment in soil. Chemosphere 20:253–265.
3. Paul, E. A., and F. E. Clark. 1989. Soil microbiology and biochemistry. Academic Press, New York.
4. Schmidt, E. L. 1979. Initiation of plant root-microbe interaction. Annu. Rev. Microbiol. 33:355–376.
5. Kloepper, J. W., and C. J. Beauchamp. 1992. A review of issues related to measuring colonization of plant roots by bacteria. Can. J. Microbiol. 38:1219–1232.
6. Dahm, D. 1984. Metabolic activity of bacteria isolated from soil, rhizosphere and mycorrhizosphere of pine. Acta. Microbiol. Polon. 33:157–158.

7. Hozore, E., and M. Alexander. 1991. Bacterial characteristics important to rhizosphere competence. Soil Biol. Biochem. 23:717–723.
8. Giddens, J., and R. L. Todd. 1979. Rhizosphere microorganisms—overview, p. 51–68. *In* Microbial-plant interactions. Special Publ. No. 47. American Society of Agronomy Madison, Wisconsin. Special Publ. No. 47.
9. Hale, M. G., L. D. Moore, and G. J. Griffin. 1978. Root exudates and exudation, p. 163–203. *In* Y. R. Dommergues and S. U. Krupa (eds.), Interactions between nonpathogenic soil microorganisms and plants. Developments in agricultural and managed forest ecology. Elsevier, New York.
10. Rovira, A. D. 1965. Interactions between plant roots and soil microorganisms. Annu. Rev. Microbiol. 19:241–266.
11. Whipps, J. M., and J. M. Lynch. 1986. The influence of the rhizosphere on crop productivity. Adv. Microb. Ecol. 9:187–244.
12. Parke, J. L. 1991. Root colonization by indigenous and introduced microorganisms, p. 33–42. *In* D. L. Kleister and P. B. Cregan (eds.), The rhizosphere and plant growth. Kluwer Academic, Dordrecht.
13. Colwell, R. R., P. R. Brayton, D. J. Grimes, D. B. Roszak, S. A. Hug, and L. M. Palmer. 1985. Viable but nonculturable *Vibrio cholerae* and related pathogens in the environment: implications for release of genetically engineered micro-organisms. Bio/Technology 3:817–820.
14. Vancura, V., and K. Kunc (eds.). 1989. Interactions between microorganisms and plants in soil, p. 492. Elsevier, Amsterdam.
15. Papavizas, G. C., and C. B. Davey. 1961. Extent and nature of the rhizosphere of Lupinus. Plant Soil 14:215–236.
16. Rouatt, J. W., and H. Katznelson. 1961. A study of the bacteria on the root surface and in the rhizosphere soil of crop plants. J. Appl. Bacteriol. 24:164–171.
17. Schmidt, E. L. 1991. Methods for microbial autoecology in the soil rhizosphere, p. 81–90. *In* D. L. Keister and P. B. Cregan (eds.), The rhizosphere and plant growth. Kluwer Academic, Dordrecht.
18. Atlas, R. M., and R. Bartha. 1992. Microbial ecology: fundamentals and applications, 3rd ed. Benjamin/Cummings, Menlow Park, California.
19. Foster, R. L. 1986. The ultrastructure of the rhizosplane and rhizosphere. Annu. Rev. Phytopathol. 24:211–234.
20. Lynch, J. M. 1990. Microbial metabolites, p. 177–206. *In* J. M. Lynch (ed.), The rhizosphere. John Wiley & Sons, Chichester.
21. Kloepper, J. W. 1991. Plant growth-promoting rhizobacteria as biological control agents of soil borne diseases, p. 142–152. *In* J. Bay Peterson (ed.), Biological control of plant diseases. FFTC Book Series No. 42. Food and Fertilizer Technology Center, Taipei, Taiwan.
22. Milus, E. A., and C. S. Rothrock. 1992. Rhizosphere colonization of wheat by selected soil bacteria over diverse environments. Can. J. Microbiol. 39:335–341.
23. Weller, D. M. 1988. Biological control of soil borne plant pathogens in the rhizosphere with bacteria. Annu. Rev. Phytopathol. 26:379–407.
24. Loper, J. E., and C. A. Ishimara. 1991. Factors influencing siderophore-mediated biocontrol activity of rhizosphere *Pseudomonas* sp., p. 253–261. *In* D. L. Keister

and P. B. Cregan (eds.), The rhizosphere and plant growth. Kluwer Academic, Dordrecht.
25. Klein, D. A., J. L. Salzwedel, and F. B. Dazzo. 1990. Microbial colonization of plant roots, p. 189–218. *In* J. P. Nakas and C. Hagedorn (eds.), Biotechnology of plant-microbe interactions. McGraw-Hill, New York.
26. Knaebel, D. B., and J. R. Vestal. 1992. Effects of intact rhizosphere communities on the mineralization of surfactants in surface soil. Can. J. Microbiol. 38:129–135.
27. Walton, B. T., and T. A. Anderson. 1990. Microbial degradation of trichloroethylene in the rhizosphere: Potential application to biological remediation of waste sites. Appl. Environ. Microbiol. 56:1012–1016.
28. Lappin, H. M., M. P. Greaves, and J. H. Slater. 1985. Degradation of the herbicide mecoprop [2-(2-methyl-4-chlorophenoxy) propionic acid] by a synergistic microbial community. App. Environ. Microbiol. 49:429–433.
29. Sandermann, E. R. I. C., and M. A. Loos. 1984. Enumeration of 2,4-D degrading microorganisms in soils and crop plants rhizospheres using indicator media: High populations associated with sugarcane (*Saccharum officinarum*). Chemosphere 13: 1073–1084.
30. Angle J. S., M. A. Levin, J. V. Gagliardi, M. S. McIntosh, and J. G. Glew. 1994. *Pseudomonas aureofaciens* in soil: survival and recovery efficiency. Microbial Releases 2:247–254.
31. McCormick, D. 1986. Detection technology: the key to environmental biotechnology. Bio/Technology 4:419–422.
32. Trevors, J. T., and J. D. van Elsas. 1989. A review of selected methods in environmental microbial genetics. Can. J. Microbiol. 35:895–902.
33. Zuberer, D.A. 1994. Recovery and enumeration of viable bacteria, p. 119–144. *In*: R. W. Weaver, J. S. Angle, and P. J. Bottomley (eds.), Methods of soil analysis. Biochemical and microbiological properties. American Society of Agronomy, Madison, Wisconsin.
34. Woolmer, P. 1994. The most probable number technique, p. 59–78. *In*: R. W. Weaver, J. S. Angle, and P. J. Bottomley (eds.), Method of soil analysis. Biochemical and microbiological properties. American Society of Agronomy, Madison, Wisconsin.
35. Schmidt, E. L., and E. A. Paul. 1982. Microscopic methods for soil microorganisms, p. 803–814. *In* A. L. Page, R. H. Miller, and D. R. Keeney (eds.), Methods of soil analysis. Part 2. Chemical and microbiological properties. American Society of Agronomy, Madison, Wisconsin.
36. Ocio, J. A., and P. C. Brookes. 1990. An evaluation of methods for measuring the microbial biomass in soils following recent additions of wheat straw and the characterization of the biomass that develops. Soil Biol. Biochem. 22:685–694.
37. Binnerup, S. J., D. F. Jensen, H. Thordal-Christenson, and J. Sorensen. 1993. Detection of viable, but non-culturable *Pseudomonas fluorescence* DF57 in soil using microcolony ep. fluorescense technique. FEMS Microbiol. Ecol. 12:97–105.
38. Katznelson, H. 1946. The rhizosphere effect of mangels on certain groups of microorganisms. Soil Sci. 62:343–354.

39. Timonin, M. I. 1946. Microflora of the rhizosphere in relation to the manganese deficiency disease of oats. Soil Sci. Soc. Am. Proc. 11:284–292.
40. Rovira, A. D., and C. B. Davey. 1974. Biology of the rhizosphere, p. 153–204. *In* E. W. Carson (ed.), The plant root and its environment. University of Virginia Press, Charlottesville.
41. Jiang, H. Y., and K. Sato. 1992. Fluctuations in bacterial populations on the root surface of wheat (*Triticum aestivum* L.) grown under different soil conditions. Biol. Fertil. Soils 14:246–252.
42. Tedla, T., and M. E. Stanghellini. 1992. Bacterial population dynamics and interactions with *Pythium aphanidermatum* in intact rhizosphere soil. Phytopathology 82: 652–656.
43. Christiansen-Weniger, C., and J. A. van Veen. 1991. Nitrogen fixation by *Azospirillum brasilense* in soil and the rhizosphere under controlled environmental conditions. Biol. Fertil. Soils 12:100–106.
44. Harley, J. L., and J. S. Waid. 1955. A method of studying active mycelia on living roots and other surfaces in the soil. Trans. Br. Mycol. Soc. 38:104–118.
45. Shaw, J. J., F. Dane, D. Geiger, and J. W. Kloepper. 1992. Use of bioluminescence for detection of genetically engineered microorganisms released into the environment Appl. Environ. Microbiol. 58:267–273.
46. Richaume, A., E. Smit, G. Faurie, and J. D. van Elsas. 1992. Influence of soil type on the transfer of plasmid RP4p from *Pseudomonas flourescens* to introduced recipient and to indigenous bacteria. FEMS Microbiol. Ecol. 101:281–291.
47. Kremer, R. J., M. F. T. Begonia, L. Stanley, and E. T. Lanham. 1990. Characterization of rhizobacteria associated with weed seedlings. Appl. Environ. Microbiol. 56: 1649–1655.
48. Seong, K. Y., M. Höfte, and W. Verstraete. 1991. Acclimatization of plant growth promoting *Pseudomonas* strain 7NSK2 in soil: effect on population dynamics and plant growth. Soil Biol. Biochem. 24:751–759.
49. Berge, O., T. Heulin, and J. Balandreau. 1991. Diversity of diazotroph populations in the rhizosphere of maize (*Zea mays* L.) growing on different French soils. Biol. Fertil. Soils 11:210–215.
50. Barraquio, W. L., A. Dumont, and R. Knowles. 1988. Enumeration of free-living aerobic N_2-fixing H_2-oxidizing bacteria by using a heterotrophic semisolid medium and most-probable number technique. Appl. Environ. Microbiol. 54:1313–1317.
51. Senoo, K., and H. Wada. 1990. Accurate and sensitive method of enumeration of a specific microorganism, *Pseudomonas paucimobilis* 5586, in nonsterile soil. Soil Sci. Plant Nutr. 36:389–395.
52. Balkwill, D. L. 1990. Deep-aquifer microorganisms, p. 183–211. *In* D. P. Labeda (ed.), Isolation of biotechnological organisms from nature. McGraw-Hill, New York.
53. van Beelen, P., and A. K. Fleuren-Kemila. 1989. Enumeration of anaerobic and oligotrophic bacteria in subsoils and sediments. J. Contam. Hydrol. 4:275–284.
54. Shafer, S. R. 1992. Responses of microbial populations in the rhizosphere to deposition of simulated acidic rain onto foliage and/or rain. Environ. Pollut. 76:267–278.

55. Sugimoto, E. E., H. A. J. Hoitink, and O. H. Tuovinen. 1990. Enumeration of oligotrophic rhizosphere pseudomonas with diluted and selective media formulations. Biol. Fertil. Soils 9:226–230.
56. Inbar, J., and I. Chet. 1991. Detection of chitinolytic activity in the rhizosphere using image analysis. Soil Biol. Biochem. 23:239–242.
57. Kloepper, J. W., W. Mahaffee, J. A. McInroy, and P. A. Boehman. 1991. Comparative analysis of methods for recovering plant growth-promoting rhizobacteria from roots. Can. J. Microbiol. 37:953–957.
58. Wollum, A. G. 1994. Soil sampling for microbiological analysis, p. 1–13. *In*: R. W. Weaver, J. S. Angle, and P. J. Bottomley (eds.), Methods of soil analysis. Biochemical and microbiological properties. American Society of Agronomy, Madison, Wisconsin.
59. Li, R., and I. C. MacRae. 1992. Specific identification and enumeration of *Acetobacter diazotrophicus* in sugarcane. Soil Biol. Biochem. 24:413–419.
60. Young, C. S., K. A. Cook, G. Lethbridge, and R. G. Burns. 1992. Survival of introduced bacteria in rhizosphere and non-rhizosphere soils, p. 145–147. *In* D. E. S. Stewart-Tull and M. Sussman (eds.), The release of genetically modified microorganisms—REGEM 2. Plenum Press, New York.
61. Kumar, B. S. D., and H. C. Dube. 1992. Seed bacterization with fluorescent *Pseudomonas* for enhanced plant growth, yield and disease control. Soil Biol. Biochem. 24:539–542.
62. DeFreitas, J. R., and J. J. Germida. 1992. Growth promotion of winter wheat by flourescent pseudomonads under growth chamber conditions. Soil Biol. Biochem. 224:1127–1135.
63. Fredrickson, J. K., and L. F. Elliott. 1985. Colonization of winter wheat roots by inhibitory rhizobacteria. Soil Sci. Soc. Am. J. 49:1172–1177.
64. Howie, W. J., and E. Echandi. 1983. Rhizobacteria: influence of cultivar and soil type on plant growth and yield of potato. Soil Boil. Biochem. 15:127–132.
65. Kluepfel, D. A., and D. W. Tonkyn. 1992. The ecology of genetically altered bacteria, p. 407–413. *In* E. S. Tjamos, G. C. Papavizas, and R. J. Cook (eds.), Biological control of plant diseases. Plenum Press, New York.
66. Thompson, I. P., C. S. Young, K. A. Cook, G. Lethbridge, and R. G. Burns. 1992. Survival of two ecologically distinct bacteria (*Flavobacterium* and *Arthrobacter*) in unplanted and rhizosphere soil: field studies. Soil Biol. Biochem. 24:1–14.
67. Walter, M. V., L. A. Porteous, V. J. Prince, L. Ganio, and R. J. Seidler. 1991. A microcosm for measuring survival and conjugation of genetically engineered bacteria in rhizosphere environments. Current Microbiol. 22:117–121.
68. Smit E., J. D. van Elsas, J. A. van Veen, and W. M. De Vos. 1991. Detection of plasmid transfer from *Pseudomonas fluorescens* to indigenous bacteria in soil by using bacteriophage ΦR2f for donor counterselection. Appl. Environ. Microbiol. 57: 3482–3488.
69. Kloepper, J. W., J. Leong, M. Teintze, and M. N. Schroth. 1980. Enhanced plant growth by siderophores produced by plant growth-promoting rhizobacteria. Nature 286:885–886.

70. Kloepper, J. W., M. N. Schroth, and T. D. Miller. 1980. Effects of rhizosphere colonization by plant growth-promoting rhizobacteria on potato plant development and yield. Phytopathology 70:1078–1082.
71. Hartel, P. G., and B. L. Haines. 1992. Effects of potential plant CS_2 emissions on bacterial growth in the rhizosphere. Soil Biol. Biochem. 24:219–224.
72. Kolb, W., and P. Martin. 1988. Influence of nitrogen on the number of N_2-fixing and total bacteria in the rhizosphere. Biol. Biochem. 20:221–225.
73. Alexander, M. 1977. The soil environment, p. 3–15. *In* M. Alexander (ed.), Introduction to soil microbiology. John Wiley & Sons, New York.
74. Liddell, C. M., and J. L. Parke. 1989. Enhanced colonization of pea taproots by a fluorescent pseudomonad biocontrol agent by water infiltration into soil. Phytopathology 79:1327–1332.
75. Coles, G. D., D. J. Abernethy, M. C. Christey, A. J. Conner, and B. K. Sinclair. 1991. Monitoring hairy-root growth by image analysis. Plant Molec. Biol. Rep. 9: 13–20.
76. Zoon, F. C., and P. H. Van Tienderen. 1990. A rapid quantitative measurement of root length and root branching by microcomputer image analysis. Plant Soil 126: 301–308.
77. Hirano, S. S., E. V. Nordheim, D. C. Arny, and C. D. Upper. 1982. Lognormal distribution of epiphytic bacterial populations on leaf surfaces. Appl. Environ. Microbiol. 44:695–700.
78. Loper, J. E., T. V. Suslow, and M. N. Schroth. 1984. Lognormal distribution of bacterial populations in the rhizosphere. Phytopathology 74:1454–1460.

7

Bacterial Movement Through Soil

Stephanie L. Murphy and Robert L. Tate III Rutgers University, New Brunswick, New Jersey

I. INTRODUCTION

Considering that the highest concentrations of organic carbon are usually found in the surface horizon (layer) of soils and that it is this fixed carbon resource that provides the energy that drives the primarily heterotrophic microbial community of soils, it is reasonable to assume a priori that maximal microbial populations are to be found in topsoils. This logic suggests that greatest microbial activity could be anticipated to occur in the rhizosphere of soils, since root exudates, sloughed cells, and decaying dead roots are major energy islands in soil. A potentially comparable soil ecosystem to the rhizosphere and its associated biological activity pattern is observed in soil ecosystems receiving major inputs of decomposing organic matter (e.g., composts, crop residues, and societal wastes).

Microbial energy metabolism in soil is driven by influxes of photosynthetically fixed carbon as plant debris (e.g., litter and thatch, as well as composts), and by a variety of belowground inputs from roots. Less obvious energy sources are derived from catabolism of humus and turnover of soil microbial biomass (see [1] for a review of this topic). Even with these organic matter reservoirs, the surface microbial community in soil is considered to be generally carbon limited. This restriction of microbial activity is accentuated in subsoils, where external fixed carbon inputs are minimal and native soil organic reservoirs are nil to minimal. The logical conclusion from this observation is that insignificant heterotrophic bacterial activity can be anticipated in subsoils and the underlying regolith.

Yet bacteria are found in subsoils, regoliths, and aquifers—sometimes in numbers generally considered to be "typical" for surface soils (e.g., [2]). These populations indicate either movement from surface soils or presence during deposition of sediment-derived materials.

This demonstration of occurrence of major populations of bacteria in subsoils alone is sufficient to stir interest in evaluating the potential for seeding of subsoils with propagules from surface bacterial populations through leaching from overlying soil strata. The intensity of such interest is accentuated by recent studies of the potential for introduction of axenically grown bacterial propagules into surface and subsurface sites that have been contaminated with biodegradable xenobiotics. The potential for movement of bacteria within soil profiles is also a realistic concern in evaluating the possibility for microbial pollution of groundwater by land disposal of wastes. In addition, Bengtsson et al. [3] indicated that bacteria or their polysaccharides may act as vectors of organic contaminants, sorbing and carrying them through subsurface materials. Whereas bacterial xenobiotic chemical cotransport may only be significant under eutrophic conditions with high microbial populations, it is notable that the bacterial isolates evaluated by Bengtsson et al. [3] in their studies were the most efficient sorbent tested for three organic compounds (acetophenone, 4-chloroaniline, and pentachlorophenol).

Factors that affect the potential for bacterial movement through soil have been recently reviewed [4]. A variety of soil and microbial properties affecting microbial mobility in soil was evaluated. For accurate evaluation of the importance of microbial mobility in soil ecosystems, compilation of these environmental and biological properties into mathematical models would be useful in determining the potential for protection and remediation of subsurface ecosystems. Thus, the objective of this review is to present a summarization and analysis of microbial transport models in soil and associated habitats.

II. FACTORS AFFECTING BACTERIAL MOVEMENT IN SOIL SYSTEMS

To relate the mathematical representation of microbial movement in soil with relevant soil properties, it is necessary to build a foundational understanding of the basic soil and microbial properties that influence bacterial movement in soil. Thus, literature pertaining to active mobility, passive transport of bacterial propagules (e.g., soil faunal transport of bacterial populations, distribution of microbial propagules by plant roots, transport with water) and environmental properties contributing to microbial dispersion will be summarized. See the review of Gammack et al. [4] for a more in-depth review of the literature relating to these topics.

A. Active Movement of Microbial Propagules in Soil Ecosystems

Although some soil bacteria are readily shown to be motile in vitro, it should not be assumed that bacteria are actually motile in situ [5,6]. Potentially motile bacteria would have to overcome the high viscosity and surface tension of water in unsaturated soil that result from the attraction of water for the mineral surfaces [5,7]. It has been questioned whether motility would provide an ecological advantage to the bacteria [6,7].

Although it is not universally accepted that flagella-driven motility occurs in soil ecosystems (e.g., [5]), a competitive advantage for motile bacteria in some soils and quasi-soil systems has been demonstrated. For example, Arora and Gupta [8] observed a chemotactic response (i.e., active movement in response to chemical gradient) of four common soil bacteria (*Agrobacterium radiobacter, Bacillus subtilus, Pseudomonas fluorescens*, and *Xanthomonas malvacearum*) in soil to a variety of fungal spores. The chemoctactic response was greatest in sandy soil, followed by sandy loam. The least response occurred in clay loam. Similarly, motility has been shown to facilitate root-to-root travel of *Azospirillum* on wheat and soybean roots in sand and coarse-textured soil [9]. Although flagella impeded the movement of rhizobial cells through small pore necks [10], possession of flagella became a benefit for mobility of those cells in soils with water-filled pores greater than the cell size [11]. Efficient spreading of rhizobia in autoclaved soils apparently requires motility and chemotaxis (e.g., [12]). Although packed sand or soil columns and autoclaved soils do not represent true soil ecosystems, the above cited studies and related work using such artificial soil systems do indicate a potential for the biological property of flagellar motion to function in systems with physical and chemical properties related to native soils.

A selective advantage of cellular motility is reliant upon the water status of soil since motile organisms must move in the water-filled capillaries, or micropores, and water films existent on soil particle surfaces. Water content (kg water kg^{-1} soil, or m^3 water m^{-3} soil) itself gives a relative measure of wetness for a given soil, but does not provide information about the activity of the water or allow an estimate of water film thickness. The physicochemical condition of soil water is characterized by its free energy per unit mass, termed the potential. One component of the total potential of soil water is the matric potential, the tenacity with which soil water is held by the soil matrix. The matric potential results from the affinity of the water to the whole matrix of the soil, including pores (capillarity) and particle surfaces (adsorptive forces). It is expressed as a negative pressure potential (i.e., a suction or tension). A water content approaching saturation corresponds to a water potential near 0.0 MPa, and decreasing water content results in lower (more negative) values of potential. The matric potential (ψ, dyne cm^{-2}) also allows an estimate of size of pores filled with water by the equation of capillarity, $\psi = 2\gamma r^{-1}$, where γ (dyne cm^{-1}) is surface tension of water and r (cm) is the maximum radius of water-filled pores [13]. Another measure of soil water status, percent water-filled porespace, has been used to quantify the relationship of soil moisture to various biological activities (e.g., [14–16]), but its utility for modeling or predicting microbial mobility in soil is unproven.

Soil wetness was the most critical factor for active movement of the rhizosphere bacteria *Azospirillum brasilense* and *P. fluorescens* in sand and clay (Terra rosa) soils [17]. *A. brasilense* demonstrated strong chemotaxic migration through soil in a distinct band toward wheat roots, glycine, and aspartic acid. In the sand at 0.20 $m^3\ m^{-3}$ water content, which was approximately saturation, *A. brasilense* moved up to 8 cm in 96 h, and this distance was linearly proportionately less at lower water contents. Migration decreased with increasing soil "weight" (i.e., fineness) when comparing the sand and clay soils.

For flagella-driven motility and colony development to be important contributors to dissemination of propagules in soil, flooded conditions with adequate nutrient supplies for microbial growth must occur. Yet in most soil systems, saturated conditions are transient and infrequent, and availability of growth substrates is low. For example, Griffin and Quail [18] provided evidence that the physical regime of a soil limits appreciable movement of bacteria by flagellar activity. They pointed out that active bacterial movement in soil requires continuity of water pathways, which would be dependent on both water content and pore size distribution. Decreasing soil water content corresponds to decreasing number and small size of water-filled pores and thinner water films in the larger pores.

Discontinuity of water pathways may be the major limitation in many soils, however. In a silt loam, Hamdi [19], determined that restriction of bacterial (*Rhizobium trifolii*) movement at potentials less than –0.015 MPa was related to discontinuity of water films rather than to size of water-filled pore necks. Similarly, Issa et al. [11] found that migration of motile bacteria (*Rhizobium* spp.) up to 4 cm in 6 days occurred with matric potentials of –0.1 MPa or greater (i.e., maximum size of water-filled pores was greater than cell size). At potentials of –1.5 MPa and –5 MPa, flagellar activity and degree of motility were not important because active migration was severely restricted by thin water films.

Griffin and Quail [18] also pointed out that water lenses at points of soil contact could provide a pathway if they were connected, but they expected that such a pathway would be limited in continuity. However, they did not consider the lenses formed along the contact of a linear structure such as fungal mycelia. Wong and Griffin [20] showed that the presence of dead fungal mycelium can enhance bacterial spread at water potentials greater than –0.015 MPa by providing a continuous water film along the hyphae. Such tensions represent water contents near saturation, however. At potentials of –0.015 MPa and less, bacterial spread along mycelia was restricted because the cross-sectional area of the water film was greatly reduced, that is, the size of the pathway was limiting rather than the continuity. (Live fungal mycelia were not as effective in facilitating bacterial spread, possibly as the result of an inhibitory accumulation of fungal metabolites in the water films.)

In a practical sense, this interaction with fungal mycelia may not be significant, as the threshold potential for active bacterial movement in soil has also been established at approximately –0.015 MPa. Wong and Griffin [21] found that the movement of motile bacteria in water films of soil pores was severely restricted at water potentials less than –0.015 MPa. They observed that adsorption to soil particles also appeared to hamper the movement of bacteria. Although nutrient amendment and the bacterial strain (three genera were represented) were significant variables in their study, they concluded that active movement of bacteria would not be important to their widespread distribution in soil [21]. Similarly, Griffin and Quail [18] determined the critical volume of water necessary for appreciable bacterial movement and concluded that active bacterial movement would be very restricted at water contents less than field capacity (i.e., at water potentials less than –0.033 MPa).

A number of factors other than soil water content and bacterial motility contribute to movement or lack thereof in soil. These include adsorption processes and pore neck size limitations. In soils with water potentials below –1.5 MPa, bacterial cells would probably be limited to their microsites as the result of the adsorptive forces exerted by soil colloids [22]. Active movement of rhizobia in a clay soil was restricted compared with that in a sandy loam soil at the same water potential, and adsorption tests in vitro indicated that clay soil adsorbed rhizobia more strongly than sandy loam soil [11]. While the mechanisms of adsorption of bacteria to soil particles have not been substantiated, numerous studies report its apparent occurrence in situ (e.g., [11,21,23]). For further information and discussion of bacterial adsorption in soil, see Stotzky [5,24] and Marshall [25]. Bacteria–soil particle interactions are not the only possible explanation for retention of bacteria in soil. Wan et al. [26] discovered that bacteria (with either hydrophobic or hydrophilic surface characteristics) are preferentially sorbed at air-water interfaces, which account for differences in bacterial retention in column flow studies.

Griffin and Quail [18] theorized that bacterial movement would be negligible in water-filled pores remaining at soil water potentials below –0.1 MPa because of small pore diameter (less than 3 μm). Movement of flagellated cells requires pore spaces larger than cell size for maximum active movement. The radius of gyration of the motile cells is the critical factor in restriction of bacterial movement by pore neck size, not simply the absolute dimensions of the pore neck and the cell [19,27]. Issa et al. [11] concluded that the effectiveness of flagella for the motility of cells in soil is limited to the range of soil wetness in which the size of water-filled pores is at least as large as the dimensions required for operation of flagella.

If some soil bacteria are indeed motile in the soil environment, the distances that they are actively able to transport themselves in the soil is minute (tens of millimeters) on the scale of the soil profile (usually meters). In a silt loam soil, Hamdi [19] observed rhizobial movement of 5 mm at a soil water potential of –0.02 MPa and up to 19 mm at –0.003 MPa. Reynolds et al. [28] provided an analysis of microbial movement in sand cores under static saturated, anaerobic conditions. Migration rate of the bacterial cells varied with type of motility (tumbling versus smoothly swimming) and the presence or absence of substrate gradient. Maximum penetration rates of motile *Escherichia coli* were 0.5 cm h^{-1}. From their study of the literature relative to colonization of plant roots, Bowen and Rovira [27] concluded that flagella are relatively unimportant for bacterial movement in soil.

Research on fungal zoospores may provide additional relevant information for elucidation of the limitations of microbial movement in soil. In saturated sand, decreasing particle size continually reduced motility of zoospores of *Phytophthora* spp. compared to water (no sand) [29]. It was suggested that more frequent contact with solid surfaces, such as in the smaller pores of the finer sand, may have induced encystment, resulting in a shorter motile period. In fine-textured, unsaturated soil, the motile period of zoospores is severely shortened compared to that in coarse-textured soil [30]. Other zoospores (e.g., of *Olpidium brassicae*) [31] may not be as sensitive to stimuli for encystment but would still be severely limited in mobility by water potentials as high as –0.004 MPa. These reports support the potential for directed movement, albeit large pores and wet conditions are required, and point out the influence of an organism's life cycle on active movement.

Bacterial growth may have been responsible for observed movement in some studies. For example, at soil water potentials of –1.5 MPa and –5 MPa, a nonflagellated strain was found farther from the inoculation point than motile strains [11]. Nutrient utilization and growth rates were more important factors than chemotaxis for motile strains in a study of *E. coli* movement in sand columns [28], suggesting that microbial growth in situ contributed to apparent microbial movement. Dispersal of nonmotile bacteria in adsorbed films of water have been observed in nonsoil situations [32,33], but the occurrence of such mechanisms in the soil environment is unknown. Under natural soil conditions, dispersal of introduced bacteria by simple growth of microcolonies is probably not appreciable [34,35].

B. Passive Transport of Bacterial Propagules

The preceding analyses demonstrated the potential for contribution of active motility to microbial transport in soil, but not necessarily the reality, and certainly not the prediction of any significant impact of bacterial motility on distribution of bacterial propagules in subsoils. Flagellar-propelled movement (or any other cell-based motility mechanisms) may contribute to bacterial migration on a microscale, and an expansion to a scale including a range of a few millimeters (at most), but definitely not at the range of meters or more required to explain a surface soil contribution to subsoil microbial communities. More extensive, passive carriers are necessarily operative at that scale. Examples of vehicles of passive transport for bacteria in the soil are migrating soil fauna, invading roots, and moving water.

Soil Fauna

Earthworms serve as dispersal agents of bacteria and other microorganisms as an indirect result of their feeding procedures. They ingest soil during feeding and burrowing and later cast it out. The potential of worm-mediated transport of microorganisms was demonstrated by Hutchinson and Kamel [36], who reported dispersal of fungal spores in sterile soil by worms. Thornton [37] found that earthworms transport spores of soil-dwelling, aquatic fungal species both internally and externally. An earthworm species, *Lubricus rubellus*, has been shown to be a transporting agent of *Pseudomonas* and *Rhizobium* [34]. Not only can preexisting populations of soil organisms be transported by earthworms, but microbial dispersal in soil may be accentuated by microbial growth in the worm gut. Parle [38,39] found that numbers of bacteria increased logarithmically during their passage through earthworm guts, indicating growth. On the other hand, Henschke et al. [40] noted that the numbers of genetically altered bacteria that were pulse fed to the earthworm, *L. terrestris*, decreased continuously within excreted casts, reaching the lower limit of detection within 50 days.

Although these studies suggest an earthworm contribution to microbial migration in soil ecosystems, their overall impact may be limited compared to other mediators of microbial movement. For example, Madsen and Alexander [34] concluded that the direct effect of burrowing earthworms carrying bacteria through the soil must be small relative to bacterial transport by percolating water in the biochannels created by the earthworms.

Other possible vectors are nematodes, insects, and other soil animals. Dispersal of nematodes may be constrained by pore size and/or water film thickness on soil particles [41], as with motile bacteria and fungal zoospores. Armstrong et al. [42] demonstrated microbial dispersal by insect larvae and differential survival and growth of bacterial species in the gut of the insect larvae. Further studies are needed to place the contributions of these populations to distribution of microbial populations into perspective with other dispersal mechanisms. Furthermore, since this animal-mediated translocation of microbial populations will be necessarily limited to the primary habitat of the transporting organism, any impact on subsoil microbial distribution would necessarily be indirect.

Plant Roots

Plant roots grow through soil sometimes to great depths and lateral extent (i.e., penetration of subsoil environs is to be anticipated). It is tempting to think that plant roots may carry bacteria of the soil surface population deeper into the soil profile as the root system expands. Consideration of the root growth process should suppress that image of roots carrying bacteria large distances through the soil, however. Root extension through soil is caused by elongation of the young cells produced in the meristematic region of the root, within a few centimeters of the root tip. The new cells at the root tip are sterile [43], and actual movement of the root surface is limited to the zone of cell elongation. The root cap moves through the soil, but its cells are constantly being sloughed in the process. Theoretically then, direct transport of bacteria by root must be minimal, but as will be shown below, roots may contribute to microbial redistribution in soil.

Worral and Roughley [44] were not able to detect an effect of growing roots on movement of rhizobia in sand columns at various water contents. Parke et al. [45] showed minor movement (0.5 cm) of inoculum with root growth in the absence of water addition. Colonization occurred at the point on the root where the apices had been at the time of inoculation, but root apices exhibited no colonization 2 days after inoculation [45]. Addition of water affected dispersal of inoculant however; the marked strain was found 3.5 cm below the inoculation point on the root, suggesting an indirect role of roots' creation of channels along the root surface for water mediated microbial transport.

Mawdsley and Burns [46] showed that cells of a nonmotile *Flavobacterium* species applied to germinating seeds moved with the expanding root in the absence of downward water flow. Percolating water distributed the bacterium throughout the soil column and in leachate. After 30 days, bacterial populations on lower (younger) root sections were greater for the percolated columns.

These studies lead to the conclusion that the soil channels formed by root extension are of greater significance to microbial movement than are the roots themselves. The channels formed by plant roots become major avenues of water flow under conditions of high matric potentials (above −0.033 MPa) or rapid infiltration. This is the probable explanation of enhanced bacterial dispersal in the presence of (*Glycine max* and *Phaseolus vulgaris*) roots [34]. Although the presence of plant roots improved the transport of two bacterial species by water flow, the growth of plant roots itself was not a vector of bacterial movement [34].

Trevors et al. [47] found that wheat roots growing in vertical soil columns facilitated transport of a *P. fluorescens* strain with percolating water through a loam, but the presence of roots contributed only slightly to bacterial movement in a loamy sand. The roots apparently formed preferential flow paths for water movement in the loam, whereas pathways were not a limitation in the loamy sand. That is, the more open pore structure of the loamy sand was already sufficient (prior to root invasion) for maximal microbial transport. Diurnal patterns of root shrinking and swelling, which creates gaps in the root-soil interface [48,49], would enhance movement of water and microorganisms along the roots.

The work of Bowen and Rovira [27] indicated that bacteria could move along roots in a water film that envelops the roots [48]. Rapid dispersal of microbes on the surface of a water film [33] could account for movement along roots.

Transport of Microbial Propagules with Water

Moving water appears to be the most important vector of passive bacterial movement in terms of distance moved and number of bacteria transported. Peterson and Ward [6] considered advection (transport by moving water) to be the primary bacterial transport mechanism in coarse soils. Dispersion was considered to be primarily an effect of the heterogeneity of a medium, with possible additional effects of bacterial diffusion (i.e., Brownian motion) and motility.

As with active movement, increasing water content improves passive transmittal of bacteria through soil. For example, Bitton et al. [50] compared bacterial movement (two strains of *Klebsiella aerogenes*) into packed columns of soils of different texture and at different water contents. After inoculating and flushing an initially dry soil, the distribution of the bacteria in a sandy soil corresponded closely to the water content above the –0.033 MPa water potential (water content = 0.14 kg kg^{-1}), but movement was impeded even at that high potential in a sandy clay loam (water content = 0.18 kg kg^{-1}), possibly by filtration in smaller pores or by adsorption. In saturated columns, retention of bacteria was least in the sandy soil, allowing the greatest movement.

Transport of root nodule bacteria in the field has been attributed to percolating water and unsaturated flow of soil water [51]. A laboratory study [44] indicated that movement of rhizobia in sand relies on saturated conditions or downward movement of water; movement did not occur in a static sand core even at a matric potential of –0.009 MPa.

Based on many sets of data in the literature, Hagedorn et al. [52] concluded that survival and movement of bacteria were more probable under conditions of saturated flow, whereas percolation of water through unsaturated soil promoted filtering, adsorption, and death of microorganisms in soil. Prevention of contamination of surface water or groundwater by septic effluents relies on the development of a clogged zone, that is, a "biological mat," at the base of the drainfield trench to filter out fecal bacteria [52]. Similar situations may occur in cases where soils receive other societal wastes.

Studies on viruses have shown that saturated flow maximized virus movement into a loamy sand column compared with unsaturated flow, and the actual flow rates in the saturated soil columns did not affect viral distribution [53].

Unsaturated flow greatly limited virus penetration compared with saturated flow (40 cm versus 160 cm depth). Apparently adsorption of viruses to soil particles is promoted as the water films become thinner with decreasing water content. Bacteria-mineral interactions should also increase as water content and water activity decrease.

A primary question relating to water-mediated bacterial transport relates to the magnitude of the process; that is, how far could flowing water move bacterial propagules within the soil profile? Corapcioglu and Haridas [54] approached water flow in unsaturated soil by matrix flow theory and concluded that microbial movement of about 0.2 m in 2 weeks was the maximum rate of transport in soil. However, greater transport can occur under different conditions. For example, Smith et al. [55] compared chloride elution to bacterial flow-through in a study of the movement of *E. coli* in unsaturated soils. All evidence indicated that bacterial transport was occurring in the macropores of the intact soils under conditions of moderate to high water flow (5 to 40 mm h^{-1} for up to 12 h), resulting in bacterial concentrations in the effluent of 22% to 96% of the applied solution within one pore volume of effluent collection from 28 cm cores. The effluent curves showed that a fraction of the water moves through the matrix, removing *E. coli* but not Cl^-, while at these rates of water addition, macropores become functional in transporting water with its solutes and suspended bacteria, which have little interaction with the soil [55]. Bitton et al. [50] also found that only part of the total soil water, the mobile phase, was responsible for transporting bacteria in soil.

Studies by White [56] expanded on the preferential flow of water and solutes or tracers in macropores. Water movement in soils with macropores does not conform to models that assume homogenous media and Darcian flow, that is, where resident water of the soil matrix is displaced by incoming water (displacement, piston flow), and the size of transmitting pores is limited by the water potential of the bulk soil. Naturally structured soils with well-developed macropores (e.g., biochannels) may display rapid conductance of water in macropores even at water contents less than field capacity (water potentials less than −0.033 MPa). This phenomenon has been referred to as partial displacement, preferential flow, bypass flow, channeling, or piping. White [56] concluded that macropores in undisturbed soils can greatly decrease the time for (increase the apparent velocity of) dissolved or suspended matter applied at the soil surface to reach subsurface drains or groundwater.

Partial displacement, in which a major portion of water movement occurs rapidly through macropores while slower movement (displacement) occurs through the soil matrix [57], has been found to be responsible for significant bacterial movement [58]. With rapid movement through macropores, there was little filtering or dilution of bacteria by the soil. A lower proportion, but still a significant amount, of bacteria was traced through a soil of finer texture and no visible macropores. Thomas and Phillips [57] listed factors that determine the proportion of water involved in macropore flow, for example, the rate of water addition and soil properties such as structure, pore size distribution, water content, and clay orientation. These are dynamic properties—water content changing constantly, aggregates reforming and pore sizes changing over months and seasons [59,60], and clay orientation increasing over a time scale in years.

Partial displacement also has indirect implications for the growth and transport of bacteria in soil. If nutrient enrichment occurs in macropores, partial displacement would leach those nutrients. On the other hand, nutrients held within soil aggregates (matrix) are preserved during macropore (bypass) flow [57]. White [56] stated that the amount of solute leached depends on the location of the solute, the ratio of macropore to matrix flow, the saturated hydraulic conductivity of the matrix, the antecedent water content of the soil, the contact area between the bypass flow and relatively static water of the matrix, and the rate of solute diffusion between mobile and immobile water volumes.

Germann et al. [61] proposed a quantitative model of water movement and bacterial transport by preferential flow using kinematic wave theory. According to this model, water movement was considered to occur primarily in macropores, while the unsaturated soil matrix sorbed water from the macropores. Sorbance, that is, the rate of water uptake from the macropore system into the adjacent soil matrix (with unit s^{-1}), was described as occurring at a rate proportional to the initial volumetric water content of the macropore system. With water conductance of 0.1 or 1.0 m s^{-1} and sorbance in the range 10^{-4} to 10^{-5} s^{-1} (typical for the experimental soils), maximum depths of transport were then calculated assuming certain conditions of input. The minimum water content of the profile was considered to be an important factor for transport of microorganisms since water content determines continuity and thickness of water films. The water content of the macropores had to be greater than 0.015 m^3 m^{-3} and macropore conductance had to exceed 0.1 m s^{-1} for significant transport of the cells of *E. coli* to occur. Typical rainfall conditions (2-year storm characteristics) for the United States east of the Rocky Mountains and for the tropical storm region of the Gulf of Mexico were not sufficient to move microorganisms beyond the depth of the soil profile (several meters). However, prolonged input or concentrated flow was sufficient to increase the depth of penetration 10- to 100-fold [61]. Such conditions might occur with a leaking pipe, with flood irrigation, or with stem flow in forests. Assumptions in this model included homogenous macroporous soils of infinite depth.

McCoy and Hagedorn [62] demonstrated rapid transport of *E. coli* by translatory channel (pipe) flow in soil, which occurs commonly in sloping landscapes. Such flow essentially circumvented the soil filtration mechanism.

Under saturated conditions, flow of water and transport of bacteria in the soil is most easily understood and modeled. Under saturated conditions, water flow is driven predominantly by gravity, which simplifies models to the one-dimensional case (downward). In saturated soil, even the largest pores are filled with water and may contribute to the movement of water and bacteria. Macropores tend to be more continuous and less tortuous than micropores. Water in large pores is affected less by matric forces than water in micropores, and moves more rapidly than micropore water (flow rate being related to the fourth power of the radius, according to Poiseuille's law). When the soil is saturated, large pores transmit a greater total volume of water and, therefore, can be considered the major throughway of water and bacteria, and micropores can be considered as merely a modifying influence. The terms macropore and micropore are vague with respect to actual size measurement, but instead refer to relative sizes and provide a functional description, as described here (see e.g., [63,64]).

Water-facilitated transport of bacteria in the soil matrix may be smaller in extent, but may be equally or more important than rapid macropore movement in some situations. Dispersal of an inoculum of *B. japonicum* was closely associated with water movement [35]. Downward movement and lateral dispersal increased in response to an increased amount of water moving through the soil. Water-facilitated movement of *B. japonicum* was not limited to gravitational flow of soil water through large channels. The bacteria were readily transported in sand and silt loam soils by an advancing water front as well as by freely percolating water, suggesting that capillary flow of soil water can transport organisms into protective microenvironments within the soil matrix. These observations suggest that capillary flow caused by water intake by plant roots may contribute to movement of bacteria toward root surfaces.

Wallace [65] stated that water-facilitated dispersal of bacteria is likely to be limited to a few centimeters because soil hydraulic conductivity is small relative to the short-term motility rates of flagellated cells. However, all other evidence presented here suggests that soil water movement is a much more important factor than bacterial motility. Perhaps the short time period of bacterial motility (if present) was not considered in relation to the constant movement of water resulting from ever-changing gradients of water potential.

A practical example of the significance of microbial dispersal via water flow is found in remediation of petroleum-contaminated soil. Cyclical fluctuations of the water table level have been shown to enhance biodegradation of diesel fuel in contaminated soil [66]. Although the alternating exposure to water and air may have been responsible for greater metabolism of the contaminant by the bacterial consortium [66], greater dispersal of the bacteria by movement with the water table should also be considered as a possible explanation.

Bandoni and Koske [33] described rapid microbial dispersal at the air-water interface of spreading water films. Preferential sorption of bacterial cells at air-water interfaces has been demonstrated [26]. Floating cells were transported as well as submerged cells, presumably within the narrow boundary layer dragged by the spreading surface monolayer [33]. Such a situation might occur as dry soil is wetted.

A final caveat in interpreting the cited research relates to discrepancies between laboratory and field studies. Much of the research described above was conducted in the laboratory using soil columns. The question of the applicability of such studies to field situations must be examined. For example, Bitton et al. [67] produced a critique of soil column experiments for studying virus transport in soils that probably also applies to studies of bacterial transport in soil columns. They concluded that soil column studies lack standardization in experimental conditions, yet have been instrumental in providing valuable information concerning transport characteristics (of viruses) in soils. Results of some field studies were consistent with laboratory column studies. It was recommended that standardized procedures for column transport studies be established [67]. Large soil columns (12 cm in diameter and 50 cm in length) of intact and disturbed soil may offer "a degree of realism absent in small cores and permit experimental flexibility" [68]. Despite the problems of scale and variability, Bentjen et al. [69] concluded that intact soil-core microcosms maintained some of the complexities of the natural ecosystem and, therefore, were useful at least for initial valuation of the fate of (plant-associated, genetically engineered) bacteria. Perhaps the most important qualifier in this statement is "some." Interpretation of such data must always be predicated on the knowledge that soil column studies represent only a small part of the complexities of field soils.

C. Environmental Conditions Controlling Microbial Movement

Aside from the method of transport, there are other factors that influence the degree of bacterial transport in soil. Those other factors to be discussed here are properties of the water, properties of the soil, and properties of the bacteria.

Rinsing Solution

The previous analysis documents the conclusion that movement of the liquid phase (water or aqueous solution) is the primary vector for microbial movement in the soil profile. Variables related to the liquid phase impacting microbial transport that must be evaluated are flow rate, number of flushes, and its composition and ionic strength.

Chemical composition of the soil solution can impact the stability of the colloidal suspension of the bacterial cell. For example, variability in divalent cation concentrations alter cell-cell interactions as well as adsorption of the cells onto soil particles. Increasing cation loadings may decrease the charge deficiency (i.e., net negative charge) of clay surfaces sufficiently to reduce repulsion, allow van der Waals or electrostatic interactions between clay and bacterial cells, and thereby, increase adsorption [70].

Similarly, Gannon et al. [71] evaluated the transport of a *Pseudomonas* species through water-saturated aquifer sand under various conditions. Using inoculant concentrations of 10^8 cells ml^{-1}, deionized water greatly increased the number of cells eluted from the columns when compared with a 0.01 M NaCl solution. Increasing the flow rate of the eluent increased the amount of flow-through of bacterial cells for both the NaCl solution and deionized water. However, increasing the inoculant cell density to 10^9 ml^{-1} increased the elution of cells with NaCl but did not change the elution rate with deionized water.

The importance of solution ion composition can be placed in perspective by the observation that the ionic strength of the transporting solution and the cell size of the bacteria are of approximately equal importance [72]. The effect of ionic strength would logically be more important in soil relative to sand. The differential role is derived from the occurrence of variable-charged components (such as iron oxide coatings and humic substances) in more complex soil systems and their role in soil dispersion versus flocculation and aggregation [72]. Huysman and Verstraete [73] found that $MgCl_2$ solution increased adhesion of bacteria to a clay loam soil but did not change adhesion to sandy soil, compared to NaCl solution of equal molarity. Again, greater cationic charge of the soil solution allows greater interaction between negatively charged soil particles and bacteria.

Pulsing of soil water flow may also increase the microbial transport. For example, Trevors et al. [47] found that flow rate of water and number of flushes were the important factors for transport of genetically altered *P. fluorescens* in soil. High flow rates (9 ml h^{-1} versus 3 ml h^{-1}) resulted in larger numbers of cells being eluted from the columns. The first flush rinsed a greater number of cells from the columns than subsequent flushes (following 4-, 7-, and 10-day incubation periods), and a series of flushes moved a greater total number of cells through the columns compared with one flush following the same incubation period. Others have found that the first flush established the vertical distribution pattern of introduced bacteria, and little change occurred with subsequent flushes [68]; however, time periods between flushes was much shorter (in the range of hours) in the latter study.

Soil Properties

Much evidence has already been cited that demonstrated the effect of soil water content on active and passive translocation of bacteria in soil. For example, van Elsas et al. [68] introduced bacteria to soil columns at initial water contents of 0.053 m^3 m^{-3} and 0.13 m^3 m^{-3} (29% and 72% of field capacity, respectively; no matric potential given), and a larger fraction of the cells added to the wetter soil were transported to lower soil layers. Worral and Roughley [44] found a similar effect of initial water content on rhizobia movement for clay loam soil columns, and movement in the initially saturated columns was similar to movement in sand columns. Bacteria added to a relatively wet soil would be found in macropores to a greater extent than would be cells added to a dry soil [74], and water flow in macropores is more likely to transport bacterial cells than flow in micropores.

Although soil water is the primary mediator of bacterial transport in soil, soil texture (i.e., the composition in terms of sand, silt, and clay fractions) is the primary determinant of the kinetics manifested in the system. Soil texture affects the degree of bacterial adsorption (presumably as a result of the clay content) and filtration (as a result of differing pore size distributions). Pore size distribution is also dependent on soil structure, which accounts for biochannels and macroporespace between soil aggregates. Analysis of the impact of soil physical properties on bacterial transport is further complicated by the effect of the chemical environment on interactions of soil particles. For example, soil pH relates to the physicochemical aspects of bacterial interactions with soil minerals.

Breitenbeck et al. [35] investigated the effect of soil particle size on transport of added *B. japonicum* by percolating water; movement was greater in soils consisting primarily of sand-sized particles (Crevasse sand) compared with soils with a large proportion of silt-sized particles (Crowley silt loam). They also found that additions of kaolin to a sand markedly reduced downward movement of surface-applied inoculum of *B. japonicum*, apparently as the result of adsorption by kaolin. Adsorption tests in vitro indicated that kaolin sorbed more *B. japonicum* than the Crevasse or Crowley soils, but less than montmorillonite clay. Pore size differences may have contributed to these effects of texture as well as adsorption. Bitton et al. [50] also considered differences in retention of bacteria in soils of different textures to be the result of filtration in smaller pores and/or adsorption by clay. Furthermore, Issa et al. [11] found that clay-textured soil retarded active bacterial movement more relative to a sandy loam; simultaneous laboratory (in vitro) tests with the soils and bacteria in question indicated that the clay-textured soil adsorbed (removed from suspension) more bacteria than the sandy loam.

Peterson and Ward [6] did not consider adsorption of bacteria to be significant in soils of the textural classes sandy loam, loamy sand, or sand; physical entrapment at particle contact points was proposed to be the major mechanism of bacterial retention in these soils. Matthess and Pekdeger [70] stated that the filtration process may have a critical pore diameter of 72 μm. For sediments of uniform grain size distribution, this would correspond to a coarse loam, and for natural heterogeneous sediments, some percentage of the pores (depending on the texture of the sediment) would interfere with bacterial transport. However, Huysman and Verstraete [75] found that a large proportion of water-mediated bacterial transport occurred in pores of the size range 16 to 30 μm, and clay-bacteria aggregates could be filtered out of soil water by pores in that size range

[76]. Huysman and Verstraete [73,75] found significant adhesion of bacteria to sandy soil, although it was always less than adhesion to clay loam soil. Adhesion to soil particles, as well as filtering by soil pores, was considered important for transport of bacteria in soil columns [76].

Grain size of sand columns has been shown to be the major controlling factor in a study of microbial transport [72]. Transport of cells through the 14-cm-long sand columns varied greatly, with 1% to 90% of the inoculated cells being eluted from the columns. Heterogeneity in the porous media resulted in two peaks in the breakthrough curve of bacteria being eluted from the columns; large pores of a coarse-sand vein provided preferred pathways of movement for water and bacteria, causing the first peak, and water moving through the smaller pores of the fine-sand matrix took longer to pass through the column and caused the second peak of bacterial elution.

Again, it must be stressed that packed soil columns are poorly representative of native soils—except perhaps in field situations where soil aggregation has been totally disrupted by poor management procedures. Differences in the proportion of bacterial cells transported in packed and native soils is exemplified by the studies of Smith et al. [55]. At initial water contents of 23% to 37%, sieved and repacked soils were relatively effective filters of an *E. coli* strain, whereas intact structured soil transmitted up to 96% of added cells under conditions of moderate to high water flow (5 to 40 mm h^{-1}). Features of the structured soil allowed preferential flow of water and transport of bacteria. Similarly, van Elsas et al. [68] found that the heterogeneities of intact soil columns enhanced water-mediated movement of bacteria introduced at the surface, compared to the structure of repacked columns. Furthermore, transport in the repacked columns was dependent on the bulk density at which they were packed, with greater transport at low bulk density. Higher bulk density in the repacked cores would correspond to reduced porosity, especially for the pore size range in which bacterial transport occurs [75].

Chemical properties of soils must also be considered as a factor in bacterial transport. Bitton et al. [50] found retention of *Klebsiella aerogenes* in sandy soil to be greater at pH 3.0 to 3.5 compared to pH 6.5 to 6.7. Santoro and Stotzky [77] found that bacteria were able to sorb to montmorillonite at pH below 4.0, the range at which the bacteria cells would be neutral or positively charged, allowing Coloumbic forces between bacteria and the clay, which has a net negative charge. However, soil pH values below 4.0 are neither common nor desirable for most organisms.

Bacterial Properties Affecting Transport Kinetics

From the foregoing analyses, the impression that the microbes are passive participants in this transport process, except for possibly a minor contribution by active motility and chemotaxis, could possibly be retained. This conclusion is far from the true situation. Even a superficial consideration of the complexity of the situation leads to the conclusion that cell size and morphology (as it affects the effective radius of the cell) as well as cell surface properties affecting sorptive processes are of prime importance.

Gannon et al. [23] screened 19 bacterial strains for relative transportability in the soil matrix. Under saturated conditions at 3°C, elution of bacteria from 5-cm (long) columns of Kendaia loam soil (ground, sieved, sterilized, and packed) ranged from 0.01% to 15% with addition of four pore volumes of water. Total recovery (including extraction from soil) was between 4.3% and about 100%. Adsorption of bacteria to clay particles correlates (negatively) with recovery; viability and filtration were probably other factors in controlling percent recovery of bacterial propagules. Furthermore, Camper et al. [78] found the adsorption rate coefficient, that is, the tendency to adsorb to surfaces of the porous medium, to be a better predictor of microbial transport than bacterial size, motility, or porous medium hydrodynamics.

Huysman and Verstraete [73] found cell surface hydrophobicity to be the major bacterial characteristic affecting adhesion to sandy soil (i.e., greater hydrophobicity, more adhesion). Electrostatic interaction between cells and soil particles was an additional factor in a clay loam soil. In transport studies [75,76], hydrophobicity of bacterial cells corresponded to greater retention in clay loam and sandy soil columns.

Gannon et al. [79], however, found that size was the only bacterial property of those examined that was significantly related to retention or transport of the bacterial cells in Kendaia loam soil columns, with bacteria shorter than 1 μm exhibiting greater total transport. Neither hydrophobicity, net surface electrostatic charge, capsulation (19 strains representing 6 genera), nor possession of flagella (10 strains examined) were statistically related to the extent of movement for a variety of bacteria.

Although total numbers of cells eluted may be inversely related to size, large bacteria may move through soil faster than small bacteria at low flow velocity [78,80] as the result of exclusion of large bacteria from smaller pores, which are more tortuous, are less continuous, and have slower pore velocities. Matthess and Pekdeger [70] considered diffusion of bacterial cells in soil increasingly more important with declining size below 1 μm, and interception became the most effective filtering mechanism for cells larger than 1 μm. The effective radius of bacteria is not only increased by flagella but also by surficial polysaccharide deposits, that is, slime layers, glycocalyx, or capsules. For example, Bitton et al. [67] found that a noncapsulated strain was retained more than a larger, encapsulated strain in a range of soils and across a range of pH. Growth period significantly affected retention of encapsulated bacteria, possibly as the result of the polysaccharide content of the capsule (negative relationship) or of the size of the encapsulated cell (negative relationship).

Cell size of the bacteria and the ionic strength of the transporting solution had approximately equal impact on mobility, but were of lesser importance than grain size of sand columns [72]. The cell size comparison was confounded by the use of bacteria of different shape (rods versus spheres). Although the two strains of bacteria had similar hydrophobicity, which might indicate similar adsorption capabilities, the retention of cells could also have been strongly influenced by their shapes, with rods being subject to more mechanical entrapment than spheres. The cells used were in a resting state, which facilitated the quantification of cell transport; however, actively metabolizing cells may have different transport characteristics.

Again, the extent of transport of a microbial cell in soil is the product of more than predicting the potential of an effective sphere to pass through a constricted soil pore. Mobility can and will be reduced by cell sorption. The extent of these situations is controlled in part by system properties as presented above, but it is also determined by intrinsic cell properties, for example, surface charge. Differences in cell wall composition and morphology result from the metabolic condition of the bacteria. Cell walls of *Arthrobacter globiformis* rods (pleomorphic) were less polymerized and more charged than those of cocci, and walls of cocci whose morphology was nutritionally controlled were more polymerized and less charged than walls of cocci in the stationary stage [81].

Lappin-Scott and Costerton [82] suggested that the dispersal potential of bacteria improves with changes induced by starvation (the normal condition in soil environments), such as size reduction, morphogenesis to cocci, and reduction of exopolysaccharide production. The starvation response of bacterial cells includes an initial phase of dwarfing (fragmentation and continuous size reduction) followed by a more stable period of starvation [83]. Cell hydrophobicity increased during starvation, and the change in surface properties may allow increased adhesion [83].

Usually microbial attachment to soil particles is initially a loose association susceptible to disruption by washing, and firm attachment may require several months, and the production and/or polymerization of extracellular polysaccharide slime and a degree of drying [84]. The loose association, or equilibrium adsorption, results from physical forces (van der Waals and Coloumbic forces), and is reversible and characterized by low specificity; in contrast, "binding" is a firm attachment and probably involves a chemical bond as well as physical forces, and exhibits more specificity [24,60]. Drying cycles are believed to aid sorption by decreasing the electrokinetic potentials to permit hydrogen bonding and van der Waals forces [77]. Gerba and Bitton [80] further discuss filtration and adsorption as it relates to bacterial movement in soil.

A third property of bacterial cells controlling transport in soils is the total microbial biomass. Simply stated, the higher the biomass, the greater the probability of soil pore plugging. Reductions in hydraulic conductivity of sand columns has been related to the rate of biomass accumulation in bacterial colonies in the columns [85]. One bacterial strain was relatively ineffective in clogging the sand columns, apparently for two reasons: it had a lower biomass yield for the limiting nutrient (O_2), and a larger proportion of cells were washed from the column compared to the strain that clogged the columns effectively. Lower levels of cell transport and subsequent clogging of sand columns were related to the tendency of cell aggregates to form as the microorganisms multiplied. Two different clogging mechanisms may have been involved, as two exponential curves were required to describe the relationship between saturated hydraulic conductivity and biomass density. The initially slow decline in conductivity may have been a simple result of an increase in biomass volume, and the subsequent rapid decrease was hypothesized to be the result of detachment of biomass flocs that then blocked pore constrictions. Hence, biofilm growth may initially enhance preferential flow by coating pore walls, but then biomass detachment and clogging of pores may prevent preferential flow. Matthess and Pekdeger [70] described the water/sediment boundary of filtration beds, where a biologically active layer exists that helps prevent pollutants from entering groundwater. At high microbial populations, flocculation and aggregation occur at that layer and limit transport through it. Moderate populations and moderate levels of chemical contamination in a medium are potentially the least desirable situation because disturbances in the biological layer can lead to chemical breakthrough and groundwater contamination [70].

Reduction in the permeability of sand column reactors (for biodegradation of contaminants) was observed to correlate well with biomass density for values less than 0.4 mg cm^{-3} and exhibited independence at higher densities [86]. Steady-state biofilms developed at a specific substrate level for a given porous media. Permeability was reduced up to 1000-fold in some cases, but there was a limit on permeability reduction. The rate of flow, duration of flow, and substrate loading affected the transport and subsequent distribution of biomass in the column. Taylor and Jaffé [86] hypothesized (the open pore model) that biofilms coating the pore walls increase dispersivity and reduce permeability until some point where increased fluid velocity and shear stress (caused by the constriction) remove biofilm and thus maintain open pores. While soils do not normally present the nutritional status of these model systems, the potential for pore plugging by biomass growth at soil microsites must be considered.

A variety of other cell-associated properties have been found to alter cell transport potential. For example, Reynolds et al. [28] recorded that motile bacteria penetrated four times faster than nonmotile strains, but nonchemotactic motile strains penetrated more than chemotactic motile strains. For the nonmotile strains, gas production significantly improved their penetration. Growth rate was important for penetration of motile bacteria. Similarly, in a stopped-flow chamber, Mercer et al. [87] investigated the effect of growth rate on the transport of bacteria by random motility and chemotaxis. Random movement was constant over the range of growth rates, but low growth rates led to increased sensitivity and attraction to and movement toward D-fucose. Peterson and Ward [6] hypothesized that copiotrophic bacteria (requiring abundant nutrients) introduced to soils have a greater potential for transport than the indigenous, oligotrophic population as the result of the copiotrophic bacteria's reduction in size resulting from the carbon-limited conditions, possible short-term increase in number, and reduced competitive ability at soil particle surfaces. Bacterial survival must also be considered a factor. Lovins et al. [88] introduced genetically engineered microorganisms (GEMs) to the surface of large intact soil columns and added water at low rates to maintain unsaturated flow conditions. The GEMs leached only to a depth of 30 to 40 cm in 70 days, and its limited extent of transport was related to its linear decline in population with time. The GEMs apparently were not well adapted to the soil environment. However, a marker plasmid was transferred to indigenous bacteria, and the transconjugants appeared to be well adapted to the environment and more mobile than GEMs.

It can be concluded from these studies that bacterial movement in soil is primarily a result of passive transport in soil water, but soil animals and plant root channels may make minor contributions to transport processes. Water-mediated microbial movement is controlled primarily by flow rate and ionic strength, as well as by the size and shape of microbial cells.

III. CONCEPTS AND MODELS OF THE MOVEMENT OF BACTERIA IN SOIL

Dickinson [89] describes the ideal model for the movement of microbes in soil to include suspended as well as adsorbed microorganisms, to allow for porosity changes as clogging and declogging occur, to consider potential ranges of values of all parameters, and to possibly be statistically based, generating a probability distribution for output solution from spatially and temporally distributed data. The data set used to develop and evaluate the model would need to include multiple measurements and boundary conditions. Hydraulic conductivity, hydraulic gradient, and porosity measurement would be necessary for saturated conditions, and the soil-water characteristic curve relating water content to water potential would also be needed for unsaturated conditions. A variety of models have been developed that meet these standards to varying degrees.

Matthess and Pekdeger [70] discussed the life and transport of bacteria and viruses in groundwater. Native groundwater populations may develop densities much greater than 10^3 ml^{-1} under favorable conditions. Contaminant pathogenic microorganisms are usually eliminated, but under oligotrophic conditions they may survive or even increase slightly in number, due to fragmentation, during the first 1 to 7 days. Afterwards, an exponential function describes their elimination from groundwater:

$$C_t = C_0 e^{-\lambda(t-t_0)} \tag{1}$$

where C_0 is the initial concentration (ml^{-1}), and C_t is the concentration at time t. The elimination rate, λ, is specific for each species (and must be measured) and environment. This model predicts very rapid elimination of microbial cells from soil cores initially and then a sharp decline, but small quantities of microorganisms remain even after long periods. The process of bacterial adsorption and desorption to and from soil particles retards microorganisms in their movement through soil, allowing greater elimination time. The extent of retardation (R_d) depends on properties of the microorganisms and the porous material:

$$R_d = V_w/V_m = 1 + (\rho_b/n)\, K_d \tag{2}$$

where V_w is mean groundwater velocity (m d^{-1}), V_m is average velocity of the microorganisms, ρ_b is bulk density (g cm^{-3}), and n is porosity (m^3 m^{-3}). The distribution coefficient, K_d (ml g^{-1}), describes the affinity of the media for the microorganisms. The retardation values, R_d, are in the range of 1 to 2 (dimensionless) for bacteria but may approach 500 for viruses. Native bacteria attach to soil particles to a greater extent than alien organisms during their exponential growth phase. Enteric bacteria entering soil systems grow very little under groundwater conditions, and therefore active attachment is minimal. Sedimentation of microorganisms was not included in this model as the result of their density being very close to 1 g cm^{-3}, and therefore, sedimentation is considered to be insignificant. The transport equation of Matthess and Pekdeger [70] took the form

$$\partial C/\partial t = \nabla\, [(D/R_d)\ \text{grad}\ C - (V_w/R_d)\ \text{grad}\ C] - \lambda C \tag{3}$$

where D is the coefficient of hydrodynamic dispersion (m^2 d^{-1}), C is concentration of the microorganisms (ml^{-1}), and grad C is the concentration gradient.

Corapcioglu and Haridas [90] developed a model of fate and movement of microorganisms in saturated porous media using a mass conservation approach; that is, the change in number of cells in a volume of soil is equal to the inputs by transport processes and cell division minus the losses by transport, filtration, adsorption, sedimentation, and cell death. The presence of bacteria and viruses in wastewater added to soil is the initiation point of the model. The mass-dispersive and mass-diffusion fluxes were admittedly neglected, with the mass conservation equation [see Equation (4)] referring simply to J, the specific mass discharge of suspended particles, to represent the transport processes. Still their contribution is relevant to this discussion in that sources and sinks of bacteria (or viruses) in saturated media are thoroughly summarized. Removal of bacteria from water by such processes as filtration, adsorption, and sedimentation were mathematically described. Addition by bacterial replication, reduction as the result of death, and enhancement of transport by Brownian motion, chemotaxis, and tumbling were also considered.

Corapcioglu and Haridas [54,90] expanded this mass conservation equation into a series of equations useful for estimating the three-dimensional spatial and temporal dynamics of microbial populations in water-saturated porous media. The summary mass conservation equation,

$$R_a + \partial(\theta C)/\partial t = -\nabla J + R_{df} + R_{gf} \tag{4}$$

includes microbial deposition on grains (R_a, interception or net mass transfer rate which considers clogging versus unclogging), population growth (R_{gf}) and decay (R_{df}), as well as transport (J, specific mass discharge of "particles"). C is concentration of suspended particles, θ is water content, and t is time. The actual movement (mass discharge) of microorganisms was summed up in the term J:

$$J = -D_d\,(n - \sigma)\,\nabla C + (n - \sigma)v_f C + (n - \sigma)v_g C + J_B + J_{CT} \tag{5}$$

where D_d is dispersion, $(n - \sigma)$ represents water content θ, V_f is flow velocity, V_g is gravitational settling velocity, J_B is diffusion (Brownian motion), and J_{CT} is chemotactic and random motility. The decay term takes into account straining, adsorption, sedimentation (for cells larger than 5 µm), and death. The governing equation for microbial transport is coupled with a transport equation for the bacterial nutrient. To apply this modeling technique to unsaturated flow, the air phase and the water phase would be treated individually in terms of the mass balance.

Although Corapcioglu and Haridas [54] took into account changes in porosity resulting from bacterial growth, they did not consider the accompanying loss of permeability and increase in dispersivity. Taylor and Jaffé [91] developed a bacterial transport model that accounts for reductions in porosity and permeability and increases in dispersivity that would occur with significant biofilm growth in saturated porous media. The macroscopic transport equation for active biomass suspended in the water phase is

$$\partial(n^w C^w{}_b)/\partial t = \partial[(n^w D_{ij})\partial C^w{}_b/\partial x_j]/\partial x_i - \partial(V_i C^w{}_b)/\partial x_i + n^w \rho^w r^w{}_b + n^w \rho^w [e^w(\rho\omega_b) + \Gamma^w{}_b] \tag{6}$$

where $C^w{}_b$ is active biomass per unit volume of water phase, n^w is the volume fraction of the water phase, D_{ij} is the hydrodynamic dispersion tensor, V_i is the macroscopic Darcy velocity, ρ^w is the water density, $r^w{}_b$ is the rate of biomass production in the water phase, $n^w\rho^w e^w(\rho\omega_b)$ is the transfer of biomass from the biofilm phase to the water phase as the result of movement of the water-biofilm phase boundary, and $n^w\rho^w\Gamma^w{}_b$ is the transfer of biomass from the biofilm phase to the water phase by diffusion across the water-biofilm phase boundary. A numerical solution was found, and then the model was calibrated and validated with experimental data.

Huysman and Verstraete [76] found that bacterial transport by water in soil columns could be described by a colloid filtration model if irrigation was started immediately after inoculation of the soil surface.

$$N = N_0 \times 10^{-\lambda l} \tag{7}$$

N and N_0 are the number of bacteria (CFU g) at a certain depth—1 cm—and at the soil surface, respectively. The filtration coefficient, λ (cm^{-1}), is inversely related to D_{90}, the depth (cm) at which the number of bacteria is reduced by 90%, $D_{90} = \lambda^{-1}$. The filtration coefficient was greater for the more hydrophobic bacterial strains, and it decreased with increasing irrigation rate. Deviations from the colloid filtration model were attributed to conditions that promoted sorptive interactions between bacterial cells and soil particles.

McDowell-Boyer et al. [92] evaluated three filtration processes that particles might encounter in porous media. Surface filtration is considered to occur when the particles are too large to enter the media. Straining filtration is particle capture by the media for which there is a critical ratio of the diameters of media to particle (bacteria). The third type is physical-chemical filtration, where particles are much smaller than the size of the media and are retained only if attractive forces dominate when particles collide with the media. Assuming that the retention (presumably from adsorption) is reversible, the net result of physical-chemical filtration is retardation of the particles in their movement through the media. Linear chromatography theory is used to describe this process. Although this model was suggested for use with microorganisms, its use in soil presents a number of problems. Adsorption is probably not reversible in all cases. In addition, straining filtration of bacteria occurs to some extent in most soils, and surface filtration may even occur at the faces of soil aggregates. The model is also not designed to address soil heterogeneity. Other problems associated with this model center on the fact that the subjects of interest are living "particles," that is, microorganisms that may divide, metabolize, change form, and/ or die. It cannot be assumed that the input bacterial population will equal the eventual output. Without additional factors in the equation, the model is inadequate to predict with any accuracy the microbial population at a particular place and time during transport through the soil matrix.

Germann and Douglas [93] were critical of the filtration theory model of particle transport because it is based on Darcy's law. Darcy's law for fluid flow through porous media is a macroscopic and averaging theory, whereas particles tend to flow along preferred flow patterns. Germann et al. [61] proposed a quantitative model of bacterial transport by preferential flow using kinematic wave theory. According to this model, "unsaturated flow is treated as a wave equation composed of diffusive and gravitational components. The water content profile is considered to be a series of waves advancing through the soil. The waves are kinematic in the sense that they are described by an expression for velocity as a function of water content plus a differential equation of mass continuity" [94]. Kinematics involves the study of motion exclusive of the influence of mass and force. In this case, flux (q) is assumed to be a function of water content (θ) alone, and the wave equation is

$$\partial\theta/\partial t + (dq/d\theta)\,(\partial\theta/\partial z) = 0 \tag{8}$$

[94] where z is distance and t is time. For a given θ, movement of a location on the water content profile has velocity

$$V_c(\theta) = dz/dt\,|_{\theta} = dq/d\theta \tag{9}$$

[94]. Germann et al. [61] added a term to the kinematic wave theory to account for the sorbance of water into soil micropores and used the model to predict the transport of microorganisms in soil macropores. Water flow parameters of the soils, sorbance (r, s^{-1}) and conductance (b, m s^{-1}), were calculated from experimental data. The minimum water content of the profile [i.e., macropore water content at the bottom of the core, $w(Z)$] was also determined, because it was anticipated that a critical water content exists for transport of microorganisms. Increasing soil water sorbance decreased the depth of penetration, but greater macropore conductance increased the depth. Microbial size, shape, and affinities to soil particles were not considerations in the model. Assumptions included homogenous macroporous soils (constant b and r) of infinite depth. In reality, most soils could not be considered homogenous even within the typical profile depth, much less to the depths calculated by Germann et al. [61].

In the kinematic wave model of Germann et al. [61], water movement was considered to occur primarily in macropores, with the unsaturated soil matrix (micropores) allowed to sorb water from the macropores. Like partial displacement models, this model does not conform to the traditional concept of water movement in unsaturated soil. Water moving through unsaturated soil was usually described as occurring in the matrix via the micropores, whereas the macropores theoretically remained unfilled until a critical water content was reached.

Peterson and Ward [6] developed a model for bacterial transport in coarse-textured soils. Bacterial diffusion, sedimentation, and random or chemotactic motility were ignored because they were considered to be minor factors in microbial movement relative to advection, that is, movement with flowing water. Straining and adsorption are handled with the distribution coefficient (k_a) to represent the partitioning of bacteria between soil solution and soil matrix. This model is based on the assumption that the filtration mechanisms are fast, reversible, and follow a linear isotherm. Growth and death of attached bacteria were assumed to be balanced, and growth of bacteria in the soil solution was assumed to be negligible. The resulting equation is

$$[(\rho k_a/\theta) + 1]\partial C/\partial t = (1/\theta)\partial[D\theta(\partial C/\partial z)]/\partial z - (q_w/\theta)\partial C/\partial z - k_d C \quad (10)$$

where z represents the depth in the soil, q_w is the soil water flux, and k_d is the die-off rate constant. The model was validated by a comparison with published data. Simulations were run with randomly generated input data to create probability distributions of bacterial transport to specific depths. The soil water retention and hydraulic conductivity models are a critical but underemphasized component of this bacterial transport model. In coarse-textured soil, where gravitational water had just drained (which is consistent with the authors' focus on wastewater drainfields), this model may be adequate if all the assumptions are reasonable, except that drainfields usually have a biological mat that act as a filter, whereas biofilm development was considered beyond the scope of this model. Use of this model with loamy to clayey soils at water contents less than field capacity (i.e., the threshold of gravitational water) is inappropriate because matric potential and unsaturated hydraulic conductivity become major factors in water movement.

To predict the adsorption and transport of bacteria in unsaturated soil in the absence of straining, Tan et al. [95] start with the one-dimensional continuity equation

$$\partial\mu/\partial t = -\,\partial J_m/\partial x - R_d + R_g \quad (11)$$

where μ is the number of microorganisms per volume of soil, including those in the matrix as well as in liquid ($\mu = \theta C + \rho q$, where q is the number of bacteria adsorbed per unit mass of solid particles); J_m is the flux of microorganisms in the direction of flow, R_d is the rate of decay, and R_g is the rate of growth. Sedimentation and chemotactic movement is considered to be negligible. The flux of bacteria, J_m, is the sum of fluxes resulting from advective transport of water, random motility, diffusion, and hydrodynamic dispersion:

$$J_m = -\,\mu D(u)\,[\partial C/\partial x] + JC \quad (12)$$

where, $D(u)$ is the dispersion coefficient of microorganisms, which accounts for active movement, diffusion, and hydrodynamic dispersion, J is the flux of water, and x is distance. The general equation for one-dimensional transport of bacteria in porous soils becomes

$$\partial(\rho q)/\partial t + \partial(\theta C)/\partial t = \partial[\theta D(u)\,\partial C/\partial x]/\partial x - \partial(JC)/\partial x - R_d + R_g \quad (13)$$

Then Tan et al. [95] ignore R_d and R_g for short-period results. Assuming the bulk density of soil to be constant and invoking the continuity equation of water, then

$$[\theta + \rho\,(dq/dC)]\,\partial C/\partial t = \partial[\theta D(u)\,\partial C/\partial x]/\partial x - J\,\partial C/\partial x \quad (14)$$

Moving coordinates for bacterial transport are introduced to relate the bacterial front to space and time. Similarity variables (distance divided by the square root of time) are used to eliminate x (distance) and t (time) from the one-dimensional horizontal transport equation for water and allows unique values of volumetric water content for given x and t.

McCaulou et al. [96] used a model (previously used for solutes) for describing colloid transport in saturated porous medium with two types of sites for attachment: fast and kinetically limited. With the concentration of bacteria sorbed at the fast sites represented by S_1 and that at the kinetically limited sites by S_2, the governing equation is

$$\theta(\partial C/\partial t) + \rho_b(\partial S_1/\partial t + \partial S_2/\partial t) = \theta D(\partial^2 C/\partial z^2) - u\theta(\partial C/\partial z) \quad (15)$$

with

$$S_1 = K_{p1}C \quad (16)$$

and

$$\rho_b(\partial S_2/\partial t) = \theta k_1 C - \rho_b k_2 S_2 \quad (17)$$

C is the concentration of bacteria in the pore water, θ is the porosity, that is, the volumetric water content at saturation, ρ_b is soil bulk density, D is the dispersion coefficient, u is the average pore water velocity, and K_{p1} is the distribution coefficient for the Freundlich-type linear isotherm of the fast sites. Rate constants for attachment and detachment at kinetically limited sites are k_1 and k_2, respectively. Experiments were performed and data fitted to evaluate related dimensionless parameters including the retardation factor,

$$R = 1 + [\rho_b(K_{p1} + K_{p2})/\theta] = 1 + (\rho_b K_{p1}/\theta) + (k_1/k_2) \quad (18)$$

the Péclet number,

$$P = Lu/D, \text{ where } L \text{ is the length of the column;} \quad (19)$$

the mass transfer coefficient,

$$\omega = (L/u)/(1/k_1) = k_1 L/u \quad (20)$$

and the ratio of equilibrium to total adsorption,

$$\beta = (\theta + \rho_b K_{p1})/[\theta + \rho_b K_{p1} + \theta(k_1/k_2)] = 1 - [k_1/k_2 R)] = 1 - [\omega u/(k_2 LR)] \quad (21)$$

The experiments consisted of short-pulse inputs of hydrophilic and hydrophobic bacteria in continuous, upward, unsaturated flow columns of quartz (negatively charged), hematite-coated quartz (positively charged), and polymer-coated quartz (hydrophobic). With the specified variables of bacteria and porous media characteristics, the fitted values determined for R, P, ω, and β were used to make conclusions regarding the model and factors of bacterial transport. For instance, retardation factors, R, for hydrophilic bacteria were much larger than for hydrophobic bacteria on positively charged and hydrophobic surfaces, indicating that hydrophilic bacteria were attached for a longer period under those conditions and hydrophobic bacteria moved faster and further in the columns. Retardation in quartz columns was the same for both (hydrophobic and hydrophilic) bacterial strains, but low compared to coated quartz. The model was close to first order (β values were close to $1/R$), and most of the attachment sites were kinetically controlled (β values were low). Rates of attachment (k_1) were similar for all materials, but rates of detachment (k_2) varied within two orders of magnitude, the greatest being for hydrophilic bacteria on quartz (–) and the lowest being for hydrophilic bacteria on hematite-coated quartz (+). The sticking efficiency (α), which is related to the rate coefficient for attachment (k_1), was also evaluated by estimating the single-collector removal efficiency (η) in the equation

$$\eta\alpha = (2/3)(k_1 d/u)[1/(1-\theta)] \tag{22}$$

where d is the bacterial diameter.

Loehle and Johnson [97] combined two approaches to modeling microbial transport, that consider the different scales of processes that affect transport and allow rescalings as variables change. Regions are modeled as chemostats. Percolation models are used at the fine scales to allow for soil heterogeneity and the dynamics of the microbial community, and to derive the parameters for the coarse-scale model. Percolation models are used at the coarse scale to define flow paths.

Sarkar et al. [98] modeled transport and growth of bacteria in porous media using a fractional flow approach. The shape of the fractional flow curve is selected based on the mechanisms of bacterial retention to be included in the model. Comparison of results with experimental data suggests that the physical processes incorporated in the model are reasonably represented.

McInerney [99] addressed modeling of bacterial movement in groundwater aquifers. He categorized the movement into three general situations depending on flow rates and the mineralogy of the formation:

1. Relatively unimpeded flow of bacteria would occur with high flow rates and small amounts of clay and dissolved solids, especially when secondary pore structure provided preferential flow paths.
2. Slower and discontinuous movement would occur with lower flow rates and greater amounts of clay and/or dissolved solids because of interactions between the bacteria and solid surfaces. Self-propulsion by bacteria was considered unimportant relative to advective flow.
3. Bacterial chemotaxis would be important only at very low flow rates and would also provide a mechanism for permeation of the bacteria through the matrix.

IV. CONCLUSIONS

Bacterial movement in soil is a complex process involving a variety of interrelated factors. Thus, prediction of the extent and perhaps even the potential for bacterial movement in soils is very difficult at best. Numerous mathematical models have been developed, most of which focus on the transport of bacteria (or viruses or microorganisms in general) by water, that is, advection. That is consistent with the literature, which indicates that water is the major mediator of bacterial transport in soil. The models also contain a range of other variables that represent soil and microbiological properties that have been shown to affect transport of microorganisms, that is, dispersion, sedimentation, chemotactic or random motility, growth and death, clogging and unclogging, filtration, and adsorption. The greater the number of soil and microbial traits accounted for in a model, the more accurate generally is the depiction of the process. Unfortunately this accuracy leads to cumbersome and complicated solutions. Because of possible error in the estimation of associated parameters, eliminating minor factors from the model may not reduce the accuracy of the solution for some situations. Therefore, the simpler models may be adequate to predict changes in microbial distributions as long as the major factors are represented.

The assumptions used in developing the transport models and solving the equations may not be valid for every situation. In selecting a model, the assumptions must be examined together with the input data set and estimated parameters required to ensure that the most appropriate model is chosen. For example, one-dimensional flow downward (under the influence of gravity) is a common assumption that greatly simplifies the transport equations. However, gravity is not the primary driving force in unsaturated soil (matric potential gradient is), and the error associated with that assumption would increase with increasingly finer soil texture.

Even with an increasing understanding of the complexity of the interactions controlling microbial dispersion in soil and a growing capability to represent this knowledge mathematically, it is still not possible to explain the origin of subsoil and aquifer bacterial populations. Exploration of this intriguing question awaits augmentation of the understanding of the long-term interactions (decades to centuries) determining distribution of soil bacteria and of the induction of phenotypic microbial properties expressed in soil that are not evident in controlled cultures. As this basic information is assembled, extended mobility of native and alien microbial properties within the soil system will become more predictable.

REFERENCES

1. Tate, R. L., III. 1986. Soil organic matter: biological and ecological effects. John Wiley & Sons. New York.

2. Sinclair, J. T., and W. C. Ghiorse. 1989. Distribution of aerobic bacteria, protozoa, algae, and fungi in deep subsurface sediments. Geomicrobiol. J. 7:15–31.
3. Bengtsson, G., R. Linqvist, and M. D. Piwoni. 1993. Sorption of trace organics to colloidal clays, polymers, and bacteria. Soil Sci. Soc. Am. J. 57:1261–1270.
4. Gammack, S. M., E. Paterson, J. S. Kemp, M. S. Cresser, and K. Killham. 1992. Factors affecting the movement of microorganisms in soils, p. 263–305. *In* G. Stotzky and J.-M. Bollag (eds.), Soil biochemistry, Vol. 7. Marcel Dekker, New York.
5. Stotzky, G. 1986. Influence of soil mineral colloids on metabolic processes, growth, adhesion, and ecology of microbes and viruses, p. 305–428. *In* P. M. Huang and M. Schnitzer (eds.), Interaction of soil minerals with natural organics and microbes. Special Publ. No. 17. Soil Science Society of America, Madison, Wisconsin.
6. Peterson, T. C., and R. C. Ward. 1989. Development of a bacterial transport model for coarse soils. Water Res. Bull. 25:349–357.
7. Purcell, E. M. 1977. Life at low Reynolds number. Am. J. Phy. 45:3–11.
8. Arora, K. D., and S. Gupta. 1993. Effect of different environmental conditions on bacterial chemotaxis toward fungal spores. Can. J. Microbiol. 39:922–931.
9. Bashan, Y., and G. Holguin. 1994. Root-to-root travel of the beneficial bacterium *Azospirillum brasilense*. Appl. Environ. Microbiol. 60:2120–2131.
10. Issa, S., L. P. Simmonds, and M. Wood. 1993. Passive movement of chickpea and bean rhizobia through soils. Soil Biol. Biochem. 25:959–965.
11. Issa, S., M. Wood, and L. P. Simmonds. 1993. Active movement of chickpea and bean rhizobia in dry soil. Soil Biol. Biochem. 25:951–958.
12. Soby, S., and K. Bergman. 1983. Motility and chemotaxis of *Rhizobium meliloti* in soil. Appl. Environ. Microbiol. 46:995–998.
13. Hillel, D. 1982. Introduction to soil physics. Academic Press, San Diego.
14. Doran, J. W., L. N. Mielke, and S. Stamatiadis. 1988. Microbial activity and N cycling as regulated by soil water-filled pore space. Paper No. 132. Proc. Int. Soil Tillage Res. Org. ISTRO, July 11–15, 1988.
15. Linn, D. M., and J. W. Doran. 1984. Effect of water-filled pore space on carbon dioxide and nitrous oxide production in tilled and nontilled soils. Soil Sci. Soc. Am. J. 48:1267–1272.
16. Skopp, J., M. D. Jawson, and J. W. Doran. 1991. Steady-state aerobic microbial activity as a function of soil water content. Soil Sci. Soc. Am. J. 54:1619–1625.
17. Bashan, Y. 1986. Migration of rhizosphere bacteria *Azospirillum brasilense* and *Pseudomonas fluorescens* towards wheat roots in the soil. J. Gen. Microbiol. 132:3407–3414.
18. Griffin, D. M., and G. Quail. 1968. Movement of bacteria in moist, particulate systems. Aust. J. Biol. Sci. 21:579–582.
19. Hamdi, Y. A. 1971. Soil-water tension and the movement of rhizobia. Soil Biol. Biochem. 3:121–126.
20. Wong, P. T. W., and D. M. Griffin. 1976. Bacterial movement at high matric potentials—II. In fungal colonies. Soil Biol. Biochem. 8:219–223.
21. Wong, P. T. W., and D. M. Griffin. 1976. Bacterial movement at high matric potentials—I. In artificial and natural soils. Soil Biol. Biochem. 8:215–218.

22. Marshall, K. C. 1971. Sorptive interactions between soil particles and microorganisms, p. 409–445. *In* A. D. McLaren and J. Skujins (eds.), Soil biochemistry, Vol. 2. Marcel Dekker, New York.
23. Gannon, J. T., U. Mingelgrin, M. Alexander, and R. J. Wagenet. 1991. Bacterial transport through homogenous soil. Soil Biol. Biochem. 23:1155–1160.
24. Stotzky, G. 1980. Surface interactions between clay minerals and microbes, viruses and soluble organics, and the probable importance of these interactions to the ecology of microbes in soil, p. 231–249. *In* R. C. W. Berkeley, J. M. Lynch, J. Melling, P. R. Rutter, and B. Vincent (eds.), Microbial adhesion to surfaces. Ellis Horwood, Chichester.
25. Marshall, K. C. 1980. Adsorption of microorganisms to soil and sediments, p. 317–329. *In* G. Bitton and K. C. Marshall (eds.), Adsorption of microorganisms to surfaces. John Wiley & Sons, New York.
26. Wan, J., J. L. Wilson, and T. L. Kieft. 1994. Influence of the gas-water interface on transport of microorganisms through unsaturated porous media. Appl. Environ. Microbiol. 60:509–516.
27. Bowen, G. D. and A. D. Rovira. 1976. Microbial colonization of plant roots. Annu. Rev. Phytopath. 14:121–144.
28. Reynolds, P. J., P. Sharma, G. E. Jenneman, and M. J. Mclnerney. 1989. Mechanisms of microbial movement in subsurface materials. Appl. Environ. Microbial. 55:2280–2286.
29. MacDonald, J. D., and J. M. Duniway. 1978. Influence of soil texture and temperature on the motility of *Phytophthora cryptogea* and *P. megasperma* zoospores. Phytopath. 68:1627–1630.
30. Duniway, J. M. 1976. Movement of zoospores of *Phytopthora cryptogea* in soils of various textures and matric potentials. Phytopathology 66:877–882.
31. Westerlund, F. V., R. N. Campbell, R. G. Grogan, and J. M. Duniway. 1978. Soil factors affecting the reproduction and survival of *Olpidium brassicae* and its transmission of big vein agent to lettuce. Phytopathology 68:927–935.
32. Henrichsen, J. 1972. Bacterial surface translocation: a survey and a classification. Bacteriol. Rev. 36:478–503.
33. Bandoni, R. J., and R. E. Koske. 1974. Monolayers and microbial dispersal. Science 183:1079–1081.
34. Madsen, E. L., and M. Alexander. 1982. Transport of *Rhizobium* and *Pseudomonas* through soil. Soil Sci. Soc. Am. J. 46:557–560.
35. Breitenbeck, B. A., H. Yang, and E. P. Dunigan. 1988 Water-facilitated dispersal of inoculant *Bradyrhizobium japonicum* in soils. Biol. Fertil. Soils 7:58–62.
36. Hutchinson, S. A., and M. Kamel. 1956. The effects of earthworms on the dispersal of soil fungi. J. Soil Sci. 7:213–218.
37. Thornton, M. L. 1970. Transport of soil-dwelling aquatic Phycomycetes by earthworms. Trans. Br. Mycol. Soc. 55:391–397.
38. Parle, J. N. 1963. Micro-organisms in the intestines of earthworms. J. Gen. Microbiol. 31:1–11.
39. Parle, J. N. 1963. A microbiological study of earthworm casts. J. Gen Microbiol. 31:13–22.

40. Henschke, R. B., E. Nucken, and F. R. J. Schmidt. 1989. Fate and dispersal of recombinant bacteria in a soil microcosm containing the earthworm *Lumbricus terrestris*. Biol. Fertil. Soils 7:374–376.

41. Wallace, H. R. 1958. Movement of eel worms. I. The influence of pore size and moisture content of the soil on the migration of larvae of the beet worm, *Heterodera schachtii* Schmidt. Ann. Appl. Biol. 46:74–85.

42. Armstrong, J. L., L. A. Porteous, and N. D. Wood. 1994. The cutworm *Peridroma saucia* (Lepidoptera: Noctuidae) supports growth and transport of pBR322-bearing bacteria. Appl. Environ. Microbiol. 55:2200–2205.

43. Rovira, A. D., and R. Campbell. 1974. Scanning electron microscopy of microorganisms on the roots of wheat. Microb. Ecol. 1:15–23.

44. Worral, V., and R. J. Roughley. 1991. Vertical movement of *Rhizobium leguminosarum* bv. trifolii in soil as influenced by soil water potential and water flow. Soil Biol. Biochem. 23:485–486.

45. Parke, J. L., R. Moen, A. D. Rovira, and G. D. Bowen. 1986. Soil water flow affects the rhizosphere distribution of a seed-borne biological control agent, *Pseudomonas fluorescens*. Soil Biol. Biochem. 18:583–588.

46. Mawdsley, J. L., and R. G. Burns. 1994. Root colonization by a *Flavobacterium* species and the influence of percolating water. Soil Biol. Biochem. 26:861–870.

47. Trevors, J. T., J. D. van Elsas, L. S. van Overbeek, and M.-E. Starodub. 1990. Transport of a genetically engineered *Pseudomonas fluorescens* strain through a soil microcosm. Appl. Environ. Microbiol. 56:401–408.

48. Huck, M. G., B. Klepper, and H. M. Taylor. 1970. Diurnal variations in root diameter. Plant Physiol. 45:529–530.

49. Faiz, S. M. A., and P. E. Weatherley. 1982. Root contraction in transpiring plants. New Phytol. 92:333–343.

50. Bitton, G., N. Lahav, and Y. Henis. 1974. Movement and retention of *Klebsiella aerogenes* in soil columns. Plant Soil 40:373–380.

51. Lowther, W. L., and H. N. Patrick. 1993. Spread of *Rhizobium* and *Bradyrhizobium* in soil. Soil Biol. Biochem. 25:607–612.

52. Hagedorn, C., E. L. McCoy, and T. M. Rahe. 1981. The potential for ground water contamination from septic effluents. J. Environ. Qual. 10:1–8.

53. Lance, J. C., and C. P. Gerba. 1984. Virus movement in soil during saturated and unsaturated flow. Appl. Environ. Microbiol. 47:335–337.

54. Corapcioglu, M. Y., and A. Haridas. 1985. Microbial transport in soils and groundwater: a numerical model. Adv. Water Res. 8:188–200.

55. Smith, M. S., G. W. Thomas, R. E. White, and D. Ritonga. 1985. Transport of *Escherichia coli* through intact and disturbed soil columns. J. Environ. Qual. 14:87–91.

56. White, R. E. 1985. The influence of macropores on the transport of dissolved and suspended matter through soil. Adv. Soil Sci. 3:95–120.

57. Thomas, G. W., and R. E. Phillips. 1979. Consequences of water movement in macropores. J. Environ. Qual. 8:149–152.

58. Rahe, T. M., C. Hagedorn, E. L. McCoy, and G. F. Kling. 1978. Transport of antibiotic-resistant *Escherichia coli* through western Oregon hill slope soils under conditions of saturated flow. J. Environ. Qual. 7:487–494.

59. Kay, B. D. 1990. Rates of change of soil structure under different cropping systems. Adv. Soil Sci. 12:1–52.

60. Staricka, J. A., R. R. Allmaras, W. W. Nelson, and W. E. Larson.1992. Soil aggregate longevity as determined by the incorporation of ceramic spheres. Soil Sci. Soc. Am. J. 56:1591–1597.

61. Germann, P. F., M. S. Smith, and G. W. Thomas. 1987. Kinematic wave approximation to the transport of *Escherichia coli* in the vadose zone. Water Resource Res. 23:1281–1287.

62. McCoy, E. L., and C. Hagedorn. 1980. Transport of resistance-labeled *Escherichia coli* strains through a transition between two soils in a topographic sequence. J. Environ. Qual. 9:686–691.

63. Luxmoore, R. J. Micro-, meso-, and macroporosity of soil. 1981. Soil Sci. Soc. Am. J. 45:671–672.

64. Bouma, J. 1981. Comment on "Micro-, meso-, and macroporosity of soil." Soil Sci. Soc. Am. J. 45:1244–1245.

65. Wallace, H. R. 1978. Dispersal in time and space: soil pathogens, p. 181–202. *In* J. G. Horsfall and E. B. Cowling (eds.), Plant disease: an advanced treatise, Vol. 2. Academic Press, New York.

66. Rainwater, K., M. P. Mayfield, C. Heintz, and B. J. Claborn. 1993. Enhanced *in situ* biodegradation of diesel fuel by cyclic vertical water table movement—preliminary studies. Water Environ. Res. 65:717–725.

67. Bitton, G., J. M. Davidson, and S. R. Farrah. 1979. On the value of soil columns for assessing the transport pattern of viruses through soils: a critical outlook. Water Air Soil Pollut. 12:449–457.

68. van Elsas, J. D., J. T. Trevors, and L. S. van Overbeek. 1991. Influence of soil properties on the vertical movement of genetically marked *Pseudomonas fluorescens* through large soil microcosms. Biol. Fertil. Soils 10:249–255.

69. Bentjen, S. A., J. K. Fredrickson, P. van Voris, and S. W. Li. 1989. Intact soil-core microcosms for evaluating the fate and ecological impact of the release of genetically engineered microorganisms. Appl. Environ. Microbiol. 55:198–202.

70. Matthess, G., and A. Pekdeger. 1981. Concepts of a survival and transport model of pathogenic bacteria and viruses in groundwater. Sci. Total Environ. 21:149–159.

71. Gannon, J., Y. Tan, P. Baveye, and M. Alexander. 1991. Effect of sodium chloride on transport of bacteria in a saturated aquifer material. Appl. Environ. Microbiol. 57:2497–2501.

72. Fontes, D. E., A. L. Mills, G. M. Hornberger, and J. S. Herman. 1991. Physical and chemical factors influencing transport of microorganisms through porous media. Appl. Environ. Microbiol. 57:2473–2481.

73. Huysman, F., and W. Verstraete. 1993. Effect of cell surface characteristics on the adhesion of bacteria to soil particles. Biol. Fertil. Soils. 16:21–26.

74. Postma, J., S. Water, and J. A. van Veen. 1989. Influence of different initial soil moisture contents on the distribution and population dynamics of introduced *Rhizobium leguminosarum* biovar *trifolii*. Soil Biol. Biochem. 21:437–442.

75. Huysman, F., and W. Verstraete. 1993. Water-facilitated transport of bacteria in unsaturated soil columns: influence of cell surface hydrophobicity and soil properties. Soil Biol. Biochem. 25:83–90.

76. Huysman, F., and W. Verstraete. 1993. Water-facilitated transport of bacteria in unsaturated soil columns: influence of inoculation and irrigation methods. Soil Biol. Biochem. 25:91–97.
77. Santoro, T., and G. Stotzky. 1968. Sorption between microorganisms and clay minerals as determined by the electrical sensing zone particle analyzer. Can. J. Microbiol. 14:299–307.
78. Camper, A. K., J.T. Hayes, P. J. Sturman, W. L. Jones, and A. B. Cunningham. 1993. Effects of motility and adsorption rate coefficient on transport of bacteria through saturated porous media. Appl. Environ. Microbiol. 59:3455–3462.
79. Gannon, J. T., V. B. Manilal, and M. Alexander. 1991. Relationship between cell surface properties and transport of bacteria through soil. Appl. Environ. Microbiol. 57:190–193.
80. Gerba, C. P., and G. Bitton. 1984. Microbial pollutants: their survival and transport pattern to groundwater, p. 65–88. *In* G. Bitton and C. P. Gerba (eds.), Groundwater pollution microbiology. John Wiley & Sons, New York.
81. Gillespie, D. C. 1963. Cell wall carbohydrates of *Arthrobacter globiformis*. Can. J. Microbiol. 9:509–514.
82. Lappin-Scott, H. M., and J. W. Costerton. 1990. Starvation and penetration of bacteria in soils and rocks. Experientia 46:807–812.
83. Kjelleberg, S., B. A. Humphrey, and K. C. Marshall. 1983. Initial phases of starvation and activity of bacteria at surfaces. Appl. Environ. Microbiol. 46:978–984.
84. Balkwill, D. L., and L. E. Casida, Jr. 1979. Attachment to autoclaved soil of bacterial cells from pure cultures of soil isolates. Appl. Environ. Microbiol. 37:1031–1037.
85. Vandevivere, P., and P. Baveye. 1992. Relationship between transport of bacteria and their clogging efficiency in sand columns. Appl. Environ. Microbiol. 58:2523–2530.
86. Taylor, S. W., and P. R. Jaffé. 1990. Biofilm growth and the related changes in the physical properties of a porous medium. 1. Experimental investigation. Water Resource Res. 26:2153–2159.
87. Mercer, J. R., R. M. Ford, J. L. Stitz, and C. Bradbeer. 1993. Growth rate effects on fundamental transport properties of bacterial populations. Biotechnol. Bioengin. 42:1277–1286.
88. Lovins, K. W., J. S. Angle, J. L. Wievers, and R. L. Hill. 1993. Leaching of *Pseudomonas aeruginosa*, and transconjugants containing pR68.45 through unsaturated, intact soil columns. FEMS Microbiol. Ecol. 13:105–112.
89. Dickinson, R. A. 1991. Problems with using existing transport models to describe microbial transport in porous media, p. 21–47. *In* C. J. Hurst (ed.), Modeling the environmental fate of microorganisms. American Society for Microbiology, Washington, D.C.
90. Corapcioglu, M. Y., and A. Haridas. 1984. Transport and fate of microorganisms in porous media: a theoretical investigation. J. Hydrol. 72:149–169.
91. Taylor, S. W., and P. R. Jaffé. 1990. Substrate and biomass transport in a porous medium. Water Resource Res. 26:2181–2194.
92. McDowell-Boyer, L. M., J. R. Hunt, and N. Sitar. 1986. Particle transport through porous media. Water Resource Res. 22:1901–1921.

93. Germann, P. F., and L. A. Douglas. 1987. Comments on "Particle transport through porous media" by Laura M. McDowell-Boyer, James R. Hunt, and Nicholas Sitar. Water Resource Res. 23:1697–1698.
94. Smith, R. E. 1983. Approximate soil water movement by kinematic characteristics. Soil Sci. Soc. Am. J. 47:1–8.
95. Tan, Y., W. J. Bond, and D. M. Griffin. 1992. Transport of bacteria unsteady unsaturated soil water flow. Soil Sci. Soc. Am. J. 56:1331–1340.
96. McCaulou, D. R., R. C. Bales, and J. F. McCarthy. 1994. Use of short-pulse experiments to study bacteria transport through porous media. J. Contam. Hydrol. 15:1–14.
97. Loehle, C., and P. Johnson. 1994. A framework for modeling microbial transport and dynamics in the subsurface. Ecol. Model. 73:31–49.
98. Sarkar, A. K., G. Georgiou, and M. M. Sharma. 1994. Transport of bacteria in porous media. 2. A model for convective transport and growth. Biotechnol. Bioengin. 44:499–508.
99. McInerney, M. J. 1991. Use of models to predict bacterial penetration and movement within a subsurface matrix, p. 115–135. *In* C. J. Hurst (ed.), Modeling the environmental fate of microorganisms. American Society for Microbiology, Washington, D.C.

8

A Critical Analysis of Methods for Determining the Composition and Biogeochemical Activities of Soil Microbial Communities In Situ

Eugene L. Madsen Cornell University, Ithaca, New York

I. INTRODUCTION

A. Scope of this Chapter

This chapter scrutinizes methodologies designed to address four long-standing questions in environmental microbiology:

1. What microorganisms are present in terrestrial and aquatic habitats that comprise the biosphere?
2. What are these microorganisms actually doing in situ—that is, in the habitats where the microorganisms reside?
3. Can the microorganisms responsible for particular biogeochemical processes in the field be identified?
4. Can accurate measurements of in situ microbiological processes be integrated in time and space?

Because this chapter is a contribution to the Soil Biochemistry Series, efforts to answer these questions will emphasize the soil habitat, hence soil microbiology and biochemistry whenever possible. However, the issues and approaches discussed here generally apply to all habitats of concern in the broader field of environmental microbiology (i.e., soils, sediments, freshwater, groundwater, and other systems); therefore, information relevant to many of these habitats, especially aquatic microbiology, will be included. Because uncertainties in understanding microbial activities in field habitats where microorganisms actually reside (in situ) have plagued environmental microbiology since its inception, revolutionary insights generating fully satisfying answers to the questions of interest should not be expected. However, by drawing on history and on both established and recently developed methodologies, progress toward answers to the above questions will be outlined. The overall goal is to strive for a heuristic essay by attempting to synthesize information from a variety of sources.

To assess the understanding of microbial activity in soil and how to measure it, it will be necessary to (1) appreciate why processes catalyzed by microorganisms in soil, sediments, and waters are of interest (Section I); (2) review historical issues and trends in soil microbiology and environmental microbiology (Section II); (3) scrutinize the basis of methodological problems that have impaired the understanding of soil microbial communities (Section III); (4) scrutinize state-of-the-art methodologies that provide accurate glimpses into the composition, general physiological status, and in situ biogeochemical activity of terrestrial and aquatic microorganisms (Section IV); (5) scrutinize procedures for identifying the specific microbial agents responsible for in situ biogeochemical changes in field habitats (Section V); (6) discuss procedures that integrate activity measurement methodologies in time and space (Section VI); and (7) speculate on future developments that may provide avenues for achieving instantaneous real-time analysis of in situ biogeochemical activity in soil (Section VII). Addressing these seven objectives in a single book chapter reveals a certain degree of optimism. To convey the universality and steadfastness of many issues of environmental microbiology, especially goals and methodological limitations of this discipline, historical quotes from pertinent scholars throughout this century will appear. These quotes are included to give weight to the arguments presented by demonstrating that precedent for the logic in this chapter has been set historically.

B. Impetus for Studying Microbial Metabolism in Soil and Other Habitats

Microorganisms that reside in the soil habitat have critical roles both in maintaining the biosphere and advancing knowledge of biochemistry [1–6]. In 1945, Waksman ([7], p. 341) described the significance of soil microorganisms as follows:

> The liberation, through the activities of microorganisms, of the nutrient elements essential for plant life in available forms; namely, the carbon as CO_2, the nitrogen as ammonia, the phosphorus as phosphate, the sulfur as

> sulfate, and others: Without these microbial activities, the limited supplies of these elements in the surface of the earth would soon become locked up in the form of plant and animal life, thus bringing to a stand-still the development of new life. The varied activities of the soil-inhabiting microorganisms keep these elements in continuous circulation. These processes take place in a series of reactions, many of which are brought about by different organisms with different physiological mechanisms. In studying these reactions, the soil microbiologist thus contributes to a knowledge of not only soil processes but microbial physiology as a whole.

The above message, delivered by Waksman for the soil habitat, applies to all other habitats that comprise the biosphere. The small size, ubiquitous distribution, high specific surface area, potentially high rate of metabolic activity, genetic malleability, potentially rapid growth rate, and unrivaled enzymatic and nutritional versatility of microorganisms cast them in the role of recycling agents for the biosphere.

The vast diversity and quantities of inorganic (e.g., S°, NH_3, H_2, CH_4) and organic (e.g., carbohydrates, fats, proteins, lipids, nucleic acids, hydrocarbons, etc.) materials present on the Earth's surface are disseminated between a matching diversity of habitats whose physical and chemical characteristics span wide ranges of pH, temperature, salinity, oxygen tension, redox potential, water potential, et cetera [8–10]. It is appealing to speculate that this distribution of thermodynamically unstable resources between a variety of environments was the source of selective pressure during the evolutionary diversification of microorganisms. The end product of this evolution is a microbial world capable of exploiting virtually all of the naturally occurring (and many of the synthetic) metabolic resources on Earth. Physiological exploitation of thermodynamically unstable resources by microorganisms allows them to survive and grow. This simple growth and survival of microorganisms drives biogeochemical cycling of the elements and simultaneously maintains the conditions required for life by other inhabitants of the biosphere. Understanding the detailed microbiological mechanisms of the maintenance of ecosystems provides both practical and intellectual challenges for inquiries into soil and environmental microbiology.

C. Definition of Soil Microbiology

Soil microbiology is concerned with the nature, activities, and characteristics of growth of microorganisms that inhabit soil. In prefacing both editions of his classic textbook, Alexander [1,11] has emphasized that

> Soil microbiology is not a pure discipline. Its origins may be traced through bacteriology, mycology, and soil science; biochemistry and plant pathology have also made their mark . . . each (microbial) transformation is viewed (1) as a reaction of importance to soil and crop production, (2) as a biological

process brought about by specific microorganisms whose habitat is the earth's crust, and (3) as a sequence of enzymatic steps.

This definition is strengthened by a passage in S. A. Waksman's 1945 essay entitled, "Soil Microbiology as a Field of Science" [7], paraphrased as follows:

> Of particular interest are the interrelationships of the various elements of microbial populations, the reactions with which they are concerned, the influence of soil conditions on the microorganisms comprising the soil population, and the effects of the activities of the population on the soil and on the plants which develop in it.

The major fields or interests of a soil microbiologist are listed in Table 1. As defined in Section I.B, the intellectual and practical concerns of soil microbiol-

Table 1 Major Fields of Interest in Soil Microbiology[a]

1. The microbiological populations of the soil including bacteria, fungi, actinomycetes, protozoa, algae, and other lower forms of plant and animal life.
2. The numerous interrelationships (*including competition, antagonism, parasitism, antibiosis, symbiosis*) of the soil populations, as well as the relationships of these organisms to higher plants.
3. The decomposition of plant and animal residues on the surface of the soil, as in composts, as well as in the soil itself and in the formation of humus.
4. The liberation, through activities of microorganisms, of the nutrient elements essential for plant life in available forms.
5. Nitrogen-fixing bacteria and the mechanism of the fixation process; no other group of microorganisms and no other microbiological process are so characteristic of the soil as the reactions involving the fixation of gaseous nitrogen.
6. The process of nitrification: the soil microbiologist has largely initiated and has been chiefly responsible for the development of the knowledge gained on autotrophic bacteria as a whole and the nitrifying forms in particular.
7. Various other processes of oxidation and reduction brought about by soil microorganisms (*especially those concerning sulfur*).
8. The role of microorganisms in the conservation of the soil.
9. The influence of the soil microbiota on the effectiveness and persistence and degradation of pesticides.
10. The critical role of the soil microbiota in modifying or destroying environmental pollutants.
11. The potential of soil microbial populations to form toxic substances.
12. The development of microbial ecology as an independent discipline with implications for applied issues, such as survival and activity of microorganisms important to food production, commerce, and pathology, as well as for issues of pure science, such as biochemistry, genetics, and species diversity.

[a] Items 1 through 8 are from Waksman [7]. Items 9 through 11 are from Alexander [1,11]. Item 12 is a synthesis from Alexander [1]; Atlas and Bartha [2]; Stolp, [5]; Brock [8,13]; Brock *et al.* [14]; Hobbie [15]; Hobbie and Ford [16]; Pace *et al.* [17]; Paul [18]; Mitchell [19]; and Woese [20].

ogy are generally shared by all environmental microbiologists. The key metabolic processes are almost universal. However, details of the methodologies, motivations, and goals vary amongst the habitats under scrutiny (e.g., soils, oceans, lakes, rivers, sediments, hot springs) and biogeochemical reactions that are of concern.

II. HISTORICAL OVERVIEW OF SOIL AND ENVIRONMENTAL MICROBIOLOGY

One approach for grasping the history and traditions of soil and environmental microbiology is to recognize the contributions from various centers of training. These necessarily evolve around various investigators and the institutions where they have been based. Major advancements in soil microbiology, microbial ecology, and related disciplines have been made by the following "schools" or prominent individual investigators.

S. Winogradsky is regarded by many as the founder of soil microbiology [2]. Working in the latter part of the nineteenth and early decades of the twentieth centuries, Winogradsky's career contributed immensely to our knowledge of soil and environmental microbiology, especially regarding microbial metabolism of sulfur, iron, and nitrogen. In 1949, much of Winogradsky's work was published as a major treatise entitled, *Microbiologie du sol, problèmes et méthodes; cinquante ans de recherches. Oeuvres complètes* [21].

Additional significant contributions to early soil microbiology [22] were made by Löhnis [23], Conn and Conn [24], Pochon [25], Omelianskii [26], Russell at the Rothamsted Experiment Station in England [27], and by Allison [28] and Clark [4,29] among others at the United States Department of Agriculture's laboratory in Beltsville, Maryland.

At the University of Delft near the end of the nineteenth century, M. W. Beijerinck founded the Delft School traditions of elective enrichment techniques that allowed Beijerinck's crucial discoveries including microbiological transformations of nitrogen and sulfur [2,14,30]. The helm of the Delft School changed hands from Beijerinck to Kluyver, and the traditions have been continued in Germany, the Netherlands, and other parts of Europe through to the present.

After training in Delft with Beijerinck, C. B. van Niel established a research program at Stanford University's Hopkins Marine Station, where R. Y. Stainer, R. Hungate, M. Doudoroff, and many others were trained, later establishing their own research programs at other institutions [30].

In the early portion of this century at Rutgers University, Selman A. Waksman was perhaps the foremost American scholar in the discipline of soil microbiology. Many of the Rutgers traditions in soil microbiology were initiated by J. Lipman, Waksman's predecessor (R. Bartha, personal communication; [22]).

Waksman produced numerous treatises that summarized the history, status, and frontiers of soil microbiology, often in collaboration with R. Starkey. Among the prominent works published by Waksman are "Soil microbiology in 1924: An attempt at an analysis and a synthesis" [22], *Principles of Soil Microbiology* [31], "Soil microbiology as a field of science" [7], and *Soil Microbiology* [6]. In the latter part of the twentieth century, a steady flow of Rutgers-based contributions to soil microbiology and microbial ecology continue to be published [2].

In the 1920s and 1930s, E. B. Fred and collaborators, Baldwin and McCoy, comprised a unique cluster of investigators whose interests focused on the *Rhizobium*-legume symbiosis [32]. Several decades later at the University of Wisconsin, T. D. Brock and his students made important contributions to microbial ecology, anaerobic metabolism, thermophily, and general microbiology.

For several decades beginning in the 1920s, patterns of mentorship between researchers in the central United States were established (G. Stotzky, and M. Alexander, personal communication). These produced significant advances in soil microbiology and biochemistry. Among the prominent contributors in these areas were T. L. Martin (Brigham Young University), A. G. Norman (University of Michigan), J. P. Martin (University of Idaho), W. P. Martin and E. L. Schmidt (University of Minnesota), and W. V. Bartholomew, J. M. Bremner, and L. R. Fredrick (Iowa State University).

M. Alexander moved with a recently granted Ph.D. from the University of Wisconsin to Cornell University in 1955. During nearly four decades before Alexander's arrival, soil microbiological research was conducted at Cornell by J. K. Wilson and F. Broadbent. From 1955 to the present, Alexander's contributions to soil microbiology have examined a broad diversity of phenomena, which include various transformations of nitrogen, predator-prey relations, most probable number methodologies, microbial metabolism of pesticides and environmental pollutants, and advancements in environmental toxicology. Many environmental microbiologists have received training with M. Alexander and become prominent investigators.

Other schools and individuals in Britain, Italy, France, Belgium, other parts of Europe, Japan, Russia, other parts of Asia, Africa, Australia, the United States, other parts of the Americas, and other nations certainly have contributed in significant ways to advancements in soil microbiology. An insightful review of the history of soil microbiology, with special emphasis on eastern European and Russian developments was written by Macura in 1974 [33].

The many historical milestones in the development of soil microbiology (most of which are shared with broader fields of biology and microbiology) have been reviewed by Atlas and Bartha [2], Brock [34], Brock et al. [14], LeChevalier and Solotorovsky [35], Macura [33], van Niel [30], Waksman [6,22,31], and others. Some of the highlights include the first visualization of microscopic life by van Leeuwenhoek in 1683; the role of microorganisms as causative agents of fermentations by Pasteur in 1857; the use of gelatin plates for enumeration of soil microorganisms by Koch in 1881; nitrogen fixation by nodules on the roots of legumes by Hellriegel and Wilfarth in 1885; the use of elective enrichment methods, by Beijerinck and Winogradsky, in the isolation of single organisms able to carry out ammonification, nitrification, and both symbiotic and nonsymbiotic nitrogen fixation; recognition of the diverse populations in soil (e.g., bacteria, fungi, algae, protozoa, nematodes, insect larvae); documentation of anaerobic cellulose decomposition by Omelianskii in 1902; the study of sulfur-utilizing bacteria by van Niel and others; the specificity of legume-nodulating bacteria [32]; the discovery and development of antibiotics; direct microscopic methods of examining soil microörganisms via staining and contact slide procedures; the development of radiotracer techniques; a diversity of advancements in analytical chemistry for detecting and quantifying biochemically and environmentally relevant compounds; developments in molecular phylogeny [20,36]; and the application of molecular methods to environmental microbiology [17, 37–42].

As this historical treatment reaches toward the end of the twentieth century, the branches and traditions in environmental microbiology become so complex that patterns of individual contributions become difficult to discern. A complete list of schools, individual investigators, and their respective discoveries extending into the 1990s is beyond the scope of this chapter. The author apologizes for his biases, limited education, and any and all omissions that readers may notice in this brief historical overview.

III. THE CHALLENGE OF UNDERSTANDING IN SITU MICROBIAL METABOLIC ACTIVITY IN SOIL

A. Impediments to Measuring In Situ Soil Microbial Activities

Soil, especially as it occurs as a portion of the landscape in field sites, poses a variety of obstacles that impede measuring the in situ activities of resident microbial communities (Table 2). Many of these obstacles and their impact on ecological interpretations of microbial activity measurements have been discussed by Nannipieri et al [43]. The first two items listed in Table 2 recognize the complexity of the environmental context where soils occur in landscapes. Soils are natural bodies, whose lateral and vertical boundaries usually occur as gradients between mixtures of materials of atmospheric, geologic, aquatic, and/or biotic origin. Soils, similar to many other habitats of interest to microbiologists, are open systems subject to fluxes in energy (e.g., sunlight, wind) and materials (e.g., aqueous precipitation, soil erosion and deposition, inputs of organic compounds from activities of plants, human beings, and other animals). Furthermore, soil is intrinsically complex (characteristic 3 in Table 2). Soil is an

Table 2 Characteristics of the Soil Habitat that Influence Determination of In Situ Metabolic Activity of Resident Microorganisms

1. Soils are situated in landscapes. Soils often support plant growth and are underlain by deeper soil horizons and geologic strata that are usually poorly characterized.
2. Soil is subject to complex variations in climatic processes that influence plant growth, temperature, and mass transfer of solids, liquids and gases.
3. The soil matrix defies complete compositional analysis. It is a three-dimensional porous array consisting of inorganic solids (sand, silt, and clay) intermingled with humus and organic and inorganic chemical coatings (organic matter and amorphous oxides), as well as organisms (micro-, meso- flora and fauna). The pore spaces are shared in variable proportions by gases (whose composition reflects a balance between atmospheric diffusion and biotic activity) and aqueous soil solution (whose composition reflects complex equilibria between inorganic, organic, and biotic reactions).
4. There is a diverse array of complex microbial activities in soil that cause in situ geochemical changes by producing and consuming solid, liquid, dissolved, or gaseous substances that are indicative of the geochemical process of interest.
5. The microbiological processes (point 4) occur simultaneously with abiotic (e.g., photochemical, organic, inorganic, and physical transport) and animal- or plant-mediated processes that also contribute to geochemical change in soil.
6. The complexities of soil characteristics in points 1 through 5 are compounded by being dynamic in time and inhomogeneously distributed in space.

assemblage of solid, liquid, gaseous, organic, inorganic, and biological constituents whose chemical composition and random three-dimensional structure have not been completely characterized.

The fourth characteristic listed in Table 2 emphasizes that even if it were possible to focus only on microorganisms (bacteria, fungi, algae, protozoa, and viruses; thus ignoring biotic geochemical changes effected by nonmicroscopic inhabitants of soil), the physiological processes and the multitude of interactions are dauntingly complicated. This is particularly the case for efforts in documenting geochemical changes caused by consortia of microbial species in which the metabolites (e.g., ammonia, CO_2, methane, or organic acids derived from fermentation of organic matter) produced by some microbial populations serve as the substrates for others (e.g., nitrifiers, other autotrophs, methanotrophs, and methanogens, respectively). When such food web linkages are spatially and/or temporally coincidental, then measurements of metabolite fluxes through bulk volumes of soil only describe net geochemical changes. Net geochemical change is certainly important, but interpretations and predictions of such changes are probably best based on separating the "source" and "sink" components of the microbial processes, so as to understand their independent influences and regulation (see Section IV.E).

Compounding the challenge of understanding in situ microbial soil processes is the fact that many nonmicrobial biotic reactions also influence the concentrations of geochemical parameters of interest (characteristic 5 of Table 2). In a field setting, it is seldom the case that microorganisms are the unique biological agents responsible for the geochemical change of interest.* Plants and animals of a multitude of types and sizes may produce, consume, influence, or transform the same substrates as microorganisms. Abiotic factors must also be considered when attempting to measure geochemical change in a field setting. For instance, efforts to measure in situ biodegradation of organic pollutants released into soils in field sites should ideally account for all of the mechanisms of pollutant attenuation. Among the abiotic attenuating mechanisms are dilution, advection, dispersion, volatilization, sorption, formation of bound residues, photolysis, alteration of the pollutants by clays or other inorganic materials, and various inorganic and organic chemical equilibria reactions (see Section IV.E). Thus, distinguishing between biotic and abiotic mechanisms of attenuation often becomes the key concern when documenting in situ biodegradation processes [3].

The last characteristic in Table 2 brings attention to the fact that all of the prior entries in the table are subject to dynamic changes in time and space. No field setting is homogeneous or static. Regarding spatial inhomogeneity, the physical, chemical, nutritional, and ecological conditions for soil biota undoubtedly vary at the micrometer or submicrometer scale. This has been documented by microelectrode-based investigations [44–46]. However, acknowledging microscale heterogeneities may have little practical bearing at other operational scales (millimeter, centimeter, meter) where investigators more readily perform measurements. Thus, measurements conducted at millimeter, centimeter, and meter scales necessarily average the contributions of many microenvironments contributing to the process(es) of interest. The problems of spatial heterogeneity of field sites also arise from the variabilities found when examining gram- to kilogram-sized samples of soils. In this regard, the scale of spatial heterogeneities in landscapes may vary continuously from micrometer to kilometer, as governed by a variety of influences such as climate, water availability, soil type, and vegetation. Some of the issues of scaling in soil biogeochemistry will be described in Sections VI and VII. These have also been discussed by Groffman [47–49], Groffman et al. [50], Parkin [51], and Tiedje et al. [52].

Regarding temporal variability (the second issue within characteristic 6 of Table 2), in situ processes that directly and indirectly influence fluxes of materi-

* Investigators that examine processes unique to microorganisms are the most likely to succeed in their attempts to measure microbial activity in field sites. Some of the most promising microbiological processes in this regard are nitrogen fixation, methanogenesis in the absence of geologically derived methane, and perhaps denitrification. Such measurements and their pitfalls are discussed in Section IV.E.

als into, out of, and within soil are dynamic [53]. Climate-related influences (such as temperature, sunlight, evaporation, and precipitation [53,54]) are probably major variables that cause temporal variations in biogeochemical processes in soil. The biota and their respective physiological (e.g., growth rate, substrate utilization, excretion, exudation, death) and behavioral (e.g., movement, predation, competition, parasitism, symbiosis) activities undoubtedly respond to climate-induced and/or other temporal environmental changes. In turn, these biotic responses almost certainly impose a variety of second-order interactions between the organisms themselves. Thus, the activities of microorganisms (the primary agents of biogeochemical change) are regulated by the temporal and spatial feedback loops among the microorganisms themselves and between their surroundings.

B. Limitations for Determining the Composition and Function of Microbial Communities: The Heisenberg Uncertainty Principle

As is evident from the quotes from prominent scholars assembled in Table 3, methodological limitations have plagued the overlapping disciplines of soil and environmental microbiology from their inception to the present. Fundamental questions such as "Who is there?" and "What are they doing?", have yet to be fully answered. Chief among the obstacles to obtaining satisfactory answers to these two questions are (1) microbiologists cannot take an accurate census of the members of a given naturally occurring microbial community; (2) even if an accurate census were possible, discerning between dormant and active cells is difficult; (3) determining the types and rates of biogeochemical metabolic processes being expressed by active cells in situ is extremely challenging; and (4) microbial communities contained within soil and other environmental samples withdrawn from field sites have a propensity to change rapidly in response to altered environmental and physiological conditions imposed by sampling and laboratory incubation.

Regardless of the microbiological habitat examined, counts of viable cells on agar media usually are one to three orders of magnitude lower than microscopic (total) counts performed on the same sample [56,58–62]. This discrepancy is cause for two important conceptual and operational distinctions: between cultured versus uncultured microorganisms and between culturable versus nonculturable microorganisms. These distinctions reflect both the physiological requirements and the metabolic status of microorganisms in nature. Each culture medium and set of incubation condition is selective to some degree and, therefore, is able to meet the physiological requirements of only a small proportion of the populations within a given microbial community. All of the microorganisms grown under laboratory conditions have been cultured because the ingenuity of microbiologists in media design has been successful. Correspondingly, all other microorganisms that have not yet been cultured in the laboratory are classified as uncultured. Attempts to explain why this uncultured component of the microbial community has resisted growth in the laboratory have led to the terms culturable and nonculturable. For a given culture medium, the majority of organisms that do not grow are probably culturable but fail to grow because their requisite physiological needs have not been provided. However, a subset of the uncultured microorganisms may be unculturable. Unculturability may have, at least, two distinct physiological bases: cells may be dormant (i.e., they may require special resuscitative events before regaining an ability to grow) or they may be nonviable (i.e., intact and, therefore, detectable microscopically but unable to reproduce regardless of the growth conditions provided by the microbiologist) [60, 63–67]. Nonviability does not consider the possibility that bacterial cells may maintain some of their enzymatic activities and physiological functions but lose the ability to reproduce. Thus, the concept of nonviability and death among bacterial cells in any natural setting is somewhat arbitrary (see Henis [64] for further discussion of the arbitrary nature of the terms, survival, dormancy, and death, as applied to bacteria).

Based on the above discussion, it is clear that knowing "Who is there?" in soil microbial communities is an evasive goal, because of imperfect methodological, physiological, and conceptual tools. Similarly, knowing "What they are doing?" represents a major challenge.

The complicated characteristics of soils as they occur in landscapes in the field (Table 2) and the implications of these for measuring in situ microbial activities (Section III.A) frequently force the investigator to perform laboratory studies. These are appealing, because control over a large number of experimental variables is gained. Instead of contending with the vicissitudes of soil as a segment of the open landscape that is subject to almost countless uncharacterized influences, the investigator becomes able to redefine the experimental context and objectives. Typical procedures often involve the examination of soil samples enclosed within laboratory-incubated vessels. These laboratory preparations are commonly subject to a variety of replicated treatments—often the most important of which is an abiotic (sterile or poisoned) control that distinguishes biotic from abiotic processes [68].

The transition from in situ to in vitro investigations of microbial activities in soil solves many of the complications discussed in Section III.A. Experiments become tidy and interpretable. Microbially mediated geochemical processes are documented by applying techniques of chemical and biochemical analysis (e.g., manometry, spectrophotometry, chromatography, enzymology, radioisotopic tracing) to each of the replicated treatments. The metabolic activities and their regulation by a variety of experimental variables can be readily measured and contrasted with results from abiotic treatments, which are often inactive. Experimental manipulations of single variables frequently elicit unambiguous responses from the naturally occurring microorganisms contained in the

Table 3 Historical Excerpts Documenting Methodological Limitations in Soil and Environmental Microbiology

Excerpt	Reference
We have as yet no science of soil microbiology proper, although we possess a great deal of information on various groups of soil microorganisms and what they can do when grown in artificial culture media, said S. Winogradsky, whose exact contributions to our understanding of certain soil microbiological processes can hardly be equaled (Waksman, 1927, p. 203).	[31]
A knowledge of the activities of certain organisms isolated for the soil is certainly necessary, but that is not a knowledge of the extent to which the processes take place in the soil itself (Waksman, 1927; Preface, p. vii).	[31]
One may be led, then, to make a hasty, but wrong conclusion that the action of the microorganisms in the soil is the same as in pure culture upon artificial culture media. In the latter case, free from stimulative and competitive influences of other microorganisms, an organism may manifest certain activities, which would not take place in the soil or *vice versa* (Waksman, 1925, p. 202).	[22]
The lack of proper methods for a microbiological investigation of soils, the fact that our present methods are far from adequate in giving us an idea of what is taking place in the soil, is widely recognized (Waksman, 1925, p. 234–235).	[22]
We possess, at the present time, considerable information on the organisms inhabiting the soil and on the chemical processes of many of these organisms, under controlled laboratory conditions; but little is known of the processes carried out in the soil itself, by the numberless representatives of the soil flora and fauna (Waksman, 1927, p. 842).	[31]
When I stand on the litter of a forest floor, or on the thatch of a grassland, I like to look down to ask myself just what do I really know about the microorganisms in that ecosystem. Immediately a whole host of categorical relationships, or definitions and generalizations from textbooks of ecology, come to mind. I feel a certain satisfaction in knowing how to define symbiosis, and fungistasis, and biocenosis, and in knowing that moisture and temperature influence microbial activity, but nevertheless I still feel unable to comprehend the total functional ecology in that ecosystem. I know with reasonable assurance what measurements I can make or what microbes I can isolate or enumerate or identify within my own limited area of competence, and I realize what an exceedingly small area that competence embraces (Clark, 1973, p. 13).	[55]
But, however the organism is obtained, whether from nature or from the laboratory, it is clear that it is not sufficient that the organism or	[56]

Table 3 Continued

Excerpt	Reference
consortium of organisms act . . . in the laboratory. Action must occur in the world at large. If we have learned nothing else from research in microbial ecology, we have learned that microbes do not do the same things in the laboratory that they do in nature (Brock, 1987).	
The composition of the microbial community is still uncertain. The old explanations for the low recoveries of bacteria by plate counts and the platitudes concerning the role of particular morphological or physiological groups in soil processes remain (Alexander, 1991).	[57]
The stunning impact that new methods are having on aquatic microbial ecology points out that the field is still methods limited. Despite tremendous progress over the last 25 years, we lag well behind other ecological research areas. We have only a primitive ability to describe the organisms present in nature, what these organisms are actually doing, and what controls their activity and growth (Hobbie, 1993, p. 1).	[15]
Both the activity and biomass measurements can be artifacts of the experimental methods employed. To some degree, the whole field is still methods limited. There are many successful methods but we still lack some crucial techniques that would allow a complete picture. Much of the effort in the field has been spent on investigations of the physiology and potentialities of many types of bacteria in the laboratory, yet in the field we do not even know the identity of the species carrying out many of the important processes, or the correct rate of growth of bacteria in water and sediments (Hobbie and Ford, 1993, p. 3).	[16]
We need information on the species carrying out the microbial processes, the controls of these processes, the importance of grazing by protozoans or infection by viruses as controls of the microbial biomass, and microbial growth rates in nature. Although there has been tremendous progress, key techniques for measurement are still needed before we can have faith in the results and begin to apply the techniques over a range of habitats. These techniques will allow a quantitative view of the ecology of aquatic bacteria rather than the largely qualitative and descriptive view we have today (Hobbie and Ford, 1993, p. 14).	[16]
The development of new methodologies to understand microorganisms in their environments has become a science unto itself. The need for these methods comes from the fact that classic methodologies of medical and industrial microbiology developed over the last hundred years produce artifacts when applied to natural populations in the environment (Paul, 1993, p. 38).	[18]

enclosed soil samples. Thus, the interpretability of the data, and hence the information yield, often increase dramatically. However, in altering the context of the measurements from field to laboratory, the experimental objectives frequently change. Rather than directly acquiring the information of interest, an estimation is obtained. Yet, this change in objectives may not be acknowledged by the investigator(s) conducting the measurements. In vitro determinations of soil microbial activities are not and cannot be considered valid substitutes for true field measurements of in situ activities. In vitro determinations are approximations—they are model systems—that provide information on the metabolic potential of the soil microbial communities that, according to most experimental protocols, have been (1) structurally disturbed when sampled; (2) removed from their physical, chemical, biological, climatic, and other connections to the landscape; (3) placed into a container; (4) transported to a laboratory; (5) stored in sealed or open containers under variable conditions of light, temperature, and humidity; (6) often subjected to drying and mixing or some other homogenization process; (7) dispensed to replicate vessels; and (8) incubated for metabolic assays under conditions of light, temperature, oxidation-reduction potential, soil matrix geometry, etcetera that may depart radically from the intricate and balanced situation found in situ in the field.

The degree to which sample handling procedures impose artifacts on the results of subsequent physiological assays is both uncertain and controversial [47,52,69] (see Section III.C and Table 4). In light of the success of Beijerinckian elective-enrichment techniques (see Section II), probably every environmental microbiologist would agree that component populations within native microbial communities in soil samples have the potential to respond to artificially imposed laboratory conditions [56,61,70,71] (Table 3). Given time, those populations most favored by the incubation conditions will grow and come to dominate the community, both numerically and physiologically. This dilemma, which acknowledges the linkage between performing measurements on soil microorganisms and imposing artifacts on resultant data, is analogous to the Heisenberg uncertainty principle in quantum chemistry [3,72–75]. Accordingly to quantum theory, accurate measurements of the position and momentum of an electron are mutually exclusive. In soil microbiology, the precision of reductionistic procedures and the relevance of information derived therefrom to field biogeochemical processes are mutually exclusive.

The uncertainty principle is not limited to soil microbiology and quantum chemistry. Many, perhaps most, methods of scientific measurements have the potential to influence the resultant data. Indeed, the powerful tools of reductionism (from smashing atoms, to breaking microbial cells, to atomic-force microscopy) that are responsible for many scientific advancements of the twentieth century are often plagued by artifactual results [76] and by difficulties in integrating data obtained from system components back to the whole [77–79].

C. Scrutinizing Soil Microbiology's Uncertainty Principle

The basis for the above-described uncertainty principle simply reflects the fact that when a sample of soil is removed from the field, the intricate balance in activities of accompanying microorganisms can change rapidly in response to sampling-related influences, such as oxygen concentration, water flow, temperature, and nutrient availability. However, admitting that the composition of soil microbial communities and their physiological activities may change does not mean that both of these profound community characteristics always do change when soil is removed from the field and incubated in the laboratory. Soil microbiology's uncertainty principle should be recognized constantly, but evoked conditionally. This is because the degree to which artifacts will be imposed upon laboratory-based experiments depends upon (1) the degree to which a specific set of environmental conditions have changed between in situ and laboratory settings; (2) the duration of the change; and (3) the specific responses of particular microbial populations that carry out the biogeochemical process(es) of interest.

For example, if the objective of an investigation is to determine microbial respiratory activity in recently tilled agricultural field soil, it is likely that reasonably accurate data on in situ activity can be obtained by measuring fluxes of oxygen and carbon dioxide caused by intact peds of field-moist soil taken immediately into the laboratory and incubated for 2 h (for instance) at the ambient field temperature. Artifacts are unlikely to be imposed upon this determination of respiration because (1) the laboratory and field conditions (e.g., soil structure, moisture, temperature) are virtually identical; and (2) the incubation period is brief, thus minimizing changes in populations and physiologies during the incubation. On the other hand, if the investigation's objective changes to determining nitrogen-fixation activity in an organic muck soil supporting a dense network of plant roots, complications may develop (1) because laboratory and field conditions (e.g., soil structure; microbial growth substrates, such as root exudates released by sampling procedures; oxygen tension which may be toxic to anaerobic microorganisms, generally, and nitrogen-fixing enzyme systems, specifically) may be very different; and (2) because the nitrogen fixation assay procedure may require a long assay period, thus accentuating changes in the microbial populations and physiologies during the incubation.

The contrast between the valid (respiration in agricultural soil) and invalid (nitrogen fixation in muck soil) scenarios above makes it clear that three questions need to be posed whenever laboratory experiments are being contemplated as a substitute for field determinations:

1. Can field sampling and laboratory incubation conditions be designed to match in situ field conditions so as to minimize alterations in the composition and activity of the microbial community?

2. How long does it take for the changes in the microbial community to occur? More pragmatically, is there an artifact-free period when valid measurements can be performed before the onset of the changes?
3. Is another approach available that bypasses the uncertainties inherent in removal of samples from the field and subsequent laboratory incubations?

Answers to these three questions reside at the heart of the uncertainty principle of soil and environmental microbiology. The first two questions will be discussed briefly in this section. A major portion of this chapter (Sections IV, V, and VI) is devoted to answering the third question, by discussing field-based methodologies.

Artifacts and Laboratory Incubation Conditions

Can field sampling and laboratory incubation conditions be designed to match in situ field conditions, so as to minimize alterations in the composition and activity of the community?

As alluded to above, the answer to this question is literally conditional, that is i.e., it depends. It depends on the investigator's ability to understand the environmental context and the microbial process of interest sufficiently well to avoid artifacts. It is clear from the above two soil incubation scenarios that the conditions of managed systems (i.e., the "valid" determination of respiration in agricultural soil) can probably be readily duplicated in laboratory settings. It is also clear (from the above "invalid" determination of nitrogen fixation by a muck soil sample) that conditions of undisturbed natural systems may be fragile, complex, very difficult to characterize fully, and therefore, the probability of matching the laboratory to field conditions becomes unlikely. This is true even for intact soil cores (used by many soil microbiologists) that have been disturbed minimally by driving a rigid cylinder into the ground at field sites. In such intact cores, disturbance is certain, and its quantification as minimal may be impossible to achieve. Some conservative, perhaps purist, environmental microbiologists would advocate that laboratory incubations of field samples can never be artifact-free because the investigator can never understand the in situ context and its associated microbial processes sufficiently well to be certain that incubation conditions have not departed from those in situ.* This purist view is meritorious, because it recognizes that extrapolation of information (especially quantitative extrapolation) from laboratory incubations to in situ field processes is risky. This risk is simply a manifestation of environmental microbiology's

* Even the sometimes superior procedure of leaving the intact soil core in situ for periodic headspace gas sampling, for instance, is subject to artifacts, because this approach relies on isolating a portion of the soil from exchanges of water and gases that support and regulate the microbiological processes of interest.

uncertainty principle and its corollaries: if in situ and in vitro conditions are matched, then laboratory metabolic activity measurements may accurately reflect field processes; furthermore, if in situ and in vitro conditions are not matched, then measurements, especially those of long duration, will be inaccurate.

Thus, it is clear that when microorganisms in environmental samples are disturbed for a sufficiently long period of time, a broad spectrum of enzymatic, anabolic, catabolic, gene regulatory, cellular, population, and community (among others) alterations often develop. For example, the respiratory response of microorganisms in a Collamer silt loam to physical disturbance is shown in Figure 1. Carbon dioxide production was monitored in the headspace of sealed tests tubes containing equal weights of freshly gathered soil in which the soil peds had been left intact, had been left intact but vertically agitated during the initial minute of the experiment, or had been sieved (2-mm mesh) before measurement of CO_2 concentration. It is clear from this simple experiment (Madsen, unpublished data) that physical disturbance can have a marked stimulatory effect on microbial respiration in soil.

An insightful comparative study by Kelly and Chynoweth [69] examined rates of methanogenesis in situ (as determined from methane trapped in field-

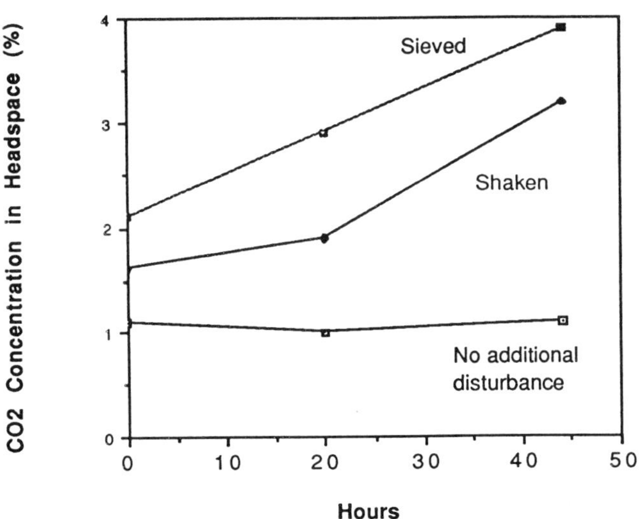

Figure 1 Effect of disturbance (sieving and shaking) on respiration by microorganisms in 10-g portions of soil. CO_2 was produced from soil organic matter. The sieved treatment was passed through a 2-mm mesh before being sealed in glass tubes with a large headspace of air. The shaken treatment was agitated for 30 s after the tubes were sealed. Each point represents the average CO_2 concentration found in the headspace of four replicate tubes (Madsen, unpublished data).

installed collection devices) and in laboratory-incubated freshwater sediments from the same site. The nearly identical rates that Kelly and Chynoweth [69] found for in situ and in vitro determinations demonstrated that these researchers were able to gather their sediment samples carefully and that the laboratory-imposed anaerobic conditions matched those in situ. Similarly, Ryden et al. [80] have achieved remarkable agreement between denitrification rates measured in field chambers and in field-incubated, glass-enclosed soil cores. Such success in mimicking in situ conditions may be most probable when dealing with a water-saturated milieu, which is likely to possess fewer structure-dependent microsite heterogeneities than soil. However, despite good quantitative agreement with one set of sediment samples (0–3 cm), drastic overestimates of the in situ methanogenesis rate were obtained when deeper (0–20 cm) sediments were incubated in vitro [69]. Table 4 presents a compilation of additional studies pertinent to assessing the impact of sampling and/or incubation procedures on the measurements of microbial activity in soil, other environmental samples, and pure culture. As is evident from the spectrum of reports in Table 4, responses of microorganisms in environmental samples and pure culture to environmental perturbations can be subtle or drastic.

Given the broad range of possible laboratory artifacts, the conservative view seems to be most prudent. Without being certain that all artifacts have been eliminated, it seems wise, whenever possible, to adopt the viewpoint that measurements of biogeochemical activity generated in the laboratory provide insights into "metabolic potential," that is, into what might be, but not what is (see Table 3).

Artifacts and Incubation Time

Related to understanding in situ field environmental conditions and duplicating them in the laboratory is the issue of rates of change of sampled microbial communities. After soil and environmental microbiologists have admitted that every laboratory incubation will (sooner or later) lead to artifacts, two crucial questions remain: (1) How long does it take for changes to occur in the environmental samples and in the resident microbial cells? and (2) Is there an artifact-free period when valid measurements can be performed or when samples can be fixed before the onset of changes?* The dilemma that arises from posing these two related questions is depicted graphically in Figure 2. Two main schools of thought have developed in attempting to answer these questions. The conservative (or purist) view, which carries over directly from attempts at understanding true in situ field conditions, would contend that there is no "safe period" before

* It should be noted that in discussing these two questions, performance of metabolic activity measurements on laboratory-incubated field samples will be treated in this section and Section VI; fixation of samples will be discussed in Sections IV and VI.

Table 4 Survey of Studies Examining Sampling and/or Disturbance-Induced Alterations of Microorganisms or Microbial Activities in Soil and Other Environments

Action	Effect and evidence	References
Physical Disturbance and Sample Storage		
Tillage of soil in agricultural field plots	Increased respiration of soil organic matter as indicated by diminished organic matter content of cultivated soils relative to adjacent, uncultivated ones	[81–84]
Sieving and laboratory incubation of soil	Burst of respiratory activity by soil microorganisms, which gradually diminishes with time	[85]; Figure 1
Removal of bacterial cells from surface film of a pond	Structural alteration of cell morphology was diminished as the period between sample removal and fixation diminished from hours to less than 1 s	P. Hirsch, personal communication
In situ and in vitro incubation of methanogenic freshwater sediments	Comparison between in situ and in vitro methanogenesis rates agreed well for shallow sediments but poorly for deep ones	[69]
Sieving of an aggregated silt loam soil	Laboratory assay showed an increase in denitrification rate relative to intact cores	[86]
Storage of intact soil cores at 4°C for up to 19 days	No effect on measured rates of denitrification measured in intact cores	[86]
Sieving, storage, and incubation of soil at various temperatures	Phospholipid fatty acid, CO_2, evolution, and principal component analyses revealed modest microbiological changes in agricultural soil	[87]
Storage (for up to 56 days) of three different sieved soils at −20°C, 4°C, and 25°C on biomass	Initial bursts of CO_2 evolution were observed consistently. Biomass estimates (levels of ATP, mineralized N, and CO_2) varied with soil type, storage time, and temperature	[88]
Soil denitrification rates were measured using an in situ cover method and also using an isolated core technique in which the core was enclosed in glass and then repositioned in the ground	Both the in situ soil cover procedure and the in situ core incubation procedure determined denitrification rates by measuring N_2O in acetylene-inhibited treatments. Procedures gave nearly identical results for well-drained soils but widely different rates for poorly drained soils	[80]
Sieving, drying, and repacking of soil columns	Dissipation (via chemical and microbiological mechanisms) of metolachlor and atrazine in laboratory incubated soil was unaffected by sieving, drying, and repacking of soil columns	[89]

Table 4 Continued

Action	Effect and evidence	References
Measurement of chlorophyll a, ATP, $^{14}CO_2$ assimilation, ammonia, and microscopically determined abundances of ocean plankton before and after prefiltration and 24-h incubation of water samples in the laboratory	Storage decreased chlorophyll a and ATP concentrations, markedly altered the composition of the phytoplankton community, and increased measured rates of photosynthetic activity. Variance (noisiness) of replicated measurements for many parameters increased when samples were stored	[90]
Outdoor stockpiling of field soil used in vegetation projects	Both microbial biomass and its respiratory responses to added glucose were reduced compared with freshly gathered, sieved top soil	[91]
Removal of subsurface sediments in cores from the field and distribution of 2-g portions to test tubes for metabolic activity measurements	Radiotracer time-course experiments over estimated rates organic carbon oxidation, sulfate reduction, and biomass production by a factor of 10^3–10^6 compared to estimates of the same rates calculated from in situ geochemical measurements	[92]
Inject conserved tracers, propane and sodium chloride, into a contaminated stream to measure in situ toluene biodegradation. Compare field rates to rates obtained from laboratory incubation of water, rocks, and sediment in both flow-through column and batch-incubated microcosms	Laboratory biodegradation rates were dependent on the mass of sediment or rocks included in test vessels; when normalized to the weight of stream solids, the in situ toluene biodegradation rate constant was similar to that obtained from batch laboratory incubation but 100-fold higher than that from the laboratory incubated flow-through column microcosms	[93,94]
Change in Soil Water Tension		
CO_2 evolution was monitored from sieved soil samples held at a variety of water tensions and periodically rewetted	Rapid bursts of CO_2 evolution occurred with each rewetting event	[95,96]
Desiccation of soil samples before incubation	Diminished numbers of viable and culturable bacteria as assessed by growth of microcolonies on filter membranes	[97]
Addition of water to soil prior to incubation	Increased number of viable and culturable bacteria as assessed by growth of microcolonies on filter membranes	[97]

Table 4 Continued

Action	Effect and evidence	References
Physiological Studies		
Transfer of *S. typhimurium* from minimal-glucose medium to nutrient broth	Onset of DNA and RNA synthesis commences within minutes in laboratory grown cultures	[98]
Introduction of molecular oxygen into anaerobically grown cultures of bacteria	Oxygen toxicity, induced by a variety of cellular mechanisms, occurs within 1 min	[99,100]

imposing artifacts on measurements of microbial activity in samples removed from field sites (the line in the left-hand graph in Figure 2 passes through the origin). However, the alternative prevailing approach, implicit in statements appearing in review articles by Karl [71], Staley and Konopka [61], and Tunlid and White [101], among others, is to hypothesize that accurate qualitative and quantitative microbial activity measurements indicative of in situ processes can be performed within a safe period before artifacts begin to develop (right-hand graph of Figure 2). This latter approach has been applied with varying degrees of caution in a broad array of investigations on soils and/or sediments, including determination of sulfate reduction [45,102,103], denitrification [86,104-106], nitrification ([107], methanotrophy [108], methanogenesis [102], and selenium reduction [109] (see also, studies listed in Table 4).

From a theoretical or pure science standpoint, assumptions about a safe period have very little basis. For instance, the amount of time required for

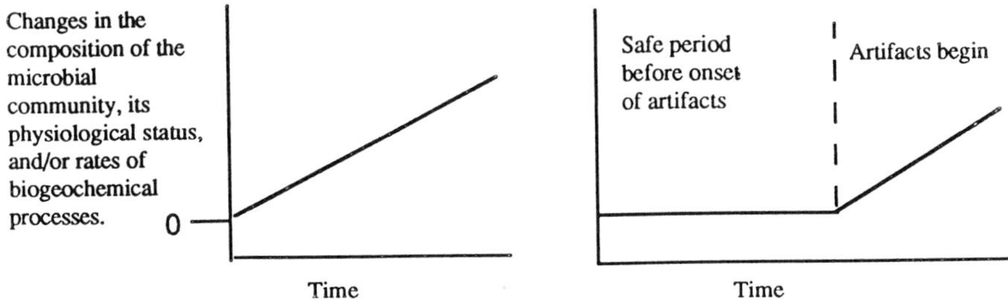

Figure 2 Uncertainties in seeking data on in situ biogeochemical processes from samples removed from the field and incubated in the laboratory. The two graphs describe the quantitative and/or qualitative influence of sampling and incubation on biogeochemical processes of interest. Changes in environmental samples may begin the instant an environmental sample is disturbed in a field site (left) or after some uncertain "safe period," during which valid measurements may be completed (right).

microorganisms to respond to cellular and environmental perturbations, while highly variable, can be nearly instantaneous [light-catalyzed biochemical reactions occur within picoseconds; enzyme-catalyzed reactions require micro- to milliseconds; entire proteins can be synthesized in tens of seconds; transcription of messenger RNA occurs at a rate of 30 to 50 base pairs per second; after transcription is complete, the average half-life of mRNA transcripts is approximately 1.3 min; cell sporulation can occur within 1 to 3 h; and chromosomal DNA replication may require as little as 40 min, although the generation time of microorganisms may require minutes to months or years (depending on growth rates as governed by environmental conditions)] [14,98,110–112]. Furthermore, the magnitude of perturbation-induced microbial responses can range from the lethality of oxygen for obligate anaerobes [99,100] to virtually imperceptible changes in the combined biotic and abiotic dissipation of microbially resistant pesticides [89] (Table 4). Using an empirical approach, Moriarty [113] has discussed time dependence of disturbance effects, such as sediment homogenization and duration of incubation, on incorporation of both ^{3}H-thymidine into DNA and ^{32}P-phosphate into phospholipids. Moriarty [113] has stated that "one can disturb sediments quite considerably without immediately changing rates of DNA synthesis." However, Moriarty [113] also cautions investigators about the possible revival of dormant cells in environmental samples and about the possibility that disturbance of samples may affect such processes as synthesis of phospholipids and proteins, as well as uptake of organic nutrients and oxygen.

It must be concluded, then, that from both practical and theoretical considerations, the Heisenberg uncertainty principle prevails in soil and environmental microbiology. The prudent way to minimize the influence of this uncertainty is to take steps that minimize contrasts between in situ conditions and those of sample handling, storage, and incubation. But, no matter how much care is taken in gathering and incubating environmental samples, this author is aware of no convincing evidence that a safe period exists (Figure 2, right-hand side) that guarantees measurements performed on laboratory-incubated environmental samples to indisputably serve as a basis for extrapolation to in situ metabolic activity.

The primary reasons for uncertainty about a safe period for metabolic activity measurements may be that the time required for completing the activity assays may be long relative to rates of microbial responses to sampling-imposed environmental changes. The safe period may only be micro-or milliseconds in duration rather than the minutes, hours, days, or weeks presumed by many investigators. As will be discussed in Sections IV.B and VI, rapid fixation of environmental samples may be far less likely to impose artifacts on the microbial community of interest.

IV. STRATEGIES FOR AVOIDING LABORATORY ARTIFACTS IN DESCRIBING NATURALLY OCCURRING SOIL MICROBIAL COMMUNITIES

In light of soil and environmental microbiology's uncertainty principle and other methodological limitations (Section III), it becomes essential to develop an answer to the third question posed in Section III.C: "Is another approach available that bypasses the uncertainties inherent in removal of samples from the field and subsequent laboratory incubations?" The answer to this question is assuredly yes.

A. Lessons from Geological Sciences

As has been pointed out by Chapelle [114], one approach to developing an answer to the above question is to compare the entire discipline of microbiology (which encompasses medical microbiology, industrial microbiology, food microbiology, microbial physiology, molecular microbiology, microbial genetics, soil microbiology, and environmental microbiology, among others) with geological sciences, in general, and structural geology, in particular. Virtually all areas of microbiology rely on reductionism to achieve insights and progress. Structural geologists, however place a strong emphasis on field observations and inference rather than reductionism. Structural geologists can seldom, if ever, capture or reproduce field-relevant processes in laboratory experiments [115]. For example, in contrast to microbial processes, plate tectonic activity cannot be chipped away from exposed rock faces in the Himalayan mountain range (created by the collision of two of the Earth's crustal plates) so that this dynamic geologic force can be duplicated in laboratory flasks, examined, transferred, dissected, and then understood at the molecular level. With no such temptation to obtain information from laboratory incubations, structural geologists are forced to gather, compile, and analyze data taken directly from the field. Using strictly deductive reasoning, structural geologists assemble logical cases for advancing the mature scientific discipline that describes the developmental history of the Earth.

For decades some field-oriented soil and environmental microbiologists (see Tables 5, 6, 7, and 8, and related discussion) have adopted expectations and practiced strategies similar to those of structural geologists. The resultant body of knowledge and methodologies constitute the state of the art in assessing biogeochemical processes in the field. By emphasizing data that can be obtained by direct field observation, discerning the status of microorganisms as they occur in the field has been a major goal. Furthermore, by deemphasizing the behavior and activity of microorganisms in artifact-riddled laboratory incubations of environmental samples, the tendency to be intellectually consumed and misled by the artifacts has been diminished.

Philosophically, the objective of state-of-the-art field methods seems to have been to advocate a shift in emphasis in soil and environmental microbiology. The shift seems to have three simultaneous themes: (1) focus much more strongly on the kind of field-oriented research that has allowed structural geology to develop and flourish; (2) be mindful of the artifacts and intellectual detours that are characteristic of laboratory incubations; yet at the same time (3) exploit, to the greatest extent possible, those pure-science disciplines (biochemistry, enzymology, molecular biology, taxonomy, physiology, genetics, etc.) that rely almost exclusively on reductionism made possible by laboratory incubations.

B. Viewing Fixed Environmental Samples as "Snapshots" Describing the In Situ Status of Microbial Communities

What follows this section is an overview of four key aspects of methodologies for field-oriented environmental microbiology. These emphasize methods for determining in situ community composition (Section IV.C); methods for determining the general in situ physiological status of microbial cells (Section IV.D); methods for determining in situ biogeochemical metabolic activity (Section

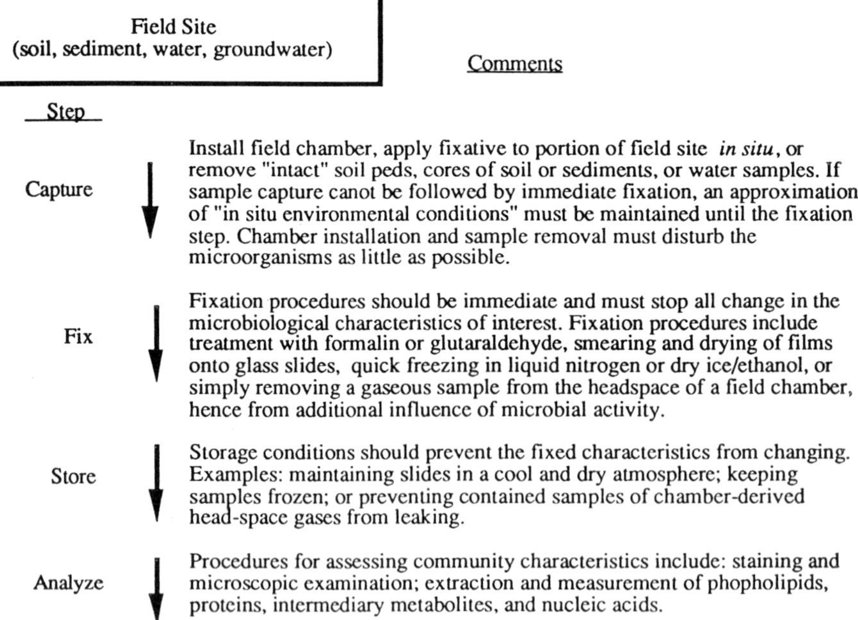

Figure 3 General methodological scheme for taking "snapshots," which allow extraction of information describing microbial populations as they occur in their habitats.

IV.E), and methods for determining the identity of particular microorganisms responsible for specific biogeochemical processes (Section V). An idealized general strategy that is common to all four types of methodologies is depicted in Figure 3. In the "capture" step of the general strategy, theory usually yields to practicality because very few sampling events can truly be instantaneous. Under rare circumstances the investigative protocol might call for freezing the sample in situ (perhaps by inundation in liquid nitrogen) before removing it from the field study site. But for most investigations, a major compromise is required: the assumption of a safe period (discussed in Section III.C) before the onset of artifacts (see Figure 2) must be evoked because time passes while samples are gathered from field sites. But the validity of the safe period becomes far more tenable because the absolute amount of time required for sample capture and fixation can be brief (Figure 3) relative to the period required for microorganisms to respond to many sampling-imposed environmental disturbances and laboratory incubations.

As noted in Figure 3, and as will be discussed below, the capture step seeks to eliminate physiological or other alterations of microbial cells present in the environment under study. This is theoretically achieved by installation of chambers in the field (for determination of fluxes of volatile materials) and by removal and immediate fixation* of soil, sediment, or water samples from a field site. The fixation and storage step of Figure 3 are designed to prevent or minimize changes from occurring in the samples before the analysis step. In this way, the information revealed by the analyses can be related directly back to the composition and metabolic characteristics of the microbial community at the time of sample removal. Thus, sets of data obtained in the manner depicted in Figure 3 represent snapshots of the microorganisms as they occur in their habitats. The usefulness of snapshot-derived data for studies seeking to integrate biogeochemical reactions over space and time will be discussed in Section VI.

A large number of methodological innovations in environmental microbiology have been advanced by those studying microbial behavior in aquatic environments [16,18,71,116–119]. The reasons for advancements by aquatic microbiologists are uncertain, but they may be related to the absence of the confoundingly complex influences of a matrix of solid particles present in soil. Many of the methodologies described in the remainder of this chapter (and listed in Tables 5, 6, 7, and 8) have great promise for use in studying the soil habitat, but they may not yet have been successfully adapted since their initial develop-

* Under exceptional circumstances for particular assays, a brief incubation period sometimes precedes the fixation step. Implicit in the brief incubation step for many assays (see Section III.C, Figure 2, and Section VI) is the assumption that a safe period prior to the development of artifacts is valid. But, according to a central theme of this chapter, artifacts will necessarily be imposed. Therefore, such incubation steps should be eliminated or minimized whenever possible.

ment by aquatic microbiologists. Several comprehensive reviews pertinent to describing naturally occurring microbial populations and their activities have recently been published. These include Paul [18], Carter and Lynch [42] Nannipieri et al. [43] Staley and Konopka [61], Karl [71], Kemp et al [117] Geesey and White [120], Hurst et al. [121], Grigorova and Norris [122], Burlage [123], Weaver et al. [124], Pepper et al. [125], Alef and Nannipieri [126], and Tate [127]. Older classic compilations of soil and environmental microbiological methods by Parkinson et al. [128], Pramer and Schmidt [129] and Rosswall [130] are also noteworthy. To this author's knowledge, previous reviews of methods in soil and environmental microbiology have not presented the notion of viewing fixed environmental samples as snapshots of the in situ status of microbial communities.

C. Methods for Determining Community Composition

Representative methodologies for answering the question "Who is there" in microbial communities present in a variety of habitats are listed in Table 5. As stated by Karl [71]:

> Without doubt, the optical microscope is one of the most important and useful instruments in (environmental) microbiology. In principle, microscopic observations of cell numbers, in conjunction with estimates of cell size, should provide the absolute standard with which to compare the results derived from more indirect methods. The three major criticisms of direct microscopy are that the method is (1) highly subjective, (2) labor intensive, and (3) relatively insensitive.

In this manner, Karl [71] both praises the unique insights that can be obtained via microscopic procedures and mentions their limitations. This careful evaluation of methodological strengths and weaknesses was applied to virtually all environmental microbiology techniques available at the time that Karl published his review [71]. Valid implementation and application of every measurement technique can only be successful when the methodological limitations (third column of Table 5) have been thoroughly understood.

As alluded in the scheme in Figure 3, discerning the composition of microbial communities present in soil and other habitats usually requires capture, fixation, storage, and analysis steps. Obviously, the procedures used to capture and fix field specimens are designed to be compatible with the analytical procedures that ultimately provide the information sought. Thus, the variety of procedures listed in Table 5 are geared toward making the microorganisms and their diversity of cellular and subcellular characteristics discernible when the microorganisms and portions of their native milieu (e.g., accompanying soil particles and/or water components) are examined via a variety of microscopic

Table 5 Methods for Determining In Situ Community Composition

Technique	Information provided	Methodology limitations	Reference
Microscopic examination of fixed, stained samples	Total cell numbers and cell morphologies	A large number of microscopic fields need to be examined to push minimum detection limits below 10^5 cells/g of soil; cell morphology may reveal little information about individual cells	[61,71, 131–135]
—may be combined with immunofluorescent visualization if cell-specific fluorescently tagged antibodies are available	—presence and abundance of immunologically recognizable populations	Cells must be culturable; mono- or polyclonal antibodies must be prepared, and their specificity tested before use; antigenic cross-reactions may lead to ambiguous results	[132, 136–140]
—may be combined with nucleic acid hybridization if gene-specific fluorescently tagged oligonucleotides are available	—presence and abundance of cells with nucleic acid homology	Nucleic acid probes must penetrate cells, must not cross-react with nontarget genomes, must be detectable when target cells and/or ribosome numbers per cell may be very low	[37,38, 131, 141–146]
Extraction and analysis of biomarkers			
—nucleic acids	—cloning and sequencing of 16S rRNA genes allows "uncultured" organisms to be discerned	Very insightful, but selectivities and efficiencies of nucleic acid isolation, cloning, characterization, and probing procedures are uncertain	[41,42, 62, 147–150]
	—hybridization, reassociation kinetics, and restriction length polymorphism analyses may be used as "diversity indices"	Very insightful, but interpretations rely on uncertain efficiencies in cell lysis, DNA purification, DNA shearing and require extensive positive and negative controls for differing hybridization conditions	[59,148, 151–159]

Table 5 Continued

Technique	Information provided	Methodology limitations	Reference
—phospholipid fatty acids (PLFA)	Estimates of total microbial biomass, and community composition based on interpretation of group-specific PLFA biomarkers	Presumes 100% extraction efficiency of PLFA from all environmental samples; requires specialized equipment and facilities; demands specialized knowledge of GC/MS data in relation to libraries of fatty acid profiles from pure cultures and microbial communities	[101, 160–162]
—muramic acid, chlorophyll a, chitin	Indicative of prokaryotes, algae, and fungi, respectively	Useful primarily in aquatic environments because of uncertain efficiencies of extraction from soils	[71,101]
Estimation of microbial biomass using fumigation and incubation	Estimates of seasonal fluctuations in total biomass using postfumigation microbial release of microbial carbon and phosphorus	Presumes the fumigation with $CHCl_3$ efficiently releases carbon (C) and phosphorus (P) and that these cellular components can be efficiently retrieved from the soil matrix in a manner that distinguishes microbial from nonmicrobial C and P	[128,163]
Plate counts and/or most probable number procedures	Viability of microorganisms able to grow on selective media	Selectivity of growth media may overlook as much as 99% of the populations	[61, 164–167]

(e.g., bright field, epifluorescence, electron, confocal laser scanning microscopy) and other methods.

Procedures utilizing microscopic techniques (Table 5), similar to those utilizing virtually all other techniques, may be viewed as having synecological or autecological [61] goals that guide an investigator in deciding which particular method to implement. For instance, when synecological information such as total microscopic counts is sought, a fluorescent stain such as acridine orange (which binds to nucleic acids) can be employed. In contrast, autecological monitoring of the fate, abundance, and/or characteristics of only a single population of microorganisms within the entire community can be achieved when specific stains (reliant upon immunological binding to cell-specific surface antigens or the annealing of fluorescently labeled oligonucleotide probes to DNA or RNA known to be present in the target cells) are employed.

Before applying immunofluorescent or nucleic acid probing techniques to environmental samples, every detail of the staining, destaining, and visualization procedures must be refined and proven effective using laboratory-prepared controlled mixtures of microorganisms (Table 5). These laboratory validation tests are designed to simulate the spectrum of negative and positive assay reactions expected in naturally occurring communities [168–170]. There are at least three drawbacks to the immunofluorescent visualization approach of identifying individual cells in environmental samples. The first is that the organism of interest must be grown in pure culture, in order to prepare and develop confidence in the utility and specificity of fluorescently tagged antibodies. The second drawback is the improbability that the organism being sought will be present in sufficient abundance in the habitat of interest to be detectable within a reasonable number of microscopic fields. Lastly, the antigens on the surface of targeted cells may fail to be expressed in the soil environment.

The newer molecular techniques for visualizing and identifying individual cells [37,38,42,142,144,169–171] offer several advantages in discerning the composition of naturally occurring communities (Table 5). Most important of the advantages is the elimination of the need to culture the organism of interest before development of molecular probes. Rather than isolating and growing the entire organism whose abundance in environmental samples is of interest, only the nucleic acid sequence for genes coding for phylogenetically (and taxonomically) significant rRNA needs to be obtained. Procedures for acquiring this sequence information (e.g., via PCR amplification of 16S ribosomal DNA (rDNA) [37,38] or reverse transcriptase-based synthesis of cDNA derived from 16S rRNA [40,62,150], followed by a cloning step that enables individual copies of rDNA molecules to be separated and isolated [40,172]) are potentially more reliable and less biased than are procedures dependent upon selective growth of particular members of mixed microbial communities. After the rDNA molecules retrieved from naturally occurring communities have been sequenced, the sequences can be deposited within the growing libraries of computerized databases, especially the Ribosomal Database Project [173] and GenBank [174]. Access to these and other computerized databases allows computer-assisted comparisons and phylogenetic analyses to be performed on the sequences of interest [175,176]. Such procedures have been effective in identifying previously uncultured microorganisms from aquatic environments [e.g., 37,38,62,147]. However, when applied to microorganisms native to soil, the fluorescently tagged oligonucleotide visualization procedures did not result in significant hybridization signals, possibly because of low rRNA contents of inactive soil bacteria [146].

In addition to microscopic procedures, Table 5 also lists extraction and analysis of key biological compounds (or biomarkers) as a means for characterizing microbial communities. Carter and Lynch [42] have discussed a variety of biomarker methodologies suited to soil. When applied to samples from a variety of habitats, the nucleic acid extraction and sequence analysis techniques described above have provided many unique insights. Unknown 16S rRNA gene sequences indicative of the diversity of uncultured microorganisms have been retrieved from aquatic [147] and hot spring [149] environments. Another innovative, more synecological approach has characterized total community genomic DNA, to assess the genetic diversity of naturally occurring microbial assemblages from both soil and aquatic environments. Using DNA reassociation curves, Torsvick et al. [158,159] reported highly heterogeneous bacterial DNA from a forest soil that corresponded to approximately 4000 different genomes—a diversity that was 200-fold higher than obtained by traditional cultivation and strain isolation techniques. Lee and Fuhrman [59,154], in characterizing saltwater bacterioplankton assemblages, have also utilized whole-community DNA hybridization measures; these investigators discovered broadly varying species compositions in naturally occurring communities gathered from different locations, depths, and at different times. Moyer et al. [156] used restriction fragment length polymorphism (RFLP) patterns of PCR-amplified and cloned genes that encode 16S rRNA to investigate total community diversity and structure of a Pacific Ocean hydrothermal vent mat community near Hawaii. Moyer et al. [156] concluded that the deep-sea hydrothermal vent system consisted of at least 12 taxa, 2 of which were dominant.

Although the molecular phylogenetic approach to assessing the composition of naturally occurring microbial communities has provided new and incisive information, limitations of the procedures are already becoming evident. Each of the steps in the chain of events leading from sampling the intact environment through cell lysis to nucleic acid sequencing (nucleic acid extraction, nucleic acid purification, enzymatic reactions in gene cloning and amplification, screening of clone libraries, and computer-assisted sequence alignment and phylogenetic analysis) has its own inefficiency, selectivity, and/or bias [40,148, 152,153,177,178]. In this regard, several investigators [179,180] have reported that chimeric artifacts can be formed when the PCR is used to retrieve 16S rDNA sequences from naturally occurring microbial communities. Thus, molecular approaches alone will probably never be able to overcome all of the methodological problems that confront environmental microbiologists in taking a census of microbial communities. Nonetheless, the ecological and phylogenetic insights provided by molecular techniques are unique and powerful [38].

Various additional biomarker approaches are also represented in Table 5. Of the available non–nucleic acid biomarkers, perhaps the phospholipid fatty acids (PLFAs) provide the richest insights into community composition. These insights are derived from the fact that various taxonomic microbial groups synthesize PLFAs of distinctive architecture. Structurally discernible characteristics of PLFAs [based on gas chromatography/mass spectrometry (GC/MS) analyses] include such traits as total number of carbon atoms per molecule, number of double bonds, position of the double bond relative to the omega end of the molecule, and whether or not the stereochemistry across the double bond is cis or trans. The profiles of PLFA biomarkers from aquatic, sedimentary, and soil environments have been used to discern relative abundances of a variety of distinctive eukaryotic (e.g., plant, animal, microeukaryote) and prokaryotic (e.g., *Desulfobacter*, anaerobic desaturase pathway, *Bacillus*-type Gram positive) groups [101,160,162,181]. The procedures used to obtain and interpret PLFA profiles are elaborate and perhaps best learned by direct instruction in laboratories where the techniques have been established.

Several other biomarker molecules mentioned in Table 5 (e.g., muramic acid, chitin, chlorophyll a) can be extracted from soil or other environmental samples to provide information about the abundance of specific groups of microorganisms (i.e., bacteria, fungi, and algae, respectively). In utilizing these types of procedures, certain assumptions about the content of each biomarker per cell and their extraction efficiency must be carefully evaluated. Growth of viable cells on various liquid or solid media is also a valuable technique whose strengths and limitations have been discussed in Section II.B, as well as by Karl [71,182] and Staley and Konopka [61], among others.

Despite the promising potential of the above-discussed procedures for accurately assessing "Who is there?" in naturally occurring microbial communities, a complete census has yet to be successfully accomplished in any environment.

D. Methods for Determining the General Physiological Status of Microbial Cells In Situ

There is another broad class of measurements that can be applied to microorganisms present in soil or other environmental samples. These measures ask "What is the general physiological status of microorganisms in field sites?" by addressing such issues as growth, viability, dormancy, heat generation, and starvation. Many such techniques have been critically evaluated for the soil habitat by Nannipieri et al. [43]. As listed in Table 6, measures for determining the general in situ metabolic status of microbial communities when constituent cells were captured and fixed are often refinements of procedures listed in Table 5. Thus, the microscopic preparations used in census-taking procedures also provide opportunities for observing whether cells are dividing (indicative of in situ growth), small in size (which often implies a nutrient limitation), or enriched

Table 6 Methods for Determining the General In Situ Physiological Status of Microbial Cells

Technique	Information provided	Methodology limitations	Reference
Microscopic examination			
—of fixed, stained cells	—frequency of dividing cells, poly-β-hydroxyalkanoate (PHA) content, cell size, cell brightness, and color	Presumes time between the initiation of cell constriction and cell separation is constant	[61,134]
—of fixed samples stained with 4',6-diamidino-2-phenylindole (DAPI) and 16S rRNA oligonucleotide probes	—protein synthesis activity and perhaps growth rate, as inferred from rRNA content	Data from field samples are difficult to interpret because relationships between RNA, DNA, cell growth, and protein synthesis are probably species specific; requires sophisticated microscopy and photometer systems	[134,183]
—of samples briefly incubated with respiratory stains	—frequency of respiring cells as discerned by production of the reduced dark red formazan product, from 2-(*p*-iodophenyl)-3-)*p*-nitrophenyl)-5-phenyl tetrazolium chloride (INT) or 5-cyano-2,3-ditolyl tetrazolium chloride (CTC)	Community composition may change during the assay, which is selective for cells that respond to both added growth substrate and respiratory mechanism	[61,71, 97,134]
—of samples briefly incubated with naladixic acid, which inhibits DNA replication and cell division	—distinguishes viable from nonviable cells	Incubation with added carbon source to stimulate growth may be selective; naladixic acid may not be inhibitory to all microorganisms	[61,71,184]
—colonization and growth of cells on membranes, glass slides, or other substrates inserted into terrestrial and aquatic habitats	—In situ colonization and/or growth	Substitutes an artificial colonization substrate (e.g., a glass slide) for naturally occurring solid surfaces; in situ growth may be confused with in situ cell transport	[44,120, 185]

Table 6 Continued

Technique	Information provided	Methodology limitations	Reference
Microcalorimetry	Heat output of microorganisms that have colonized a substratum inserted into habitats and subsequently removed	Method is not truly an in situ one; colonization substrates may not mimic those that are naturally occurring	[120,186]
Extraction and analysis of phospholipid fatty acids: comparison of *trans* to *cis* isomers	Starvation may be indicated by an increase in the ratio of *trans* to *cis* monoenoic PLFA	The significance of relative abundance of *cis* and *trans* PLFAs has not been universally established among a broad selection of microorganisms	[101,120, 187]
Extraction of PHA storage granules	Nutritional imbalance may be indicated by intracellular accumulation of the endogenous PHAs	Many microorganisms fail to accumulate PHAs; PHA accumulation indicates a nonspecific response of assimilation of a carbon source relative to uncertain other essential nutrients	[101,120]
Uptake and metabolism of radiolabeled substrates by microorganisms incubated in environmental samples	A variety of anabolic (e.g., tritiated thymidine) and catabolic (e.g., glucose or organic pollutants) can be added to environmental samples incubated in the laboratory to reveal "metabolic potential"	Cannot be performed in situ, laboratory artifacts are imposed by the incubation period (see Section III)	[61,71, 184, 188–190]
Extraction and quantification of adenosine-5' triphosphate (ATP) and related nucleotides	Highly sensitive assay for short-lived universal cellular energy currency (ATP); allows estimation of total biomass and physiological status of community via adenylate energy charge index	Extraction of ATP, especially from soil and sediments may be inefficient; interpretation of data may rely on uncertain assumptions about cellular ATP contents; modified versions of procedure rely on laboratory incubations	[98,101, 182, 190–192]

in poly-β-hydroxyalkanoate (PHA) storage granules (an indication of carbon sufficiency but nutrient imbalance). Microscopy of acridine orange (AO)-stained cells offers additional insights into the in situ status of microbial cells: active cells exhibit brighter, more orange images relative to the faint green images of inactive cells. This differential staining is caused by proportionately higher amounts of RNA in the active cells, but such observations are reliable only when consistent (e.g., 0.1%) AO concentrations are employed (W. C. Ghiorse, personal communication [134]). Furthermore, recent refinements using 16S rRNA fluorescence procedures suggest that signal intensity from individual cells can be quantified to assess protein synthesis activity as indicated by rRNA content [183].

The procedure in Table 6 involving the frequency of respiring cells is controversial according to the terms defined in Section III, because an incubation (albeit as short as 20 min) of the sample is required before the fixation step. Despite the possibility that artifacts may be induced during the incubation period, this procedure is capable of differentiating nonrespiring from respiring cells, whose electron transport systems reduce 2-(*p*-iodophenyl)-3-(*p*-nitrophenyl)-5-phenyl tetrazolium chloride (INT) or 5-cyano-2,3-ditolyl tetrazolium chloride (CTC) to the corresponding formazan red dye.

Another approach that utilizes laboratory incubation of environmental samples is the naladixic acid technique (Table 6). This is designed to distinguish viable from nonviable cells, because the added naladixic acid (which among other things is an inhibitor of DNA gyrase) reportedly prevents cells constriction and division. Thus, cells that respond to an added growth substrate become elongated, as discerned through microscopic examination. This technique not only physically disturbs environmental samples but also encourages population shifts via incubation and addition of selective growth substrates.

A variety of procedures have developed around the strategy of inserting an initially sterile substrate into the habitat of interest and then periodically removing and inspecting the substrate for growth and colonization by microorganisms (Table 6). One manifestation of this approach, the "buried slide technique" (developed by Rossi et al., 1936, as cited by Parkinson et al. [128]), fixes and stains adherent soil microorganisms for subsequent microscopic examination. Other procedures may be designed to allow metabolic activities, such as colonization and corrosion of metal surfaces, to be discerned from substrates temporarily installed within engineered bioreactor systems [120].

Microcolorimetry has been employed by several investigators as a means of assessing general metabolic activity of microorganisms colonizing various habitats (Table 6). Heat outputs from cells colonizing substrates inserted into a variety of habitats can provide insights into relative activities of microbial communities. However, the in situ relevance of microcolorimetry data is suspect because procedures require that an initially sterile colonization substrate be inserted into the habitat of interest, followed by transfer to a laboratory-based calorimeter for incubation.

The insightful PLFA extraction and analysis procedures, discussed above (Section IV.C) as a means for defining the composition of microbial communities, can also theoretically reveal information about the general in situ physiological status of the cells (Table 6). D.C. White and colleagues, studying pure cultures of *Pseudomonas atlantica* [187] and other organisms, have speculated that a marked shift between cis to trans monoenoic PLFA may confer resistance to nutrient deprivation [101]. Should this shift be shown to be applicable to a significant portion of the microbial world, then much information about the in situ nutrient status of microbial communities may be gleaned from the detailed GC/MS analyses of PLFA cis and trans profiles extracted from environmental samples. As indicated above, microscopically detectable internal storage granules composed of PHAs are indicative of carbon sufficiency, but a deficiency of other nutrients. PHAs can also be extracted from environmental samples and analyzed, similar to PLFAs [101].

The problematic aspects of procedures that utilize laboratory incubation of environmental samples exposed to a variety of radioactive tracers of microbial metabolic reactions (Table 6) were discussed in Section III. For the purposes of this chapter, even brief laboratory incubations were deemed undesirable because of potential artifacts. These are most likely to occur in soil samples because of disruption of the solid matrix of the soil and uncertainties about the physical and chemical conditions provided for resident microorganisms (Section III). For aquatic habitats, Karl [190] advocates the use of incubation procedures in isolating and quantifying ATP and total adenine nucleotides (TAN) as a means for estimating carbon and energy flux in microbial communities. A related approach, the adenylate energy charge procedure (Table 6) relies on extraction of adenine nucleotides and estimates the physiological state of communities by calculating the proportion of total adenine nucleotides as energy-rich ATP [98, 191,192].

E. Methods for Determining In Situ Biogeochemical Metabolic Activity

Measurement procedures of greatest prevalence and, possibly, with the greatest practical impact in biogeochemistry are those discussed in this section. These are designed to focus beyond general in situ physiological status (discussed in Section IV.D) by answering the question, "Specifically which chemical reactions are microorganisms in soil (and other) field habitats actively catalyzing?" (i.e., What are microorganisms really doing in situ?). Representative methodologies for answering this question and related ones [such as "What are the rates of reaction?" (discussed at length in Section VI)] are listed in Table 7. There are three themes that unify the studies selected for inclusion in Table 7. The investi-

Table 7 Methods for Determining In Situ Biogeochemical Metabolic Activity

Technique	Information provided	Methodology limitations	Reference
Deployment of chambers that cover portions of aquatic or terrestrial field sites; periodic monitoring of changes in concentrations of gaseous reactants and products within the enclosures	Respiratory activity, as reflected by CO_2 evolution and/or O_2 consumption	Contributions of root and microbial activity cannot be separated.[a] Substrates being respired are uncertain. Chambers are often in place between 1 and several days and, therefore, may alter in situ conditions; CO_2 fixation is assumed to be negligible	[54, 193–196]
	Methanogenesis, as reflected by CH_4 production	Net methanogenesis may overlook simultaneous removal of methane by methanotrophs	[69,102, 197,198]
	Denitrification to dinitrogen as determined by N_2O evolved within chambers with or without acetylene amendment	Acetylene inhibitor is unlikely to reach all points in the soil profile; thus, much flux may continue as N_2; acetylene may inhibit nitrification, thus altering supply of nitrite; acetylene may alter microbial community by serving as a carbon source	[52,80,195 199–203]
	Methanotrophy, as reflected by consumption of atmospheric CH_4	Only net fluxes across soil and water-air interfaces can be determined (see entry for methanogenesis)	[108,204]
	Nitrogen fixation as reflected by conversion of field-released acetylene to ethylene	Presumes (1) complete saturation of nitrogenase enzyme by added acetylene; (2) known and consistent stoichiometric conversion between moles of N_2 fixed and moles of acetylene reduced	[205,206]

[a] Separation may sometimes be achieved through stable isotope analysis (see appropriate entry in this Table).

Table 7 Continued

Technique	Information provided	Methodology limitations	Reference
Isolation of a soil plot using concrete walls, installation of piezometers for monitoring surface application of ammonium- and nitrite-amended water	Nitrification	Data from field study was limited to depth profiles of ammonium, nitrite, and nitrate; mass balances were difficult to achieve; plot needed to be amended with the water and solutes	[207]
Deployment and periodic retrieval of crop residues in field site	Decomposition and nitrogen dynamics of crop residues placed in fiberglass bags at two depths under several irrigation regimes and periodically retrieved from field site	Placement of crop residues in bags in field may diminish contact between soil and microorganisms; retrieval of crop residues from field may be biased toward large fragments that remain within bags	[208]
Release of stable isotopes in field plots followed by periodic analysis of soil, water, and atmospheric samples (with or without deployment of field chambers)	Denitrification as revealed by:		
	—$^{15}N_2$ and $^{15}N_2O$ evolved into the headspace of a sealed chamber after $^{15}NO_3$ was added to field plot	Nonrandom distribution of ^{14}N and ^{15}N atoms may violate assumptions for isotope dilution calculations; nitrogen mineralization-immobilization exchange may occur; long residence times of chamber may impose environmental artifacts	[195,199, 200,209]
	—differences in changes of nitrate concentrations in soil plots amended with $^{15}NO_3^-$ and $^{14}NO_3^-$	Assumes (1) no leaching of NO_3^-; (2) assimilatory nitrate reduction to be negligible; (3) disturbance caused in insertion of hollow cylinder in soil to have no effects	[210]
	—In situ CO_2 and N_2O concentrations and	Uncertain efficiency in disseminating denitri-	[195,211]

Table 7 Continued

Technique	Information provided	Methodology limitations	Reference
	by in situ conversion of $^{15}NO_3^-$ to $^{15}N_2O$ or $^{15}N_2$ recovered from field piezometers. Treatments included field release of C_2H_2 to block complete denitrification and of glucose or ethanol to overcome carbon limitation	fication reactants (C_2H_2, $^{15}NO_3$, glucose) and in recovering denitrification products (labeled and unlabeled N_2 and N_2O); N_2 immobilization may have occurred	
	Nitrogen fixation as revealed by		
	—conversion of $^{15}N_2$ to organic ^{15}N in plant biomass in a field-installed chamber	chamber may alter environmental and physiological conditions; method presumes complete recovery of root biomass	[205,206, 209]
	—dilution of $^{15}NO_3^-$ incorporated into legumes relative to a non-N_2-fixing plant	requires growth of both N_2-fixing and nonfixing crops; presumes identical root geometry, nitrogen demand, and $^{15}NO_3^-$ uptake for treatment and control plants	[212–214]
Fractionation of naturally occurring stable isotopes measured in field samples	In situ biodegradation of hydrocarbon contaminants at three field sites, as deduced from $\delta\ ^{13}C$ stable isotopic signatures in soil gases	Requires mass spectral evidence for distinctive $\delta\ ^{13}C$ signatures of native organic matter, the organic contaminants, and mixtures of CO_2 derived therefrom	[215]
	Cycling of acetate in sulfate-reducing and methanogenic sediments as determined by measuring $\delta\ ^{13}C$ values for acetate and its methyl- and	Data difficult to interpret because both site heterogeneity and isotope-fractionation physiology are complex	[216]

Table 7 Continued

Technique	Information provided	Methodology limitations	Reference
	carboxyl-moieties at various depths		
	CO_2 production in deep coastal plain aquifer sediments as shown by (1) increases in bicarbonate, (2) detection of anaerobic bacteria, and (3) preferential metabolism of lighter (^{12}C) carbon isotopes along a known hydrologic flow path	Hydrologic flow path uncertain, abiotic processes may have contributed to increased CO_2 concentration, many anaerobic isotope fractionation processes are poorly understood	[217]
	Nitrogen fixation by chickpea plants in the field was revealed by $\delta\,^{15}N$ in plant tissue after laboratory tests calibrating $\delta\,^{15}N$ with varying percentages of fixed N	Method loses accuracy at low rates of N_2 fixation; extensive experimentation in the laboratory is required for assembling the calibration curve for plant cultivars; recalibration may be required with different *Rhizobium* endosymbionts	[218]
	Sulfur cycling as indicated by vertical profiles of $\delta\,^{34}S$ in forest soils; partitioning of ^{32}S into soil organic matter and ^{34}S into inorganic materials suggests both mineralization and immobilization processes by microorganisms	Physiological principles underlying microbial S isotope fractionation need to be strengthened; several alternative hypotheses could explain the vertical profiles reported	[219]
	Methanogenesis in anoxic groundwater sediments as revealed by $\delta\,^{13}C$ values of inorganic and organic	Data interpretation depends on the validity of a variety of assumptions about sample handling and analysis	[220]

Table 7 Continued

Technique	Information provided	Methodology limitations	Reference
	carbon compounds and their abundances in well water samples	procedures, as well as isotope fractionation and hydrologic flow processes	
	Soil organic matter turnover rates can be inferred using ^{13}C natural abundance techniques	Limited to situations where vegetation changes (from C_3 to C_4 or C_4 to C_3-type photosynthetic pathways) leave characteristic $\delta\,^{13}C$ signatures; best suited to processes enduring up to thousands of years; interpretation of field data may require complex models; microbial decomposition may be an additional confounding fractionation process; requires specialized sample handling and mass spectrometry equipment	[221]
Chemical analysis of field samples without chamber deployment	Denitrification, as reflected by changes in		
	—N_2O profiles above the soil surface (without installation of chambers) and interpreted using micrometerological theories of turbulent diffusion	Lack of analytical sensitivity limits this procedure to soils with very high denitrification rates; flux to N_2 is masked by high ambient N_2 in the atmosphere; N_2O may also be produced during nitrification	[199,200]
	—nitrate concentrations in field samples —nitrate:chloride ratios in field samples	Many processes besides denitrification may consume nitrate; spatial heterogeneity may impede interpretation of data	[200,209]

Table 7 Continued

Technique	Information provided	Methodology limitations	Reference
	—total mass balance for nitrogenous substances added and removed from agricultural soil; method presumes that difference between initial and final content is due to denitrification	Limitations include those listed above [200], also analytical errors in total nitrogen determinations are likely to mask small reductions caused by denitrification	[199,200, 209]
	—N_2O concentrations in piezometers installed in field	N_2O is a transient intermediate of both nitrification and denitrification and, therefore, may or may not indicate denitrification activity; the piezometers may have been screened at depths below the active zones	[104]
	Nitrogen fixation, as reflected by		
	—relative content of plant nitrogen in the form of ureides	Promising, but many uncertainties remain in the biochemistry of production, transport, and disposition of ureides within plant tissues	[205]
	—nitrogen balance of plant and soil	Lack of precision in quantifying the many pools of organic N in soil; spatial variability of the pools of nitrogenous materials	[205]
	—differences in nitrogen content of nitrogen-fixing and nonfixing legumes	Assumes nitrogen lost to the soil by roots is negligible. Requires availability of non-N_2-fixing legume with comparable physiological traits. Can only be discerned if variability imposed	[205]

Table 7 Continued

Technique	Information provided	Methodology limitations	Reference
		by soil and plant heterogeneity is low	
	The terminal electron accepting process (TEAP) (e.g., reduction of nitrate, iron, sulfate, or CO_2) that dominates microbial communities in anoxic sediments and groundwater is indicated by the in situ concentration of H_2 gas, a key microbial intermediary metabolite	Measuring nM concentrations of H_2 requires specialized equipment; the procedure has only been applied to water-saturated environments, never soil; distinctions between TEAPs may be difficult to discern	[222]
	In situ biodegradation of crude oil as revealed by shifts in ratios of biodegradable to microbially resistant hydrocarbons (which serve as conserved tracers) over time	Not always reliable, because resistant hydrocarbons may be biodegradable or may undergo abiotic reactions that differ from those of the degradable hydrocarbons	[223–225]
	—polychlorinated biphenyls (PCBs) as revealed by a suite of indicators: microbial numbers, biodegradation potential, and detection of intermediary metabolites extracted from Hudson River sediments	Requires highly sensitive extraction and analytical methods; chlorobenzoate metabolite may originate from compounds other than PCBs; microbial numbers and biodegradation potential, alone, are not definitive evidence	[226,227]
	—^{14}C-labeled herbicides; by releasing these to field plots and monitoring ^{14}C in percolation water using lysimeters	recovery efficiency of ^{14}C applied was close to 1%; no $^{14}CO_2$ was recovered; little could be concluded; release of radioactivity is seldom allowed	[228]

Table 7 Continued

Technique	Information provided	Methodology limitations	Reference
	—trichloroethylene (TCE) in an engineered aerobic groundwater system as revealed by the linkage between phenol addition and cometabolism of TCE	requires intensive instrumentation and characterization of field site; hydrology must be understood sufficiently well to establish mass balances	[229]
	—atrazine; by recovering metabolites in water draining from field plots amended with atrazine	presumes single metabolic pathway for atrazine biodegradation; procedure can recover only mobile (soluble and nonsorbing) metabolites	[230]
	—pentachlorophenol in replicated field soil plots inoculated with fungi, *Phanerochaete chrysosporium* and *P. sordida*	requires uniform site subdivided into replicated plots; uninoculated controls must be prepared	[231]
	—organic contaminants that serve as growth substrates (in this case, naphthalene and phenanthrene) by demonstrating a combination of metabolic adaptation and enhanced numbers of predatory protozoans	method relatively nonspecific; requires access to adjacent, uniform uncontaminated "control" site for determining both metabolic adaptation and enhanced numbers of protozoa	[232]
	—perchloroethylene (PCE) accidentally spilled into an anaerobic mixed-textured aquifer. Biodegradation was shown using laboratory incubations measuring metabolic potential of the sediments and/or by mapping field concentration contours	requires a relatively uniform field site so that heterogeneity does not interfere with attempts to measure and map field concentrations of metabolic intermediate compounds	[233,234]

Table 7 Continued

Technique	Information provided	Methodology limitations	Reference
	for dechlorinated metabolites over time		
	—PCE accidentally spilled into a coarse-grained, unconfined aquifer. An engineered bioremediation pumping regime was established on site; reductive dehalogenation of PCE to nontoxic dechlorinated products occurred only after driving anaerobic processes by benzoate addition.	requires a relatively uniform field site so that heterogeneity does not interfere with attempts to alter site hydrology or to measure and map field concentrations of metabolic intermediate compounds	[235]
	—oil released to replicated soil field plots with and without fertilizer amendments	requires uniform site, subdivided into replicated plots; also requires field release of oil	[236]
	—TCE in an engineered aerobic groundwater system as revealed by detection of unique intermediary metabolites and a linkage between oxygen and methane delivery required for cometabolic oxidation of TCE	requires intensive instrumentation and characterization of field site; hydrology must be understood sufficiently well to establish mass balances	[237]
	—naphthalene in surface water flowing from a coal-tar contaminated site as indicated by careful sampling and detection of the unique intermediary metabolite, 1,2-dihydroxy-1,2-dihydronaphthalene	requires rapid sample handling field procedures, a rare standard for unique metabolite, and GC/MS analytical capabilities	[238]

Table 7 Continued

Technique	Information provided	Methodology limitations	Reference
	—herbicides 2,4-D and 2-methyl-4-chloro-phenoxyacetic acid (MCPA) as reflected by need for continued application for weed control and by enhanced biodegradation potential by microorganisms in soils with previous exposure to the herbicides	extraction and chemical analysis of herbicides from treated soils were not performed; formation of bound residues was not considered	[239]
	—polycyclic aromatic hydrocarbons (PAHs) applied for two decades to soil. In situ metabolism was reflected by lack of accrual of the chemicals	mass balances uncertain; formation of unextractable "bound residues" not considered as alternative explanation for "loss" of PAHs	[240]
	—toluene in a contaminated stream, as indicated by attenuation which exceeded that predicted by tracers (NaCl and propane) released into the field study site	tracers may not mimic the transport properties of toluene; assumptions about equations predicting toluene transport warrant scrutiny	[93]
Extraction and analysis of nucleic acids	—extraction of RNA and probing for mRNA to indicate gene expression —reverse transcriptase polymerase chain reaction (rtPCR) amplification of expressed messenger RNA	Nucleic acid extractions may be selective and inefficient, mRNA is unstable and transient; specificity of gene probes is uncertain; conditions for hybridization reactions (including both positive and negative controls) must be carefully selected to avoid ambiguity in data interpretation	[241–249]
Microelectrodes	Small-scale in situ profiles of O_2, H_2S,	Biotic versus abiotic and dynamic causes	[44,120, 250]

Table 7 Continued

Technique	Information provided	Methodology limitations	Reference
	NO_3^-, pH, etc. in microbial habitats are measured directly	of concentration gradients may be difficult to discern; many studies are conducted in laboratory microcosms rather than field sites; calculations of metabolic rates based on diffusion models may or may not be accurate	
Laboratory incubation of field samples simulating in situ conditions (see discussion in Section VI)	denitrification rates in intact soil cores	Intact soil cores were removed from field site and incubated in the laboratory using the acetylene block assay; acetylene may not have been adequately mixed; samples were disturbed; artifacts may appear (see discussion in Section VI)	[52,251]

gators have (1) emphasized field approaches designed to eliminate artifact-plagued laboratory incubation techniques; (2) struck a balance between both simplifying the complexities of field sites (so that interpretable measurements can be obtained) and minimizing the artifacts of the simplification and measurement procedures; and (3) pursued a central goal of distinguishing biotic from abiotic [252–256] processes which govern fluxes of material in field sites. Because laboratory incubations are strongly entrenched in environmental microbiological practices, some of these appear in Table 7 as well.

One of the most direct approaches for measuring in situ biogeochemical metabolic activity in soil is to place a chamber on the soil surface and periodically remove and analyze gaseous samples from the chamber headspace (Table 7). This approach will only address processes that affect volatile materials (e.g., CO_2, O_2, N_2O, CH_4). Another requirement is that the flux rate of these volatile materials be relatively rapid so that the chamber can be installed for only brief periods of time, thus minimizing alterations in environmental conditions (temperature, light, gaseous exchange, etc.) that affect the microbial processes under scrutiny. Furthermore, as has been mentioned in Section III.A, the metabolic products of many physiological processes serve as reactants for other processes. Thus, in most instances, field rates of microbial metabolism discerned via soil cover procedures overlook the contributions of individual processes.

A strategy common to a variety of both laboratory and field biogeochemical assays (Table 7) is to separate two competing processes via the application of selective inhibitors. In this regard, acetylene is useful in competing for active sites on enzymes involved in both nitrogen fixation [leading to the production of ethylene, which is readily measured by gas chromatography (GC)] [206] and in denitrification (leading to the truncation of the denitrification respiratory chain and consequent production of N_2O, which is also readily determined by GC) [202,257]. Methylfluoride has been proposed as a selective inhibitor of the methane monooxygenase enzyme of methanotrophic bacteria that is responsible for methane oxidation [258]. When applied in field settings, protocols utilizing selective inhibitors may produce data that are difficult to interpret because the compounds may fail to reach the targeted populations and/or may have unknown secondary effects (such as serving as carbon and energy sources) for some members of the microbial community (Table 7). In field studies of denitrification, denitrification to N_2 has received the greatest, if not exclusive, attention; thus, the significance of the other major anaerobic pathway of nitrate respiration, denitrification to ammonia [257,259], is uncertain.

Nitrification represents an important segment of nitrogen cycling processes, yet none of the major chemical indicators of nitrification (ammonium, nitrite, nitrate) is volatile. Thus, field documentation of nitrification processes (Table 7) is difficult because of mass balance and other problems [207]. In this regard the obstacles to measuring nitrification in field sites resemble the challenges encountered in measuring biodegradation of chemical pollutants (see Table 7 and text below).

The study by Schomberg et al. [208] represents a recent example of the widespread practice of assessing decomposition processes in the field (Table 7). Changes in crop or forest litter layer residues can be monitored by enclosing these materials in open mesh bags and periodically retrieving them from field sites. Both microbiological and physical changes characteristic of the decay processes are recorded.

Nearly artifact-free insights into in situ biogeochemical processes can be obtained by methodologies that capitalize on the mass spectrometric tools that analytical chemists employ to distinguish between elements with more than one atomic weight. Although unstable (radioactive) isotopes (indispensable tracers of microbial metabolism in laboratory studies) have occasionally been used in intentional field releases of radioactive materials, it is the stable isotopes (which carry no radioactive health hazards) that are much more amenable to routine experimentation aimed at measuring a variety of biogeochemical activities in situ. As shown in Table 7, stable isotopic procedures have been applied to field microbial processes using two fundamentally different approaches: (1) isotopically enriched compounds of interest can be released to field sites where the fate of compounds can be monitored by periodic sampling of soil, water, and atmosphere (with or without deployment of field chambers); or (2) field samples can be analyzed for their naturally occurring relative abundances of stable isotopes whose patterns may reflect selective enzyme-mediated metabolism in situ.

Although a variety of biogeochemical processes can theoretically be measured by releasing stable isotopes into field plots, agricultural interests in fertilizer efficiency and nitrogen cycling have directed the majority of research efforts toward denitrification [195,199,209–211] and nitrogen fixation [205,206,209, 212–214; Table 7]. The data gathered from stable isotope, field release experiments rigorously document microorganism—(alone or as a plant endosymbiont) mediated conversion of the labeled compound from one form (e.g., $^{15}NO_3^-$) to another (e.g., $^{15}N_2$). However, the logistics of field experimentation often present such a challenge that unambiguous quantitative rates of biogeochemical processes are seldom obtained. Weaknesses in the methodologies become especially apparent when attempts at mass balance accounting of the released compounds are made, when dubious assumptions about experimental controls are made, and when field deployed chambers alter the in situ conditions.

Fractionation of naturally occurring stable isotopes conforms well to the snapshot paradigm for determining in situ biogeochemical activity (Figure 3). Samples from unmodified field sites can be captured, fixed, and analyzed, and the resultant data can often be used to draw inferences about in situ microbial processes. Implementation and interpretations of isotopic fractionation analyses relies on a rich and intricate body of knowledge that spans analytical chemistry and both plant and microbial physiology. The details of sample preparation, analyses, and signature ratios of stable isotopes are described by Hoefs [260] and by Wolf et al. [221]. As mentioned in Table 7, delta (δ) values, derived from high-resolution mass spectrometric determination of stable isotopic ratios for carbon, nitrogen, and sulfur, have been used to document microbial cycling of these elements. As is the case with virtually every field methodology (Section III), stable isotope fractionation procedures deliver data on biogeochemical rates that suffer somewhat from uncertainties associated with both variabilities in understanding field sites and/or in the biochemical principles used in interpretation of stable isotopic fractionation data.

A substantial portion of Table 7 lists investigations that have documented in situ biogeochemical processes simply by analyzing field-derived samples for substrates, reactants, products, and intermediary metabolic compounds indicative of the process of interests. This approach (similar to that used in stable isotopic fractionation) is attractive because artificial conditions, capable altering true in situ microbial metabolism, are imposed only in the brief sample capture and fixation steps (Figure 3). But this field authenticity is accompanied by the complexity of field conditions that often impedes data interpretation [43]. For instance, atmospheric gradients of the partially reduced gas N_2O (formed from NO_3^-) are indicative of denitrification. However, not only is N_2O difficult to measure in field samples, but interpretation of vertical N_2O profiles is subject to micrometerological uncertainties implicit in both field measurements and theories of turbulent diffusion [200]. Similarly, nitrogen fixation can theoretically be estimated in the field by assembling cumulative nitrogen balance data from all pools of organic N. However, both spatial variability of the N content of field samples and lack of precision in measuring organic N make it very difficult to document small increments of this element and attribute these to nitrogen fixation processes [205].

A recently proposed innovation for determining the terminal electron accepting process dominant in anoxic aquatic habitats (sediments and groundwaters) is based on the concentration of hydrogen gas [222] (Table 7). Hydrogen is an important intermediate in the coupling of microbial oxidation of organic matter to the reduction of inorganic electron acceptors. Lovley et al. [222] have shown that the in situ H_2 concentration ranges of 7–10, 1–1.5, 0.2, and <0.05 nM correspond to methanogenic, sulfate-reducing, Fe(III)-reducing, and Mn(IV)- or nitrate-reducing conditions, respectively. It is not yet clear how these procedures can be applied to the soil habitat.

The portion of the carbon cycle concerned with the biodegradation of pollutant organic chemicals is another process that can be addressed by chemical analysis of field samples. As outlined by Madsen [3,261,261a] and expanded by the National Research Council [261b], when organic compounds are inadvertently released to field sites, microbial processes are often the only ones capable of breaking intramolecular carbon-carbon bonds, hence achieving pollutant detoxification. However, distinguishing between microbiological and abiotic (e.g., transport, dilution, volatilization, sorption, humification [252–256]) mechanisms of pollutant attenuation in the field can be challenging. The principles for determining in situ biodegradation recommended by the National Research Council [261] are as follows: (1) assemble historical evidence documenting diminishing concentrations of contaminant compounds in site-derived samples; (2) demonstrate in laboratory-based studies that microorganisms from site-derived samples have the potential to metabolize the compounds; and (3) demonstrate that metabolic potential is actively expressed in the field. It is this latter issue that is the main focus of the in situ biodegradation studies listed in Table 7. For instance, Atlas [223] and Pritchard and Costa [224] have successfully utilized an internal tracer technique that relies on diminishing ratios of straight-chain (readily metabolized) to branched-chain (microbially resistant) hydrocarbons for demonstrating in situ biodegradation of crude oil. Other investigators have proposed that alternative internal marker compounds present in crude oil, notably hopanes, can also serve as indices for demonstrating in situ selective microbial metabolism of components of crude oil [262,263]. Additional successful strategies for distinguishing microbial from abiotic attenuation of organic contaminants from field sites include the detection of unique intermediary metabolites [226,227,230,233,234,238], inoculation of replicated soil plots with biodegrading microorganisms [231], fertilization of replicated soil plots [236], and the detection of both high numbers of protozoan predators and pollutant-adapted microbial populations inside but not outside the contaminated zone [232]. The major drawback of virtually all of the in situ biodegradation field studies listed in Table 7 is that, except for the highly engineered sites [229,235,237], they are largely qualitative. This reflects the fact that estimates of field processes must reflect both the biotic and abiotic attenuation processes simultaneously in progress, as well as the generally large variability of contaminant concentration data typically found in heterogeneous field sites.

Nucleic acid extraction and analysis techniques (discussed in Sections IV.C and D because of their usefulness in describing the composition and protein-synthesis activity of naturally occurring microbial populations) offer an additional dimension of information—that of gene expression. Several investigations in aquatic [242,244,245] and soil or sedimentary [241,246–249] habitats have successfully extracted, amplified, and/or probed naturally occurring microbial populations for the mRNA encoded by structural genes. The functional genes examined to date include those involved in photosynthesis (ribulose-1,5-bisphosphate carboxylase [244]), metabolism of environmental pollutants (naphthalene dioxygenase [241] methane monooxygenase [248] toluene monooxygenase [247]), and metal detoxification (*mer*A, specifying mercuric reductase [242]). Although sophisticated and elegant, these molecular techniques are laborious and biased to some degree (see discussion in Section IV.C, where molecular methods were first introduced). Additionally, synthesis of mRNA is only one step toward full expression of an in situ biogeochemical process. The presence of an expressed structural gene does not guarantee that the corresponding enzyme has been synthesized, has successfully contacted the targeted substrate, and that the intermediary compounds have been fully integrated into cellular metabolism. Indeed, an instance of metabolic activity being uncoupled from gene expression has been recently documented by Nazaret et al. [264] in examining both transcription of *mer*A and mercury volatilization from field and laboratory incubated samples.

Microelectrode technology has contributed substantially to documentation of the small-scale spatial gradients of key regulators and end products of aerobic and anaerobic microbial activity (Table 7). Microelectrode measurements have been applied widely in intracellular physiological studies as well as in laboratory assays utilizing environmental samples. As is the case with many other habitat-description procedures, the precise contribution of microbial metabolism to the measured gradients may be difficult to discern. The current strategy of using computer-based diffusion calculations to attain estimated rates of microbial processes (e.g., [250]) seems reasonably successful, albeit seldom applied to field settings.

The final entry in Table 7 relies on laboratory incubation of field samples under conditions simulating those found in situ. Perhaps the most extensively investigated application of this approach is the estimation of field denitrification rates using intact soil cores [52,86,251] although methanogenesis [69] and methanotrophy [108], among others, have also been studied (see Table 4 and Section III.C). As has been extensively discussed in Section III.C, the author's view of laboratory incubated assays is that they represent model systems that must change over time, cannot be artifact-free, and whose accuracy in estimating true field rates of microbial processes is variable. However, because perfect, artifact-free procedures have yet to be developed in soil and environmental microbiology (see Section VI), the information provided by laboratory incubated assays often remains unrivaled, hence essential.

V. METHODS FOR IDENTIFYING MICROORGANISMS RESPONSIBLE FOR BIOGEOCHEMICAL PROCESSES IN SITU

It is appropriate, at this point, to recall the remarks from various scholars shown in Table 3 about methodological limitations in soil and environmental microbiology. Given the limited tools for asking "Who is there?" and "What are they doing?" (Sections III and IV), it may seem glaringly optimistic to ask "Can the microorganisms responsible for particular biogeochemical processes in the field be identified?" After all, the means for simply demonstrating in situ biogeochemical activity (without discerning the particular responsible party) are both demanding and rare (Section IV.E; Table 7). Thus, studies pursuing the goal of identifying the microorganisms responsible for particular biogeochemical field processes are perhaps the most ambitious of those yet examined in this chapter.

Using a combination of experience, common sense, and the logic which Koch developed in his studies of pathogenic microorganisms, Waksman [31] expressed the following criteria for attributing a given biogeochemical process in soil to a particular member of the soil community:

> Before we can conclude that a microorganism is active in the soil and that certain chemical transformations are produced by this organism under ordinary soil conditions, certain requirements must be satisfied. The following postulates, applied by Koch to pathogenic bacteria, and modified by Conn in

their application to soils should hold true for soil microorganisms: (1) The organism must be shown to be present in the soil in an active form when the chemical transformation under investigation is taking place. (2) The organism must be shown to be present in larger numbers in such soil than in similar soil in which the chemical change is not taking place. (3) The organism must be isolated from the soil and studied in pure culture. (4) The same chemical change must be produced by the organism in experimentally inoculated soil, making the test, if possible, in unsterilized soil. (5) The organism must be found in the inoculated soil.

Implicit in Waksman and Conn's criteria are variations of almost all of the concerns discussed in Section III of this chapter. Postulate 1 raises the issue that presence of microorganisms is not equivalent to in situ activity in soil and environmental microbiology. Postulate 2 attributes in situ growth to in situ activity. The remaining three postulates show biases from medical microbiology, which presume that the organism of interest can be grown under laboratory conditions (postulates 3 and 5) and that the agent(s) of chemical change (hence the change, itself) is either absent from some soils or will function in sterilized soil (postulate 4). While the logic of Waksman and Conn is sound (and quite effective in studying cases of unusual metabolic function, such as plant disease or symbiotic nitrogen fixation), there clearly are a variety of situations in which the applicability of the five postulates breaks down. These include situations where the microorganisms of interest resist laboratory cultivation, where the process of interest is catalyzed widely by many members of the community, and where the soil sterilization procedure renders the medium toxic or nonconductive to the physiological function under scrutiny.

With the advent of the combination of three technologies (improved microscopy, immunolabeling, and radiotracer procedures) near the end of the twentieth century, new progress has been made in addressing this long-standing issue of determining the identity of particular microorganisms responsible for specific biogeochemical processes (Table 8). A technique common to two of the three main methodological categories shown in Table 8 is autoradiography. This highly sensitive procedure (pioneered in microbial ecology by Brock and Brock [270] and still in current practice [168,268]) utilizes a photosensitive coating on glass slides to achieve the microscopic visualization of single cells that have incorporated radioactive tracers into their cellular structure.

Taking advantage of the fact that autotrophic ammonium-oxidizing bacteria (*Nitrobacter* spp.) are rather rare in soil, Fliermans and Schmidt [265] carried out an elegant set of experiments that allowed metabolically active cells (i.e., those carrying out $^{14}CO_2$ fixation, as determined by microautoradiography) to be distinguished within a population of soil microorganisms bearing *Nitrobacter*-specific surface antigens (as determined by fluorescence microscopy; Table 8). The study by Fliermans and Schmidt [265] illustrated that important insights

Table 8 Methods for Determining the Identity of Particular Microorganisms Responsible for Specific Biogeochemical Processes

Technique	Information provided	Method	Limitations	Reference
Combination of microscopy, immunofluorescence, and autoradiography	Identification of active nitrifying bacteria present in laboratory incubated soil samples	Fluorescently tagged antibody for *Nitrobacter* in combination with microautoradiography allowed *Nitrobacter* cells active in $^{14}CO_2$ fixation to be detected in soil samples previously enriched with NO_2^- and exposed to $^{14}CO_2$ in laboratory incubations	Not a field study. *Nitrobacter* cells were isolated, grown; fluorescent antibodies were prepared, and their specificity tested. The detection assay required laboratory incubation of soil, which involved both enrichment for nitrifying microorganisms and exposure to $^{14}CO_2$	[265]
	Relative contribution of individual members of an artificial mixture of algae to photosynthetic activity	Incorporation of $^{14}CO_2$ into cell constituents was detected by deposition of silver grains on slides coated with photographic emulsion	A laboratory, rather than a field study. Experimental design required the three test algae to be morphologically distinctive	[266]
	Identification of active ammonium-oxidizing bacteria in seawater samples	Fluorescently labeled antibodies specific for *Nitrosomonas oceanus* and *N. marina* were used to identify the cells that incorporated $^{14}CO_2$ into their cellular material	$^{14}CO_2$ assimilation assay required a 24-h laboratory incubation. A large number of microscopic fields (100) needed to be counted	[267]
Combination of microscopy, autoradiography, and hybridization of intracellular nucleic acid probes to messenger RNA	Cytoplasmic localization of photosynthetic and nitrogen-fixation gene expression in laboratory-grown cyanobacteria	Intracellular hybridization of ribulose-1,5-bisphosphate carboxylase and nitrogenase gene transcripts was achieved using ^{35}S-labeled oligonucleotide probes visualized using a photosensitive emulsion and light microscopy	Cells were selected, enriched, and grown in the laboratory rather than being derived from natural samples	[268]

Table 8 Continued

Technique	Information provided	Method	Limitations	Reference
Immunofluorescence microscopy	Presence of translated nitrogenase gene in whole cyanobacterial cells grown in the laboratory	Intracellular binding of fluorescently tagged antibodies recognizing the Fe-S component of the nitrogenase enzyme complex	May underestimate number of N_2 fixers due to inadequate cell permeability; some nitrogenases may lack requisite antigens; controls for nonspecific antibody binding are complex	[269]

could be attained using microscopic visualization methods for both metabolic function and cell identity; however, the procedures were implemented using laboratory grown cells added to laboratory incubated soil. Such a contrived situation has little bearing on in situ field nitrification processes.

One decade later, Ward [267] carried out what emerges as possibly the most impressive study to date using a combination of autoradiography and immunofluorescence microscopy (Table 8). As with Fliermans and Schmidt [265], nitrification was the process of interest. But unlike Fliermans and Schmidt [265], Ward [267] examined freshly gathered, naturally occurring microbial populations (in this case those found in seawater). The dual labeling procedure allowed Ward [267] to report the relative autotrophic activity of individual ammonium-oxidizing bacterial cells, as discerned by the number of silver grains formed in the photosensitive emulsion by β particles released from the ^{14}C tracer. However, using strict criteria for avoiding incubation-induced artifacts (described in Section III), even the study by Ward [267] was flawed because shipboard laboratory incubation for ^{14}C labeling was required. Ward [267] acknowledged that study was conducted under simulated rather than actual in situ conditions.

Gutel'makker [266; Table 8] provided the methodological basis for the studies by both Fliermans and Schmidt [265] and Ward [267] by showing, in an artificial mixture of algal cells, that autoradiographic detection of $^{14}CO_2$ fixation activity was possible. Similarly, an innovative physiological investigation using pure cultures of cyanobacteria [268; Table 8] also lays the groundwork for future applications of microautoradiography in microbial ecology. The study by Madan and Nierzwicki-Bauer [268] identified the location of photosynthetic gene expression in the structurally differentiated heterocysts of *Anabaena*. The advancement made by Madan and Nierzwicki-Bauer [268] was to substitute ^{35}S nucleic acid mRNA probing for Ward's [267] incubation step that was required for ^{14}C labeling of cells. Madan and Nierzwicki-Bauer's [268] strategy eliminated the need for the incubation-requiring metabolic activity assay ($^{14}CO_2$ incorporation in cell biomass) by, instead, detecting the mRNA transcribed from two ribulose-1,5-biphosphate structural genes (*rbcL* and *rbcS*). Thus, it has only recently become feasible to use radiolabeled nucleic acid probing procedures to detect gene expression in individual cells. It should also be noted that the recent report by Hudson et al. [268a] of intracellular amplification of naphthalene catabolic genes and transcripts may have significant bearing on our ability to identify active cells.

The final entry in Table 8 [269] goes one step further in the sequence of gene expression than that of Madan and Nierzwicki-Bauer [268] by using immunofluorescent labeling to detect a translated functional enzyme (in this case, nitrogenase) within individual cells. The procedures developed by Madan and Nierzwicki-Bauer [268] and Currin et al. [269] have not yet been tested on members of naturally occurring microbial communities. Nonetheless, these approaches have great promise if they can be successfully applied to individual cells present in environmental samples. Such samples of soil, sediment, or water theoretically may be gathered and fixed using the artifact minimizing snapshot approach described in Section IV.B and Figure 3.

VI. METHODS FOR INTEGRATING IN SITU SOIL METABOLIC ACTIVITY MEASUREMENTS IN TIME AND SPACE

As outlined in Section I.A, the fourth major question to be addressed by this chapter is "Can accurate measurements of in situ microbiological processes be integrated in time and space?" Although discerning the particular agents responsible for specific biogeochemical reactions (Section V) represents a major conceptual goal within the disciplines of soil and environmental microbiology, this fourth question carries a far more expansive and pragmatic theme. Microbial ecologists, global biogeochemical modelers, and ecosystem scientists are driven by concerns over potentially catastrophic human-induced developments, such as global warming caused by atmospheric accumulation of greenhouse gases [271]. The obvious role that microorganisms often have as the primary agents of biogeochemical change makes integration of their in situ activities a critical issue in both understanding and managing the biosphere.

For soil and environmental microbiologists who have long been conducting field-oriented biogeochemical studies, the above question may seem unnecessary, perhaps even insulting. But the objective here is simply to continue the central theme of this chapter by critically reevaluating what the tools of soil and environmental microbiology can and cannot reveal. Of particular interest are the related questions, "Which measurements are most accurate?" and "Can these accurate measurements be integrated in time and space?"

A. Mass Balance Lessons from Ecosystem Ecology

Among the central conceptual rules for understanding global as well as watershed biogeochemical cycles [272,273] are the boundary conditions imposed by mass balance considerations. For a given ecological system (regardless of its size), if it is in steady state, the inputs must equal the outputs. By the same token, if "source" and "sink" terms within a system are not in balance, the biogeochemical parameters of interest (such as organic carbon, nitrogen, or sulfur species) may be accruing or diminishing with time. Net loss or gain is dependent on relative rates of consumption and production. These simple truths have major implications for assessing the accuracy of microbiologically catalyzed geochemical processes measured at field sties.

The essential feature of mass balance approaches in ecosystem studies is their uncompromising link to field site characteristics (to reality, if you will). Misleading laboratory-imposed artifactual data can be readily dismissed because the study site in the real world demands that things make sense. The study site is there to confirm or deny any conclusions drawn from both field and laboratory data. It is historical mass balance data from field sites that provide the most rigorous check on the accuracy of estimates of biogeochemical processes. The mass-balance approach is far more convincing than even a set of convergent, independently conducted alternative assays (such as laboratory microcosm and field chamber methods discussed in Section III.C and Table 4). Even if non-mass balance techniques confirm one another, they may suffer from inaccuracies imposed by shared or different methodological weaknesses.

B. Examining the Accuracy of Laboratory-Based Biogeochemical Extrapolations: Three Examples

In certain subdisciplines of biogeochemistry, it is a challenging yet routine practice to extrapolate data from laboratory incubated soil cores to the scale of field plot, landscape, and region [51]. Indeed, Groffman [49] has insightfully recognized that the hierarchical scales for understanding biogeochemical processes range from gene and organism to the field, landscape, region, and globe. Corresponding to each of these scales, Groffman [49] has recognized an appropriately scaled controlling factor for best approaching the management and understanding of biogeochemical processes. For instance, at the organism scale, denitrification is regulated primarily by oxygen, nitrate, and available carbon. At the field level, denitrification is controlled by the supply of soil water, nitrate, and organic carbon. At the landscape level, soil type and plant community are of major concern. At the regional level, geomorphology and land use are the key issues. And global concerns about denitrification focus on biome type and climate [49].

The science of integrating highly reductionistic information of molecular mechanisms through the hierarchies to global processes is still in its infancy [16, 274,274a] Among the theoretical and practical uncertainties about extrapolating small-scale measurements upwards are whether (1) the environmental samples used to obtain biogeochemical rate data are a valid "unit cells" that adequately incorporate all spatial variability factors found in the field; (2) the duration and timing of the laboratory incubation period (needed for rate data) accurately represent the average hourly or daily temporal variability for the microbial process under scrutiny; (3) adequate care has been taken in designing statistically valid procedures for both sampling field sites and estimating the extrapolation factors used to integrate rates in space and time; and (4) Heisenberg uncertainty-type artifacts have been imposed on the experimental system used to generate raw biogeochemical rate data.

To illustrate the weak links in the series of measurements and assumptions necessary for integrating laboratory microcosm data over space and time, details from three examples are presented. The first two examples were chosen because they represent sound, state-of-the-art studies on important biogeochemical processes, denitrification, and methanogenesis. The third, treated only briefly, examines rates of carbon turnover in an aquifer system. All three examples are particularly rigorous, because in addition to implementing laboratory microcosm assays, they also assess the validity of their respective results using mass balance approaches typical of ecosystem scientists.

Denitrification in North Temperate Forest Soils

Standard procedures for using laboratory-derived rates of microbial activity as a basis for extrapolating in time and space are illustrated by Groffman et al. [50] and Groffman and Tiedje [275] in their studies of denitrification. The primary objective of these investigations was to analyze denitrification in north temperate forest soils at the regional level. To do this, the following steps were taken:

1. Nine categories of soil type were identified representing three soil textural classes spanning three drainage regimes.
2. For each of the nine soil categories, 20 intact soil cores (15 cm deep × 2 cm in diameter) were gathered at weekly to monthly intervals.
3. Each core was stored overnight at 4°C, then warmed to 22°C for 2 h. After the cores were exposed to acetylene (to truncate dissimilatory nitrate reduction at N_2O), a 6-h incubation was initiated and the change in N_2O concentration during the last 4 h of the incubation was considered the linear in situ denitrification rate for each core.

4. The denitrification rates were corrected using a Q_{10} factor for in situ temperatures.
5. Rates measured in soil cores were expressed on an areal basis using bulk density values and extrapolated over the periods of time that elapsed between sampling dates.
6. Annual denitrification losses were calculated by integrating the areal extent of the soils over space and time.

Groffman and Tiedje [275] reported that annual nitrogen losses by denitrification ranged from <1 kg. ha^{-1} to over 40 kg. ha^{-1}. The higher rates were deemed "unsustainable from a long-term nutrient balance standpoint."

Methanogenesis in Peatlands

A study by Wieder et al. [102] used an approach similar to that of Groffman and Tiedje [275] to examine methanogenesis in two Appalachian peatlands. The procedures employed by Wieder et al. [102] for obtaining biogeochemical rate data and extrapolating over time and space were as follows: (1) duplicate (35 cm deep × 10 cm in diameter) cores were removed from each peatland at eight sampling dates; (2) under N_2 flow in the laboratory, three 75-cm^3 subsamples from four depths from each core were placed into Erlenmeyer flasks, and covered with site-derived interstitial water; (3) the flasks were capped, flushed with He, and incubated at in situ temperatures; (4) at 24- to 48-h intervals over a 9-day period, headspace gases were analyzed for concentrations of CO_2 and CH_4; (5) rates of CH_4 and CO_2 production were calculated on a millimolar per liter per day basis, and the mean rate values were integrated across all depths; (6) the integrated rates were plotted as a function of time, and then the annual rates were obtained by integrating he seasonal data. To assess the accuracy of the results of their work, Wieder et al. [102] accompanied their estimates of CH_4 and CO_2 production for the peatlands under study with other carbon-budget parameters, including above ground and below ground net primary production and aerobic decay processes. Wieder et al. [102] concluded that "carbon-mineralization exceeds annual input of carbon into the anaerobic zone, placing the peat deposits into a present state of negative carbon balance, so that these peat deposits are being degraded rather than accumulating."

Turnover of Organic Carbon in a Deep Aquifer

The third selected example of a study using field observations to test the accuracy of laboratory data was carried out by Chapelle and Lovley [276]. These investigators demonstrated that their laboratory microcosm assays for the conversion of ^{14}C-labeled acetate to $^{14}CO_2$ in anaerobic groundwater sediments were two to four orders of magnitude too high to be sustainable in the deep aquifer system being studied. In discussing the discrepancies between laboratory derived rates and sustainable in situ rates derived from hydrologic field modeling, Chapelle and Lovley [276] considered a variety of causes. These ranged from in situ versus in vitro contrasts in hydrostatic pressure, substrate concentrations, and sample heterogeneity—all of which fall within fall within the range of laboratory artifacts discussed in Section III of this chapter.

C. A Proposed General Strategy for Measuring Biogeochemical Rates In Situ

In Part A of this section, it was emphasized that the mass balance approach to biogeochemical processes serves as a rigorous check for the accuracy of estimates of in situ microbiological processes. In Part B of this section, examples were discussed in which results of time- and space-integrated laboratory data were believable but uncertain [102,275] or clearly misleading [276]. Given these unsettling experimental results, it becomes necessary to ask, "What really is the status of the abilities of soil and environmental microbiologists to integrate in situ metabolic activity in time and space?"

This author's view of the state of the art for integrating in situ microbial metabolic activity is itemized as follows:

1. The most reliable data on in situ biogeochemical process are those gathered according to the time-tested guidelines of ecosystem scientists. Field observations and mass balances represent the best available approximations of truth in biogeochemistry (Section VI.A). However, large-scale watershed data usually represent net changes in complex, open systems in which simple cause-and-effect linkages are difficult to decipher. Indeed, the intricate microscale interactions between individual biotic and abiotic field processes are often masked in data gathered in large-scale systems. Thus, ecosystem-level biogeochemical data may often fail to satisfy a desire of microbiologists for details of the processes of interest.
2. Many of the procedures for measuring specific biogeochemical processes described in Section IV.E and Table 7, especially those developed for application in agriculture, overlap with those utilized by ecosystem scientists. The stable isotopic procedures (both field release and those based on fractionation of naturally occurring isotopes; Table 7) provide reliable field data that are virtually artifact free yet successfully integrate microbially mediated fluxes of materials (e.g., carbon and nitrogen) in time and space. Similarly, the time-integrated approaches reliant on chemical analysis of field samples for determining both nutrient fluxes and in situ biodegradation (Section IV.E and Table 7) often provide accurate estimates of rates of biogeochemical processes in situ.

3. When the field approaches mentioned in points 1 and 2 fail to provide the required information, alternative field- or laboratory-based procedures are often required. In implementing these alternatives to in situ biogeochemical investigations that integrate microbial activity data in time and space, realism should be the goal that is relentlessly pursued. However, this goal of realism must be tempered by practical concerns of obtaining interpretable data.
4. Soil and environmental microbiology's Heisenberg uncertainty principle (described in Section III) prompts an investigator toward handling environmental samples using the rapid capture-and-fix snapshot procedures described in Sections IV.B and IV.E. These probably are the best way to minimize artifacts imposed both by removing environmental samples from field sites and by the propensity of naturally occurring microbial population to shift subsequent to sampling.
5. There are several variations of the snapshot approach to gathering data on the status of microorganisms in field sites. When the soil-cover procedure is employed (Section IV.E and Table 7), the snapshots capture and fix headspace samples of volatile gases whose flux is regulated by microbial activity. Analysis of the sequentially gathered headspace samples allows rates of in situ biogeochemical processes to be estimated. Although subject to some limitations (see Table 7 and Section IV.E), these rates are probably reasonably accurate.
6. However, in another variation of the snapshot approach, the entire environmental sample (rather than gaseous metabolites produced therefrom) is captured and fixed. This fixation step, while creating a snapshot of the status of the microbial community at the moment it was sampled, necessarily eliminates the dimension of time from additional consideration. Thus, rates of biogeochemical processes cannot be discovered from fixed environmental samples. The autoradiographic, nucleic acid hybridization, and immunofluorescence determinations that can be applied to fixed samples (see Sections IV and V) offer qualitative information about whether the process of interest was in progress at the time of sample fixation. But, unless reliable, elaborate calibration schemes are completed (which show quantitative, reproducible relationships between actual geochemical change and the concentration of metabolic intermediates, enzymes, or mRNA), the prospects of inferring in situ rates from these static measurements are dim. Even if snapshot assays could be conducted sequentially in time, heterogeneity of field sites would probably impede interpretation of data derived from different samples. Furthermore, some type of elaborate calibration procedure would still be required for inferring rates of activity.

7. In contrast to captured, fixed, and intact snapshot samples (see point 6 above), unfixed samples that are removed from field sites are perfectly amenable to laboratory incubation-based metabolic activity assays. These necessarily incorporate the dimension of time; thus, rate data are implicit when geochemical changes are monitored in environmental samples. The strength of performing geochemical assays on laboratory incubated environmental samples is that specific microbially mediated chemical changes are directly measured. Thus, the data can be unequivocal because experimental designs often include abiotic controls, and also because the need for using indirect biochemical or molecular probing surrogates (see point 6) is eliminated. The irony of laboratory metabolic activity assays is, of course, that the very characteristic that allows rate data to be obtained (i.e., time) also allows artifacts to develop (see Section III.C and Figure 2).
8. Perhaps the soundest approach for integrating in situ microbial activity data in time and space is to synthesize results from all of the appropriate items listed above (and all others available) into a consistent, coherent picture consisting of convergent lines of independent evidence.

VII. CONCLUSIONS AND FUTURE PROSPECTS

It is the privilege and responsibility of soil and environmental microbiologists to advance and steward knowledge of microbially mediated ecological processes. Yet despite more than a century of effort, soil and environmental microbiologists have been unable to definitively answer what might considered to be the most fundamental of questions.

1. What microorganisms are present in terrestrial and aquatic habitats that comprise the biosphere?
2. What are these microorganisms actually doing in situ—in the habitants where the microorganisms reside?
3. Can the microorganisms responsible for particular biogeochemical field processes be identified?
4. Can accurate measurements of in situ microbiological processes be integrated in time and space?

The motivations for posing each of these questions, the methodological obstacles to delivering unequivocal answers, and possible technological innovations for improving the answers have been discussed in the appropriately titled sections of this chapter. A summary of the state-of-the-art methods capable of providing answers to the above four questions appears below.

1. Recently developed procedure that do not require the cultivation of microorganisms, especially those based on 16S rRNA sequencing and PLFA analyses, have provided and will continue to provide major new insights into the composition of naturally occurring microbial communities (see Section IV.C). Although often passed over, novel and innovative methods of culturing microorganisms offer promise in this area as well. Despite marked progress, a complete census of any single microbiological habitat remains a distant goal.
2. Much progress toward understanding both the general in situ physiological status (Section IV.D) and the specific in situ biogeochemical process catalyzed by microorganisms (Section IV.E) has recently been made. Key advancements have been linked to new knowledge of the molecular and biochemical mechanisms of microbial processes. When probed using procedures from both analytical chemistry (e.g., GC, GC/MS, isotopic fractionation) and biochemistry (e.g., nucleic acid hybridization and immunolabeling), partial but unambiguous evidence describing the activities of microorganisms as they occur in field sites can be obtained.
3. The ability of soil and environmental microbiologists to attribute particular biogeochemical field processes to specific microbial cells is still limited (Section V). Nonetheless, new combinations of microscopy, immunolabeling, and nucleic acid probing offer avenues for approaching this evasive objective.
4. The key to integrating in situ biogeochemical processes in time and space is in understanding the accuracy of the methods used to determine rates of the respective processes. Ecosystem-level integration procedures are unquestionably accurate yet vague. Laboratory incubation of environmental samples provides detailed information, but this is often at the expense of realism. As recommended in Section VI.C, a multidisciplinary approach providing convergent lines of independent evidence is probably the best strategy for integrating microbiological processes in time and space.

To conclude this chapter, I would like briefly to address two issues that may be important in the future developments pertinent to determining the composition and in situ biogeochemical activities of soil microbial communities.

A. Exploring the Biochemical and Molecular Diversity of Microbiological Processes

Figure 4 provides an overview of the cellular processes responsible for biogeochemical changes in field sites. Each microbially mediated biogeochemical reaction is preceded by a series of specific intracellular events. Genes must be

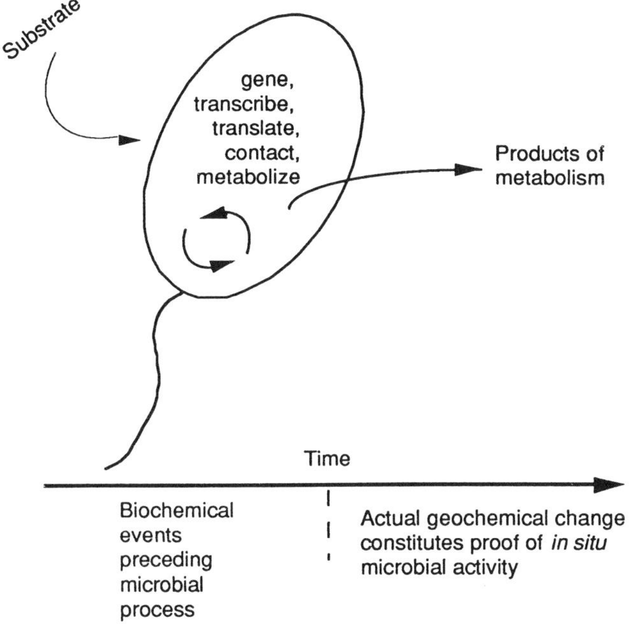

Figure 4 Generalized view of the sequence of biochemical events required for biogeochemical reactions. Intracellular events are essential for and precede the geochemical change.

present, transcription and translation must occur, enzymes and enzyme systems must contact their substrates, and the specific metabolic pathways for each substrate must be both carried through to completion and integrated into the cell's anabolic and catabolic physiological function. Because of the growing availability of information on nucleic acid sequences, on metabolic pathways, and on the enzymes responsible for each metabolic step, all three of these key parameters can theoretically be measured in field site-derived samples using the variety of analytical procedures (see Sections IV, V, and VI). Thus, to document qualitatively whether a given biogeochemical process of interest was in progress at the time that an environmental sample was fixed, all that one must do is analyze the cellular components in the sample (by extraction or microscopy) for process-specific biochemical indicators (mRNA, enzymes, or unique intermediary metabolites). This strategy for documenting in situ biogeochemical activity, in combination with historical mass balance and other site-derived measurements (Section VI.C), is likely to be increasingly influential in documenting in situ biogeochemical processes.

However, it is necessary to scrutinize at least one major assumption underlying the use of process-specific biochemical indicators for analyzing the in situ biogeochemical activity of microorganisms. Whenever such analyses are undertaken, one needs to know which genes, enzymes, and metabolites to seek. This type of sophisticated a priori information (gene sequence, enzymology, intermediary metabolite) is only available for a limited number of laboratory cultured microorganisms able to carry out a limited number of physiological processes. Although soil and environmental microbiologists have no choice but to use existing knowledge to discover new knowledge, it must be recognized that the universality of particular metabolic pathways and their encoding genes is uncertain. In this regard, Kluyver [277], Kluyver and van Neil [278], and Neidhardt et al. [98] have discussed the astonishing uniformity of biochemical mechanisms throughout the biological world. There is no doubt that nucleic acids are presently the dominant information storage motif for inheritance. Furthermore, significant alternatives to ATP as the universal cellular energy currency have not been discovered. For less universal metabolic functions, however, there is clear evidence that genes coding for particular processes have diverged and converged over evolutionary time [279–281]. Indeed, the aerobic biodegradation of toluene proceeds via at least five different metabolic pathways [282].

Thus, it becomes essential to ask how many metabolic pathways exist for a given biogeochemical process? It is important to ask this question because it reveals an additional limitation to molecular and biochemical approaches to solving age-old methodological dilemmas in environmental microbiology (see Section III). The answer to this question about metabolic diversity will gradually unfold as comparative studies examining newly isolated microorganisms and their genetic, physiological, thermodynamic, and molecular reaction capabilities are completed. Rigorous discussion about the manifestations of evolutionary pressures and both convergence and divergence in physiological pathways [283] are beyond the scope of this chapter. It will suffice to state simply that progress made in the area of the diversity of metabolic mechanisms will almost surely have major implications for both advancing the discipline of molecular phylogeny and for interpreting data derived from captured and fixed environmental samples.

B. Real-Time Analysis of In Situ Microbial Metabolism in Soil

The term real-time analysis has its origin in data processing and computer-controlled systems. The McGraw-Hill Dictionary of Scientific and Technical Terms [284] defines real-time as "pertaining to a data processing system that controls an on-going process and delivers its outputs (or controls its inputs) no later than the time when these are needed for effective control." Real time is a commonly used term in chemical engineering [285–287] and analytical chemistry [288–290], but it also has applications in wastewater analysis [291] and even in microbial behavior [292].

Regarding future developments in soil and environmental microbiology, it seems prudent to speculate on the possibility of developing systems for real-time analysis of soil microbial communities. The notion of real-time analysis of in situ microbial metabolism in soil is designed to be provocative; to formally recognize both the dynamic aspect of processes catalyzed by microorganisms in situ and the practical daily impact of such processes on the biosphere. The services rendered by soil microbial communities are highly diverse and essential for continued functioning of the biosphere. As these services (e.g., cycling of nitrogen and carbon, pollutant detoxification) become increasingly critical for protecting both the health of human beings and for supporting ecosystems, it may become increasingly important to refine, understand, monitor, and manage these processes in real time. Indeed, it is clear from three recent symposia [53,293,294] that the concept of in situ real-time analysis has come of age in soil science. However, the primary emphasis in these symposia was soil-water relationships, rather than the microbial activities in soil discussed here. Real-time analysis of soil microbial communities should aspire toward knowing (1) the physiological potential of laboratory-incubated microorganisms contained in environmental samples (soil, water, sediments); (2) the rich, detailed rules of population genetics, molecular biology, biochemistry, and evolution that underlie that physiological potential; (3) the strategies for demonstrating that the physiological potential, easily measured in the laboratory, is expressed in situ; (4) the composition of soil microbial communities; (5) the particular members of soil populations that are responsible for specific metabolic processes in situ; and (6) how the biogeochemical activities of microorganisms in situ are expressed over widely varying dimensional units of time and space.

Obviously, achieving an ability to analyze the in situ composition and biogeochemical activities of soil microorganisms in real time will depend on significant methodological advancements. In this regard, Paul [18] whimsically proposed the use of a satellite-based "tricorder" that needs only be pointed at an environment of interest to obtain the relative abundances and taxonomic identities of microorganisms, their metabolic rates, and the rates of ecological and geochemical functions that they catalyze. Without dreams, we would be lost.

ACKNOWLEDGEMENTS

The author would like to express his gratitude to W. C. Ghiorse, J. B. Yavitt, J. B. Herrick, and S. H. Zinder for stimulating insightful discussions pertinent to microscopy, field measurements, molecular biology, and prokaryotic diversity, respectively. Patience and expertise in manuscript preparation on the part of P. Lisk is greatly appreciated. During the time that this manuscript was in preparation, research in the author's laboratory was supported by a grant from the

Cornell Biotechnology Program which is sponsored by the New York State Science and Technology Foundation (grant no. NYS CAT 92054), a consortium of industries, and the National Science Foundation. Additional support was provided by the Air Force Office of Scientific Research (grant nos. AFOSR-91-0436, F49620-93-1-0414, and F49620-95-1-0346), Merck and Co., Inc. (grant no. PY-641308), the New York State Center for Hazardous Waste Management, and the Hughes Foundation Undergraduate Scholars Program (awarded by the Howard Hughes Medical Institute to the Division of Biological Sciences, Cornell University).

REFERENCES

1. Alexander, M. 1977. Introduction to soil microbiology, 2nd ed. John Wiley & Sons, New York.
2. Atlas, R. M., and R. Bartha. 1993. Microbial ecology: fundamentals and applications, 3rd ed. Benjamin Cummings, Menlo Park, California.
3. Madsen, E. L. 1991. Determining in situ biodegradation: facts and challenges. Environ. Sci. Technol. 25:1662–1673.
4. Paul, E. A., and F. E. Clark (eds.). 1989. Soil Microbiology and biochemistry. Academic Press, San Diego.
5. Stolp, H. 1988. Microbial ecology: organisms, habitats, activities. Cambridge University Press, New York.
6. Waksman, S. A. 1952. Soil microbiology. John Wiley & Sons, New York.
7. Waksman, S. A. 1945. Soil microbiology as a field of science. Science 102:339–344.
8. Brock, T. D. 1980. Environmental biogeochemistry, p. 93–103. *In* Biogeochemistry of ancient and modern environments. Proc. Fourth Int. Symp. Environ. Biogeochem. (ISEB). Australian Academy of Sciences, Canberra, Australia.
9. Schlegel, H. G., and H. W. Jannasch. 1992. Prokaryotes and their habitats, p. 75–125. *In* A. Balows, H. G. Trüper, M. Dworkin, W. Harder, and K.-H. Schleifer (eds.), The Prokaryotes, 2nd ed. Springer-Verlag, New York.
10. Zehnder, A. J. B., and W. Stumm. 1988. Geochemistry and biogeochemistry of anaerobic habitats, p. 1–38, *In* A. J. B. Zehnder (ed.), Biology of anaerobic microorganisms. John Wiley & Sons, New York.
11. Alexander, M. 1961. Introduction to soil microbiology. John Wiley & Sons, New York.
12. Alexander, M. 1971. Microbial ecology. John Wiley & Sons, New York.
13. Brock, T. D. 1966. Principles of microbial ecology. Prentice Hall, Englewood Cliffs, New Jersey.
14. Brock, T. D., M. T. Madigan, J. M. Martinko, and J. Parker. 1994. Biology of microorganisms, 7th ed. Prentice Hall, Englewood Cliffs, New Jersey.
15. Hobbie, J. E. 1993. Introduction, p. 1–5. *In* P. F. Kemp, B. F. Sherr, E. B. Sherr, and J. J. Cole (eds.), Handbook of methods in aquatic microbial ecology. Lewis Publishers Boca Raton, Florida.

16. Hobbie, J. E., and T. E. Ford. 1993. A perspective on the ecology of aquatic microbes, p. 1–14. *In* T. E. Ford (ed.), Aquatic Microbiology. Blackwell Scientific, Boston.
17. Pace, N. R., D. A. Stahl, D. J. Lane, and G. J. Olsen. 1986. The analysis of natural microbial populations by ribosomal RNA sequences. Adv. Microbial Ecol. 9:1–55.
18. Paul, J. H. 1993. The advances and limitations of methodology, p. 15–46. *In* T. E. Ford (ed.), Aquatic microbiology. Blackwell Scientific, Boston.
19. Mitchell, R. (ed.) 1992. Environmental microbiology. Wiley-Liss, New York.
20. Woese, C. R. 1992. Prokaryotic systematics: the evolution of a science, p. 3–18. *In* A. Balows, H. G. Trüper, M. Dworkin, W. Harder, and K.-H. Schleifer (eds.), The prokaryotes, 2nd ed. Springer-Verlag, New York.
21. Winogradsky, S. 1949. Microbiologie du sol, problémes et méthodes; cinquante ans de recherches. Oeuvres completes. Masson, Paris.
22. Waksman, S. A. 1925. Soil microbiology in 1924: an attempt at an analysis and a synthesis. Soil Sci. 19:201–249.
23. Löhnis, F. 1925. Bacterial nitrogen fixation. Agron. J. 17:445–450.
24. Conn, H. W., and H. J. Conn. 1926. Bacteriology. A study of microorganisms and their relation to human welfare, 3rd ed. Williams and Wilkins, Baltimore, Maryland.
25. Pochon, J. 1954. Manuel technique d'analyse microbiologique du sol [par]. J. Pochon avec la collaboration de J. Augier [et al.]. Monigraphies de I'Institut Pasteur, Masson, Paris.
26. Omelianskii, V. L. 1953. Izbrannye trudy. Moskva, Izd-vo Akademii nauk SSSR. Institut Mikrobiologie.
27. Russell, E. W. 1988. Soil conditions and plant growth, 11th ed. Longman, Essex, UK, and John Wiley & Sons, New York.
28. Allison, F. E. 1973. Soil organic matter and its role in crop production. Elsevier, Amsterdam, and New York.
29. Bartholomew, W. V., and F. E. Clark (eds.) 1965. Soil nitrogen. American Society of Agronomy, Madison, Wisconsin.
30. van Niel, C. B. 1967. The education of a microbiologist: some reflections. Ann. Rev. Microbiol. 21:1–30.
31. Waksman, S. A. 1927. Principles of soil microbiology. Williams & Wilkins, Baltimore, Maryland.
32. Fred, E. B., I. L. Baldwin, and E. McCoy. 1932. Root nodule bacteria and leguminous plants. University of Wisconsin Studies in Science No. 5. University of Wisconsin, Madison.
33. Macura, J. 1974. Trends and advances in soil microbiology from 1924 to 1974. Geoderma 12:311–329.
34. Brock, T. D. 1961. Milestone in microbiology. Prentice-Hall, Englewood Cliffs, New Jersey.
35. Lechevalier, H. A., and M. Solotorovsky. 1965. Three centuries of microbiology. McGraw-Hill, New York.
36. Woese, C. R. 1987. Bacterial evolution. Microbiol. Rev. 51:221–271.
37. Amann, R., N. Springer, W. Ludwig, H.-D. Görtz, and K.-H. Schleifer. 1991. Identification in situ and phylogeny of uncultured bacterial endosymbionts. Nature 351:161–164.

38. Amann, R. I., W. Ludwig, and K.-H. Schleifer. 1995. Phylogenetic identification and in situ detection of individual microbial cells without cultivation. Microbiol. Rev. 59:143–169.

39. Olsen, G. J., D. J. Lane, S. J. Giovannoni, and N. R. Pace. 1986. Microbial ecology and evolution: a ribosomal RNA approach. Annu. Rev. Microbiol. 40:337–365.

40. Ward, D. M., M. M. Bateson, R. Weller, and A. L. Ruff-Roberts. 1993. Ribosomal RNA analysis of microorganisms as they occur in nature. Adv. Microb. Ecol. 12: 219–286.

41. Sayler, G. S., K. Nikbakht, J. T. Fleming, and J. Packard. 1992. Application of molecular techniques to soil biochemistry, p. 131–172. *In* G. Stotzky and J.-M. Bollag (eds.), Soil Biochemistry, Vol. 7. Marcel Dekker, New York.

42. Carter, J. P., and J. M. Lynch. 1993. Immunological and molecular techniques for studying the dynamic of microbial populations and communities in soil, pp. 249–272. *In* J.-M. Bollag and G. Stotzky (eds.), Soil biochemistry, Vol. 8. Marcel Dekker, New York.

43. Nannipieri, P., S. Grego, and B. Ceccanti. 1990. Ecological significance of the biological activity in soil, pp. 293–356. *In* J.-M. Bollag and G. Stotzky (eds.), Soil biochemistry, Vol. 6. Marcel Dekker, New York.

44. Costerton, J. W., Z. Lewandowski, D. DeBeer, D. Caldwell, D. Korber, and G. James. 1994. Biofilms, the customized microniche. J. Bacteriol. 176:2137–2142.

45. Elsgaard, L., D. Prieur, G. M. Mukwaya, and B. B. Jørgensen. 1994. Thermophilic sulfate reduction in hydrothermal sediment of Lake Tanganyika, East Africa. Appl. Environ. Microbiol. 60:1473–1480.

46. Revsbech, N.P., and B. B. Jørgensen. 1986. Microelectrodes: their use in microbial ecology. Adv. Microb. Ecol. 9:293–352.

47. Groffman, P. M. 1985. Nitrification and denitrification in conventional and no-tillage soils. Soil Sci. Soc. Am. J. 49:329–334.

48. Groffman, P. M. 1991. Ecology of nitrification and denitrification in soil evaluated at scales relevant to atmospheric chemistry, p. 201–217. *In* J. Rogers and W. B. Whitman (eds.), Microbial production and consumption of greenhouse gases: methane, nitrogen oxides and halomethanes. American Society for Microbiology, Washington, D.C.

49. Groffman, P. M. 1993. Soil microbiology: Contributions from the gene to the global scale, p. 19–26. *In* Agriculture Research in the northeastern United States: critical review and future perspectives. American Society of Agronomy, Madison, Wisconsin.

50. Groffman, P. M., J. M. Tiedje, D. L. Mokma, and S. Simkins. 1992. Regional scale analysis of denitrification in north temperate forest soils. Landscape Ecol. 7:45–53.

51. Parkin, T. B. 1993. Spatial variability of microbial processes in soil—a review. J. Environ. Qual. 22:409–417.

52. Tiedje, J. M., S. Simkins, and P. M. Groffman. 1989. Perspectives on measurement of denitrification in the field including recommended protocols for acetylene based methods. Plant Soil 115:261–284.

53. Bouma, J. 1993. Soil behavior under field conditions: differences in perception and their effects on research. Geoderma 60:1–14.

54. Buyanovsky, G. A., G. H. Wagner, and C. J. Gantzer. 1986. Soil respiration in a winter wheat ecosystem. Soil Sci. Soc. Amer. J. 50:338–344.

55. Clark, F. E. 1973. Problems and perspectives in microbial ecology, p. 13–16. *In* T. Rosswall (ed.), Modern methods in the study of microbial ecology, Vol. 17. Bull. Ecol. Res. Comm. Stockholm, Sweden.

56. Brock, T. D. 1987. The study of microorganisms in situ: progress and problems, p. 1–17. *In* Ecology of microbial communities. Forty-first Symposium of the Society for General Microbiology, University of St. Andrews. Cambridge University Press, New York.

57. Alexander, M. 1991. Soil microbiology in the next 75 years: fixed, flexible, or mutable? Soil Sci. 151:35–40.

58. Ferguson, R. L., E. N. Buckley, and A. V. Palumbo. 1984. Response of marine bacterioplankton to differential filtration and confinement. Appl. Environ. Microbiol. 47:49–55.

59. Lee, S., and J. A. Fuhrman. 1991. Spatial and temporal variation of natural bacterioplankton assemblages studied by total genomic DNA cross-hybridization. Limnol. Oceanogr. 36:1277–1287.

60. Roszak, D. B., and R. R. Colwell. 1987. Survival strategies of bacteria in the natural environment. Microbiol. Rev. 51:365–379.

61. Staley, J. T., and A. Konopka. 1985. Measurement of in situ activities of nonphotosynthetic microorganisms in aquatic and terrestrial habitats. Annu. Rev. Microbiol. 39:321–346.

62. Ward, D. M., R. Weller, and M. M. Bateson. 1990. 16S rRNA sequences reveal numerous uncultured microorganisms in a natural community. Nature 345:63–65.

63. Fuhrman, J. A., and F. Azam. 1982. Thymidine incorporation as a measure of heterotrophic bacterioplankton production in marine surface waters: evaluation and field results. Mar. Biol. 66:109–120.

64. Henis, Y. (ed.). 1987. Survival and dormancy of microorganisms. John Wiley & Sons, New York.

65. Hoppe, H.-G. 1978. Relationships between active bacteria and heterotrophic potential in the sea. Neth. J. Sea Res. 12:78–98.

66. Magariños, B., J. L. Romalde, J. L. Barja, and A. E. Toranzo. 1994. Evidence of a dormant but ineffective state of the fish pathogen *Pasteurella piscicida* in seawater and sediment. Appl. Environ. Microbiol. 60:180–186.

67. Nilsson, L., J. D. Oliver, and S. Kjelleberg. 1991. Resuscitation of *Vibrio vulnificus* from the viable but nonculturable state. J. Bacteriol. 173:5054–5059.

68. Brock, T. D. 1978. The poisoned control in biogeochemical investigations. p. 717–726. *In* W. V. Krumbein (ed.), Environmental biogeochemistry and geomicrobiology, Vol. 3. Ann Arbor Science Publishers, Ann Arbor, Michigan.

69. Kelly, G. A., and D. P. Chynoweth. 1980. Comparison of in situ and in vitro rates of methane release in freshwater sediments. Appl. Environ. Microbiol. 40:287–293.

70. Bull, A. T. 1980. Biodegradation: some attitudes and strategies of microorganisms and microbiologists, p. 107–136. *In* D. C. Ellwood et al. (eds.), Contemporary microbial ecology. Academic Press, London.

71. Karl, D. M. 1986. Determination of in situ microbial biomass, viability, metabolism, and growth, p. 85–176. *In* J. S. Poindexter and E. R. Leadbetter (eds.), Bacteria in nature. Plenum Press, New York.
72. Castellan, G. W. 1983. Physical chemistry, 3rd ed. Addison-Wesley, Reading, Massachusetts.
73. Gibbs, C. F. 1976. Methods and interpretation in measurement of oil biodegradation rate, p. 127–140. *In* J. M. Sharpley and A. M. Kaplan (eds.), Proceedings of the third international biodegradation symposium. Applied Science, London.
74. Madsen, E. L., and W. C. Ghiorse. 1993. Ground water microbiology: subsurface ecosystem processes, p. 167–213. *In* Ford, T. (ed.), Aquatic microbiology: an ecological approach. Blackwell Scientific, Cambridge, Massachusetts.
75. Zumdahl, S. S. 1986. Chemistry. D. C. Heath, Lexington, Massachusetts.
76. Aldrich, H. C., and H. H. Mollenhauer. 1986. Secrets of successful embedding, sectioning, and imaging, p. 101–132. *In* H. C. Aldrich and W. J. Todd (eds.), Ultrastructure techniques for microorganisms. Plenum Press, New York.
77. Caplan, A. L. 1987. Why the problem of reductionism in biological science will not go away. Growth 51:22–34.
78. Krebs, C. J. 1986. Ecology, 3rd ed. Harper Collins, New York.
79. Levins, R., and R. Lewontin. 1985. The dialectical biologist. Harvard University Press, Cambridge.
80. Ryden, J. C., J. H. Skinner, and D. J. Nixon. 1987. Soil core incubation system for the field measurement of denitrification using acetylene-inhibition. Soil Biol. Biochem. 19:753–757.
81. Beare, M. H., P. F. Hendrix, and D. C. Coleman. 1994. Water-soluble aggregates and organic matter fractions in conventional- and no-tillage soils. Soil. Sci. Soc. Amer. J. 58:777–786.
82. Brady, N. C. 1990. The nature and properties of soil, 10th ed. MacMillan, New York.
83. van Veen, J. A., and E. A. Paul. 1981. Organic carbon dynamics in grassland soil. I. Background information and computer simulations. Can. J. Soil. Sci. 61:185–201.
84. Waksman, S. A., and R. L. Starkey. 1931. The soil and the microbe: an introduction to the study of the microscopic population of the soil and its role in soil processes and plant growth. John Wiley & Sons, New York.
85. Elmholt, S. 1992. Effect of propiconazole on substrate amended soil respiration following laboratory and field application. Pestic. Sci. 34:139–146.
86. Parkin, T. B., H. F. Kaspar, A. J. Sexstone, and J. M. Tiedje. 1984. A gas-flow soil core method to measure field denitrification rates. Soil Biol. Biochem. 16:323–330.
87. Petersen, S. O., and M. J. Klug. 1994. Effects of sieving, storage, and incubation temperature on the phospholipid fatty acid profile of a soil microbial community. Appl. Environ. Microbiol. 60:2421–2430.
88. Ross, D. J., K. R. Tate, A. Cairns, and K. F. Meyrick. 1980. Influence of storage on soil microbial biomass estimated by three biochemical procedures. Soil Biol. Biochem. 12:369–374.
89. Topp, E., W. N. Smith, W. D. Reynolds, and S. U. Khan. 1994. Atrazine and metolachlor dissipation in soils incubated in undisturbed cores, repacked cores, and flasks. J. Environ. Qual. 23:693–700.

90. Venrick, E. L., J. R. Beers, and J. F. Heinbokel. 1977. Possible consequences of containing microplankton for physiological rate measurements. J. Exp. Mar. Biol. Ecol. 26:55–76.
91. Visser, S., J. Fujikawa, C. L. Griffiths, and D. Parkinson. 1984. Effect of topsoil storage on microbial activity, primary production and decomposition potential. Plant Soil 82:41–50.
92. Phelps, T. J., E. M. Murphy, S. M. Pfiffner, and D. C. White. 1994. Comparison between geochemical and biological estimates of subsurface microbial activities. Microb. Ecol. 28:335–349.
93. Kim, H., H. F. Hemond, L. R. Krumholz, and B. A. Cohen. 1995. In-situ biodegradation of toluene in a contaminated stream. 1. Field studies. Environ. Sci. Technol. 29:108–116.
94. Cohen, B. A., L. R. Krumholz, H. Kim, and H. F. Hemond. 1995. In-situ biodegradation of toluene in a contaminated stream. 2. Laboratory studies. Environ. Sci. Technol. 29:117–125.
95. Kieft, T. L., E. Soroker, and M. K. Firestone. 1987. Microbial biomass response to a rapid increase in water potential when dry soil is wetted. Soil Biol. Biochem. 19: 119–126.
96. Orchard, V. A., and F. J. Cook. 1983. Relationships between soil respiration and soil moisture. Soil Biol. Biochem. 15:447–454.
97. Winding, A., S. J. Binnerup, and J. Sørensen. 1994. Viability of indigenous soil bacteria, assayed by respiratory activity and growth. Appl. Environ. Microbiol. 60: 2869–2875.
98. Neidhardt, F. C., J. I. Ingraham, and M. Schaechter. 1990. Physiology of the bacterial cell. Sinauer, Sunderland, Massachusetts.
99. O'Connor, L. T., and D. C. Savage. 1993. Characterization of an activity from the strict anaerobe *Roseburia cecicola* that degrades DNA when exposed to air. J. Bacteriol. 175:4681–4687.
100. Morris, J. G. 1975. The physiology of obligate anaerobiosis, p. 169–246. *In* A. H. Rose and D. W. Tempest (eds.), Advances in microbial physiology, Vol. 12. Academic Press, New York.
101. Tunlid, A., and D. C. White. 1992. Biochemical analysis of biomass, community structure, nutritional status, and metabolic activity of microbial communities in soil, p. 229–262. *In* G. Stotzky, and J.-M. Bollag (eds.), Soil biochemistry, Vol. 7. Marcel Dekker, New York.
102. Wieder, R. K., J. B. Yavitt, and G. E. Lang. 1990. Methane production and sulfate reduction in two Appalachian peatlands. Biogeochem. 10:81–104.
103. Howarth, R. W. 1993. Microbial processes in salt-marsh sediments, p. 239–259. *In* T. E. Ford (ed.), Aquatic Microbiology. Blackwell Scientific, Boston.
104. Schnabel, R. R., and W. L. Stout. 1994. Denitrification loss from two Pennsylvania floodplain soils. J. Environ. Qual. 23:344–348.
105. Groffman, P. M., C. W. Rice, and J. M. Tiedje. 1993. Denitrification in a tallgrass prairie landscape. Ecology 74:855–862.
106. Sexstone, A. J., T. B. Parkin, and J. M. Tiedje. 1985. Temporal response of soil denitrification rates to rainfall and irrigation. Soil Sci. Soc. Amer. J. 49:99–103.
107. Bramley, R. G. V., and R. E. White. 1990. The variability of nitrifying activity in field soils. Plant Soil 126:203–208.

108. Keller, M., M. E. Mitre, and R. F. Stallard. 1990. Consumption of atmospheric methane in soils of Central Panama: effects of agricultural development. Global Biogeochem. Cycles 4:21–27.
109. Oremland, R. S., N. A. Steinberg, A. S. Maest, L. G. Miller, and J. T. Hollibaugh. 1990. Measurement of in situ rates of selenate removal by dissimilatory bacterial reduction in sediments. Environ. Sci. Technol. 24:1157–1164.
110. Johnson, J. L. 1991. Isolation and purification of nucleic acids, p. 1–19. *In* E. Stackbrandt and M. Goodfellow (eds.), Nucleic acid techniques in bacterial systematics. John Wiley & Sons, New York.
111. Neidhardt, F. C., J. L. Ingraham, K. B. Low, B. Magasanik, M. Schaechter, and H. E. Umbarger (eds.). 1987. *Escherichia coli* and *Salmonella typhimurium:* cellular and molecular biology, Vols. 1 and 2. American Society for Microbiology, Washington, D.C.
112. Stryer, L. 1988. Biochemistry, 3rd ed. W. H. Freeman, New York.
113. Moriarty, D. J. W. 1990. Techniques for estimating bacterial growth rates and production of biomass in aquatic environments, p. 211–234. *in* T. Grigorova and J. R. Norris (eds.) Techniques in microbial ecology: Methods in microbiology, Vol. 22. Academic Press, New York.
114. Chapelle, F. H. 1993. Ground-water microbiology and geochemistry. John Wiley & Sons, New York.
115. Larson, E. E., and P. W. Birkeland, 1982. Putnam's geology, 4th ed. Oxford University Press, New York.
116. Austin, B. 1988. Methods in aquatic bacteriology. John Wiley & Sons, New York.
117. Kemp, P. F., B. F. Sherr, E. B. Sherr, and J. J. Cole (eds.). 1993. Handbook of methods in aquatic microbial ecology. Lewis Publishers, Boca Raton, Florida.
118. Litchfield, C. D. (ed.). 1976. Marine microbiology. Dowden, Hutchinson & Ross, Stroudsburg, Pennsylvania.
119. Zobell, C. 1946. Marine microbiology. Chronica Botanica, Waltham, Massachusetts.
120. Geesey, G. G., and D. C. White. 1990. Determination of bacterial growth and activity at solid-liquid interfaces. Annu. Rev. Microbiol. 44:579–602.
121. Hurst, C., G. Knudsen, M. McInerney, L. Stetzenbach, and M. Walter (eds.) 1996. Manual of environmental microbiology. American Society for Microbiology, Washington, D.C.
122. Grigorova, R., and J. R. Norris (eds.). 1990. Techniques in microbial ecology. Methods in microbiology, Vol. 22. Academic Press, New York.
123. Burlage, R. S. (ed.). 1996. Techniques in microbial ecology. Oxford University Press, New York.
124. Weaver, R. W., S. Angle, P. Bottomley, D. Bezdicek, S. Smith, A. Tabatabai, and A. Wollum (eds.). 1994. Methods of soil analysis, Part 2: Microbiological and biochemical properties. Soil Science Society of America, Madison, Wisconsin.
125. Pepper, I. L., C. P. Gerba, and J. W. Brendecke. 1995. Environmental Microbiology: a laboratory manual. Academic Press, San Diego.
126. Alef, K., and P. Nannipieri (eds.). 1995. Methods in applied soil microbiology and biochemistry. Academic Press, San Diego.
127. Tate, R. L. 1995. Soil microbiology. John Wiley & Sons, New York.

128. Parkinson, D., T. R. G. Gray, and S. T. Williams. 1971. Methods for studying the ecology of soil micro-organisms. Blackwell Scientific, Oxford, U.K.

129. Pramer, D., and E. L. Schmidt. 1964. Experimental soil microbiology. Burgess Publishing, Minneapolis, Minnesota.

130. Rosswall, T. 1973. Modern methods in the study of microbial ecology, Vol. 17. Bull. Ecol. Res. Comm. Stockholm, Sweden.

131. Bottomley, P. J. 1994. Light microscopic methods for studying soil microorganisms, p. 81–105. *In* R. W. Weaver, S. Angle, P. Bottomley, D. Bezdicek, S. Smith, A. Tabatabai, and A. Wollum (eds.), Methods of soil analysis, Part 2. Soil Science Society of America, Madison, Wisconsin.

132. Schmidt, E. L., and E. A. Paul. 1982. Microscopic methods for soil microorganisms. p. 803–814. *In* A. L. Page, R. H. Miller, and D. R. Keeney (eds.), Methods of soil analysis, Part 2. American Society of Agronomy, Madison, Wisconsin.

133. Turley, C. M. 1993. Direct estimates of bacterial numbers in seawater samples without incurring cell loss due to sample storage, p. 143–148. *In* P. F. Kemp, B. F. Sherr, E. B. Sherr, and J. J. Cole, Handbook of methods in aquatic microbial ecology. Lewis Publishers, Boca Raton, Florida.

134. McFeters, G. A., F. P. Yu, B. H. Pyle, and P. S. Stewart. 1995. Physiological assessment of bacteria using fluorochromes. J. Microbiol. Meth. 21:1–13.

135. Schmidt, E. L., and R.O. Bankole. 1962. Detection of Aspergillus flavus in soil by immunofluorescent staining. Science 136:776–777.

136. Bohlool, B. B., and E. L. Schmidt. 1973. A fluorescent antibody technique for determination of growth rates of bacteria in soil. Bull. Ecol. Res. Comm. (Stockholm) 17:67–76.

137. Bohlool, B. B., and E. L. Schmidt. 1973. Persistence and competition aspects of *Rhizobium japonicum* observed in soil by immunofluorescence microscopy. Soil Sci. Soc. Am. Proc. 37:561–564.

138. Campbell, L. 1993. Immunofluorescence method for the detection and characterization of marine microbes, p. 295–302. *In* P. F. Kemp, B. F. Sherr, E. B. Sherr, and J. J. Cole, Handbook of methods in aquatic microbial ecology. Lewis Publishers, Boca Raton, Florida.

139. Hoff, K. A. 1993. Total and specific bacterial counts by simultaneous staining with DAPI and fluorochrome-labeled antibodies, p. 149–154. *In* P. F. Kemp, B. F. Sherr, E. B. Sherr, and J. J. Cole, Handbook of methods in aquatic microbial ecology. Lewis Publishers, Boca Raton, Florida.

140. Schmidt, E. L. 1974. Quantitative autoecological study of microorganisms in soil by immunofluorescence. Soil Sci. 118:141–149.

141. Amann, R. I., B. J. Binder, R. J. Olson, S. W. Chisholm, R. Devereux, and D. A. Stahl. 1990. Combination of 16S rRNA-targeted oligonucleotide probes with flow cytometry for analyzing mixed microbial populations. Appl. Environ. Microbiol. 56: 1919–1925.

142. Amann, R. I., B. Zarda, D. A. Stahl, and K.-H. Schleifer. 1992. Identification of individual prokaryotic cells by using enzyme-labeled, rRNA-targeted oligonucleotide probes. Appl. Environ. Microbiol. 58:3007–3011.

143. DeLong, E. F. 1993. Single cell identification using fluorescently labeled, ribosomal RNA-specific probes, p. 285–294. *In* P. F. Kemp, B. F. Sherr, E. B. Sherr, and J. J.

Cole, Handbook of methods in aquatic microbial ecology. Lewis Publishers, Boca Raton, Florida.

144. DeLong, E. F., G. S. Wickham, and N. R. Pace. 1989. Phylogenetic stains: ribosomal RNA-based probes for the identification of single cells. Science 243:1360–1363.
145. Giovannoni, S. J., E. F. DeLong, G. J. Olsen, and N. R. Pace. 1988. Phylogenetic group-specific oligodeoxynucleotide probes for identification of single microbial cells. J. Bacteriol. 170:720–726.
146. Hahn, D., R. I. Amann, W. Ludwig, A. D. L. Akkermans, and K.-H. Schleifer. 1992. Detection of micro-organisms in soil after in situ hybridization with rRNA-targeted, fluorescently labelled oligonucleotides. J. Gen. Microbiol. 138:879–887.
147. Giovannoni, S. J., T. B. Britschgi, C. L. Moyer, and K. G. Field. 1990. Genetic diversity in Sargasso Sea bacterioplankton. Nature 345:60–65.
148. Moré, M. I., J. B. Herrick, M. C. Silva, W. C. Ghiorse, and E. L. Madsen. 1994. Quantitative cell lysis of indigenous microorganisms and rapid extraction of microbial DNA from sediment. Appl. Environ. Microbiol. 60:1572–1580.
149. Ruff-Roberts, A. L., J. G. Kuenen, and D. M. Ward. 1994. Distribution of cultivated and uncultivated cyanobacteria and *Chloroflexus*-like bacteria in hot spring microbial mats. Appl. Environ. Microbiol. 60:697–704.
150. Weller, R., and D. M. Ward. 1989. Selective recovery of 16S rRNA sequences from natural microbial communities in the form of cDNA. Appl. Environ. Microbiol. 55: 1818–1822.
151. Atlas, R. M., A. Horowitz, M. Krichevsky, and A. K. Bej. 1991. Response of microbial populations to environmental disturbance. Microb. Ecol. 22:249–256.
152. Holben, W. E. 1994. Isolation and purification of bacterial DNA from soil, p. 727–752. *In* R. W. Weaver, S. Angle, P. Bottomley, D. Bezdicek, S. Smith, A. Tabatabai, and A. Wollum (eds.), Methods of soil analysis, Part 2. Soil Science Society of America, Madison, Wisconsin.
153. Herrick, J. B., E. L. Madsen, C. A. Batt, and W. C. Ghiorse. 1993. Polymerase chain reaction amplification of naphthalene-catabolic and 16S rRNA gene sequences from indigenous bacteria. Appl. Environ. Microbiol. 59:687–694.
154. Lee, S., and J. A. Fuhrman. 1990. DNA hybridization to compare species compositions of natural bacterioplankton assemblages. Appl. Environ. Microbiol. 56:739–746.
155. Ogram, A. V., and D. F. Bezdicek. 1994. Nucleic acid probes, p. 665–687. *In* R. W. Weaver, S. Angle, P. Bottomley, D. Bezdicek, S. Smith, A. Tabatabai, and A. Wollum (eds.), Methods of soil analysis, Part 2. Soil Science Society of America, Madison, Wisconsin.
156. Moyer, C. L., F. C. Dobbs, and D. M. Karl. 1994. Estimation of diversity and community structure through restriction fragment length polymorphism distribution analysis of bacterial 16S rRNA genes from a microbial mat at an active, hydrothermal vent system, Loihi Seamount, Hawaii. Appl. Environ. Microbiol. 60:871–879.
157. Pepper, I. L., and S. D. Pillai. 1994. Detection of specific DNA sequences in environmental samples via polymerase chain reaction, p. 707–725. *In* R. W. Weaver, S. Angle, P. Bottomley, D. Bezdicek, S. Smith, A. Tabatabai, and A. Wollum (eds.),

Methods of soil analysis, Part 2. Soil Science Society of America, Madison, Wisconsin.

158. Torsvik, V., K. Salte, R. Sørheim, and J. Goksøyr. 1990. Comparison of phenotypic diversity and DNA heterogeneity in a population of soil bacterial. Appl. Environ. Microbiol. 56:776–781.

159. Torsvik, V., J. Goksøyr, and F. L. Daae. 1990. High diversity in DNA of soil bacteria. Appl. Environ. Microbiol. 56:782–787.

160. Findlay, R. H., and F. C. Dobbs. 1993. Quantitative description of microbial communities using lipid analysis, p. 271–284. *In* P. F. Kemp, B. F. Sherr, E. B. Sherr, and J. J. Cole (eds.), Handbook of methods in aquatic microbial ecology. Lewis Publishers, Boca Raton, Florida.

161. White, D.C. 1988. Validation of quantitative analysis for microbial biomass, community structure, and metabolic activity. Adv. Limnol. 31:1–18.

162. Zelles, L., Q. Y. Bai, T. Beck, and F. Beese. 1992. Signature fatty acids in phospholipids and lipopolysaccharides as indicators of microbial biomass and community structure in agricultural soils. Soil Biol. Biochem. 24:317–323.

163. Buchanan, M., and L. D. King. 1992. Seasonal fluctuations in soil microbial biomass carbon, phosphorus, and activity in no-till and reduced-chemical-input maize agroecosystems. Biol. Fertil. Soils 13:211–217.

164. Alexander, M. 1982. Most probable number method for microbial populations, p. 815–820. *In* A. L. Page, R. H. Miller, and D. R. Keeney (eds.), Methods of soil analysis, Part 2. American Society of Agronomy, Madison, Wisconsin.

165. Wollum, A. G., II. 1982. Cultural methods for soil microorganisms, p. 781–802. *In* A. L. Page, R. H. Miller, and D. R. Keeney (eds.), Methods of soil analysis, Part 2. American Society of Agronomy, Madison, Wisconsin.

166. Woomer, P. L. 1994. Most probable number counts, p. 59–79. *In* R. W. Weaver, S. Angle, P. Bottomley, D. Bezdicek, S. Smith, A. Tabatabai, and A. Wollum (eds.), Methods of soil analysis, Part 2. Soil Science Society of America, Madison, Wisconsin.

167. Zuberer, D. A. 1994. Recovery and enumeration of viable bacteria, 119–143. *In* R. W. Weaver, S. Angle, P. Bottomley, D. Bezdicek, S. Smith, A. Tabatabai, and A. Wollum (eds.), Methods of soil analysis, Part 2. Soil Science Society of America, Madison, Wisconsin.

168. Ghiorse, W. C., D. N. Miller, R. L. Sandoli, and P. L. Siering. 1995. Applications of laser scanning microscopy for analysis of aquatic microhabitats. Microscopy Res. Techniques. 32:1–14.

169. Stahl, D. A., and R. I. Amann. 1991. Development and application of nucleic acid probes, p. 205–248. *In* E. Stackebrandt and M. Goodfellow (eds.), Nucleic acid techniques in bacterial systematics. John Wiley & Sons, New York.

170. Wagner, M., R. Amann, H. Lemmer, and K.-H. Schleifer. 1993. Probing activated sludge with oligonucleotides specific for proteobacteria: inadequacy of culture-dependent methods for describing community structure. Appl. Environ. Microbiol. 59:1520–1525.

171. Manz, W., R. Amann, W. Ludwig, M. Wagner, and K.-H. Schleifer. 1992. Phylogenetic oligodeoxynucleotide probes for the major subclasses of proteobacteria: problems and solutions. Sys. Appl. Microbiol. 15:593–600.

172. Schmidt, T. M., E. F. DeLong, and N. R. Pace. 1991. Analysis of marine picoplankton community by 16S rRNA gene cloning and sequencing. J. Bacteriol. 173:4371–4378.

173. Larsen, N., G. J. Olsen, B. L. Maidak, M . J. McCaughey, R. Overbeek, T. J. Macke, T. L. Marsh, and C. R. Woese. 1993. The ribosomal database project. Nucleic Acids Res. 21:3021–3023.

174. Benson, D., D. J. Lipman, and J. Ostell. 1993. GenBank. Nucleic Acids Res. 21: 2963–2965.

175. Swofford, D. L., and G. J. Olsen. 1990. Phylogeny reconstruction, p. 411–501. *In* D. M. Hillis and C. Moritz (eds.), Molecular systematics. Sinauer, Sunderland, Massachusetts.

176. Sackin, M. J., and D. Jones. 1993. Computer-assisted classification, p. 281–309. *In* M. Goodfellow and A. G. O'Donnell (Eds.), Handbook of new bacterial systematics. Academic Press, New York.

177. Atlas, R. M. 1993. Extraction of DNA from soils and sediments, p. 261–266. *In* P. F. Kemp, B. F. Sherr, E. B. Sherr, and J. J. Cole (eds.), Handbook of methods in aquatic microbial ecology. Lewis Publishers, Boca Raton, Florida.

178. Farrelly, V., F. A. Rainey and E. Stackebrandt. 1995. Effect of genome size and *rrn* gene copy number on PCR amplification of 16S rRNA genes from a mixture of bacterial species. Appl. Environ. Microbiol. 61:2798–2801.

179. Robison-Cox, J. F., M. M. Bateson, and D. M. Ward. 1995. Evaluation of nearest-neighbor methods for detection of chimeric small-subunit rRNA sequences. Appl. Environ. Microbiol. 61:1240–1245.

180. Liesack, W., H. Weyland, and E. Stackebrandt. 1991. Potential risks of gene amplification by PCR as determined by 16S rDNA analysis of a mixed-culture of strict barophilic bacteria. Microb. Ecol. 21:191–198.

181. Hedrick, D. B., J. B. Guckert, and D. C. White. 1991. Archaebacterial ether lipid diversity analyzed by supercritical fluid chromatography integration with a bacterial lipid protocol. J. Lipid Res. 32:659–666.

182. Karl, D. M. 1993. Total microbial biomass estimation derived from the measurement of particulate adenosine-5'-triphosphate, p. 359–368. *In* P. F. Kemp, B. F. Sherr, E. B. Sherr, and J. J. Cole (eds.), Handbook of methods in aquatic microbial ecology. Lewis Publishers, Boca Raton, Florida.

183. Kemp, P. F., S. Lee, and J. LaRoche. 1993. Evaluating bacterial activity from cell-specific ribosomal RNA content measured with oligonucleotide probes, p. 415–422. *In* P. F. Kemp, B. F. Sherr, E. B. Sherr, and J. J. Cole (eds.), Handbook of methods in aquatic microbial ecology. Lewis Publishers, Boca Raton, Florida.

184. Tabor, P. S., and R. A. Neihof. 1984. Direct determination of activities for microorganisms of Chesapeake Bay populations. Appl. Environ. Microbiol. 48:1012–1019.

185. Marshall, K. C. 1986. Microscopic methods for the study of bacterial behavior at inert surfaces. J. Microbiol. Meth. 4:217–227.

186. Lock, M. A., and T. E. Ford, 1986. Microcalorimetric approach to determine relationships between energy supply and metabolism in river epilithon. Appl. Environ. Microbiol. 49:408–412.

187. Guckert, J. B., D. B. Ringelberg, and D. C. White. 1987. Biosynthesis of trans fatty acids from acetate in the bacterium *Pseudomonas atlantica*. Can. J. Microbiol. 33: 748–754.

188. Carman, K. R. 1993. Microautoradiographic detection of microbial activity, p. 397–404. *In* P. F. Kemp, B. F. Sherr, E. B. Sherr, and J. J. Cole (eds.), Handbook of methods in aquatic microbial ecology. Lewis Publishers, Boca Raton, Florida.

189. Deming, J. W. 1993. ^{14}C tracer method for measuring microbial activity in deep-sea sediments, p. 405–414. *In* P. F. Kemp, B. F. Sherr, E. B. Sherr, and J. J. Cole (eds.), Handbook of methods in aquatic microbial ecology. Lewis Publishers, Boca Raton, Florida.

190. Karl, D. M. 1993. Adenosine triphosphate (ATP) and total adenine nucleotide (TAN) pool turnover rates as measures of energy flux and specific growth rate in natural populations of microorganisms, p. 483–494. *In* P. F. Kemp, B. F. Sherr, E. B. Sherr, and J. J. Cole (eds.), Handbook of methods in aquatic microbial ecology. Lewis Publishers, Boca Raton, Florida.

191. Kaplan, L. A., and T. L. Bott. 1985. Adenylate energy charge in streambed sediments. Freshwater Biol. 15:133–138.

192. Kieft, T. L., and L. L. Rosacker. 1991. Application of respiration- and adenylate-based soil microbiological assays to deep subsurface terrestrial sediments. Soil Biol. Biochem. 23:563–568.

193. Bitton, G., B. G. Volk, D. A. Graetz, J. M. Bossart, R. A. Boylan, and G. E. Byers. 1985. Effect of acid precipitation on soil microbial activity: II. Field studies. J. Environ. Qual. 14:69–71.

194. Kowalenko, C. G., K. C. Ivarson, and D. R. Cameron. 1978. Effect of moisture content, temperature and nitrogen fertilization on carbon dioxide evolution from field soils. Soil. Biol. Biochem. 10:417–423.

195. Weier, K. L., I. C. MacRae, and R. J. K. Myers. 1993. Denitrification in a clay soil under pasture and annual crop: losses from ^{15}N-labelled nitrate in the subsoil in the field using C_2H_2 inhibition. Soil Biol. Biochem. 25:999–1004.

196. Zibilske, L. M. 1994. Carbon mineralization, p. 853–863. *In* R. W. Weaver, S. Angle, P. Bottomley, D. Bezdicek, S. Smith, A. Tabatabai, and A. Wollum (eds.), Methods of soil analysis, Part 2. Soil Science Society of America, Madison, Wisconsin.

197. Conrad, R., and H. Schütz. 1988. Methods of studying methanogenic bacteria and methanogenic activities in aquatic environments, p. 301–346. *In* B. Austin (ed.), Methods of soil analysis, Part 2. John Wiley & Sons, New York.

198. Strayer, R., and J. Tiedje. 1978. In situ CH_4 production in a small, hypereutrophic, hardwater lake. Limnol. Oceanogr. 23:1201–1206.

199. Duxbury, J. M. 1986. Advantages of the acetylene method of measuring denitrification, p. 73–92. *In* R. D. Hauck and R. W. Weaver (eds.), Field measurement of dinitrogen fixation and denitrification. Special Publ. No. 18. Soil Science Society of America, Madison, Wisconsin.

200. Hauck, R. D. 1986. Field measurement of denitrification—An overview, p. 59–72. *In* R. D. Hauck and R. W. Weaver (eds.), Field measurement of dinitrogen fixation and denitrification. Special Publ. No. 18. Soil Science Society of America, Madison, Wisconsin.

201. Keeney, D. R. 1986. Critique of the acetylene blockage technique for field measurement of denitrification, p. 103–115. *In* R. D. Hauck, and R. W. Weaver (eds.), Field measurement of dinitrogen fixation and denitrification. Special Publ. No. 18. Soil Science Society of America, Madison, Wisconsin.
202. Mosier, A. R., and L. Klemedtsson. 1994. Measuring denitrification in the field, p. 1047–1065. *In* R. W. Weaver, S. Angle, P. Bottomley, D. Bezdicek, S. Smith, A. Tabatabai, and A. Wollum (eds.), Methods of soil analysis, Part 2. Soil Science Society of America, Madison, Wisconsin.
203. Rolston, D. E. 1986. Limitations of the acetylene blockage technique for field measurement of denitrification, p. 93–101. *In* R. D. Hauck and R. W. Weaver (eds.), Field measurement of dinitrogen fixation and denitrification. Special Publ. No. 18. Soil Science Society of America, Madison, Wisconsin.
204. Yavitt, J. B., G. E. Lang, and A. J. Sexstone. 1990. Methane fluxes in wetland and forest soils, beaver ponds, and low-order streams of a temperate forest ecosystem. J. Geophysic. Res. 95:22,463–22,474.
205. Weaver, R. W. 1986. Measurement of biological dinitrogen fixation in the field, p. 1–10. *In* R. D. Hauck and R. W. Weaver (eds.), Field measurement of dinitrogen fixation and denitrification. Special Publ. No. 18. Soil Science Society of America, Madison, Wisconsin.
206. Weaver, R. W., and S. K. A. Danso. 1994. Dinitrogen fixation, p. 1019–1045. *In* R. W. Weaver, S. Angle, P. Bottomley, D. Bezdicek, S. Smith, A. Tabatabai, and A. Wollum (eds.), Methods of soil analysis, Part 2. Soil Science Society of America, Madison, Wisconsin.
207. Ardakani, M. S., R. K. Schulz, and A. D. McLaren. 1974. A kinetic study of ammonium and nitrite oxidation in a soil field plot. Soil Sci. Soc. Am. Proc. 38: 273–277.
208. Schomberg, H. H., J. L. Steiner, and P. W. Unger. 1994. Decomposition and nitrogen dynamics of crop residues: residue quality and water effects. Soil Sci. Soc. Amer. J. 58:372–381.
209. Hauck, R. D., J. J. Meisinger, and R. L. Mulvaney. 1994. Practical considerations in the use of nitrogen tracers in agricultural and environmental research, p. 907–950. *In* R. W. Weaver, S. Angle, P. Bottomley, D. Bezdicek, S. Smith, A. Tabatabai, and A. Wollum (eds.), Methods of soil analysis, Part 2. Soil Science Society of America, Madison, Wisconsin.
210. Parkin, T. B., A. J. Sexstone, and J. M. Tiedje. 1985. Comparison of field denitrification rates determined by acetylene-based soil core and nitrogen-15 methods. Soil Sci. Soc. Am. J. 49:94–99.
211. Weier, K. L., J. W. Doran, A. R. Mosier, J. F. Power, and T. A. Peterson. 1994. Potential for bioremediation of high nitrate irrigation water via denitrification. J. Environ. Qual. 23:105–110.
212. Phillips, D. A., M. B. Jones, and K. W. Foster. 1986. Advantages of the nitrogen-15 dilution technique for field measurements of symbiotic dinitrogen fixation in legumes, p. 11–22. *In* R. D. Hauck and R. W. Weaver (eds.), Field measurement of dinitrogen fixation and denitrification. Special Publ. No. 18. Soil Science Society of America, Madison, Wisconsin.
213. Rennie, R. J. 1986. Advantages and disadvantages of nitrogen-15 isotope dilution to quantify nitrogen fixation in field-grown legumes—a critique, p. 43–58. *In* R. D.

Hauck and R. W. Weaver (eds.), Field measurement of dinitrogen fixation and denitrification. Special Publ. No. 18. Soil Science Society of America, Madison, Wisconsin.

214. Vose, P. B., and R. L. Victoria. 1986. Re-examination of the limitations of nitrogen-15 isotope dilution technique for the field measurement of dinitrogen fixation, p. 23–42. *In* R. D. Hauck and R. W. Weaver (eds.), Field measurement of dinitrogen fixation and denitrification. Special Publ. No. 18. Soil Science Society of America, Madison, Wisconsin.
215. Aggarwal, P. K., and R. E. Hinchee. 1991. Monitoring in situ biodegradation of hydrocarbons by using stable carbon isotopes. Environ. Sci. Technol. 25:1178–1180.
216. Blair, N. E., and W. D. Carter, Jr. 1992. The carbon isotope biogeochemistry of acetate from methanogenic marine sediment. Geochim. Cosmochim. Acta 56:1247–1258.
217. Chapelle, F. H., J. L. Zelibor, Jr., D. J. Grimes, and L. L. Knobel. 1987. Bacteria in deep coastal plain sediments of Maryland: a possible source of CO_2 to groundwater. Water Resour. Res. 23:1625–1632.
218. Doughton, J. A., I. Vallis, and P. G. Saffigna. 1992. An indirect method for estimating ^{15}N isotope fractionation during nitrogen fixation by a legume under field conditions. Plant Soil 144:23–29.
219. Fuller, R. D., M. J. Mitchell, H. R. Krouse, B. J. Wyskowski, and C. T. Driscoll. 1986. Stable sulfur isotope ratios as a tool for interpreting ecosystem sulfur dynamics. Water Air Soil Pollut. 28:163–171.
220. Grossman, E. L., B. K. Coffman, S. J. Fritz, and H. Wada. 1989. Bacterial production of methane and its influence on ground-water chemistry in east-central Texas aquifers. Geology 17:495–499.
221. Wolf, D. C., J. O. Legg, and T. W. Boutton. 1994. Isotopic methods for the study of soil organic matter dynamics, p. 865–905. *In* R. W. Weaver, S. Angle, P. Bottomley, D. Bezdicek, S. Smith, A. Tabatabai, and A. Wollum (eds.), Methods of soil analysis, Part 2. Soil Science Society of America, Madison, Wisconsin.
222. Lovley, D. R., F. H. Chapelle, and J. C. Woodward. 1994. Use of dissolved H_2 concentrations to determine distribution of microbially catalyzed redox reactions in anoxic groundwater. Environ. Sci. Technol. 28:1205–1210.
223. Atlas, R. M. 1981. Microbial degradation of petroleum hydrocarbons: an environmental perspective. Microbiol. Rev. 45:180–209.
224. Pritchard, P. H., and C. F. Costa. 1991. EPA's Alaska oil spill bioremediation project. Environ. Sci. Technol. 25:372–379.
225. Bragg, J. R., R. C. Prince, E. J. Harner, and R. M. Atlas. Effectiveness of bioremediation for the Exxon Valdez oil spill. Nature 368:413–418.
226. Flanagan, W. P., and R. J. May. 1993. Metabolite detection as evidence for naturally occurring aerobic PCB biodegradation in Hudson River sediments. Environ. Sci. Technol. 27:2207–2212.
227. Harkness, M. R., J. B. McDermott, D. A. Abramowicz, J. J. Salvo, W. P. Flanagan, M. L. Stephens, F. J. Mondello, R. J. May, J. H. Lobos, K. M. Carroll, M. J. Brennan, A. A. Bracco, K. M. Fish, G. L. Warner, P. R. Wilson, D. K. Dietrich, D. T. Lin, C. B. Morgan, and W. L. Gately. 1993. In situ stimulation of aerobic PCB biodegradation in Hudson River sediments. Science 259:503–507.

228. Führ, F. 1985. Application of ^{14}C-labeled herbicides in lysimeter studies. Weed Sci. 33:11–17.

229. Hopkins, G. D., J. Munakata, L. Semprini, and P. L. McCarty 1993. Trichloroethylene concentration effects on pilot field-scale in-situ groundwater bioremediation by phenol oxidizing microorganisms. Environ. Sci. Technol. 27:2542–2547.

230. Jayachandran, K., T. R. Steinheimer, L. Somasundaram, T. B. Moorman, R. S. Kanwar, and J. R. Coats. 1994. Occurrence of atrazine and degradates as contaminants of subsurface drainage and shallow groundwater. J. Environ. Qual. 23:311–319.

231. Lamar, R. T., and D. M. Dietrich. 1990. In situ depletion of pentachlorophenol from contaminated soil by *Phanaerochaete* spp. Appl. Environ. Microbiol. 56:2092–3100.

232. Madsen, E. L., J. L. Sinclair, and W. C. Ghiorse. 1991. In situ biodegradation: microbiological patterns in a contaminated aquifer. Science 252:830–833.

233. Major, D. W., E. W. Hodgins, and B. J. Butler. 1991. Field and laboratory evidence of in situ biotransformation of tetrachloroethene to ethene at a chemical transfer facility in North Toronto, p. 147–172. *In* R. E. Hinchee and R. F. Olfenbuttel (eds.), On-site Bioreclamation. Butterworth-Heinemann, Stoneham, Massachusetts.

234. Semprini, L., P. K. Kitanidis, D. H. Kampbell, and J. T. Wilson. 1995. Anaerobic transformation of chlorinated aliphatic hydrocarbons in a sand aquifer based on spatial chemical distributions. Water Resour. Res. 31:1051–1062.

235. Beeman, R. E., J. Howell, S. H. Shoemaker, E. A. Salazar, and J. R. Buttram. 1994. A field evaluation of in situ microbial reductive dehalogenation by the biotransformation of chlorinated ethenes, p. 14–27. *In* R. E. Hinchee, A. Leesen, L. Semprini, and S. K. Ong (eds.), Bioremediation of chlorinated and polycyclic aromatic hydrocarbon compounds. Lewis Publishers, Boca Raton, Florida.

236. Raymond, R. L., J. O. Hudson, and V. W. Jamison. 1976. Oil degradation in soil. Appl. Environ. Microbiol. 31:522–535.

237. Semprini, L., P. V. Roberts, G. D. Hopkins, and P. L. McCarty. 1990. A field evaluation of in-situ biodegradation of chlorinated ethenes: Part 2, results of biostimulation and biotransformation experiments. Ground Water 28:715–727.

238. Wilson, M. S., and E. L. Madsen. 1995. Recovery of 1,2-dihydroxy-1,2-dihydronaphthalene from a contaminated field site indicates *in situ* naphthalene biodegradation. American Society for Microbiology Annual Meeting, Washington, D.C. Abst. Annu. Meet. Amer. Soc. Microbiol. Q126, p. 421.

239. Smith, A. E., L. Hume, G. P. LaFond, and V. O. Biederbeck. 1991. A review of the effects of long-term 2,4-D and MCPA applications on wheat production and selected biochemical properties of a Black Chernozem. Can. J. Soil Sci. 71:73–87.

240. Wild, S. R., J. P. Obbard, C. I. Munn, M. L. Berrow, and K. C. Jones. 1991. The long-term persistence of polynuclear aromatic hydrocarbons (PAHs) in an agricultural soil amended with metal-contaminated sewage sludges. Sci. Total Environ. 101:235–253.

241. Fleming, J. T., J. Sanseverino, and G. S. Sayler. 1993. Quantitative relationship between naphthalene catabolic gene frequency and expression in predicting PAH degradation in soils. Environ. Sci. Technol. 27:1068–1074.

242. Jeffrey, W. H., S. Nazaret, and R. Von Haven. 1994. Improved method for recovery of mRNA from aquatic samples and its application to detection of *mer* expression. Appl. Environ. Microbiol. 60:1814–1821.

243. Nuovo, G. J. 1992. PCR in situ hybridization: protocols and applications. Raven Press, New York.

244. Pichard, S. L., and J. H. Paul. 1991. Detection of gene expression in genetically engineered microorganisms and natural phytoplankton populations in the marine environment by mRNA analysis. Appl. Environ. Microbiol. 57:1721–1727.

245. Pichard, S. L., and J. H. Paul. 1993. Gene expression per gene dose, a specific measure of gene expression in aquatic microorganisms. Appl. Environ. Microbiol. 59: 451–457.

246. Tsai, Y.-L., M. J. Park, and B. H. Olson. 1991. Rapid method for direct extraction of mRNA from seeded soils. Appl. Environ. Microbiol. 57:765–768.

247. Ogram, A., W. Sun, F. J. Brockman, and J. K. Fredrickson. 1995. Isolation and characterization of RNA from low-biomass deep-subsurface sediments. Appl. Environ. Microbiol. 61:763–768.

248. Fleming, J. T., and D. J. Zhang. 1994. Quantitation of soluble methane monooxygenase mesenger RNA by RT-PCR from trichloro ethylene contaminated sediments. Abstr. Annu. Meet. Am. Soc. Microbiol. Q-161, p. 416.

249. Lamar, R. T., B. Schoenike, A. Vanden Wymelenberg, P. Stewart, D. M. Dietrich, and D. Cullen, 1995. Quantitation of fungal mRNAs in complex substrates by reverse transcription PCR and its application to *Phanerochaete chrysosporium*-colonized soil. Appl. Environ. Microbiol. 61:2122–2126.

250. Jensen, K., N. P. Sloth, N. Risgaard-Petersen, S. Rysgaard, and N. P. Revsbech. 1994. Estimation of nitrification and denitrification from microprofiles of oxygen and nitrate in model sediment systems. Appl. Environ. Microbiol. 60:2094–2100.

251. Parkin, T. B., and J. A. Robinson. 1989. Stochastic models of soil denitrification. Appl. Environ. Microbiol. 55:72–77.

252. Macalady, D. L., P. G. Tratnyek, and T. J. Grundl. 1986. Abiotic reduction reactions of anthropogenic organic chemicals in anaerobic systems: a critical review. J. Cont. Hydrol. 1:1–28.

253. Valentine, R. L. 1986. Nonbiological transformation, p. 223–243. *In* S. C. Hern and S. M. Malancon (eds.), Vadose zone modeling of organic pollutants. Lewis Publishers, Boca Raton, Florida.

254. Voudrias, E. A., and M. Reinhard. 1986. Abiotic organic reactions at mineral surfaces, p. 463–486. *In* J. A. Davis and K.F. Hayes (eds.), Geochemical processes at mineral surfaces. ACS Symposium Series 323. American Chemical Society, San Diego.

255. Wolfe, N. L. 1992. Abiotic transformations of pesticides in natural waters and sediments. John Wiley & Sons, New York.

256. Zepp, R. G., and L. N. Wolfe. 1987. Abiotic transformation of organic chemicals at the particle-water interface, p. 423–455. *In* W. Stumm (ed.), Aquatic surface chemistry. John Wiley & Sons, New York.

257. Tiedje, J. M. 1994. Denitrifiers, p. 245–267. *In* R. W. Weaver, S. Angle, P. Bottomley, D. Bezdicek, S. Smith, A. Tabatabai, and A. Wollum (eds.), Methods of soil analysis, Part 2. Soil Science Society of America, Madison, Wisconsin.

258. Oremland, R. S., and C. W. Culbertson. 1992. Importance of methane-oxidizing bacteria in the methane budget as revealed by the use of a specific inhibitor. Nature 356:421–423.

259. Tiedje, J. M. 1988. Ecology of denitrification and dissimilatory nitrate reduction to ammonium, p. 179–244. *In* A. J. B. Zehnder (ed.), Biology of anaerobic microorganisms. John Wiley & Sons, New York.

260. Hoefs, J. 1987. Stable isotope geochemistry, 3rd ed. Springer-Verlag, New York.

261. Madsen, E. L. 1996. Theoretical and applied aspects of bioremediation: the influence of microbiological processes on organic compounds in field sites. *In* R. Burlage (ed.) Techniques in Microbial Ecology. Oxford University Press. (In press).

261a. Madsen, E. L. 1996. Method for determining biodegradability. *In* C. Hurst et al (eds.) Manual of Environmental Microbiology. American Society of Microbiology, Washington, DC. (In press).

261b. National Research Council. 1993. In situ bioremediation: when does it work? National Academy Press, Washington, D.C.

262. Prince, R. C., D. L. Elmendorf, J. R. Lute, C. S. Hsu, C. E. Halth, J. D. Senius, G. J. Dechert, G. S. Douglas, and E. L. Butler. 1994. 17α*(H)*,21β*(H)*-hopane as a conserved internal marker for estimating the biodegradation of crude oil. Environ. Sci. Technol. 28:142–145.

263. Roques, D. E., E. B. Overton, and C. B. Henry. 1994. Using gas chromatography/mass spectroscopy fingerprint analyses to document process and progress of oil degradation. J. Environ. Qual. 23:851–855.

264. Nazaret, S., W. H. Jeffrey, E. Saouter, R. Von Haven, and T. Barkay. 1994. *merA* gene expression in aquatic environments measured by mRNA production and Hg(II) volatilization. Appl. Environ. Microbiol. 60:4059–4065.

265. Fliermans, C. B., and E. L. Schmidt. 1975. Autoradiography and immunofluorescence combined for autoecological study of single cell activity with *Nitrobacter* as a model system. Appl. Microbiol. 30:676–684.

266. Gutel'makker, B. L. 1973. Autoradiography as a method of determining the relative contribution of individual algal species to the primary production of plankton. Hydrobiol. J. 9:61–64.

267. Ward, B. B. 1984. Combined autoradiography and immunofluorescence for estimation of single cell activity by ammonium-oxidizing bacteria. Limnol. Oceanogr. 29: 402–410.

268. Madan, A. P., and S. A. Nierzwicki-Bauer. 1993. In situ detection of transcripts for ribulose-1,5-bisphosphate carboxylase in cyanobacterial heterocysts. J. Bacteriol. 175:7301–7306.

268a. Hudson, R. E., W. A. Dustman, R. P. Garg, and M. A. Moran. 1995. In situ PCR for visualization of microscale distribution of specific genes and gene products in prokoryotic communities. Appl. Environ. Microbiol. 61:4074–4082.

269. Currin, C. A., H. W. Paerl, G. K. Suba, and R. S. Alberte. 1990. Immunofluorescence detection and characterization of N_2-fixing microorganisms from aquatic environments. Limnol. Oceanogr. 35:59–71.

270. Brock, T. D., and M. L. Brock. 1966. Autoradiography as a tool in microbial ecology. Nature 209:734–736.
271. Rogers, J. E., and W. B. Whitman (eds.). 1991. Microbial production and consumption of greenhouse gases: methane, nitrogen oxides and halomethanes. American Society for Microbiology, Washington, D.C.
272. Bormann, F. H., and G. E. Likens. 1979. Pattern and process in a forested ecosystem. Springer Verlag, New York.
273. Likens, G. E. (ed.) 1985. An ecosystem approach to aquatic ecology: mirror lake and its environment. Springer-Verlag, New York.
274. Allen, T. F. H., and T. W. Hoekstra. 1992. Toward a unified ecology. Columbia University Press, New York.
274a. Madsen, E. L. 1995. Impacts of agricultural practices on subsurface microbial ecology. *In* D. Sparks (ed.). Advances in agronomy, Vol. 54, pp. 1–67.
275. Groffman, P. M., and J. M. Tiedje. 1989. Denitrification in north temperate forest soils: spatial and temporal patterns at the landscape and seasonal scales. Soil Biol. Biochem. 21:613–620.
276. Chapelle, F. H., and D. R. Lovley. 1990. Rates of microbial metabolism in deep coastal plain aquifers. Appl. Environ. Microbiol. 56:1865–1874.
277. Kluyver, A. J. 1924. Eenheid en verscheidenheid in de stofwisseling der microben'. Chemisch Weekblad 21:266. Translation as "Unity and diversity in the metabolism of micro-organisms" in Albert Jan Kluyver, His Life and Work, 1959, North Holland, Amsterdam (cited in [34]).
278. Kluyver, A. J., and C. B. van Niel. 1954. The microbe's contribution to biology. Harvard University Press, Cambridge.
279. Warren, L., and H. Koprowski (eds.). 1991. New perspectives on evolution: proceedings of a multidisciplinary symposium designed to interrelate recent discoveries and new insights in the field of evolution held at the University of Pennsylvania. *In* The Wistar symposium series, Vol. 4. Wiley-Liss, New York.
280. Poland, B. W., M. M. Silva, M. A. Serra, Y. Cho, K. H. Kim, E. M. S. Harris, and R. B. Honzatko. 1993. Crystal structure of adenylosuccinate synthetase from *Escherichia coli*: Evidence for convergent evolution of GTP-binding domains. J. Biol. Chem. 268:25334–25342.
281. Brown, J. R., Y. Masuchi, F. T. Robb, and W. F. Doolittle. 1994. Evolutionary relationships of bacterial and archaeal glutamine synthetase genes. J. Mol. Evol. 38: 566–576.
282. Zylstra, G. J. 1994. Molecular analysis of aromatic hydrocarbon degradation, p. 83–115. *In* S. J. Garte (ed.), Molecular environmental biology. Lewis Publishers, Boca Raton, Florida.
283. Fambrough, D. M. (ed.) 1994. Molecular evolution of physiological processes. Society of General Physiologists, 47th Annual Symposium. Rockefeller University Press, New York.
284. Parker, S. P. 1984. McGraw-Hill dictionary of scientific and technical terms, 3rd ed. McGraw-Hill, New York.
285. Alcala, J. R., I. Scott, J. Parker, B. Atwater, C. Yu, R. Fischer, and K. Bellingrath. 1993. Real time frequency-domain fiber-optic sensor for intra-arterial blood oxygen measurements. Proc. SPIE-Int. Soc. Opt. Eng. 1885:306–316.

286. Alcala, J. R., C. Yu, and G. J. Yeh. 1993. Digital phosphorimeter with frequency domain signal processing: application to real-time fiber-optic oxygen sensing. Rev. Sci. Instrum. 64:1554–1560.
287. Nore, D., A. M. Gomes, J. Bacri, and J. Cabe. 1993. Development of an apparatus for the detection and measurement of the metallic aerosol concentrations in atmospheric air in situ and in real time: preliminary results. Spectrochim. Acta. 48B: 1411–1419.
288. Berresheim, H., D. J. Tanner, and F. L. Eisele. 1993. Method for real-time detection of dimethyl sulfone in ambient air. Anal. Chem. 65:3168–3170.
289. Sauren, H., E. Gerkema, D. Bicanic, and H. Jalink. 1993. Real-time and in situ determination of ammonia concentrations in the atmosphere by means of intermodulated stark resonant carbon dioxide laser photoacoustic spectroscopy. Atmos. Environ. 27A:109–112.
290. Watanabe, F., G. M. McClella, and, H. Heinzelmann. 1993. Real-time observation of the vibration of a single absorbed molecule. Science 262:1244–1247.
291. Minunni, M., and M. Mascini. 1993. Detection of pesticides in drinking water using real-time biospecific interaction analysis (BIA). Anal. Lett. 26:1441–1460.
292. Hasegawa, K., A. Tanakadate, and H. Ishikawa. 1988. A method for tracking the locomotion of an isolated microorganism in real time. Physiol. Behav. 42:397–400.
293. Bouma, J., and A. K. Bregt. 1989. Land qualities in space and time. Pudoc, Wageningen.
294. Mausbach, M. J., and L. P. Wilding. 1991. Spatial variability of soils and land forms. Special Publ. No. 228. Soil Science Society of America, Madison, Wisconsin.

9

Soil Suppressiveness to Diseases Induced by Soilborne Plant Pathogens

Claude Alabouvette, Heinrich Hoeper, Philippe Lemanceau, and Christian Steinberg Institut National de la Recherche Agronomique, Dijon, France

I. INTRODUCTION

If a disease results from the intimate interaction between a plant and a pathogen, its severity is influenced by environmental factors that affect either the plant, the pathogen, or both. In the case of diseases induced by soilborne plant pathogens, the soil has a major role in controlling interactions between the plant, the pathogen, and the saprophytic microbiota. Since the first observations by Atkinson [1], many papers have described field situations where soil influences either disease incidence or disease severity. The existence of soils that suppress diseases provides good examples of the role of abiotic and biotic factors that affect the relationship between a plant and its pathogen: in suppressive soils, disease incidence remains low despite the presence of the pathogen, a susceptible host plant, and climatic conditions favorable for disease development. These soils have been designated as antagonistic, biologically buffered, competitive, decline, fungistatic, long-life, low pathogen, intolerant, immune, suppressive, or resistant [2,3]. Studies of suppressive soils led to a basic concept of soil receptiveness to diseases [4,5]. Soil is not a neutral milieu in which pathogenic microorganisms interact freely with the roots of host plants; on the contrary, soil interferes with the relations between and among microorganisms, pathogens, and plants. Every natural soil possesses some ability to limit diseases, and soil receptiveness has to be considered as a continuum from strongly suppressive to highly conducive. Distinct levels of disease severity can be observed following artificial infestation of different soils with the same inoculum concentrations of a pathogen. In conducive soils, disease incidence is severe even at a low inoculum density, whereas in suppressive soils, disease incidence is low even at high inoculum density. Although the term soil receptiveness better describes this concept of a continuum, the term soil suppressiveness will be used because it is more popular in the literature.

A soil may be *pathogen* suppressive, in that it limits the growth and saprophytic survival of the pathogen, or it may be *disease* suppressive. These two properties are not necessarily linked. Disease suppressive soils may contain a higher inoculum density and enable a better survival of the pathogen than a conducive soil [6]. In contrast, soils that are pathogen suppressive may not be suppressive to disease [7]. This chapter will focus on disease suppressiveness, as this phenomenon implies interactions that result in reduced disease and high crop yield and quality.

From a theoretical point of view, soil may exert its influence through its physicochemical characteristics, its biological properties, or both. It is often difficult to distinguish clearly between the role of the abiotic and biotic properties of suppressive soils. For example, Reinking and Manns [8], Walker and Snyder [9], Stover [10], and Stotzky and Martin [11] drew attention to the role of abiotic factors in soil suppressiveness to Fusarium wilts. In general, Fusarium wilts are more severe in light-textured soils, and suppressiveness seems linked to high pH and the presence of smectite-type clays. In contrast, Louvet et al. [12] and Scher and Baker [13] considered the role of the soil microbiota more important, based on observations that suppressiveness disappears after the application of biocide treatments that do not affect the abiotic characteristics of soil. In recent years there have been extensive studies on the microorganisms that may be involved in soil suppressiveness in an attempt to utilize them for biological control. Less research has been devoted to the role of abiotic factors and to the relations between microbial balance and the physicochemical properties of soil. Most physicochemical factors are not independent of each other, which makes experimentation and interpretation of data very difficult. However, the role of abiotic factors in relation to soil suppressiveness to diseases should not be underestimated.

Another problem is related to the length of time necessary to reach an equilibrium between the biotic and abiotic characteristics of soils. Hornby [14] distinguished between long-term and short-term suppressiveness; long-term suppressiveness has been observed for decades, but short-term suppressiveness results from rapid changes in the soil environment due to liming, fertilization, and tillage. The first type of suppressiveness is probably the result of stable soil properties, whereas the latter often disappears rapidly with changes in cultural practices. The best documented example of short-term suppressiveness is that of take-all decline, which is induced by wheat monoculture, but is immediately destroyed by an interruption of the monoculture [3]. It is, therefore, difficult to correlate laboratory studies on the mechanisms of soil suppressiveness with field observations on naturally suppressive soils. The longevity of laboratory induced suppressiveness is almost never studied, and the history of natural suppressiveness is generally ignored. For example, it is possible to transfer suppressiveness to a heat-treated conducive soil by supplying a small amount of suppressive soil, but it is not known how long this effect will be maintained. The soil in which suppressiveness has been induced will probably revert to the conducive state if its physicochemical properties are not suitable for the development of a disease suppressive microbiota.

Studies on the mechanisms of soil suppressiveness have shown that the activity of both the total microbial biomass and specific populations of antagonists can be responsible for disease suppression [15]. Interactions between the antagonistic populations and the pathogens involve several mechanisms, including competition for trophic resources and antibiosis resulting from the production of toxic metabolites. However, it is obvious that disease suppression is not independent of the physical and chemical properties of soils, which may have a direct or an indirect role [16]. These abiotic properties may favor the activity of the microbial biomass or influence the balance between pathogens and their specific antagonists. It has been observed that the physicochemical properties of soil influence the rapidity with which monocultures induce the decline of diseases [17], which is a suppressive phenomenon based on microbial interactions.

Soils that are suppressive to many important plant diseases have been described, but, in general, the same soil may be suppressive to only one type of disease and conducive to others. This indicates that pathogens belonging to different genera or species have different requirements for the development of their pathogenic activity. Moreover, soils suppressive to the same disease may have different characteristics, indicating that different mechanisms may be responsible for the same global effect, that is, reduction of disease incidence or severity. Exhaustive review of the literature dealing with soil suppressiveness to diseases reveals many contradictory data, probably as the result of the complexity of the interactions between biotic and abiotic soil factors and also of the diversity of experimental approaches. Therefore, the aim of this chapter is not to review all the available data, but to discuss the most significant examples to illustrate how abiotic and biotic factors, alone or together, contribute to the control of disease incidence.

II. INVOLVEMENT OF SOIL ABIOTIC FACTORS IN DISEASE SUPPRESSIVENESS

As indicated above, most of the early studies on soil suppressiveness to diseases emphasized the role of abiotic factors. For example, Walker and Snyder [9] observed that the incidence of Fusarium wilt of pea was reduced in clay soils in comparison to sandy soils. Hancock [18] suggested that suppressiveness to diseases induced by *Phthium ultimum* was associated with certain classifications of soil at the series level, with the more coarsely textured soils being more conducive. There are a few examples of a clear relation between pH and disease incidence. Soils suppressive to clubroot of crucifers, induced by *Plasmodiophora brassicae,* always have a pH higher than 7, and liming of the soil is a usual practice to control this disease [19].

However, a review of the literature gives a rather imprecise image of the relations between physicochemical soil properties and soil suppressiveness. Different plant pathogens have different physiological and ecological requirements, and even for the same disease, contradictory results have been published. Moreover, it is often difficult to compare data obtained with different experimental approaches. One approach consists of an attempt to correlate disease incidence recorded in the field, or sometimes assessed after artificial infestation of soils, with physicochemical characteristics of soil. This approach, followed by Stotzky et al. [20], led to the conclusion that clay type is an important factor in soils suppressive to Fusarium wilt. In Central America, soils that had a long lifetime of banana production, despite the presence of the pathogen, were rich in montmorillonitic clays, whereas soils with a short lifetime did not contain this type of clay. This approach was also followed by Broadbent and Baker [21,22] to determine the most important factors involved in the suppressiveness of Australian soils to diseases induced by *Phytophthora* spp. Another approach consists of experimental modifications of soil characteristics to induce variations of the level of soil suppressiveness. Modifications of soil pH has been used to demonstrate the role of pH in the mechanisms of soil suppressiveness to Fusarium wilt [13,23], and additions of different clay types have been used to induce suppressiveness in conducive soils [6,16]. Interpretation of such manipulations must be considered with caution, as many soil characteristics may be modified by such manipulations.

A. Soil Texture and Structure

The importance of texture and structure in soil suppressiveness has not been studied extensively. Soil texture may exert at least two types of influence on microorganisms: a direct interaction between soil particles and microorganisms, and an indirect effect through the influence of texture on soil structure, aeration, and water potential. In general, sand and silt, composed mainly of primary minerals (e.g., quartz and feldspars), are considered to be essentially inactive, especially as the result of their low electrokinetic potential [16,24]. However, the silt fraction may contain some structural clay minerals (e.g., micas). The clay fraction is thought to have the major role as a result of its cation exchange capacity and high internal and external surface area, as well as of the permanent negative or variable charge of clay minerals [16,24]. It is generally accepted that clays favor bacterial activity and are less favorable to fungal growth [16,25–28].

Soil texture influences soil structure, especially the distribution of different pore sizes, which determines the habitable space for bacteria, fungi, and predators. Depending on their size, different types of microorganisms can only colonize pores of a certain minimal diameter [29,30]. As a result of their structural instability, silty soils are more susceptible to compaction than are sandy or clay soils. Diseases induced by *Aphanomyces, Phytophthora, Verticillium,* and *F. solani* are more prevalent in dense or compacted soils [17,31–33]. Compacted soils are less aerated, more easily waterlogged, and show a higher resistance to the penetration of plant roots, which predisposes the roots to attack by pathogens [34]. This effect appears to be most important for species of *Aphanomyces* and *Phytophthora,* whereas the incidence of Fusarium wilts is apparently not affected by soil compaction.

The influence of soil texture on disease severity is related to the environmental conditions of the soil in situ. For example, even a sandy soil will be wet all year in a humid climate, and the larger diameter of water-filled pores in a sandy soil than in a clay soil favors the mobility of propagules with a high diameter, for example, the zoospores of *Pythium* or *Phytophthora* [17,35]. Soils with a fine texture buffer fluctuations in environmental conditions such as wet-dry weather cycles or variations in temperature.

The complexity of interactions between soil texture and other abiotic or biotic parameters makes the generalization of data impossible. More studies are needed to understand better the relation between disease suppressiveness and soil type.

B. Soil Water

Two different aspects of soil water have to be considered: (1) the total water content, which, together with the bulk density of the soil, determines the relative content of water and air in the soil; and (2) the water potential, which determines the availability of soil water to plants and microorganisms. Soil moisture is determined primarily by the climate and is thus relatively independent of other soil properties. However, under the same climatic conditions, soil type, soil texture, location of the soil (e.g., uphill, downhill, slope), and agricultural practices, such as irrigation and drainage, modify water retention and thus the soil water content, as well as fluctuations in soil moisture.

Diseases induced by *F. roseum* or *Rhizoctonia solani* are favored by low water content and are suppressed in fine-textured soils, especially in wet climates [36–38]. In contrast, diseases caused by species of *Phytophthora* or *Pythium* are more severe in wet soils [17,22]. Agricultural practices that increase soil moisture, such as irrigation, reduce the incidence of potato scab and Rhizoctonia damping-off [17,39] but enhance the incidence of southern blight of vegetables and take-all of wheat [40,41]. Compaction of subsoil and impeded drainage, resulting in a higher moisture in the surface soil, favor black root rot of tobacco and banana wilt [10,42].

Different hypotheses have been proposed to explain the effect of soil water on disease: (1) at a given water content, the pathogen has a relative competitive advantage or disadvantage compared with its antagonists [17]; (2) at high soil moisture, increased root exudation and chemotaxis intensify the host-pathogen relationship and favor pathogen activity [34]; (3) at a higher water content, pores with a larger diameter are filled with water, increasing the mobility of zoospores of *Pythium* and *Phytophthora* and favoring root contamination [17,35]; and (4) drought causes water stress in the plant and increases the osmotic potential in soil and in plant tissues [17]. Not only the water content per se but also its fluctuations were found to favor some diseases, such as damping-off of pepper, southern blight, and banana wilt, indicating that different moisture conditions may be optimal for spore germination, mycelial growth, and pathogenicity [10].

In the laboratory, the interactions between soil texture, compaction, impeded drainage resulting from an impermeable subsoil, and climate, which determine the soil moisture content and its fluctuations, are usually eliminated. Therefore, the failure to consider the influence of soil water on soil suppressiveness may be one reason for some of the discrepancies observed between field and laboratory studies of some diseases.

C. Clay Type

The role of clay type in soil suppressiveness has been studied for only a few plant diseases, and available information is mainly based on field observations.

Clay mineralogy was apparently not involved in the suppressiveness of soils to Phytophthora root rot induced by *P. cinnamomi* in eastern Australia [21,22]. However, the studied soils contained essentially kaolinites, and there was only one soil that contained montmorillonites in mixture with kaolinites. Moreover, the interaction with vegetation (culture of avocado or native rainforest) prevented reliable conclusions on the role of clay minerals on this root rot. A definitive study was done by Stotzky et al [20] and Stotzky and Martin [11] in Central America: soils suppressive to Fusarium wilt of banana contained montmorillonitic clays, whereas conducive soils did not. Both suppressive and conducive soils contained other types of clay (e.g., kaolinites and illites). The suppressive soil from Chateaurenard contains predominantly kaolinites and illites but also small amounts of montmorillonites in the clay fraction [H. Hoeper (unpublished data)]. Another clear example of the involvement of soil type is provided by the Swiss soil suppressive to black root rot of tobacco. Suppressiveness is correlated with moraine soils rich in vermiculitic clays, whereas a conducive soil contained essentially smectites (montmorillonites) and illites [43].

The role of clay type in soil suppressiveness to other plant diseases has not been studied, although clay type is known to interact with other soil properties involved in suppressiveness. The type of clay influences the relation between soil water content and water potential, the availability and viscosity of soil water, and soil aeration [16,24]. Fungal respiration was already reduced in media containing 4% montmorillonite, probably as a result of the increased viscosity of the system and reduced oxygen diffusion, whereas in media containing kaolinite this effect was only observed at concentrations above 40% clay [27]. Clay type also modifies soil structure and mediates the soil habitat for different groups of microorganisms, protecting them more or less against desiccation and predation: (1) smectite addition to a soil enhanced the survival of *Rhizobium leguminosarum* biovar *trifolii,* possibly as the result of the increase of the number of pores with diameters less than 6 μm, which protected the bacterium against predation by protozoa, whereas kaolinite had a much lesser effect [44]; and (2) shrinkage of smectite as the result of drying, and the subsequent reduction in pore size, was detrimental to *Pseudomonas solanacearum* [45]. Clays may also coat bacterial cells and fungal hyphae, and then reduce transmembrane transport [24,28]. The type of clay determines in large part the cationic exchange capacity (CEC) of soil and thus the content and availability of macro- and microelements, as well as the buffer capacity of the soil and the soil pH: addition of montmorillonite to culture media increased bacterial respiration as a result of its high content of exchangeable calcium and pH buffering capacity [46]. Montmorillonite, with its large internal and external surface, interacts with organic components, affecting the activity of enzymes, antibiotics, or siderophores: it inactivated streptomycin or siderophores, whereas kaolinite did not [28,47,48]. By the formation of clay-humus complexes, it stabilizes soil organic matter and influences its quality.

Clay type also appears to be involved in antagonism between bacteria and fungal pathogens. In soil amended with montmorillonite, *Serratia marcescens* reduced the growth rate of *Aspergillus niger*, whereas it did not in soil amended with kaolinite [49]. However, the addition of bentonite did not enhance bacterial antagonism toward *Gaeumannomyces graminis* var. *tritici* [50].

These examples demonstrate that the type of clay may influence the abundance and activity of pathogens and antagonists and therefore the level of soil suppressiveness.

The lack of specific knowledge of how the clay minerals influence disease suppressiveness of soils results from a multitude of possible interactions between clays and other soil factors. In field studies, it is difficult to separate the influence of clay type from other soil properties. In laboratory studies, the clay content can be altered, but it is difficult to maintain other soil properties constant. Therefore, experiments based on the introduction of pure clays into soil are difficult to interpret, but they are needed to test hypotheses on the role of clay type in relation to microbial balance and disease suppression [51].

D. pH and Elements Affected by pH

It is not possible to consider the role of pH in relation to disease incidence without taking into account the role of several elements, especially calcium, iron, and aluminum, whose availability greatly depends on pH values.

Two methods of determining pH have to be distinguished: pH measurement in a water suspension, expressed as $pH_{(water)}$, or in a salt solution (0.01 M $CaCl_2$ or 1 M Kcl), expressed as $pH_{(salt)}$, which is, in general, 0.6 lower than the $pH_{(water)}$ [52]. Uncertainty about the threshold value of the pH above or below which soils are suppressive to disease is often the result of the fact that the method of measurement is not indicated.

Soils with extreme pH values are often highly suppressive against several plant diseases (Table 1).

At a $pH_{(salt)}$ value below 3.8 to 4.5, diseases induced by *Streptomyces scabies, Phytophthora* spp., *G. graminis, R. solani, Thielaviopsis basicola, Verticillium* spp., and *F. solani* are suppressed [53,59,62,64,65,67,68]. Whether the high proton concentration has a direct effect on the pathogen has not been generally established. More often, the phenomenon of disease suppressiveness is associated with the presence of free aluminium (Al^{3+}) in the soil [62,69,70].

Table 1 Relationship Between Soil pH and Suppressiveness to Diseases

Disease	Pathogen	pH values	References
Black root rot of tobacco	*Thielaviopsis basicola*	<5.0–5.6	53
Clubroot of rape	*Plasmodiophora brassicae*	>7	19
Fusarium wilt of banana	*Fusarium oxysporum* f. sp. *cubense*	>7.6	54
Fusarium wilt of carnation	*F. oxysporum* f. sp. *dianthi*	>6.7	55
Fusarium wilt of tomato	*F. oxysporum* f.sp. *lycopersici*	>7	56, 57
Fusarium root rot of beam	*F. solani* f.sp. *phaseoli*	>7.9	58
Fusarium dry rot of potato	*F. solani* var. *coeruleum*	<5.3	59
Potato scab	*Streptomyces scabies*	>8	60
		<5.2	61
Rhizoctonia damping-off	*Rhizoctonia solani*	<4.2	62
Root rot of pineapple	*Phytophthora cinnamomi*	<3.8	63
Take-all of wheat	*Gaeumannomyces graminis*	<5	64
Tobacco black shank	*Phytophthora parasitica* var. *nicotianae*	<4.2–4.5	65
Verticillium wilt of potato	*Verticillium dahliae*	<3.6	66

Al^{3+} is solubilized at a $pH_{(water)}$ below 5.0 to 5.5 [52]. The quantity of aluminum solubilized at a given pH is a function of the mineralogical composition of the soil [71]. Sandy soils, rich in quartz that is very resistant to the attack of acids and contains little aluminum, or organic soils release less Al^{3+} upon acidification than clay soils [52]. The mechanism of disease suppression by Al^{3+} has been insufficiently studied. Al^{3+} may directly affect inoculum density and/or activity of the pathogen, as shown in vitro for *Verticillium* spp [69] and *F. solani* [A. Ridao (1990) Ph.D. thesis, Université de Rennes], but such demonstrations are lacking in vivo.

Alkaline soils with a $pH_{(water)}$ above 7.8 to 8.0 are highly suppressive to diseases induced by *S. scabies, P. brassicae, Sclerotium* spp., and probably *F. oxysporum* [19,40,54,60]. A chemical explanation for this threshold pH value is lacking. Generally, a pH above 7.8 indicates the presence of $CaCO_3$ [72], a high base saturation of the cation exchange complex, and a high content of divalent cations, such as calcium and magnesium, in the soil solution. High contents of exchangeable calcium and magnesium in soil were shown to control disease caused by *P. brassicae* [73]. Moreover, it has been demonstrated that a high pH favors calcium uptake by roots [73], and that calcium reinforces the wall of root cells and presumably inhibits the activity of polygalacturonase produced by several fungi [74]. However, the importance of a high calcium content in plant tissues as a major factor controlling disease has not been confirmed for Fusarium wilts [56,57]. High pH and high calcium also reduce the water solubility and thus the availability of phosphate and some micronutrients (boron, manganese, iron, zinc) to microorganisms and the host plant [52,75]. In contrast, the water solubility of other micronutrients, such as copper or molybdenum, is less affected by high pH [52]. The role of the availability of micronutrients in soil suppressiveness will be discussed later.

In rare cases, high pH is not associated with calcium but with high salinity. High electric conductivity or osmotic pressure were correlated with soil suppressiveness to Aphanomyces root rot and Fusarium wilts [76].

Little or no correlation has been found between disease suppressiveness and soil pH in the range of 5 to 7, because at intermediate pH values, the soil microbiota is little affected by soil pH.

Although soil pH and calcium content are the most studied factors in relation to soil suppressiveness, their modes of action are not clearly understood. Moreover, extreme pH, high Al^{3+} concentration, and salinity cannot be used to induce soil suppressiveness because they are unfavorable for the growth of most host plants. In the intermediate range of pH found in most agricultural soils, modification of the pH and calcium content can be used together with other soil amendments to reduce the incidence of several diseases, and liming is a commonly used practice [77].

E. Micronutrients

Many papers have reported interactions between macroelements, such as nitrogen, and disease incidence. But in relation to soil suppressiveness, only microelements have been considered.

The availability of microelements to microorganisms and the host plant is involved in soil suppressiveness to several diseases (Table 2). Nevertheless, more precise methods to determine the soil content of biologically available microelements are still needed. An often used method is based on the extraction of microelements with the weak ligand ethylenediaminetetraacetic acid (EDTA), although it may underestimate the availability of microelements for microorganisms; for example, iron for *Pseudomonas* spp. producing siderophores with a strong iron-complexing capacity [24].

Soils with a high content in one or several of the elements iron, magnesium, copper, or zinc were more suppressive to diseases induced by *S. scabies, Aphanomyces* spp., *G. graminis*, and *T. basicola* [80,81,83–85]. Although growth, zoospore formation, and spore germination of *A. euteiches* were completely inhibited in vivo by 50 mg l^{-1} copper or iron [81], a toxic effect of these elements on the pathogen at the concentrations normally found in soil solutions (e.g., 0.1–10 mg l^{-1} Fe^{2+}, 0.3–910 μg l^{-1} Mn^{2+}, 0.009–0.31 μg l^{-1} Cu^{2+}, <0.8–20 μg l^{-1} Zn^{2+}) can, in general, be excluded [52,86,87]. In contrast, a high availability of micronutrients to microorganisms and a generally high nutrient level in soil should contribute to a more diversified microbiota and thus to a more complex and efficient antagonism of pathogens [22]. Moreover, micronutrients are probably involved in defense reactions and in susceptibility of the host plant to disease [88]. Diseases may be suppressed in acid soils, because under acid conditions iron, magnesium, copper, and zinc form water-soluble

Table 2 Relationship Between Availability of Microelements in Soil and Suppressiveness to Diseases

Disease	Pathogen	Microelement	Reference
Fusarium wilt of carnation	*Fusarium oxysporum* f. sp. *dianthi*	Low Fe	23
Fusarium wilt of flax	*F. oxysporum* f. sp. *lini*	Low Fe	78
Fusarium wilt of tomato	*F. oxysporum* f. sp. *lycopersici*	Low Mn, Zn, and Fe	79
Potato scab	*Streptomyces scabies*	High Mn	80
Root rot of pea	*Aphanomyces euteiches*	High Cu and Zn	81
Root rot of avocado	*Phytophthora cinnamomi*	High Mn and Fe	21
Take-all of wheat	*Gaeumannomyces graminis* var. *tritici*	High Mn	82
		High Zn, Fe, Cu, and Mn	83

salts that are more available [52,88]. Adding microelements to alkaline, conducive soils does not always effectively control disease [89], possibly because in alkaline soils, these elements are rapidly precipitated and become unavailable to plants and microorganisms.

Some soils suppressive to Fusarium wilts are poor in microelements, especially in iron [13,23], which has been shown to be necessary for supporting the growth of germ tubes arising from chlamydospores of *F. oxysporum* [90]. The concentration of available iron controls the efficacy of bacteria antagonistic to plant pathogens by enhancing or repressing the production of toxic metabolites. This point will be discussed in part III.C.

In addition to soil pH, other soil factors affect the availability of micronutrients: (1) the microelement content of soil minerals [84]; (2) the age of soil, which affects the degree of alteration of soil minerals and the leaching of microelements from soil [43]; and (3) the redox potential. A low redox potential increases the availability of some microelements, especially of iron and magnesium, which form water-soluble salts in the reduced state. The redox potential is low in fine-textured soils under conditions of high soil moisture and high microbial activity, such as after the incorporation of large quantities of organic matter [91]. The combination of high soil moisture or low redox potential and high levels of micronutrients has been demonstrated in the suppressiveness of two diseases: potato scab is suppressed in moist soils in which the availability of magnesium and iron is high [80], whereas Fusarium wilts are enhanced by impeded drainage and high soil moisture, as these conditions lead to an increase in the availability of iron for the pathogen.

F. Organic Matter

Soils rich in organic matter content tend to be more suppressive to *Phytophthora* induced diseases and to Fusarium root rot of wheat, whereas take-all of wheat and black root rot of tobacco appear to be independent of the organic matter content [21,22,38,43,92]. Most studies on the influence of organic matter on Fusarium wilts have provided inconclusive and, more often, negative results [11,54,93,94].

The organic matter of soil is certainly important in microbial ecology, as it is both the substrate for and the product of microbial activity [95]. In addition, together with clay minerals, organic matter affects soil structure and thus moisture retention and aeration. Especially in soils with a low clay content, organic matter is mainly responsible for the water holding capacity and the CEC of the soils.

Generally, little attention has been directed to the quality of soil organic matter in relation to disease suppressiveness. The effects of different types of organic amendments and of their carbon:nitrogen ratios on disease severity and pathogen suppression have been more intensively studied [96–98], but the role of indigenous organic components on soil suppressiveness remains unclear. The carbon:nitrogen ratio of the soil organic matter was apparently not involved in soil suppressiveness to Fusarium root rot of bean induced by *F. solani* [58]. Soil contains organic matter in the form of carbohydrates, nitrogen-containing constituents (amino acids, amino sugars), aliphatic fatty acids, and complex aromatic carbon (e.g., lignin), in order of their increasing resistance to degradation. Moreover the organic matter of soil may contain active substances, such as vitamins, growth regulators, antibiotics, or toxins. The relative importance of these substances varies as a function of organic matter inputs, soil pH, soil texture, and soil moisture conditions. The composition of soil organic matter may influence the microbial balance in soil and thus also soil suppressiveness.

III. INVOLVEMENT OF MICROBIAL METABOLITES IN DISEASE SUPPRESSION

Although the physicochemical characteristics of soil have an important role, microorganisms and their metabolites are often directly or indirectly responsible for soil suppressiveness to diseases. Indeed, biocidal treatments that destroy most of the living microorganisms without altering too much the physicochemical characteristics of the soil also destroy soil suppressiveness [12,13]. Moreover, the introduction of a small amount of suppressive soil into a disinfested conducive soil may be sufficient to transfer the suppressive effect to the conducive soil [12,41]. These results, which demonstrated the involvement of the saprophytic microbiota stimulated studies devoted to the identification of antagonistic microorganisms and to the analysis of their modes of action.

According to Cook and Baker [15], two types of soil suppressiveness have to be considered: (1) general suppression caused by the activity of the total microbial biomass of the soil acting principally as a nutrient sink, and (2) specific suppression resulting from the activity of a specific population of microorganisms antagonistic to the pathogen. This specific suppression always operates against a background of general suppression, but it is based on more specific interactions between the antagonist and the target organism. As competition for nutrients is mainly responsible for general suppression, this chapter will emphasize specific suppression, which often involves the production of microbial metabolites (toxins, enzymes, antifungal compounds, siderophores) toxic to the pathogen.

There is extensive literature dealing with microorganisms and their metabolites that are antagonistic to plant pathogens. Many of the producers are soil-borne microorganisms, but only few of them have been definitely shown to be involved in the mechanisms of disease suppression in naturally occurring disease suppressive soils. Among them, the bacteria belonging to the group of fluorescent pseudomonads provide a good example of antagonists producing several types of metabolites that are more or less toxic to different types of pathogenic fungi and bacteria [99,100]. In contrast, nonpathogenic *F. oxysporum*, which also appear to be involved in the mechanisms of soil suppressiveness to Fusarium wilts, have not been shown to exert modes of action other than competition for nutrients, competition for infection sites, or induced resistance [101,102]. The presence of an active population of nonpathogenic *Fusarium* is required, but the production of specific metabolites has not been demonstrated.

Some microorganisms, such as species of *Trichoderma* [103], *Gliocladium* [104], and *Thaloromyces* [105], have been extensively studied for their ability to reduce disease incidence or severity when applied to soils or seeds. However, it has never been clearly demonstrated that they were responsible for naturally occurring suppressiveness. Nevertheless, some of these appear to be active through the production of toxic metabolites that are released in soil.

The production of metabolites and their implication in microbial antagonism have been extensively reviewed [100,106,107]. Consequently, this chapter will focus on the description of (1) the biological effects of the major metabolites described (Table 3); (2) the experimental strategies used to demonstrate the implication of these metabolites in disease suppression; and (3) the effect of the environment on the production and the activity of these metabolites.

A. Metabolites Involved in Suppression of Diseases Induced by Soilborne Plant Pathogens

The natural suppressiveness of some soils to Fusarium wilts has been related to competition for iron between the pathogens and the saprophytic microbiota [23,153]. Increases in iron availability by the introduction of the iron-containing chelate Fe-ethylenediaminetetraacetic acid (FeEDTA) to either suppressive or conducive soils increased the growth of the pathogen and disease severity, whereas a reduction in iron availability by the introduction of a strong iron chelator, ethylenediaminedi-*O*-hydroxyphenylacetic acid (EDDHA), reduced the growth of the pathogen and disease severity in both suppressive and conducive soils [23,78,154]. Species of fluorescent pseudomonads, which have been shown to have a major role in soil suppressiveness to Fusarium wilts [13,137], produce siderophores that reduce the availability of iron. Depending on the authors, these siderophores are called either pyoverdine or pseudobactin [110,155]. Numerous studies have related the suppressiveness of fungi by fluorescent pseudomonads to their production of siderophores [28,90,137,139,140,142,143,154,156–159], and several review papers are available [160-165]. Antagonism mediated by pseudobactin has been related to competition for iron because the bacterial siderophores show a higher affinity for iron than fungal siderophores. Pyoverdine or pseudobactin form complexes with iron that are called ferripyoverdine or ferripseudobactin and have a stability constant close to 10^{32} [166]. In contrast,

Table 3 Metabolites Involved in Soil Suppressiveness to Diseases

Metabolite	Antagonist	Pathogen	Disease	Reference
Agrocin 84, Agrocin 434	*Agrobacterium radiobacter* K84	*A. tumefaciens*	Crown gall	108,109
Alkyl pyrones	*Trichoderma harzianum*	*Rhizoctonia solani*	Damping-off	110
Ammonia	*Enterobacter cloacae* EcH1	*Pythium ultimum*	Damping-off	111
Cepacin	*Pseudomonas cepacia* PCI	*Phytophthora capsici*	Phytophthora blight	112
2-ceto-D-glucanate	*P. cepacia* B5	*Pseudomonas solanacearum*	Bacterial wilt	113
Chaetomin	*Chaetomium globosum*	*Pythium ultimum*	Damping-off of sugarbeet	114
Chitinases	*Aeromonas caviae*	*Fusarium oxysporum* f. sp. *vasinfectum* *R. solani* *Sclerotinia rolfsii*		115
	Gliocladium virens			116
	P. stutzeri YPL-1	*F. solani*		117
	T. harzianum P1			118,119
	Serratia marcescens	*F. rolfsii*		120
2,4-diacetyl-phloroglucinol	*P. fluorescens* PFM2	*Septoria tritici*		121
	P. aureofaciens	*Gaeumannomyces graminis* var. *tritici*	Take-all	122
	P. fluorescens CHAO	*Thielaviopsis basicola*	Black root rot of tobacco	123
	Pseudomonas sp. F113	*P. ultimum*	Damping-off	124
Gliotoxin	*Gliocladium virens* G12	*P. ultimum, R. solani*	Damping-off	125
Gliovirin	*G. virens*	*P. ultimum, R. solani*	Damping-off	126
Glucanases	*P. cepacia*	*P. ultimum, R. solani, S. rolfsii*		127
	P. stutzeri YPL-1	*F. solani*		117
	T. harzianum			118
Glucose oxidase	*Talaromyces flavus* Tf1	*Verticillium dahliae*	Verticillium wilt	128
Hydrogen cyanide	*P. fluorescens* CHAO	*T. basicola*	Black root rot of tobacco	129
	P. corrugata 2140	*G. graminis* var. *tritici*		130

Oomycin A	*P. fluorescens* Hv37	*P. ultimum*	Damping-off	131,132
Phenazines	*P. fluorescens* 2-79	*G. graminis* var. *tritici*	Take-all	133
		Tilletia laevis	Common bunt of wheat	134
	P. aeruginosa LECI	*S. tritici*	Leaf blotch	135
	P. aureofaciens Q2-87	*G. graminis.* var. *tritici*	Take-all	136
Pseudane-A	*P. cepacia* PCII	*P. capsici*	Phytophthora blight	112
Pseudobactin, Pyoverdin	*P. fluorescens* B10	*F. oxysporum*	Fusarium wilt	137
	P. putida WCS358	*F. oxysporum*	Fusarium wilt	138,139
		Minor pathogens	Yield depression in potato	140
	Pseudomonas sp. B324	*Pythium* sp.	Root rot of wheat	141
	P. fluorescens 3551	*P. ultimum*	Damping-off of cotton	142
Pyochelin	*P. aeruginosa* 7NSK2	*P. ultimum*	Damping-off of tomato	143
Pyocyanin	*P. fluorescens*	*R. solani*	Brown girdling rot of root	144
Pyoluteorin	*P. fluorescens* Pf-5	*P. ultimum*	Damping-off of cotton	145
	P. fluorescens CHAO	*P. ultimum*	Damping-off	146
Pyrrolnitrin	*P. fluorescens* Pf-5	*Pyrenophora tritici repentis*	Tan spot disease	147
		R. solani	Damping-off	148
	P. cepacia B37w	*F. sambucinum*	Potato dry root rot	149
	P. fluorescens BL915	*R. solani*	Damping-off of cotton	150
Tropolone	*Pseudomonas* sp.	*Alternaria* sp.		151
		Cladosporium sp.		
		Diplodia sp.		
		Fusarium sp.		
		Helminthosporium sp.		
		Pyricularia sp.		
		Pythium sp.		
		Rhizoctonia sp.		
Zwittermicin	*Bacillus cereus* UW85	*Phytophthora medicaginis*	Damping-off of alfafa	152

fusarinines, siderophores produced by *Fusarium*, form a complex with iron that has a stability constant of about 10^{29} [23]. Moreover, the antagonistic activity of pseudobactin in vitro and in vivo is suppressed when chelated with iron [137,158,167]. These data suggest that participation of fluorescent pseudomonads in soil suppressiveness to Fusarium wilt is related to pseudobactin-mediated competition for iron [168].

The involvement of siderophores in the suppression of the damping-off of cotton [142] and tomato [143] and in the root rot of wheat [141] has also been demonstrated. Fluorescent pseudomonads may also suppress deleterious fluorescent pseudomonads responsible for yield depression in short rotation of potato [140].

However, pyoverdines only account for a part of the antagonistic activity of fluorescent pseudomonads. Loper and Lindow [169] showed that pyoverdine was responsible for 95% of the suppression by *P. fluorescens* strain JL3551 against *P. ultimum* on cotton, but only for 5% of that by *P. fluorescens* strain Pf-5 against *P. ultimum* on cucumber.

Other metabolites have been implicated in disease suppression. Some are responsible for microbial antibiosis. According to Fravel (106), antibiosis refers to "antagonism mediated by specific or nonspecific metabolites of microbial origin, by lytic agents, enzymes, volatile compounds, or other toxic substances." A variety of different antibiotics, bacteriocins, enzymes, and volatile compounds have been described to be involved in the suppression of different pathogens and plant diseases.

Soil suppressiveness to take-all disease of wheat was associated with a high proportion of fluorescent pseudomonads that were inhibitory to *G. graminis* var. *tritici* [170,171]. The production of antimicrobial agent, especially of phenazines and 2,4-diacetylphloroglucinol, was shown to be the primary mode of antagonistic action against *G. graminis* var. *tritici* [122,133,172]. The production of phenazines has also been implicated in the suppression of *Septoria tritici* and *Tilletia laevis* on wheat [134,135]. 2,4-diacetylphloroglucinol produced by *P. fluorescens* strain CHAO was also shown to be responsible for the suppression of black root rot caused by *Thielaviopsis basicola* and damping-off caused by *P. ultimum* [123].

Soil suppressiveness to black root rot in Switzerland was correlated with fluorescent pseudomonads [173]. Suppression of black root rot by *P. fluorescens* strain CHAO, isolated from a natural suppressive soil, was shown to be related to the production of 2,4-diacetylphloroglucinol and cyanide [123,129,174]. However, depending on the host plant and the level of its production, cyanide may be either beneficial or harmful to the growth of the plant. Schippers et al. [163] showed that the deleterious effect of certain fluorescent pseudomonads on potato was associated with cyanide production, whereas Voisard et al. [129]

attributed to the production of cyanide the beneficial effect of *P. fluorescens* CHAO on tobacco.

The production of various enzymes by different antagonistic microorganisms has been suggested to be involved in the suppression of soilborne plant pathogens. Since chitin and glucans are major components of the cell wall of most fungi, chitinases and glucanases have been implicated in mycoparasitism. These enzymes are produced by different fungi (e.g., *Gliocladium virens, Trichoderma harzianum*) and bacteria (e.g., *Aereomonas caviae, Micrococcus luteus, Pseudomonas cepacia, P. stutzeri, Serratia marcescens*) [115–120]. Although these lytic enzymes have been shown to be produced by antagonistic soil fungi and bacteria, it has never been clearly established that the producer organisms are responsible for naturally occurring soil suppressiveness.

Numerous studies have been devoted to the production of volatile metabolites by bacteria that are responsible for inhibition of fungal growth and sporulation [175–178]. Despite the clear demonstration of the antagonism exerted by these volatiles on fungi, there is no evidence of their involvement in the mechanisms of soil suppressiveness. However, all the metabolites produced by soil bacteria and fungi that have some inhibitory activity on pathogens may contribute to some extent to soil suppressiveness.

As emphasized by Loper et al. [179], a single antifungal metabolite generally does not account for all the antagonistic activity of a biocontrol agent. Antagonistic strains usually produce several toxic metabolites. For example, *P. fluorescens* CHAO produces (1) siderophores: pyoverdine, salicylate, and pyochelin; (2) a volatile compound: cyanide; and (3) antibiotics: 2,4-diacetylphloroglucinol, pyoluteorin, and pyrroniltrin [100,180,181]. Synthesis of several of these metabolites (2,4-diacetylphloroglucinol, cyanide, pyoluteorin) by *P. fluorescens* CHAO has been shown to be regulated by the same operon called *gacA* (global antibiotic and cyanide control) [182]. *P. fluorescens* 2-79 produces (1) a siderophore: pyoverdine; and (2) antibiotics: phenazine-1-carboxylic acid and an iron-regulated antifungal factor identified as anthranilic acid [183–185]. This example illustrates how complex can be the modes of action of a single strain of bacterial antagonist.

The suppressiveness of natural soil to fungal diseases might be related to the diversity of the indigenous microbial populations and of their microbial metabolites. In an attempt to simulate the complex interactions responsible for soil suppressiveness, several workers have studied a combination of antagonistic microorganisms in biocontrol experiments [90,159,186–190]. Increased suppressiveness has been reported when different strains of fluorescent pseudomonads [90,186,190] have been combined or when a strain of fluorescent pseudomonad has been associated with a strain of nonpathogenic *F. oxysporum* [159,187–189]. Synergy for carbon competition by the nonpathogenic *F. oxysporum* strain Fo47 and iron competition by the *P. putida* strain WCS358 was shown to be responsible for the increased suppression achieved by the microbial association, when compared with the suppression achieved by each antagonistic microorganism separately [167]. Similarly, combining two chitinolytic enzymes from *T. harzianum* resulted in a synergistic increase of antifungal activity against a wide range of fungi [119]. The synergistic antifungal activity of an endochitinase in combination with gliotoxin from *G. virens* was attributed to a better diffusion of the antibiotic into the cells as a result of the degradation of the cell wall by the enzyme [116].

B. Strategies for the Demonstration of the Implication of Microbial Metabolites in Disease Suppression

Different approaches have been used to provide evidence for the implication of a given metabolite in the suppression of a fungal disease by a control agent.

The traditional approach is based on the demonstration of (1) the production, in vitro, of the metabolite and the antagonistic activity of the purified metabolite against the target pathogen: (2) the relation between disease suppression in vivo and the production, in vitro, of the metabolite by a collection of similar microbial strains; and (3) the ability of the metabolite introduced into soil or onto seeds to mimic the antagonistic activity of the biocontrol agent in situ.

Such an approach has been applied, for example, with bacterial siderophores. Siderophores are produced in vitro only when the concentration of available iron is low [191]. Reduction of fungal growth in vitro by different isolates of fluorescent pseudomonads was shown to occur only when siderophores were produced [137,156], and, in the presence of iron, both the synthesis of pseudobactin and bacterial antagonism were suppressed. Growth in vitro of *P. ultimum* was reduced by purified pyoverdine from a strain of *P. tolaasi* [158]. Conidial germination and mycelial growth in vitro of *F. oxysporum* were also decreased in the presence of purified pseudobactin from *P. putida* WCS358. Furthermore, a positive correlation was established between the amount of siderophores produced in vitro by different pseudomonad isolates and their ability to suppress both the germination of chlamydospores in soils and Fusarium wilts [90,157]. Introduction of partially purified pseudobactin from *P. fluorescens* B10 to soil reduced the level of germination of *Fusarium* chlamydospores [156] and the severity of Fusarium wilt [137]. The same approach demonstrated the implication of the antibiotic, pyrrolnitrin, in the biological control of diseases caused by soilborne pathogens [192].

A second approach, which has been used more recently, is based on the demonstration of (1) a reduction of the in vitro antagonism against the target pathogen and a reduction in vivo of disease suppression by mutants defective in the production of the studied metabolite; and (2) the restoration of the biocontrol ability of the defective mutants when complemented with DNA sequences from the wild-type strain.

This approach was used to demonstrate the role of siderophores and antibiotics in disease suppression by different pseudomonad strains. The elimination of the ability of *P. putida* WCS358 to synthesize pseudobactin, by single insertion of Tn5 (Sid^-), resulted in a loss of its ability to stimulate potato growth, which was suggested to be related to the production of pseudobactin [168]. Similarly, growth promotion of wheat caused by the suppression of Pythium root rot was recorded in the presence of *Pseudomonas* sp. B324 but not in the presence of a Tn5-Sid^- mutant of B324 [141]. The implication of pseudobactin in the protection by *P. fluorescens* 3551 of cotton against Pythium damping-off [142] and by *P. aeruginosa* 7NSK2 against damping-off of tomato [143], as well as the protection by *P. putida* WCS358 of carnation against Fusarium wilt in the presence [160] or absence [139] of nonpathogenic *F. oxysporum*, was also demonstrated using Tn5-Sid^- mutants.

Thomashow and Weller [133] were the first to demonstrate by genetic means the implication of antibiotic production in disease suppression. A Tn5-induced mutant of *P. fluorescens* 2-79 defective in phenazine production (Phz^-) showed both reduced in vitro antagonism against *G. graminis* var. *tritici* and reduced suppression of take-all of wheat compared with the wild-type strain. This approach was also subsequently used to assess the role of (1) oomycine in the suppression of Pythium damping-off of cotton by *P. fluorescens* Hv37a [132]; (2) cyanide in the suppression of black root rot by *P. fluorescens* CHAO [129]; (3) 2,4-diacetylphloroglucinol in the suppression of tobacco black root rot, of Pythium damping-off by *P. fluorescens* CHAO, and of take-all of wheat by *P. aureofaciens* strain Q2-87 [122,123]; and (4) pyrrolnitrin in the suppression of damping-off of radish by *P. cepacia* and of tan spot disease of wheat by *P. fluorescens* [147,193].

When Phz^- mutants of *P. fluorescens* 2-79 were complemented with homologous DNA from a 2-79 genomic library, production of phenazine, inhibition of *G. graminis* var. *tritici*, and suppression of take-all were restored [133]. Furthermore, a linear relation was established between the number of take-all lesions resulting from primary infections and root colonization by the wild-type 2-79 and the complemented strains. Such relation could not be established with the Phz^- mutant [194].

Genetic methods were also used to assess the implication of agrocin 84 and 434 in the suppression of crown gall by *Agrobacterium radiobacter* K84. This bacterium is sold to control the crown gall induced by *A. tumefasciens* in many crops. These two bacterial species are closely related, but the biocontrol agent *A. radiobacter* produces bacteriocins toxic to the pathogenic bacteria. The synthesis of agrocins being encoded by genes carried on a plasmid, plasmid-free derivatives of strain K84 deficient in the production of agrocin 84 and 434 were less efficient than in controlling the wild-type of crown fall [109].

Further evidence for the implication of different metabolites in biocontrol was obtained by genetic engineering of bacterial strains. Introduction of DNA encoding the synthesis of an antagonistic metabolite in a nonproducing strain usually resulted in increased disease suppression by the modified strain when compared with the parental strain. Several examples of such strategy have been described. Insertion of the genes responsible for the synthesis of 2,4-diacetylphloroglucinol into nonproducer strains improved the biological control of Pythium damping-off of sugar beet and the in vitro antagonism of *G. graminis* var. *tritici* by *Pseudomonas* M114 and *P. fluorescens* 2-79, respectively [122,195]. Introduction of DNA responsible for cyanide synthesis from *P. fluorescens* CHAO into *P. fluorescens* P3 resulted in better protection of tobacco against black root rot by the modified strain compared with the parental one [129]. Modification of the bacterial genome to enhance the production of a bacterial metabolite (2,4-diacetylphloroglucinol) was performed with different strains, which were usually more efficient than the parental strains in suppressing disease [146]. However, the increased production of 2,4-diacetylphloroglucinol had a deleterious effect on cress [146].

To show unequivocally the involvement of a microbial metabolite in disease suppression, its presence in the rhizosphere of plants inoculated with the biocontrol agent must be demonstrated. Phenazine was isolated from roots and rhizosphere of wheat inoculated with *P. fluorescens* 2-79 or with a Phz$^-$ mutant complemented with DNA from strain 2-79, but not from root and rhizosphere inoculated with the Phz$^-$ mutant [196]. Similarly, 2,4-diacetylphloroglucinol was isolated from the rhizosphere and roots of wheat inoculated with *P. fluorescens* CHAO but not from the rhizosphere and roots of wheat inoculated with a mutant deficient in the production of 2,4-diacetylphloroglucinol [123]. Although pseudobactin has not been isolated from soil or roots, Bakker et al. [140] indirectly demonstrated the synthesis of pseudobactin in the rhizosphere of potato inoculated with fluorescent pseudomonads. Root colonization by Tn5-Sid$^-$ mutants of fluorescent pseudomonads was significantly higher when these mutants were introduced together with wild-types that produced pseudobactins that could be used by the mutants. Pseudobactin produced by *P. fluorescens* B10 in the rhizosphere has been detected by immunoassay with monoclonal antibodies [197].

The extraction of metabolites from soil and rhizosphere is difficult. The use of reporter genes, fused to the promoter of biosynthetic operons, allows the determination, in situ, of the expression of the biosynthetic genes. Such bacterial constructs have been used to establish the expression of genes responsible for the synthesis of oomycine [132] and pyoverdine [198] in the spermosphere of cotton and bean, respectively.

Despite the numerous studies on the implication of lytic enzymes in microbial antagonism, direct evidence for their implication in disease suppression is still lacking [199]. Most of these studies were based mainly on the demonstration in vitro of the production of enzymes by the biocontrol agent and on microscopic observations that showed the lytic activities [115,117,120]. Similarly, although a large number of antibiotics and volatile compounds have been shown to be produced in vitro by *T. harzianum* [118,119], direct evidence for their implication in disease suppression is lacking.

C. Effect of the Environment on the Production and Activity of Microbial Metabolites

There is often a lack of correlation between in vitro antagonism and in situ biocontrol [106]. For example, although pyoluteorin has been shown to be responsible for the antagonism of *P. fluorescens* Pf-5 against *Pythium* in vitro, it contributes little to the suppression of Pythium damping-off of cucumber [200]. Biocontrol is also characterized by its inconsistency [99]. One explanation for these two aspects of biocontrol is related to the effects of the environment on both the production and the activity of microbial metabolites.

Various abiotic factors affect microbial metabolism. Temperature influenced siderophore production by fluorescent pseudomonads [201,202] in that synthesis of siderophores occurred only at temperatures below 28°C. Increase of temperature up to 33°C suppressed growth of some bacterial strains, but this suppression was reversed by the addition of pseudobatin [203]. Temperature also influenced the synthesis of antibiotics: for example, production of oomycine by *P. fluorescens* HV37a was highest at 20°C and was significantly reduced at 16 and 24°C [131]; production of gliotoxin by *G. virens* GL-21 increased from 15 to 30°C [125]; and production of 2,4-diacetylphloroglucinol was maximal at 12°C [124].

Soil texture may affect antibiotic production and activity. Humus and clay colloids may absorb and inactivate basic antibiotics [204]. Clay type was shown to affect the ability of *P. fluorescens* CHAO to suppress black root rot of tobacco. Under gnotobiotic conditions in artificial soil containing vermiculite, suppression was efficient, whereas when vermiculite was replaced by illite, suppression was significantly reduced. This difference was related to the higher production of cyanide in the presence of vermiculite than of illite [78]. Suppression of take-all by *P. fluorescens* 2-79, which produces phenazines, was positively correlated with the percentage of sand but negatively correlated with the percentage of silt, clay, and organic matter [205].

The pH of soil affects microbial antagonism in various ways. Soils suppressive to Fusarium wilt are usually alkaline (see section II). The availability of iron is affected by pH, the higher the soil pH the lower the concentration of Fe^{3+} in solution [206]. Therefore, high pH reduces iron availability and favors siderophore production and iron competition. In contrast, suppression of take-all by *P. fluorescens* 2-79, which was related to the production of phenazines, occurs over a wide range of pH [207]. However, the antagonistic activity of purified phenazines was higher at low than at high pH. Suppression of take-all by *P. fluorescens* CHAO was not influenced by soil pH [208].

The availability of different ions also affects production of bacterial metabolites. The most documented example is the synthesis of siderophores, which is regulated by the concentration of available iron. The synthesis of siderophores by fluorescent pseudomonads was shown to be less sensitive to the concentration of iron in the medium than that by *Fusarium* spp. [209,210]. The availability of iron being affected by pH, Duijff et al. [211] showed that *Pseudomonas* sp. WCS417r significantly reduced Fusarium wilt of carnation at pH 7.5, but not at pH 6.5 and 5.5, and suggested that this difference was related to the higher in vitro production of pseudobactin and the greater antagonism of *F. oxysporum* f. sp. *dianthi* at pH 7.5 than at pH 6.5 and 5.5. In contrast, iron is required for cyanide production [212,213], and suppression of black root rot by *P. fluorescens* CHAO only occurs when iron availability is sufficient [123]. Suppression of take-all by *P. fluorescens* 2-79 was positively correlated with the content in soil of sodium, zinc, and ammonium-nitrogen and was negatively correlated with the cation exchange capacity, exchangeable acidity, content of manganese and iron, total carbon, and total nitrogen [205]. The positive correlation between disease suppression and zinc content was suggested by Weller and Thomashow [199] to be related to a higher production of phenazines. The addition of zinc to soil low in this element improved the suppression of take-all by strain 2-79, and the addition of zinc to liquid culture medium increased the production of phenazines [214]. Suppression of take-all by *P. fluorescens* CHAO was not affected by the source of nitrogen nor by the level of manganese [208].

Microbial metabolism in the rhizosphere depends on the carbon sources available [203]. The production of oomycine is induced by glucose [215], whereas the production of 2,4-diacetylphloroglucinol by fluorescent pseudomonads was low in the presence of glucose or sorbose but high in the presence of sucrose, fructose, or mannitol [124]. Glycine is required for the synthesis of cyanide [216]. The production of siderophores by fluorescent pseudomonads occurs with a wide range of different carbohydrates [203].

In addition to abiotic factors, the soil microbiota affects the production and stability of microbial metabolites. As most microorganisms produce siderophores [191], they all contribute to competition for iron, suggesting that, as for carbon competition, the total microbiota participates in iron competition (general suppression) [153]. Consequently, the larger the soil biomass, the greater is the competition for iron and thus siderophore production. This hypothesis is consistent with the data recorded by Duijff et al. [211], who showed that the suppression of Fusarium wilt by *P. putida* WCS358, based on pseudobactin production, occurred in "raw" soil but not in steamed soil. The microbiota has also been suggested to be responsible for the degradation of antibiotics [204].

IV. PROSPECTS FOR DISEASE MANAGEMENT

One purpose for studies on soil suppressiveness is the identification of abiotic and biotic factors that are responsible for disease suppression. Once these have been identified, it may be possible to manipulate these factors to induce suppressiveness in conducive soils and, therefore, to control diseases through environmentally friendly techniques.

With the exception of liming, it is not practical to modify the physicochemical properties of soil on a large scale. The addition of different types of clay to a conducive soil only represents an experimental approach with which to analyze the modes of action of these soil constituents, although clay amendments have been suggested in some cases. However, most agricultural practices have an impact on the microbial balance in soil, either by a direct action on some microbial populations or more frequently through an indirect action on abiotic soil properties.

One of the most traditional ways to control diseases is to use crop rotations that include a nonhost plant that can "sanitize" the soil. However, continuous monocropping of a susceptible plant can also lead, after several years of high damages, to a situation of disease suppression.

Fertilization may be another method by which disease can be controlled by increasing the resistance of the host, reducing the activity of the pathogen, or favoring the development of antagonistic microorganisms. The interactions between abiotic and biotic factors are so complex that it is difficult to analyze the mechanisms by which a given cultural practice contributes to limiting disease incidence. However, specific fertilization practices can be devised to control certain diseases induced by soilborne plant pathogens.

Recent progress in the understanding of the biotic interactions involved in disease suppression enable the development of inoculation procedures that are applied to soil or seeds to control diseases through microbial antagonism. These techniques are still in their infancy, but they indicate the directions to biological control of diseases caused by soilborne plant pathogens. It is convenient to distinguish between the enhancement of an existing potential for suppression and the introduction of selected strains of antagonistic microorganisms. However, both approaches required a thorough knowledge of the soil characteristics that are necessary to ensure the survival and activity of the beneficial microorganisms.

It is not possible to review here all the literature dealing with agricultural practices in relation to disease control. Hence, only a few examples will be used to illustrate how it is possible to improve the management of disease.

A. Crop Rotation

The literature provides many examples of soils that became disease suppressive after long-term monocropping of the host plant. Weinhold et al. [217] reported a decrease in the severity of potato scab in irrigated fields: after 13 years, the severity of scab was less with continuous potatoes than with 2- or 3-year rotations. Hyakumachi et al. [218] observed a decline in Rhizoctonia root rot with monoculture of sugar beet in naturally infested fields in Japan, and Sneh et al. [219] reported the buildup of suppressiveness to Fusarium wilt in infested experimental plots used to screen for resistant varieties of melon. In addition, the phenomenon of a decline in take-all linked to monoculture of wheat has been extensively demonstrated [220].

These suppressive soils represent ideal situations with which to characterize the microbial interactions responsible for disease suppression. However, long-term monocropping to control diseases caused by soilborne plant pathogens cannot be recommended, as farmers cannot afford the several years of severe damage until the soil microbiota has reached an equilibrium state that is responsible for reduced disease incidence.

These examples of suppressiveness caused by monoculture of the host plant occulted the fact that crop rotation is probably the most efficient way to prevent the buildup of the pathogen population and the increase of disease incidence with time. Cook and Veseth [221] stated that biological control of soil-inhabiting pathogens of wheat is accomplished by not growing wheat in the same field more frequently than every second or third year.

The beneficial effect of crop rotation depends on the host range of the pathogen and its ability to survive in soil in the absence of its host plant. Crop rotation prevents the buildup of populations of specialized pathogens, but it is ineffective in controlling damping-off diseases induced by *Pythium* spp., which have a wide host range. Crop rotation may also fail to control highly specialized pathogens, such as *formae speciales* of *F. oxysporum*. These fungi are able to survive for long periods, either saprophytically or as dormant chlamydospores, in soil, and a very low inoculum density is sufficient to induce disease.

The efficacy of crop rotation depends mainly on the susceptibility of the pathogen during its saprophytic survival in soil to the competition and the antibiosis exerted by the soil microbiota, knowing that a nonhost plant may modify the microbial balance and enhance the proportion and activity of microorganisms antagonistic to the pathogen. It has been well established that root exudates stimulate the growth and activity of microorganisms, with the population density being greater in the rhizosphere than in the bulk soil [222].

The culture of a nonhost plant may affect the diversity of microbial communities and even the diversity of a given population at the infraspecific level. Lemanceau et al. [223] demonstrated that flax and tomato select some phenons of fluorescent *Pseudomonas* spp., and that the bacteria dominant in the rhizosphere and within the root tissues are rare in soil. The phenons specific to flax share in common the ability to utilize some nutrients that are probably present in the root exudates of flax. Hence, it may be possible to use either the plant or the specific nutrients to modify the diversity within populations of fluorescent *Pseudomonas* spp. and enhance the population of antagonistic pseudomonads or other plant growth promoting rhizobacteria (PGPR).

Another example is provided by a study on the influence of cover plants on soil suppressiveness to Fusarium wilt of oil palm. A leguminous plant grown under palm trees enhances soil suppressiveness, which has been correlated with an increase in the population density of the nonpathogenic fusaria in the rhizosphere of the legume plant [224].

These two examples show that the influence of the plant cannot only modify the balance between microorganisms but can also affect the diversity among microorganisms belonging to the same species. These subtle changes are difficult to assess, but they could be very important in relation to disease suppression. Much more research is needed before crop rotation can be utilized to voluntarily modify the microbial balance in a direction favorable to crop health.

B. Fertilization and Liming

As stated above, there is no evidence for an important role of major nutrients in soil suppressiveness to diseases. However, fertilization practices may help to limit disease severity in infested fields. The form and the application rate of nutrients affect plant growth and physiology and therefore defense reactions to the pathogen. Fertilizers may also directly modify some physicochemical properties of soil, such as pH and availability of different ions, which may indirectly affect the activity of the pathogen and its relations with the saprophytic microbiota. The book *Management of diseases with macro- and micro-elements* [225] reviews the most characteristic examples of disease reduction linked to appropriate fertilization.

Nitrogen has been extensively studied in relation to most of the diseases induced by soilborne plant pathogens. The form of nitrogen appears to be more important than the application rate: for example, NO_3-N contributes to reduced incidence of Fusarium wilts, but NH_4-N contributes more in soils infested with *G. graminis*. In the latter case, however, there are contradictory reports. The influence of nitrogen depends on many other soils factors such as type, pH, moisture, temperature, tillage, and availability of other nutrients [226]. Only an integrated approach can lead to the control of diseases through nutrient amendments.

Jones et al. [227] developed such an approach in Florida to control Fusarium wilts on vegetable and ornamental crops. NO_3-N associated with lime to increase the pH to 7 in the presence of low phosphorus content were the best conditions for reducing the incidence of Fusarium wilt on tomato. The mechanisms of control are not totally understood; they appear to involve a direct effect on the pathogen and an indirect effect on antagonistic microorganisms such as bacteria and actinomycetes, which are favored by high soil pH. Huber [226] and Simon and Sivasithamparam [228] discussed the possible effects of mineral nutrition on root growth, root exudation, plant susceptibility, soil pH, changes in the soil microbiota, and reduced virulence of *G. graminis*, showing the great complexity of interactions responsible for control of take-all. Sarniguet et al. [82] showed that depending on the level of natural inoculum and time of application, ammoniacal fertilizers can decrease the level of soil conduciveness and affect either the saprophytic or the parasitic phase of the development of *G. graminis*.

The mechanisms by which fertilization contributes to disease control are not fully understood. For example, even the mode of action of liming, a practice used for many years to control club root of crucifers, is still unknown [229].

C. Solarization

Solar heating of soil has been proposed as a method to control soilborne plant pathogens and has been successfully used in warm climates, such as in Israel, Greece, Italy, California, and Florida [230,231]. One effect of solar heating is the destruction of the pathogen by lethal temperatures or its weakening by sublethal temperatures; another effect is the modification of the microbial balance, which can result in the establishment of soil suppressiveness. Tjamos and Paplomatos [232] observed that *Talaromyces flavus*, a fungus antagonist to *V. dalhiae,* not only survived after solarization, but its population density increased and probably contributed to the long-lasting effect of the control induced by solar heating.

In other cases, the efficacy of solar heating has not been linked with an increase in a specific population of antagonists but with a shift in the microbial balance that favored the saprophytic microbiota, thereby making the establishment of the pathogen more difficult. This suppressiveness is probably the result of competition for nutrients. Soil solarization is a good example of a cultural practice based on a physical phenomenon that can induce suppressiveness in soil and contribute to the control of diseases through nonpolluting methods.

D. Microbial Inoculation of Soil

Inoculation of soil or seeds with antagonistic microorganisms has been used for many years to control diseases induced by soilborne plant pathogens. Evidence that naturally occurring suppressiveness is often based on microbial interactions stimulated research in this field with an attempt to select biological control agents [2,233].

Many microorganisms (e.g., different species of actinomycetes, bacteria, and fungi) have been isolated from soil and selected for their ability to antagonize pathogens in vitro or to decrease disease severity when introduced into conducive soils or disinfected suppressive soils. Research has focused on the modes of action of these microorganisms and on the technological problems involved in the mass production, formulation, and application of the biological control agents. Many recent review papers describe the current status of biological control and the need for more research in this area [234,235].

In conclusion, it must be emphasized that the introduction into a soil of a single antagonistic strain does not result in the establishment of stable suppressiveness in that soil. Although it is possible to decrease disease severity by microbial inoculation of soil, this beneficial effect is usually of short duration. Soil suppressiveness is the result of coupled interactions between the abiotic characteristics of soil and several microbial populations. Different mechanisms usually contribute to making a soil suppressive to diseases.

Two different approaches are being followed to improve the efficacy of biological control. One consists of combining several microorganisms that belong to the same or different species and that have different modes of action, mimicing the complexity of the microbial community responsible for natural suppressiveness (see section III.A). The second approach is based on genetic manipulation of the antagonistic microorganisms, with an attempt to introduce into the same strain genes that encode for several modes of action that contribute to the antagonistic activity [236]. However, to be effective, the engineered microorganism or the mixture of natural microorganisms must find in soil abiotic properties favorable not only to their survival but also to the expression of their antagonistic activity. Much more research is needed on the relations between the physicochemical properties of soil and microbial activity to be able to create stable suppressiveness in conducive soils.

V. CONCLUSION

Many field observations made in different countries under different climates have shown that some soils are naturally suppressive to plant diseases induced by soilborne pathogens. Considering all the examples of suppressive soils reported in the literature, it appears that suppressiveness is not an exceptional phenomenon; it applies to almost all the most important soilborne pathogens.

Extensive studies have been devoted to soil suppressiveness by soil microbiologists and plant pathologists willing to analyze the mechanisms responsible for soil suppressiveness in order to learn how to induce suppressiveness in conducive soils and therefore to control diseases through environmentally friendly techniques.

All studies dealing with naturally occurring disease suppressive soils emphasized the great complexity of mechanisms that involve both the physicochemical properties and the microbial communities of the soils.

Several examples clearly demonstrated the involvement of soil abiotic properties such as texture, structure, water potential, pH, nature of the clays, and content of micronutrients. These factors may directly inhibit the growth or activity of the pathogens, but more often they indirectly affect the pathogens through modifications of the microbial balance. Indeed, many species of bacteria and fungi have shown their ability to antagonize pathogens through competition for nutrients or production of metabolites toxic to other fungi or bacteria. There exists a great diversity of metabolites (siderophores, antibiotics, enzymes, volatile compounds) involved in antagonistic interactions between the pathogens and the saprophytic microbiota responsible for soil suppressiveness. Most of these interactions have been demonstrated in vitro or in microcosmes of soils, and it was difficult to assess their exact role in the mechanisms of soil suppressiveness. During the last decade, the use of molecular tools enabling the obtention of mutants deficient for the production of a given metabolite has permitted a clear demonstration of the role of these metabolites in the interactions between the antagonists and the pathogens. In a few cases, it was possible to show that the antagonistic metabolites are really produced in soil and to establish correlations between their production in soil and soil suppressiveness to diseases. However, natural suppressiveness can never be ascribed to a single mechanism or a single population of antagonistic microorganisms. It always implies complex interactions between soil abiotic factors and the saprophytic microbiota.

If more research is needed to better understand the mechanisms responsible for soil suppressiveness, actual knowledge already allows us to propose some disease management strategies based on cultural practices and microbial inoculation of soils. The recent evolution of the concept of sustainable agriculture and the necessity to better protect the environment should favor the use of these nonpolluting control strategies.

ACKNOWLEDGMENTS

The authors are thankful to G. Stotzky for critical review of the manuscript.

REFERENCES

1. Atkinson, G. F. 1892. Some diseases of cotton. Alabama Agricultural Experiment Station Bulletin 41.

2. Baker, K. F., and R. J. Cook. 1974. Biological control of plant pathogens. W. H. Freeman, San Francisco.
3. Huber, D. M., and R. W. Schneider. 1982. The description and occurrence of suppressive soils, p. 1–7. *In* R. W. Schneider (ed.), Suppressive soils and plant diseases. A.P.S. Press, St. Paul, Minnesota.
4. Alabouvette, C., Y. Couteaudier, and J. Louvet. 1982. Comparaison de la réceptivité de différents sols et substrats de culture aux fusarioses vasculaires. Agronomie 2: 1–6.
5. Linderman, R. G., L. W. Moore, K. F. Baker, and D. A. Cookrey. 1983. Strategies for detecting and characterising systems of biological control of soil-borne plant pathogens. Plant Dis. 67:1058–1064.
6. Amir, H., and C. Alabouvette. 1993. Involvement of soil abiotic factors in the mechanisms of soil suppressiveness to Fusarium wilts. Soil Biol. Biochem. 25:157–164.
7. Kobayashi, N., and W. H. Ko. 1985. *Pythium splendens*—suppressive soils from different islands of Hawaii. Soil Biol. Biochem. 17:889–891.
8. Reinking. O. A., and M. M. Manns. 1933. Parasitic and other *Fusaria* counted in tropical soils. Z. Parasitenk. 6:23–75.
9. Walker, J. C., and W. C. Snyder. 1933. Pea wilt and root rots. Wisconsin Agricultural Experiment Station Bulletin 424.
10. Stover, R. H. 1962. Fusarium wilt (Panama disease) of bananas and other *Musa* species. CMI, Phytopathol. Pap. 4. 117 p.
11. Stotzky, G., and T. Martin. 1963. Soil mineralogy in relation to the spread of Fusarium wilt of banana in Central America. Plant Soil 18:317–337.
12. Louvet, J., F. Rouxel, and C. Alabouvette. 1976. Recherches sur la résistance des sols aux maladies. I—Mise en évidence de la nature microbiologique de la résistance d'un sol au développement de la fusariose vasculaire du melon. Ann. Phytopathol. 8:425–436.
13. Scher, F. M., and R. Baker. 1980. Mechanism of biological control in a *Fusarium*-suppressive soil. Phytopathology 70:412–417.
14. Hornby, D. 1983. Suppressive soils. Annu. Rev. Phytopathol. 21:65–85.
15. Cook, R. J., and K. F. Baker. 1983. The nature and practice of biological control of plant pathogens. A.P.S. Press, St. Paul, Minnesota.
16. Stotzky, G. 1986. Influence of soil mineral colloids on metabolic processes, growth, adhesion, and ecology of microbes and viruses, p. 305–428. *In* P. M. Huang and M. Schnitzer (ed.), Interactions of soil minerals with natural organisms and microbes. Soil Science Society of America, Madison, Wisconsin.
17. Cook, R. J., and R. I. Papendick. 1972. Influence of water potential of soils and plants on root disease. Annu. Rev. Phytopathol. 10:349–374.
18. Hancock, J. G. 1979. Occurrence of soils suppressive to *Pythium ultimum,* p. 183–189. *In* B. Schippers and W. Gams (ed.). Soil-borne plant pathogens. Academic Press, London.
19. Rouxel F., and Y. Regnault. 1985. Comparaison de la réceptivité des sols à la hernie des cruciféres: application à l'évaluation des risques sur quelques sols à culture de colza oléagineux. Coll. ANPP, 375–382.
20. Stotzky, G., J. E. Dawson, R. T. Martin, and G. H. H. Ter Kuile. 1961. Soil mineralogy as a factor in the spread of Fusarium wilt of banana. Science 133:1483–1485.

21. Broadbent, P., and K. F. Baker. 1974. Behaviour of *Phytophthora cinnamomi* in soils suppressive and conducive to root rot. Aust. J. Agr. Res. 25:121–137.
22. Broadbent, P., an K. F. Baker. 1975. Soils suppressive to Phytophthora root rot in Eastern Australia, p. 152–157. *In* G. W. Bruehl (ed.), Biology and control of soil-borne plant pathogens. A.P.S. Press, St. Paul, Minnesota.
23. Scher, F. M. and K. F. Baker. 1982. Effect of *Pseudomonas putida* and a synthetic iron chelator on induction of soil suppressiveness to Fusarium wilt pathogens. Phytopathology 72:1567–1573.
24. Robert, M., and C. Chenu. 1992. Interactions between soil minerals and microorganisms, p. 307–404. *In* G. Stotzky and J. M. Bollag (ed.), Soil Biochemistry, Vol. 7. Dekker Marcel New York.
25. Stotzky, G. 1966. Influence of clay minerals on microorganisms. II. Effect of various clay species, homoionic clays and other particles on bacteria. Can. J. Microbiol. 12:831–848.
26. Marshall, K. C. 1975. Clay mineralogy in relation to survival of soil bacteria. Annu. Rev. Phytopathol. 13:357–373.
27. Stotzky, G., and L. T. Rem. 1967. Influence of clay minerals on microorganisms. IV. Montmorillonite and kaolinite on fungi. Can J. Microbiol. 13:1535–1550.
28. Lavie, S., and G. Stotzky. 1986. Interactions between clay minerals and siderophores affect the respiration of *Histoplasma capsulatum*. Appl. Environ. Microbiol. 51:74–79.
29. Filip, Z. 1979. Wechselwirkungen von Mikroorganismen und Tonmineralen - eine Übersicht. Z. Pflanzenernähr. Bodenk. 142:555–561.
30. van Elsas, J. D., A. F. Dijkstra, J. M. Govaert. and J. A. van Veen. 1986. Survival of *Pseudomonas fluorescens* and *Bacillus subtilis* introducted into two soils of different texture in field microplots. FEMS Microbiol. Ecol. 38:151–160.
31. Fulton, J. M., C. G. Mortimore, and A. A. Hildebrand. 1961. Note on the relation of soil bulk density to the incidence of Phytophthora root and stalk rot of soybeans. Can. J. Soil Sci. 41:247.
32. Lyda, S. D. 1982. Physical and chemical properties of suppressive soils, p. 9–22. *In* R. W. Schneider (ed.), Suppressive soils and plant diseases. A.P.S. Press, St. Paul, Minnesota.
33. Miller, D. E., and D. W. Burke. 1974. Influence of soil bulk density and water potential on Fusarium root rot of bean. Phytopathology 64:526–529.
34. Kuan, T. L., and D. C. Erwin. 1980. Predisposition effect of water saturation of soil on Phytophthora root rot of *Alfalfa*. Phytopathology 70:981–986.
35. Stolzy, L. H., J. Letey, L. J. Klotz, and C. K. Labanauskas. 1965. Water and aeration as factors in root decay of *Citrus sinensis*. Phytopathology 55:270–275.
36. Colhoun, J., and D. Park. 1964. Fusarium diseases of cereals. I. Infection of wheat plants with particular reference to the effects of soil moisture and temperature on seedling infection. Trans. Br. Mycol. Soc. 47:559–572.
37. Beute, M. K. and R. Rodriguez-Kabana. 1981. Effects of soil moisture, temperature and field environment on survival of *Sclerotium roffsii* in Alabama and North Carolina. Phytopathology 71:1293–1296.

38. Papendick, R. I., and R. J. Cook. 1974. Plant water stress and development of Fusarium root rot in wheat subjected to different cultural practices. Phytopathology 64:358–363.
39. Staten, G., and J. F. Cole. 1948. The effect of pre-planting irrigation on pathogenicity of *Rhizoctonia solani* in seedling cotton. Phytopathology 38:661–664.
40. Jenkins, S. F., and C. W. Averre. 1986. Problems and progress in integrated control of southern blight of vegetables. Plant Dis. 70:614–619.
41. Shipton, P. J., R. J. Cook, and J. W. Sitton. 1973. Occurrence and transfer of a biological factor in soil that suppresses take-all of wheat in eastern Washington. Phytopathology 63:511–517.
42. Johnson, J., and R. E. Hartman. 1919. Influence of soil environment on the root rot of tobacco. J. Agric. Res. 17:41–86.
43. Stutz, E., and G. Defago. 1985. Effect of parent materials derived from different geological strata on suppressiveness of soils to black-root rot of tobacco. p. 215–217. *In* C. A. Parker, A. D. Rovira, K. J. Moore, and P. T. W. Wong (ed.), Ecology and management of soil-borne plant pathogens. Proceedings of section 5 of the 4th International Congress of Plant Pathology, Melbourne, Australia, August 17–24 1983.
44. Heijnen, C. E., J. D. van Elsas, P. J. Kuikman, and J. A. van Veen. 1988. Dynamics of *Rhizobium leguminosarum* biovar *trifolii* introduced into soil, the effect of bentonite clay on predation by protozoa. Soil Biol. Biochem. 20:483–488.
45. Schmit, J., and M. Robert. 1984. Action des argiles sur la survie d'une bactérie phytopathogène *Pseudomonas solanacearum* EFS. C. R. Acad. Sci. Paris 299:733–738.
46. Stotzky, G., and L. T. Rem. 1966. Influence of clay minerals on microorganisms. I. Montmorillonite and kaolinte on bacteria. Can. J. Microbiol. 12:547–563.
47. Skinner, F. A. 1956. The effect of adding clays to mixed cultures of *Streptomyces albidoflavus* and *Fusarium culmorum* J. Gen. Microbiol. 14:393–405.
48. Soulides, D. A. 1969. Antibiotic tolerance of the soil microflora in relation to type of clay minerals. Soil Sci. 107:105–107.
49. Rosenzweig, W. D., and G. Stotzky. 1979. Influence of environmental factors on antagonism of fungi by bacteria in soil: Soil minerals and pH. Appl. Environ. Microbiol. 38:1120–1126.
50. Campbell, R., and M. Ephgrave. 1983. Effect of bentonite clay on the growth of *Gaeumannomyces graminis* var. *tritici* and on its interaction with antagonistic bacteria. J. Gen. Microbiol. 129:771–777.
51. Hoeper, H., C. Steinberg, and C. Alabouvette. 1995. Involvement of clay type and pH in the mechanisms of soil suppressiveness to Fusarium wilt of flax. Soil Biol. Biochem., in press.
52. Scheffer, F., and P. Schachtschabel. 1984. Lehrbuch der Bodenkunde. Enke, Stuttgart.
53. Doran, W. L. 1929. Effects of soil temperature and reaction on growth of tobacco infected and uninfected with black-root rot. J. Agric. Res. 39:853–872.
54. Rishbeth, J. 1957. Fusarium wilt of bananas in Jamaica. II. Some aspects of host-parasite relations. Ann. Bot. 21:215–245.

55. Yuen, G. Y., A. H. McCain, and M. N. Schroth. 1983. The relation of soil type to suppression of Fusarium wilt of carnation. Acta Hortic. 141:95–102.
56. Jones, J. P., and S. S. Woltz. 1969. Fusarium wilt (race 2) of tomato: Calcium, pH and micronutrient effects on disease development. Plant Dis. Rep. 53:276–279.
57. Jones, J. P., and S. S. Woltz. 1970. Fusarium wilt of tomato: Interaction of soil liming and micronutrients on disease development. Phytopathology 60:812–813.
58. Papavizas, G. C., J. A. Lewis, and P. B. Adams. 1968. Survival of root infecting fungi in soil. II. Influence of amendment and soil carbon to nitrogen balance on *Fusarium* root rot of beans. Phytopathology 58:365–372.
59. Tivoli, B., R. Corbiere, and E. Lemarchand. 1989. Relation entre le pH des sols et leur niveau de réceptivité à *Fusarium solani* var. *coeruleum* et *Fusarium roseum* var. *sambucinum* agents de la pourriture sèche des tubercules de pomme de terre. Agronomie 10:63–68.
60. Menzies, J. D. 1959. Occurrence and transfer of a biological factor in soil that suppresses potato scab. Phytopathology 49:648–652
61. Martin, F. N., and J. G. Hancock. 1986. Association of chemical and biological factors in soils suppressive to *Pythium ultimum*. Phytopathology 76:1221–1231.
62. Kobayashi, N., and W. H. Ko. 1985. Nature of suppression of *Rhizoctonia solani* in Hawaiian soils. Trans. Br. Mycol. Soc. 84:691–694.
63. Pegg, K. G. 1977. Soil application of elemental sulphur as a control of Phytophthora cinnamomi root and heart rot of pine apple. Aust. J. Exp. Agric. Anim. Husb. 17: 859–865.
64. Smiley, R. W., and R. J. Cook. 1973. Relationship between take-all of wheat and rhizosphere pH in soils fertilized with ammonium vs. nitrate nitrogen. Phytopathology 63:882–890.
65. Kincaid, R. R. and N. Gammon. 1954. Incidence of tobacco black shank directly related to soil pH. Plant Dis. Rep. 38:852–853.
66. Keinath, A. P. and D. R. Fravel. 1992. Induction of soil suppressiveness to Verticillium wilt of potato by successive croppings. Am. Potato J. 69:503–513.
67. Gillespie, L. J., and L. A. Hurst. 1918. Hydrogen-ion concentration—soil type—common potato scab. Soil Sci. 6:219–236.
68. Wilhelm, S. 1950. *Verticillium* in acid soils. Phytopathology 40:776–777.
69. Orellana, R. G., C. A. Foy, and A. L. Fleming. 1975. Effect of soluble aluminum on growth and pathogenicity of *Verticillium albo-atrum* and *Whetzelinia sclerotiorum* from sun-flower. Phytopathology 65:202–205.
70. Meyer, J. R., and H. D. Shew. 1991. Soils suppressive to black-root rot of burley tobacco, caused by *Thielaviopsis basicola*. Phytopathology 8:946–953.
71. Barnhisel, R., and P. M. Bertsch. 1982. Aluminum, p. 275–300. *In* A. L. Page, R. H. Miller, and D. R. Keeney (ed.), Methods of soil analysis, part 2. Chemical and microbiological properties. American Society of Agronomy, Madison, Wisconsin.
72. McLean, E. O. 1982. Soil pH and lime requirements, p. 199–224. *In* A. L. Page, R. H. Miller and D. R. Keeney (ed.), Methods of soil analysis, part 2. Chemical and microbiological properties. American Society of Agronomy, Madison, Wisconsin.
73. Myers, D. F., and R. N. Campbell. 1985. Lime and the control of clubroot of crucifers: Effects of pH, calcium, magnesium and their interactions. Phytopathology 75:670–673.

74. Corden, C. E. 1965. Influence of calcium nutrition on Fusarium wilt of tomato and polygalacturonase activity. Phytopathology 55:222–224.
75. Ulrich, B. 1961. Boden und Pflanze. Enke, Stuttgart.
76. Davey, C. B., and G. C. Papavizas. 1961. Aphanomyces root rot of peas as affected by organic and mineral soil amendments (abstr.). Phytopathology 51:131–132.
77. Engelhard, A. W., J. P. Jones, and S. S. Woltz. 1989. Nutritional factors affecting Fusarium wilt incidence and severity, p. 337–352. *In* E. C. Tjamos and C. H. Beckman (ed.), Vascular wilt diseases of plants. Heidelberg Springer, Berlin.
78. Lemanceau, P., C. Alabouvette, and Y. Couteaudier. 1988. Recherches sur la résistance des sols aux maladies. XIV—Modification du niveau de réceptivité d'un sol résistant et d'un sol sensible aux fusarioses vasculaires en réponse à des apports de fer ou de glucose. Agronomie 8:155–162.
79. Woltz, S. S., and J. P. Jones. 1972. Control of Fusarium wilt of tomato by varying the nutrient regimes in soils (abstr.). Phytopathology 62:799.
80. McGregor, A. J., and G. C. S. Wilson. 1964. The effect of applications of manganese to a neutral soil upon the yield of tubers and the incidence of common scab in potatoes. Plant Soil 20:59–64.
81. Lewis, J. A. 1973. Effect of mineral salts on *Aphanomyces euteiches* and Aphanomyces root rot of peas. Phytopathology 63:89–993.
82. Sarniguet, A., P. Lucas, and M. Lucas. 1992. Relations between take-all, soil conduciveness to the disease, populations of fluorescent pseudomonads and nitrogen fertilizers. Plant Soil 145:17–27.
83. Reis, E. M., R. J. Cook, and B. L. McNeal. 1982. Effect of mineral nutrition on take-all of wheat. Phytopathology 72:224–229.
84. Keel, C., C. Voisard, C. H. Berling, G. Kahr, and G. Defago. 1989. Iron sufficiency, a prerequisite for the suppression of tobacco black-root rot by *Pseudomonas fluorescens* strain CHAO under gnotobiotic conditions. Phytopathology 79:584–589.
85. Gusenleitner, J. 1974. Der Zusammenhang zwischen Ökologischen bzw. betriebswirtschaftlichen Gegebenheiten und dem Befall mit Kartoffelschorf (*Streptomyces scabies* und *Spongospora subterranea*). Bodenkultur 25:63–74.
86. Geering, H. R., J. F. Hodgson, and C. Sdano. 1969. Micronutrient cation complexes in soil solution: IV. The chemical state of manganese in soil solution. Soil Sci. Soc. Am. Proc. 33:81–85.
87. Hodgson, J. F., W. L. Lindsay, and J. F. Trierweiler. 1966. Micronutrient cation complexing in soil solution: II. Complexing of zinc and copper in displaced solution from calcareous soils. Soil Sci. Soc. Am. Proc. 30:723–726.
88. Reis, E. M., R. J. Cook, and B. L. McNeal. 1983. Elevated pH and associated reduced trace-nutrient availability as factors contributing to take-all of wheat upon soil liming. Phytopathology 73:411–413.
89. Keinath, A. P., and R. Loria. 1989. Management of common scab of potato with plant nutrients, p. 152–166. *In* A. W. Engelhard (ed.), Soilborne plant pathogens: Management of diseases with macro- and microelements. A.P.S. Press, St. Paul, Minnesota.
90. Sneh, B., M. Dupler, Y. Elad, and R. Baker. 1984. Chlamydospore germination of *Fusarium oxysporum* f. sp. *cucumerinum* as affected by fluorescent and lytic bacteria from *Fusarium*-suppressive soil. Phytopathology 74:1115–1124.

91. Christensen, P. D., S. J. Toth, and F. E. Bear. 1950. The status of soil manganese as influenced by moisture, organic matter and pH. Soil Sci. Soc. Am. Proc. 15:279–282.
92. Lucas, P., and M. Nignon. 1987. Influence du type de sol et de ses composantes physicochimiques sur les relations entre une variété de blé (*Triticum aestivum*, L. var. Rescher) et deux souches agressives et hypoagressives de *Gaeumannomyces graminis* (Sacc.) Von Arx et Olivier var. *tritici* Walker. Plant Soil 97:105–117.
93. Tu, C. C., Y. H. Cheng, and M. Chen. 1975. Flax Fusarium wilt suppressive soil in Taiwan. Plant Prot. Bull. 17:390–399.
94. Park, C. S., and Y. S. Cho. 1985. Properties of soil suppressiveness to cucumber wilt, caused by *Fusarium oxysporum* f. sp. *cucumerinum* Owen. Kor. J. Plant Prot. 24:85–95.
95. Stotzky, G. 1974. Activity, ecology, and population dynamics of microorganisms in soil, p. 57–135. *In* A. I. Laskin and H. Lechevalier (ed.), Microbial ecology. C.R.C. Press, Boca Raton, Florida.
96. Papavizas, G. C., and C. B. Davey. 1960. Rhizoctonia disease of bean as affected by decomposing green plant materials and associating microflora. Phytopathology 50: 516–522.
97. Papavizas, G. C. 1966. Suppression of *Aphanomyces* root rot of peas by cruciferous soil amendments. Phytopathology 56:1071–1075.
98. Huber, D. M., and R. D. Watson. 1970. Effect of organic amendment on soil-borne plant pathogens. Phytopathology 60:22–26.
99. Weller, D. M. 1988. Biological control of soilborne plant pathogens in the rhizosphere with bacteria. Annu. Rev. Phytopathol. 26:379–407.
100. Defago, G., and D. Haas. 1990. Pseudomonads as antagonists of soilborne plant pathogens: mode of action and genetic analysis. Soil Biochem. 6:249–291.
101. Alabouvette, C., and Y. Couteaudier. 1992. Biological control of fusarium wilts with nonpathogenic Fusaria, p. 415–426. *In* E. C. Tjamos, G. C. Paparizas, and R. J. Cook (ed.), Biological control of plant disease. Plenum Press, New York.
102. Mandeel, G., and R. Baker. 1991. Mechanisms involved in biological control of fusarium wilt of cucumber with strains of nonpathogenic *Fusarium oxysporum*.. Phytopathology 81:462–469.
103. Harman, G. E. 1990. Development of tactics for biocontrol fungi in plant pathology, p. 779–792. *In* R. Baker and P. E. Dunn (ed.), New directions in biological control. A. R. Liss, New York.
104. Lumsden, R. D., and K. C. Locke. 1989. Biological control of damping-off caused by *Pythium ultimum* and *Rhizoctonia solani* in soilless mix. Phytopathology 79:361–366.
105. Fravel, D. R. 1989. Biocontrol of Verticillium wilt of eggplant and potato, p. 487–492. *In* E. C. Tjamos and C. H. Beckman (ed.), Vascular wilt disease of plant. NATO ASI series. Springer Verlag, Berlin.
106. Fravel, D. R. 1988. Role of antibiosis in the biocontrol of plant diseases. Annu. Rev. Phytopathol. 26:75–91.
107. O'Sullivan, D. J., and F. O'Gara. 1992. Trait of fluorescent *Pseudomonas* spp. involved in suppression of plant root pathogens. Microbiol. Rev. 56:662–676.

108. Kerr, A. 1980. Biological control of crown gall through production of agrocin 84. Plant Dis. 64:25–30.
109. McClure, N. C., A. R. Ahmadi, and B. G. Clare. 1994. The role of agrocin 434 produced by *Agrobacterium* strain K84 and derivatives in the biological control of *Agrobacterium* biovar 2 pathogens, p. 125–127. *In* M. H. Ryder, P. M. Stephens, and G. D. Bowen (ed.), Improving plant productivity with rhizosphere bacteria. CSIRO, Adelaïde, Australia.
110. Claydon, N., M. Allan, J. R. Hanson, and A. G. Avent. 1987. Antifungal alkylpyrones of *Trichoderma harzianum*. Trans. Br. Mycol. Soc. 88:503–513.
111. Howell, C. R., R. C. Beier, and R. D. Stipanovic. 1988. Production of ammonia by *Enterobacter cloacae* and its possible role in the biological control of *Pythium* preemergence damping-off by the bacterium. Phytopathology 78:1075–1078.
112. Kim, C. H., and K. S. Park. 1994. Plant growth promotion and disease suppression by two *Pseudomonas cepacia* isolates and their antibiotics, cepacid and pseudance-A, p. 134. *In* M. H. Ryder, P. M. Stephens, and G. D. Bowen (ed.), Improving plant productivity with rhizosphere bacteria. CSIRO, Adelaïde, Australia.
113. Aoki, M., K. Uehara, K. Koseki, K. Tsuji, M. Iijima, K. Ono, and T. Samejima. 1991. An antimicrobial substance produced by *Pseudomonas cepacia* B5 against the bacterial wilt disease pathogen, *Pseudomonas solanacearum*. Agric. Biol. Chem. 55: 715–722.
114. Di Pietro, A., M. Gut-Rella, J. P. Pachlatko, and F. J. Schwinn. 1992. Role of antibiotic produced by *Chaetomium globosum* in biocontrol of *Pythium ultimum*, a causal agent of damping-off. Phytopathology 82:131–135.
115. Inbar, J., and I. Chet. 1991. Evidence that chitinase produced by *Aeromonas caviae* is involved in the biological control of soil-borne pathogens by this bacterium. Soil. Biol. Biochem. 23:973–978.
116. Di Pietro, A., M. Lorito, C. K. Hayes, R. M. Broadway, and G. E. Harman. 1993. Endochitinase from *Gliocladium virens:* isolation, characterization, and synergistic antifungal activity in combination with gliotoxin. Phytopathology 83:308–313.
117. Lim, H. S., Y. S. Kim, and S. D. Kim. 1991. *Pseudomonas stutzeri* YPL.1 genetic transformation and antifungal mechanism against *Fusarium solani*, an agent of plant root rot. Appl. Environ. Microbiol. 57:510–516.
118. Harman, G. E., C. K. Hayes, M. Lorito, R. M. Broadway, A. Di Pietro, C. Peterbauer, and A. Tronsmo. 1993. Chitinolytic enzymes of *Trichoderma harzianum:* purification of chitobiosidase and endochitinase. Phytopathology 83:313–318.
119. Lorito, M., G. E. Harman, C. K. Hayes, R. M. Broadway, A. Tronsmo, S. L. Woo, and A. Di Pietro. 1993. Chitinolytic enzymes produced by *Trichoderma harzianum:* antifungal activity of purified endochitinase and chitobiosidase. Phytopathology 83: 302–307.
120. Ordentlich, A., Y. Elad, and I. Chet. 1988. The role of chitinase of *Serratia marcescens* in biocontrol of *Sclerotium rolsii.* Phytopathology 78:84–88.
121. Levy, E., and F. J. Gough. 1991. Biocontrol of *Septoria tritici* by fluorescent pseudomonads. Phytopathology 81:1178.
122. Vincent, M. N., L. A. Harrison, J. M. Brackin, P. A. Kovacevich, P. Mukerji, D. M. Weller, and E. A. Pierson. 1991. Genetic analysis of the antifungal activity of a soilborne *Pseudomonas aereofaciens* strain. Appl. Environ. Microbiol. 57:2928–2934.

123. Keel, C., U. Schnider, M. Maurhofer, C. Voisard, J. Laville, U. Burger, P. Wirthner, D. Haas, and G. Defago. 1992. Suppression of root diseases by *Pseudomonas fluorescens* CHAO: Importance of the bacterial secondary metabolite 2,4-diacetylphloroglucinol. Mol. Plant-Microbe Interact. 5:4–13.
124. Shanahan, P., D. J. O'Sullivan, P. Simpson, J. D. Glennon, and F. O'Gara. 1992. Isolation of 2,4-diacetylphloroglucinol from a fluorescent pseudomonad and investigation of physiological parameters influencing its production. Appl. Environ. Microbiol. 58:353–358.
125. Lumsden, R. D., J. C. Locke, S. T. Adkins, J. F. Walter, and C. J. Ridon. 1992. Isolation and localization of the antibiotic gliotoxin produced by *Gliocladium virens* from alginate prill in soil and soilless media. Phytopathology 82:230–235.
126. Howell, C. R., and R. D. Stipanovic. 1983. Gliovirin, a new antibiotic from *Gliocladium virens* and its possible role in the biological control of *Pythium ultimum*. Can. J. Microbiol. 29:321–324.
127. Fridlender, M., J. Inbar, and I. Chet. 1993. Biological control of soil-borne plant pathogens by B-1,3-glucanase producing *Pseudomonas cepacia*. Soil Biol. Biochem. 25:1211–1221.
128. Kim, K. K., D. R. Fravel, and G. C. Papavizas. 1988. Identification of a metabolite produced by *Talaromyces flavus* as glucose oxidase and its role in the biocontrol of *Verticillium dahliae*. Phytopathology 78:488–492.
129. Voisard, C., C. Keel, D. Haas, and G. Defago. 1989. Cyanide production by *Pseudomonas* fluorescens helps suppress black-root rot of tobacco under gnotobiotic conditions. EMBO J. 8:351–358.
130. Ross, I. L., and M. H. Ryder. 1944. Hydrogen cyanide production by a biocontrol strain in *Pseudomoans corrugata*: evidence that cyanide antagonises the take-all fungus in *vitro,* p. 131–133. *In* M. H. Ryder, P. M> Stephens, and G. D. Bowen, Improving plant productivity with rhizosphere bacteria. CSIRO, Adelaide, Australia.
131. Gutterson N., J. S. Ziegle, G. J. Warren, and T. J. Layton. 1988. Genetic determinants for catabolite induction of antibiotic hypothesis in *Pseudomonas fluorescens* HV37a. J. Bacteriol. 170:380–385.
132. Howie, W. J., and T. V. Suslow. 1991. role of antibiotic synthesis in the inhibition of *Pythium ultimum* in the cotton spermosphere and rhizosphere by *Pseudomonas fluorescens.* Mol. Plant-Microbe Interact. 4:393–399.
133. Thomashow, L. S., and D. M. Weller. 1988. Role of a phenazine antibiotic from *Pseudomonas fluorescens* in biological control of *Gaeumannomyces graminis* var. *tritici.* J. Bacteriol. 170:3499–3508.
134. McManns, P. S., A. V. Ravenscroft, and D. W. Fulhright. 1993. Inhibition of *Tilletia laevis* tellospore germination and suppression of common bunt of wheat by *Pseudomonas fluorescens* 2-79. Plant Dis. 77:1012–1015.
135. Flaishman, M., Z. Eyal, C. Voisard, and D. Haas. 1990. Suppression of *Septoria tritici* by phenazine- or siderophore-deficient mutants of *Pseudomonas.* Curr. Microbiol. 20:121–124.
136. Pierson, E. A., and L. S. Thomashow. 1992. Cloning and heterologous expression of the phenazine biosynthetic locus from *Pseudomonas aureofaciens* 30-84. Mol. Plant-Microbe Interact. 5:330–339.

137. Kloepper, J. W., J. Leong, M. Teintze, and M. N. Schroth. 1980. *Pseudomonas* siderophores: a mechanism explaining disease suppression soils. Curr. Microbiol. 4: 317–320.
138. Lemanceau, P. 1992. Effets bénéfiques des rhizobactéries sur les plantes: exemple des *Pseudomonas* spp. fluorescents. Agronomie 12:413–437.
139. Duijff, B. J., J. W. Meijer, P. A. H. M. Bakker, and B. Schippers. 1993. Siderophore-mediated competition for iron and induced resistance in the suppression of fusarium wilt of carnation by fluorescent *Pseudomonas* spp. Neth. J. Plant Pathol. 99:277–289.
140. Bakker, P. A. H. M., J. G. Lammers, A. W. Bakker, J. D. Marugg, P. J. Weisbeek, and B. Schippers. 1986. The role of siderophores in potato tuber yield increase by *Pseudomonas putida* in short rotation of potato. Neth. J. Plant Pathol. 92:249–256.
141. Becker, J. O., and R. J. Cook. 1988. Role of siderophores on suppression of *Pythium* species and production of increased-growth response of wheat by fluorescent pseudomonads. Phytopathology 78:778–782.
142. Loper, J. E. 1988. Role of fluorescent siderophore production in biological control of *Pythium ultimum* by a *Pseudomonas fluorescens* strain. Phytopathology 78:166–172.
143. Buysens, S., J. Poppe, and M. Hofte. 1994. Role of siderophores in plant growth stimulation and antagonism by *Pseudomonas aeruginosa* 7NSK2, p. 139–141. *In* M. H. Ryder, P. M. Stephens, and G. D. Bowen (ed.), Improving plant productivity with rhizosphere bacteria. CSIRO, Adelaide, Australia.
144. Dahiya, J. S., D. L. Woods, and J. P. Tewari. 1988. Control of *Rhizoctonia solani*, causal agent of brown girdling root rot of rapeseed by *Pseudomonas fluorescens*. Bot. Bull. Acad. Sin. 29:135–142.
145. Howell, C. R., and R. D. Stipanovic. 1980. Suppression of *Pythium ultimum* induced damping-off of cotton seedlings by *Pseudomonas fluorescens* and its antibiotic, pyoluteorin. Phytopathology 70:712–715.
146. Maurhofer, M., C. Keel, U. Schnider, C. Voisard, D. Haas, and G. Defago. 1992. Influence of enhanced antibiotic production in *Pseudomonas fluorescens* strains CHAO on its disease suppressive capacity. Phytopathology 82:190–195.
147. Pfender, W. F., J. Kraus, and J. E. Loper. 1993. A genomic region from *Pseudomonas fluorescens* Pf-5 required for pyrrolnitrin production and inhibition of *Pyrenophora tritici repentis* in wheat straw. Phytopathology 83:1223–1228.
148. Howell, C. R., and R. D. Stipanovic. 1979. Control of *Rhizoctonia solani* on cotton seedlings with *Pseudomonas fluorescens* and with an antibiotic produced by the bacteria. Phytopathology 69:480–482.
149. Burkhead, K. D., D. A. Schisler, and P. J. Shininger. 1994. Pyrrolnitrin production by biological control agent *Pseudomonas cepacia* B37W in culture and in colonized wounds of potatoes. Appl. Environ. Microbiol. 60:2031–2039.
150. Hill, D. S., J. P. Stein, N. R. Torkewitz, A. M. Morse, C. R. Howell, J. P. Pachlatko, J. O. Becker, and J. M. Ligon. 1994. Cloning of genes involved in the synthesis of pyrrolnitrin from *Pseudomonas fluorescens* and role of pyrrolnitrin synthesis in biological control of plant disease. Appl. Environ. Microbiol. 60:78–85.
151. Linberg, G. O. 1981. An antibiotic lethal to fungi. Plant Dis. 65:680–683.

152. Silo-Suh, L. A., B. J. Lethbridge, S. J. Raffel, H. He, J. Clardy, and J. Handelsman. 1994. Biological activities of two fungistatic antibiotics produced by *Bacillus cereus* UW85. Appl. Environ. Microbiol. 60:2023–2030.
153. Lemanceau, P. 1989. Role of competition for carbon and iron in mechanisms of soil suppressiveness to Fusarium wilts, in vascular wilt diseases of plants, p. 385–396. *In* E. C. Tjamos and C. Beckman (ed.), Springer-Verlag, Berlin.
154. Elad, Y., and R. Baker. 1985. Influence of trace amounts of cations and siderophore-producing pseudomonads on chlamydospore germination of *Fusarium oxysporum*. Phytopathology 75:1047–1052.
155. Teintze, M., M. B. Hossain, C. L. Barnes, J. Leong, and D. Van der Helm. 1981. Structure of ferripseudobactin, a siderophore from plant growth promoting *Pseudomonas* B10, Biochemistry 20:6446–6457.
156. Misaghi, I. J., L. J. Stowell, R. G. Grogan, and L. C. Spearman. 1982. Fungistatic activity of water-soluble fluorescent pigments of fluorescent pseudomonads. Phytopathology 72:33–36.
157. Elad, Y., and R. Baker. 1985. The role of competition for iron and carbon in suppression of chlamydospore germination of *Fusarium* spp. by *Pseudomonas* spp. Phytopathology 75:1053–1059.
158. Meyer, J. M., F. Halle, D. Hohnadel, P. Lemanceau, and H. Ratefidarivelo. 1987. Siderophores of *Pseudomonas*. Biological properties, p. 189–203. *In* G. Winkelmann, D. Van der Helm, and J. B. Neilands (ed.), Iron transport in microbes, plants and animals. VCH, Weinheim, Germany.
159. Lemanceau, P., P. A. H. M. Bakker, W. J. de Kogel, C. Alabouvette, and B. Schippers. 1992. Effect of pseudobactin 358 production by *Pseudomonas putida* WCS358 on suppression of Fusarium wilt of carnations by nonpathogenic *Fusarium oxysporum* Fo47. Appl. Environ. Microbiol. 58:2978–2982.
160. Heming, B. C. 1986. Microbial-iron interactions in the plant rhizosphere. An overview. J. Plant Nutr. 9:505–521.
161. Leong, J. 1986. Siderophores: their biochemistry and possible role in the biocontrol of plant pathogens. Annu. Rev. Phytopathol. 24:187–208.
162. Neilands, J. B., and S. A. Leong. 1986. Siderophores in relation to plant growth and disease. Annu. Rev. Plant Physiol. 37:187–208.
163. Schippers, B., A. W. Bakker, and P. A. H. M. Bakker. 1987. Interactions of deleterious and beneficial rhizosphere microorganisms and the effect of cropping practices. Annu. Rev. Phytopathol. 25:339–358.
164. Bakker, P. A. H. M., R. Van Peer, and B. Schippers. 1991. Suppression of soil-borne plant pathogens by fluorescent pseudomonads: mechanisms and prospects, p. 217–230. *In* A. B. R. Beemster, G. J. Boeller, M. Gerlagh, M. A. Ruissen, B. Schippers, and A. Tempel (ed.), Biotic interactions and soil-borne diseases. Elsevier, Amsterdam.
165. Loper, J. E., and J. S. Buyer. 1991. Siderophores in microbial interactions on plant surfaces. Mol. Plant-Microbe Interac. 4:5–13.
166. Meyer, J. M., and M. A. Abdallah. 1978. The fluorescent pigment of *Pseudomonas fluorescens*: biosynthesis, purification and physicochemical properties. J. Gen. Microbiol. 107:319–328.

167. Lemanceau, P., P. A. H. M. Bakker, W. J. de Kogel, C. Alabouvette, and B. Schippers. 1993. Antagonistic effect of nonpathogenic *Fusarium oxysporum* strain Fo47 and pseudobactin 358 upon pathogenic *Fusarium oxysporum* f. sp. *dianthi.* Appl. Environ. Microbiol. 59:74–82.
168. Baker, R., Y. Elad, and B. Sneh. 1986. Physical, biological and host factors in iron competition in soils, p. 77–84. *In* T. R. Swinburne (ed.), Plenum Press, New York.
169. Loper, J. E., and S. E. Lindow. 1993. Role of competition and antibiosis in suppression of plant diseases by bacterial biological control agents, p. 144–155. *In* R. D. Lumsden and J. L. Vaughn (ed.), Pest management: biologically based technologies. American Chemical Society, Washington, D.C.
170. Smiley, R. W. 1979. Wheat rhizoplane pseudomonads as antagonists of *Gaeumannomyces graminis.* Soil Biol. Biochem. 11:371–376.
171. Weller, D. M., and R. J. Cook. 1981. Pseudomonads from take-all conducive and suppressive soils. Phytopathology 71:264.
172. Harrison, L., D. B. Teplow, M. Rinaldi, and G. Strobel. 1991. Pseumycins, a family of novel pectides from *Pseudomonas syringae* possessing broad-spectrum antifungal activity. J. Gen. Microbiol. 137:2857–2865.
173. Stutz, E., G. Defago, and H. Kern. 1986. Naturally occurring fluorescent Pseudomonads involved in suppression of black-root rot of tobacco. Phytopathology 76: 181–185.
174. Keel, C., P. Wirthner, T. H. Oberhansli, C. Voisard, U. Burger, D. Haas, and G. Defago. 1990. Pseudomonads as antagonists of plant pathogens in the rhizosphere: role of antibiotic 2,4-diacetylphloroglucinol in the suppression of black-root rot of tobacco. Symbiosis 9:327–341.
175. Moore-Landecker, E., and G. Stotzky. 1972. Inhibition of fungal growth and sporulation by volatile metabolites from bacteria. Can. J. Microbiol. 18:957–962.
176. Moore-Landecker, E., and G. Stotzky. 1973. Morphological abnormalities of fungi induced by volatile microbial metabolites. Mycologia 65:519–530.
177. Moore-Landecker, E., and G. Stotzky. 1974. Effects of concentration of volatile metabolites from bacteria and germinating seeds on fungi in the presence of selective absorbents. Can J. Microbiol. 20:97–103.
178. Stotzky, G., and S. Schenck, 1976. Volatile organic compounds and microorganisms. Crit. Rev. Microbiol. 4:333–382.
179. Loper, J. E., N. Corbell, J. Kraus, B. Nowak-Thompson, M. P. Henkels, and S. Cornegie. 1994. Contributions of molecular biology towards understanding mechanisms by which rhizosphere pseudomonads effect biological control, p. 89–96. *In* M. H. Ryder, M. P. Stephens, and G. D. Bowen (ed.), Improving plant productivity with rhizosphere bacteria. CSIRO, Adelaïde, Australia.
180. Ahl, P., C. Voisard, and G. Defago. 1986. Iron-bound siderophores, cyanic acid, and antibiotics involved in suppression of *Thielaviopsis basicola* by a *Pseudomonas fluorescens* strain. J. Phytopathol. 116:121–134.
181. Haas, D., C. Keel, L. Laville, M. Maurhofer, T. Oberkänoli, U. Schnider, C. Voisard, B. Wüthrich, and G. Defago. 1991. Secondary metabolites of *Pseudomonas fluorescens* strain CHAO involved in suppression of root diseases, p. 450–456. *In* H. Hennecke and D. P. S. Verma (ed.), Advances in molecular genetics of plant-microbe interactions. Kluwer Academic, Dordrecht, Netherlands.

182. Laville, J., C. Voisard, C. Keel, M. Maurhofer, G. Defago, and D. Haas. 1992. Global control in *Pseudomonas fluorescens* mediating antibiotic synthesis and suppression of black-root rot of tobacco. Proc. Natl. Acad. Sci. 89:1562–1566.
183. Gurusiddaiah, S., D. M. Weller, A. Sarkar, and R. J. Cook. 1986. Characterization of an antibiotic produced by a strain of *Pseudomonas fluorescens* inhibitory to *Gaeumannomyces graminis* var. *tritici* and *Pythium* spp. Antimicrob. Agents Chemother. 29:488–495.
184. Weller, D. M., W. J. Howie, and R. J. Cook. 1988. Relationship between *in vitro* inhibition of *Gaeumannomyces graminis* var. *tritici* and suppression of take-all of wheat by fluorescent pseudomonads. Phytopathology 78:1094–1100.
185. Thomashow, L. S., and L. S. Pierson. 1991. Genetic aspects of phenazine antibiotic production by fluorescent pseudomonads that suppress take-all disease of wheat, p. 443–445. *In* H. Hennecke and P. S. Verma (ed.), Advances in molecular genetics of plant-microbe interactions, Vol. I. Kluwer Academic, Dordrecht, Netherlands.
186. Weller, D. M., and R. J. Cook. 1983. Suppression of take-all of wheat by seed treatment with fluorescent pseudomonads. Phytopathology 73:463–469.
187. Park, C. S., T. C. Paulitz, and R. Baker. 1988. Biocontrol of fusarium wilt of cucumber resulting from interaction between *Pseudomonas putida* and nonpathogenic isolates of *Fusarium oxysporum*. Phytopathology 78:190–194.
188. Fuchs, J., and G. Defago. 1991. Protection of tomatoes against *Fusarium oxysporum* f. sp. *lycopersici* by combining a nonpathogenic *Fusarium* with different bacteria in untreated soil, p. 51–56. *In* C. Keel, B. Koller, and G. Defago (ed.), Plant growth-promoting rhizobacteria—progress and prospects. WPRS Bulletin XIV/8 IOBC.
189. Lemanceau, P., and C. Alabouvette. 1991. Biological control of fusarium diseases by fluorescent *Pseudomonas* and nonpathogenic *Fusarium*. Crop Protect. 10:279–286.
190. Pierson, E. A., and D. M. Weller. 1991. Recent work on control of take-all on wheat by fluorescent pseudomonads, p. 96–97. *In* C. Keel, B. Koller, and G. Defago (ed.), Plant growth-promoting rhizobacteria—progress and prospects. WPRS Bulletin XIV/8 IOBC.
191. Neilands, J. B. 1981. Iron absorption and transport in microorganisms. Annu. Rev. Nutr. 1:27–46.
192. Homma, Y. 1994. Mechanisms in biological control focussed on the antibiotic pyrrolnitrin, p. 100–103. *In* M. H. Ryder, P. M. Stephens, and G. D. Bowen (ed.), Improving plant productivity with rhizosphere bacteria. CSIRO, Adelaïde, Australia.
193. Homma, Y., and T. Suzui. 1989. Role of antibiotic production in suppression of radish damping-off by seed bacterization with *Pseudomonas cepacia*. Ann. Phytopathol. Soc. Japan 55:643–652.
194. Bull, C. T., D. M. Weller, and L. S. Thomashow. 1991. Relationship between root colonization and suppression of *Gaeumannomyces graminis* var. *tritici* by *P. fluorescens* strain 2-79. Phytopathology 81:954–959.
195. Fenton, A. M., P. M. Stephens, J. Crowley, M. O'Callaghan, and F. O'Gara. 1992. Exploitation of genes involved in 2,4-diacetylphloroglucinol biosynthesis to confer a new biocontrol capability on a *Pseudomonas* strain. Appl. Environ. Microbiol. 58: 3873–3878.

196. Thomashow, L. S., D. M. Weller, R. F. Bousall, and L. S. Pierson. 1990. Production of the antibiotic phenazin-1-carboxylic acid by fluorescent *Pseudomonas* species in the rhizosphere of wheat. Appl. Environ. Microbiol. 56:908–912.
197. Buyer, J. S., M. G. Kratzke, and L. J. Sikora. 1993. A method for detection of pseudobactin the siderophore produced by a plant growth-promoting *Pseudomonas* strain in the barley rhizosphere. Appl. Environ. Microbiol. 59:677–681.
198. Loper, J. E., and S. E. Lindow. 1994. A biological sensor for iron available to bacteria and their habitats on plant surfaces. Appl. Environ. Microbiol. 60:1934–1941.
199. Weller, D. M., and L. S. Thomashow. 1993. Microbial metabolites with biological activity against plant pathogens, p. 171–180. *In* R. D. Lumsden and J. L. Vaughn (ed.), Pest management: Biologically based technologies. American Chemical Society, Washington, D.C.
200. Kraus, J., and J. E. Loper. 1992. Lack of evidence for a role of antifungal metabolite production by *Pseudomonas fluorescens* Pf-5 in biological control of damping-off of cucumber. Phytopathology 82:264–271.
201. Mattar, J., and B. Digat. 1991. Influence of temperature on growth and siderophore production of *Pseudomonas fluorescens putida* strains, and consequence for root colonization, p. 336–339. *In* C. Keel, B. Koller, and G. Defago (ed.), Plant growth promotting rhizobacteria—progress and prospects. WPRS Bulletin XIV/8 IOBC.
202. Garibaldi, J. A. 1971. Influence of temperature on the iron metabolism of a fluorescent pseudomonad. J. Bacteriol. 105:1036.
203. Loper, J. E., and M. N. Schroth. 1986. Importance of siderophores in microbial interactions in the rhizosphere, p. 85–98. *In* T. R. Swinburne (ed.) Iron, siderophores, and plant disease. NATO ASI Series. Plenum Press, New York.
204. Williams, S. T., and J. C. Vickers. 1986. The ecology of antibiotic production. Microb. Ecol. 12:43–52.
205. Ownley, B. H., D. M. Weller, and J. R. Allredge. 1991. Relation of soil chemical and physical factors with supression of take-all by *Pseudomonas fluorescens* 2-79, p. 299–301. *In* C. Keel, B. Koller, and G. Defago (ed.), Plant growth promoting rhizobacteria—progress and prospects. WPRS Bulletin XIV/8 IOBC.
206. Lindsay, W. L, and B. Schwab. 1991. The chemistry of iron in soil and its availability to plants. J. Plant Nutr. 5:821–840.
207. Ownley, B. H., D. M. Weller, and L. S. Thomashow. 1992. Influence of *in situ* and *in vitro* pH on suppression of *Gaeumannomyces graminis* var. *tritici* by *Pseudomonas fluorescens* 2-79. Phytopathology 82:178–184.
208. Wuthrich, B., P. Haldimann, and G. Defago. 1991. Effects of pH, nitrogen source and manganese on suppression of wheat take-all by *Pseudomonas fluorescens* strain CHAO under gnotobiotic conditions, p. 340–345. *In* C. Keel, B. Koller, and G. Defago. Plant growth promoting rhizobacteria—progress and prospects. WPRS Bulletin XIV/8 IOBC.
209. Lemanceau, P., C. Alabouvette, and J. M. Meyer. 1986. Production of fusarinine and iron assimilation by pathogenic and nonpathogenic *Fusarium*, p. 251–253. *In* T. R. Swinburne (ed.), Iron siderophores and plant diseases. Plenum Press, New York.
210. Van Peer, R., A. J. Van Kuik, H. Rattink, and B. Schippers. 1990. Control of fusarium wilt in carnation growth on rock wool by *Pseudomonas* sp. strain WCS 417r and Fe-EDDHA. Neth. J. Plant Pathol. 96:119–132.

211. Duijff, B. J., J. W. Meijer, P. A. H. M. Bakker, and B. Schippers. 1991. Suppression of fusarium wilt of carnation by *Pseudomonas* in soil; mode of action, p. 152–157. *In* C. Keel, B. Koller, and G. Defago. Plant growth promoting rhizobacteria—progress and prospects. WPRS Bulletin XIV/8 IOBC.

212. Castric, P. 1975. Hydrogen cyanide a secondary metabolite of *Pseudomonas aeruginosa.* Can. J. Microbiol. 21:613–618.

213. Askeland, R. A., and S. M. Morrison. 1983. Cyanide production by *Pseudomonas fluorescens* and *Pseudomonas aeruginosa.* Appl. Environ. Microbiol. 45:1802–1807.

214. Slininger, P. J., and M. A. Jackson. 1992. Nutritional factors regulating growth and accumulation of phenazine-1-carboxylic acid by *Pseudomonas fluorescens* p. 2–79. Appl. Microbiol. Biotechnol. 37:388–392.

215. James, D. J., and N. I. Gutterson. 1986. Multiple antibiotics produced by *Pseudomonas fluorescens* HV37a and their differential regulation by glucose. Appl. Environ. Microbiol. 52:1183–1189.

216. Castric, P. 1977. Glycine metabolism by *Pseudomonas aeruginosa:* hydrogen cyanide biosynthesis. J. Bacteriol. 130:826–831.

217. Weinhold, A. R., J. W. Oswald, T. Bowman, J. Bishop, and D. Wright. 1964. Influence of green manures and crop rotation on common scab of potato. Am. Potato J. 41:265–273.

218. Hyakumachi, M., K. Kanzawa, and T. Ui. 1990. Rhizoctonia root rot decline in sugarbeet monoculture, p. 227–247. *In* D. Hornby (ed.), Biological control of soil-borne plant pathogens. C.A.B. International, Wallingford, England.

219. Sneh, B., D. Pozniak, and D. Salomon. 1987. Soil suppressiveness to Fusarium wilt of melon, induced by repeated croppings of resistant varieties of melons. J. Phytopath. 120:347–354.

220. Hornby, D. 1979. Take-all decline: a theorist's paradise, p. 133–156. *In* B. Schippers and W. Gams (ed.), Soil-borne plant pathogens. Academic Press, London.

221. Cook, R. J., and R. J. Veseth. 1991. Wheat health management. American Phytopathology Society, St. Paul, Minnesota.

222. Foster, R. C. 1985. The biology of the rhizosphere, p. 101–106. *In* C. A. Parker, A. D. Rovira, K. J. Moore, P. T. W. Wong, and J. F. Kollmorgen, Ecology and management of soilborne plant pathogens. A.P.S. Press, St. Paul, Minnesota.

223. Lemanceau, P., T. Corberand, L. Gardan, X. Latour, G. Laguerre, J. M. Boeufgras, and C. Alabouvette. 1995. Effect of two plant species of flax (*Linum usitatissinum* L.) and tomato (*Lycopersicon esculentum* Mill.) on the diversity of soilborne populations of fluorescent pseudomonads. Appl. Environ. Microbiol. 61:1004–1012.

224. Abadie, C., H. De Franqueville, and C. Alabouvette. 1994. Influence of cultural practices on level of soil receptivity to Fusarium wilt, p. 45. *In* Environmental biotic factors in integrated plant disease control. Third annual E.F.P.P. conference. Polish Phytopathology Society Poznan.

225. Engelhard, A. W. 1989. Soilborne plant pathogens: management of diseases with macro- and microelements. American Phytopathology Society, St. Paul, Minnesota.

226. Huber, D. M. 1989. The role of nutrition in the take-all disease of wheat and other small grains, p. 46–74. *In* A. W. Engelhard (ed.), Soilborne plant pathogens: man-

agement of diseases with macro- and microelements. American Phytopathology Society, St. Paul, Minnesota.

227. Jones, P. J., A. W. Engelhard, and S. S. Woltz. 1989. Management of Fusarium wilt of vegetables and ornamentals by macro- and microelement nutrients, p. 18–32. *In* A. W. Engelhard (ed.), Soilborne plant pathogens: management of diseases with macro- and microelements. American Phytopathology Society, St. Paul, Minnesota.
228. Simon, A., and K. Sivasithamparam. 1990. Effect of crop rotation, nitrogenous fertilizer and lime on biological suppression of the take-all fungus, p. 215–226. *In* D. Hornby (ed.), Biological control of soil-borne plant pathogens. CAB International, Wallingford, England.
229. Campbell, R., and A. S. Greathead. 1989. Control of clubroot of crucifers by liming, p. 90–101. *In* A. W. Engelhard (ed.), Soilborne plant pathogens: management of diseases with macro- and microelements. American Phytopathology Society, St. Paul, Minnesota.
230. Katan, J. 1981. Solar heating (solarization) of soil for control of soilborne pests. Annu. Rev. Phytopathol. 19:211–236.
231. Katan, J., J. E. Devay, and A. Greenberger. 1989. The biological control induced by soil solarization, p. 493–500. *In* E. C. Tjamos and C. H. Beckman (ed.), Vascular wilt diseases of plants. NATO ASI Series H28. Springer-Verlag, Berlin.
232. Tjamos, E. C., and E. J. Paplomatas. 1988. Long-term effect of soil solarization in controlling Verticillium wilt of globe artichokes in Greece. Plant Pathol. 37:507.
233. Burpee, L. L. 1990. The influence of abiotic factors on biological control of soilborne plant pathogenic fungi. Can. J. Plant Pathol. 12:308–317.
234. Whipps, J. M. 1992. Status of biological disease control in horticulture. Biocontrol Sci. Technol. 2:3–24.
235. Baker, R. 1991. Diversity in biological control. Crop Protect. 10:85–94.
236. Harman, G. E., C. K. Hayes, and M. Lorito. 1993. The genome of biocontrol fungi: modification and genetic components for plant disease management strategies, p. 347–354. *In* R. D. Lumsden and J. L. Vaughn (ed.), Pest management: biologically based technologies. American Chemistry Society, Washington, D.C.

10
Fractals in General Soil Science and in Soil Biology and Biochemistry

Nicola Senesi Università degli Studi di Bari, Bari, Italy

I. INTRODUCTION

Fractal geometry was introduced in the 1970s by Mandelbrot [1]. The word *fractal* comes from the Latin *fractus*, which means irregular. The dimension of a fractal object is not an integer, as it is in Euclidean geometry, but can be a noninteger. The second and complementary property of a fractal object is its description in terms of dilatational self-similarity (fractal geometry) and not in terms of usual translational symmetry (Euclidean geometry). In other words, its degree of irregularity is independent of scale, that is, increasing the magnification reveals an increasing number of smaller and smaller irregularities that are morphologically similar to those observed at the larger scale. Highly irregular objects and systems, both natural and artificial, such as coastlines, rivers, clouds, snowflakes, trees, fractures, colloidal aggregates, brain circumvolutions, and star constellations, have been shown experimentally to be fractals. General principles and techniques of fractal geometry are briefly presented in section II.

Soil, as a fragmented and porous medium, and its physical, chemical, and biological components and properties are mostly characterized by an intrinsic irregular and heterogeneous character and are amenable, in principle, to being described by fractal geometry. On these bases, fractal analysis has been usefully applied in soil physics to study pore space and particle size distribution, soil aggregates and soil structure, soil fragmentation, spatial variation of soil surface strength, and soil water retention and hydraulic conductivity. A brief summary of the results of these studies is provided in sections III.A and III.B.

More recently and more limitedly, fractal geometry has also been applied successfully to the study of soil chemical components and properties, including the evaluation of the surface area of soil particles and the nature and morphology of humic substances. A summary of these studies is provided in Section III.C.

In the last decade, fractal analysis has provided a powerful tool in biochemistry for a novel approach to the study of the structure, conformation, dynamics, surface area and reactivity, aggregation processes, reaction kinetics, and allosteric effects of proteins and enzymes. Results of these studies have provided unique and important clues on the interrelations between structural irregularity and surface roughness of proteins and enzymes, as recognized by fractal analysis, and their shape, topography, specificity, diffusion and trapping of substrates by the active sites, and interactions in biomembranes. These findings are reviewed in Section IV.

Fractal analysis has been introduced in microbiology and other biological fields of interest to soil biology only recently and with limited results. Fractal concepts have been applied to the description of growth patterns and morphological development of some microorganisms in different cultural conditions, to mechanisms of resource acquisition, to microbial aggregation processes, and to the evaluation of habitat space and spatial and temporal dynamics of microbial and other populations on plants and in soil. Results of these studies are discussed in Sections V and VI.

The purpose of this chapter is to provide a simple introduction to the general principles and concepts of fractal geometry; to introduce the applications of fractal analysis to problems in soil physics and protein and enzyme biochemistry that are related to problems in soil biology and biochemistry; and to examine in detail the few investigations that have been conducted using fractal analysis in soil biology and biochemistry.

II. PRINCIPLES AND TECHNIQUES OF FRACTAL GEOMETRY

Classical Euclidean geometry allows for perfect descriptions of simple structures, but it is an inappropriate tool for describing irregular forms, which exist everywhere in nature. Fractal geometry provides a powerful tool for the quantitative description of complex, heterogeneous, highly irregular and random systems; of the processes leading to their formation; and of their physical behavior [1]. The fractal concept addresses strongly random systems by treating "disorder" as an intrinsic rather than a perturbative feature of the system. A fractal object is basically characterized by dilatational self-similarity, or scale-invariant properties, and can be described in terms of a nonintegral dimension, that is, the fractal dimension [1,2].

Highly porous media, such as cross-linked gels and spatial networks of highly branched polymer chains that are common constituents of biological and biologically derived structures, often exhibit fractal features. Examples of phenomena that are susceptible to a fractal description include (1) the formation of complex structures by a variety on nonequilibrium processes, such as biologically controlled growth processes; and (2) small-particle (colloidal) aggregation, including the flocculation of organic and biological particles in soils and waters.

The focal point of fractal geometry is the repetition of disorder (dilatation invariance) at an appropriate length scale, that is, any real object is fractal only on a particular length of scales. The resulting picture is, therefore, radically different from the classical one. From the behavior of the system under appropriate scale transformations (dilatations), fractal geometry finds the so-called fractal dimension, D, which reflects the actual space occupied by the disordered system. Ordered, or Euclidean, nonfractal systems are also dilation invariant, but they are simultaneously translation invariant (the latter implies the former). In these systems, the fractal dimension, D, coincides with the topological dimension, D_{top}. The topological dimension is $D_{top} = 0$ for a set of disconnected points, $D_{top} = 1$ for a curve, $D_{top} = 2$ for a surface , and $D_{top} = 3$ for a solid. If D differs from D_{top}, the system is called fractal. If this is the case, translation symmetry is lost, but dilation symmetry is preserved.

For any system in which $D_{top} \leq D \leq d$, where d is the embedding dimension in the Euclidean space (usually, $d = 3$), the difference, $D - D_{top}$, is a measure of the disorder of the system. If $D = D_{top}$, the system is ordered or weakly disordered; if $D > D_{top}$, it is strongly disordered. Systems with classical dimension 1, 2, or 3 may be considered degenerate, and this degeneracy is removed in nonintegral dimensions. Fractals, therefore, have properties much richer than those expected from the classical cases, D = 1, 2, or 3, but require additional parameters for their description.

A. Natural Fractals

Natural or real fractals are random fractals that exhibit quasi or statistical self-similar structures characterized by dilation invariance within some observable length scale. This signifies that they are fractal between two finite length scales that typically differ by at least one order of magnitude. For quasi self-similar systems, the dilatations that leave the system invariant may scale the x, y, and z coordinates by the same factor (isotropic disorder) or by different factors in different directions (anisotropic disorder).

In fractal analysis, a scaling or power law of the form

$$p \propto v^{\gamma} \tag{1}$$

is sought, where p is the property, v is the variable, and the exponent, γ, can be related to a fractal dimension, D, and can be computed directly from experimental data. The existence of such a power law suggests the existence of a symmetry, under a change of scale of the variable, in the system being studied.

The length scale, or interval of length, over which a system is fractal can be estimated on the width of the interval of the variable, v, within which the property, p, follows a power law. Thus, it is important first to find this symmetry and, second, to give it a physical realization when this is not obvious. In other words, it is necessary to find which physical, chemical, or biochemical property of the system is described by a power law.

Different properties of biomolecules and microorganisms in soil, including structure, conformation, dynamics, surface properties and aggregation processes of proteins and enzymes, microbial growth patterns and morphology, and microbial populations and their distribution on soil pore surfaces, may obey scaling laws and be described by fractal concepts. Randomness and scaling are, however, only approximations. Detailed properties of specific systems or processes are averaged in fractal analysis. Further, because of the finite size of biological systems, the range of length scales that can be studied is often limited, and the validity of the fractal approach must be considered carefully.

For example, the system can be modeled as a lattice, and the lattice sites are used to count increasingly distant neighbors. The sites could be monomers in a polymer chain, adsorption sites on a surface, primary particles of a colloidal aggregate, et cetera. The quantity of interest is the number of sites located within a distance, R, from a given site, with $R >> \varepsilon$ where ε is the distance between nearest neighbor sites [3]. This quantity, $M_{sites}(R)$, is called the "mass" in a sphere of radius R, and grows as

$$M_{sites}(R) \propto R^D \tag{2}$$

for increasing R (when ε is fixed) [2]. Equation (2) is called the mass-radius relation and exhibits the macroscopic aspect of the space-filling ability of the system; that is, the larger the value of D (the fractal dimension), the more sites are in the sphere mode. For self-similar systems, Equation (2) is a power law and, thus, may serve to define D [2].

B. Mass, Surface, and Pore Fractals

A system wherein mass and surface scale the same is defined as a mass fractal. A system wherein pore space and surface scale the same is defined a pore fractal. A system for which only the surface is fractal is called a surface fractal. A lattice representation may be used to describe quantitatively the system, which is divided into mass sites (occupied sites), surface sites (occupied sites with adjacent empty sites), and pore sites (empty sites in the convex hull of the occupied sites) [2]. Then, the number of x-type sites within radius, R, from a fixed x-type site (x = mass, surface, or pore), $M_x(R)$, is counted. Each site type gives rise to a mass-radius relation of the form

$$M_{mass}(R) \propto R^{D_m} \tag{3}$$

$$M_{surf}(R) \propto R^{D_s} \tag{4}$$

$$M_{pore}(R) \propto R^{D_p} \tag{5}$$

The three types of fractals become indistinguishable in the space-filling limit, $D \rightarrow d$.

For example, when considering the surface activity of a protein or an enzyme, the major advantage of fractal analysis compared with other approaches resides in its ability to provide a measure of the surface roughness. The surface roughness is known to have a significant effect on association of different subunits, binding and diffusion on a ligand on the surface, release of adsorbed ions or molecules, and many other physical and chemical functions of the protein or enzyme. The fractal dimension of the surface, D_s, thus provides a compact parameter describing the degree of surface corrugation. The application of fractal analysis to soil particles, proteins, enzymes, and microorganisms requires, however, that the value of the dimension persist for reasonable variations in the length scale, otherwise the use of the scaling law may be questionable, as it cannot describe a local property.

C. Models and Kinetics of Formation of Fractal Objects

Considerable effort has been devoted to building various statistical models to explain the formation of fractal objects and to relate the fractal dimensions to aggregation processes. Although aggregates usually have very complicated structures, their essential aggregation processes are, in general, relatively simple and can be described by two distinct model: (1) particle-cluster aggregation, in which freely moving particles are aggregated to form a cluster; and (2) cluster-cluster aggregation (CLA), in which movable clusters collide and stick together to form larger clusters (e.g., colloidal aggregation). Aggregation can also be simulated according to the rate-limiting process into (1) ballistic aggregation; (2) diffusion-limited aggregation (DLA); and (3) reaction-limited aggregation (RLA).

Computer simulations can contribute simple models for growth, polymerization, and aggregation processes leading to the formation of fractal structures that closely resemble those found in nature. These models are helpful in developing a better understanding of complex real systems in terms of the simplest physical description possible and, in some cases, may provide a better description of a variety of apparently unrelated processes.

In the particle-cluster aggregation model simulated by a DLA process, a seed particle is fixed at the origin, and particles are released one after the other from a circle enclosing the cluster. Each particle moves in a random-walk fashion until it reaches a neighboring site of the aggregate and becomes part of the growing cluster. The cluster formed is a fractal object having a self-similarity property in all length scales. The fractal dimensions of these clusters are $D = 1.68 \pm 0.05$ for $d = 2$ and $D = 2.5 \pm 0.06$ for $d = 3$ [4].

The CLA model describes the system as consisting of a large number of dispersed particles that move in a fluid. The particles may thus come into contact with and stick to each other. The small clusters formed continue to move and collide with other clusters and particles to form larger clusters. As the aggregation process continues, clusters of larger and larger sizes are formed.

This type of model describes most colloidal aggregation processes better than the particle-cluster–DLA model and can be simulated by

1. Ballistic aggregation, where particles initially, and clusters subsequently, come together along randomly selected ballistic (linear) trajectories with a fractal dimensionality of one and are joined at their point of first contact. This model is appropriate for conditions in which the correlation length (relaxation length) of the cluster trajectories is larger than the cluster size [5].
2. Diffusion-limited aggregation (DLA), in which the trajectories followed by particles and clusters are represented by random walks with a fractal dimensionality of two (Brownian motion) better than by ballistic trajectories. In this model, particles are represented by occupied lattice sites. At the start of the simulation, the lattice contains a large number of isolated occupied lattice sites and a few small clusters, and objects (particles and clusters) are selected at random and moved by one lattice unit in a randomly selected direction on the lattice. After a particle or cluster has been moved, its perimeter is examined to determine if another object has been contacted and a new cluster has been formed. As the simulation proceeds, the clusters grow larger and larger (and fewer and fewer) until all the particles (sites) are contained in a single large cluster. At finite concentrations, this cluster has a fractal structure on short length scales and a uniform structure on larger length scales. The main advantage of this model is that it can be easily made time dependent.
3. Reaction-limited aggregation (RLA), where many collisions are required before particles and/or clusters combine. This requirement results from a small repulsive barrier potential in particle-to-particle interaction that must be overcome before aggregation can occur. In this model, the fractal dimensionality of aggregates appears to be dependent on aggregation kinetics and on the associated distributions (polydispersity) of cluster sizes [5]. The smallest value for D is found when clusters of the same size are combined. Larger values for D are found for models that lead to a broader distribution of cluster sizes. The clusters generated by this model have a fractal dimensionality of 2.11 ± 0.03 [6].

Physical aspects of the aggregation process are time dependent, which implies that a complete understanding of the process requires consideration of both the geometry and kinetics. All aggregation models here described were originally developed without regard to aggregation kinetics, but they can easily be made time dependent. Both the geometric and kinetic aspects of aggregation phenomena can, therefore, be represented by simple modifications of the models originally developed.

In real systems, there are at least two distinct regimes of colloidal aggregation, that is fast and slow aggregation, each of which has different rate-limiting physics. The resulting clusters have also been found to possess different fractal dimensions and different size distributions, as determined by the characteristic physical mechanisms that govern the kinetics of aggregation in each regime.

D. Surface Accessibility and Reactivity and Trapping Processes in Fractal Systems

The majority of natural biogeochemical and biochemical phenomena occur in heterogeneous environments at the border area of interfaces, in membranes, in pores, and in other confined zones. Most of the parameters that govern these processes are affected not only by the degree of surface irregularity of the substrate, but also by the degree of irregularity of the surroundings. A variety of molecule-surface interactions present in a wide range of natural heterogeneous systems obey empirical fractal scaling laws, which reflects an underlying fractal geometry of the surfaces. In other cases, the accessible surface that can be reached by a reactant, that is, the effective surface involved in a process and not the "purely geometric" one, is fractal. Consequently, fractal geometry may provide a useful means with which to describe the effects that complex geometries have on such interactions. In other cases, even though a scaling law is obtained, interpretations other than those from fractal considerations may apply.

Fractal geometry may offer a solution to the definition of surface area by recognizing that the concepts of length and area of irregular objects are not absolute but depend on the "yardstick" size. In fractal terms, the apparent surface area, A (in $m^2\ g^{-1}$), may be defined as [7]

$$A = Nn\sigma^{(2-D_a)/2} \tag{6}$$

where N is Avogadro's constant, n is the monolayer value (moles per unit mass), σ is the cross-sectional area of the adsorbed molecule, and D_a is the fractal dimension of the area obtained from the adsorption experiment. D_a will thus reflect the surface irregularity or roughness, that is, the effective surface geometry as seen by the specific set of adsorbent molecules used to derive it. D_a values usually fall in the theoretical range of $2 \le D_a \le 3$. Equation # (6) may be usefully applied to the evaluation of surface irregularity, flexibility, and conformation of protein and enzyme molecules.

Chemical accessibility, as well as the geometric accessibility, of the surface has to be considered in molecule-surface interactions such as those occurring in proteins and enzymes. While physical adsorption is relatively insensitive to chemical heterogeneities on the surface, chemisorption (i.e., interaction with the surface through the formation of covalent bonds) recognizes only the sites on the surface that are capable of taking part in the specific interactions. The subset of specific sites able to result in chemisorption is often fractal, and a fractal dimension, D_c, can be obtained where $D_c \leq D_a$. However, the chemical recognition of the surface may depend not only on the size of the adsorbate (i.e., the molecule being adsorbed), but also on its molecular composition. D_c will then reflect not only the geometry of the distribution of the active sites on the surface, but also the gradual change in the ability to form a chemisorbed bond with the adsorbate (i.e., the adsorbed molecule) [7].

The collection of surface points that react (i.e., the effective surface available for these processes) may coincide with all surface points (i.e., be the same as that probed by a physical adsorption experiment), or they may be different. In general, the two sets of sites accessible for physical and chemical adsorption do not coincide because of the many parameters that govern the course of a heterogenous chemical reaction with a surface. These parameters include the degree to which the reaction is diffusion controlled or chemically controlled, the relative importance of diffusion to and diffusion on the surface, and the role of various surface geometry parameters, such as its fractal dimensionality, the connectivity of the surface, and the pore size distribution [7]. A scaling relation can often be found between the activity of a surface, a (in mol g^{-1} s^{-1}), and the particle size, 2R [8]:

$$a \propto R^{D_R - 3} \tag{7}$$

where D_R is defined as the reaction dimension, that is, in its simplest interpretation, the fractal dimension of the subset of reactive surface sites.

Trapping of particles and diffusion-limited reactions are important processes occurring in micellar systems and polymers in solution, such as microorganisms, proteins, and enzymes. The dynamic properties of particles become anomalous when diffusing and being trapped in disordered systems or on fractal substrates. Also, the laws of diffusion of particles in these systems are very different from the classical laws for transport in ordered media. The trapping problem is also fundamental to the understanding of the kinetics of reactions. For instance, consider a random walker, for example, an active chemical species, in a medium that contains a finite concentration of randomly distributed static trapping centers, for example, chemically reactive sites that may combine with and deactivate the chemical species. In this case, the question of interest is the trapping rate or the survival probability of the random walker, that is, how the concentration of a random walker decays with time. In fractal space, the scaling of time, t, with the volume, V, covered by the walker is [9]

$$t \propto V^{2/D_S} \tag{8}$$

Another problem of interest is the diffusion of particles toward a fractal structure. If a particle diffuses from a homogenous solution to a solid surface and reacts to form a product when it reaches the surface, the subset of reactive sites exposed to diffusing particles is expected to have a fractal dimension $D < D_S$. In such experiments, the effective fractal dimension of the surface, as seen by the reactant molecules diffusing toward it from the bulk solution, can be measured through the reaction rate, v, according to

$$v \propto R^{D_R - 3} \tag{9}$$

where R is the reactant size and D_R is the reactant dimension. This can be interpreted as the fractal dimension of the subset of the reactive sites of the surface [10].

E. Techniques for Fractal Analysis

Scattering techniques provide the most powerful experimental means for studying fractal systems and for determining their fractal dimension. Depending on the characteristic length scale to be observed and the nature of the fractal object, scattering of visible or laser light, small-angle X-ray scattering (SAXS), and small-angle neutron scattering (SANS) can be used [11,12].

The value of the fractal dimension of an object is also accessible through the measurement of the wavelength dependence of the turbidity of the particle suspension, provided the particles are large enough. Instead of measuring the scattered light intensity at finite angles, which necessitates the use of specially designed apparatus, the complementary technique of monitoring the nonscattered or transmitted light with a conventional spectrophotometer can be used as a simpler alternative. Turbidity is a measure of the reduction in the intensity of the incident beam as the result of light scattering. The theoretical basis of this technique and examples of its application to the study of protein aggregation are provided in Ref. 13.

Transmission and scanning electron microscopy (TEM and SEM) and optical microscopy have been used successfully to study proteins, enzymes, and microorganisms. The forms and shapes observed appear to be similar to fractals described theoretically. The possible fractal character and dimension of these systems can be determined by detailed analysis of photomicrographs. A general procedure for determining the fractal dimension of an object observed in real space is based on the notion of the space density-density correlation function that can directly give the fractal dimension, D, of the object [14]. This technique has some important limitations in that it is, in general, applicable only at two dimensions because of the use of photographs. It has been shown, however, that the geometric projection of a three-dimensional aggregate onto a two-dimensional space does not change the fractal dimension, if $D < 2$ [15].

III. FRACTALS IN GENERAL SOIL SCIENCE

A. Fractal Analysis of Spatial Variations of Soil Properties

The use of fractal concepts and the fractal dimension were introduced to soil science by Burrough [16,17], who showed that the nature of the spatial variation of some soil properties, including pH, bulk density, percentages of clay, sand, and silt, sodium content, and others, could be described by examining their fractal dimension. Soil data were fractals, because increasing the scale of observation continued to reveal more and more detail. Although these data could not be referred to as "ideal" fractals, because they did not possess the property of self-similarity at all scales, the fractal parameter was a useful indicator of the presence and importance of nested scales of variation over the sampled transect or area. The fractal dimension thus gave a measure of the irregularity of a set of measurements that were independent of the magnitude of the numeric values involved. It provided a tool for examining the degree of spatial organization of soil data which could compare results from different soil properties measured at a variety of scales.

Fractal analysis has also been extended to the evaluation of spatial variation of transient soil properties, including soil surface strength along a series of transects, at 10-cm spacings, and microtopography measured by profile meter, at 1-cm intervals [18]. Subsequently, fractal analysis and multifractal concepts (e.g., probability measures with multiple fractal dimensions) were used to describe the spatial structure of soil surface strength, as measured with a micropenetrometer, and to study processes that cause soil crusting and the spatial characteristics of crust [19]. These studies suggested that the fractal dimension might be a relevant qualifier of the intrinsic variability of penetration measurements and that this variability could be modeled as a fractal process. Fluctuations in fractal dimension across sampling scales for fine-textured soils of large strength appeared to be low, as the result of the cohesiveness of the soil. For looser soils, more variability in fractal dimension was observed, especially in very dry conditions. Multifractal analysis of data suggested that at all locations, the same processes (e.g., rainfall, irrigation, dispersion, settling of fine particles) were responsible in a cumulative (multiplicative) way for the observed soil strengths, for example, via densification and crusting [19]. These various physical processes, apparently random in nature, possibly operated and interacted at closely related scales to cause the observed surface-strength features.

B. The Fractal Nature of Soil Structure and Hydraulic Conductivity

Soil is both a fragmented material and a porous medium. Thus, a fractal representation of soil may be especially appropriate, based mainly on pore space and particle size distributions and soil fragmentation, which are also useful ways of quantifying soil structure. Observations of maxima and "humps" in particle size distributions might be of diagnostic importance for soil genesis and origin.

The fractal nature of soil structure has been studied recently by several authors. The quantification of the apparent heterogeneity of soil structure by a numerical parameter, such as the fractal dimension, has important consequences for relating physical properties, such as strength, penetrability, diffusivity, hydraulic conductivity, and the spatial and temporal dynamics of soil microorganisms, to soil structure. The ultimate goal would be the incorporation of the fractal dimension of soil structure into the modeling of water and nutrient cycling and microbial dynamics in the bulk soil and, especially, in the rizosphere at the soil-plant interface.

The fractal dimension for the different pore size distributions in different soil horizons was found to range from 2.1 to 2.8 [20]. This parameter was suggested to be an intrinsic measurement of the amount of space of the different pore systems that could be used to distinguish different soil horizons and to calculate the unsaturated hydraulic conductivity of soil. However, in their study of a variety of silty and sandy soil sections by image analysis and of air-dried soil samples by mercury porosimetry, Bartoli et al. [21] suggested that soils are not pore fractals in the 10^{-9}- to 10^{-1}-m size range but are mass and surface fractals with $D_m = 1.1D_s$ when the embedding dimension is $d = 3$. These authors also suggested that soil structures could possibly be described by fractal DLA or CLA models.

The surface fractal dimension, D_s, calculated from mercury porosimetry data of transects of the organic-rich A_1 horizon of a sandy acid brown soil under fir trees, was found not to be significantly different from the D_s of the deeper aluminum-rich $A_1(B)$ horizon [22]. The surface fractal dimension was suggested to be a better discriminator than porosity of the root and nonroot zones of soil. For each horizon studied, the aggregate pore volume was not significantly larger, whereas the value of D_s was significantly smaller, for the root zone soils than the nonroot zone soils [22].

The degree of structure heterogeneity introduced by such agronomic manipulations as cultivation, tillage, irrigation, et cetera, may be quantified by the determination of the fractal dimension. The value of D will depend upon natural weathering and agronomic practices to which any particular soil may be subjected. For instance, soil moisture content will probably have a significant influence on the fractal dimension of soil structure by controlling the modes of aggregate breakdown and buildup during tillage, and this will, in turn, influence the distribution and performances of soil microbes. However, the value of the dimension may be misinterpreted if not supported by an appropriate use of fractal concepts and equations and direct measurements by adequate means of investigation.

The fractal dimension of aggregates sampled from soils under different management systems was found to range from 2.75 to 2.93 [23]. These values were higher than those of $D = 2.59$ and $D = 2.37$ previously calculated [24] in an attempt to replace empirically derived exponents in the relation between matric potential, hydraulic conductivity, and volumetric water content by the physically based fractal dimension parameter.

The fractal dimension of soil aggregates was strongly influenced by energy input (wetting treatments), cropping sequences, and aggregate size. D increased with increasing fragmentation, that is, with an increase in the number of microaggregates with respect to larger one [25,26]. The D values greater than 3 obtained in some cases are highly questionable with respect to their geometric and physical meanings, and may be artifacts of assumptions made about the density and size of the grains. Fractal dimensions of soil aggregate size distributions estimated from mass size data, D_m, were compared with those computed from actual number size data determined by counting, D_n, and then, both were compared with two fractal dimensions, D_T and D_R, obtained by a newly developed mass size-based model, with or without certain assumptions [27–29]. Estimates of D_m ranging from 0.79 to 4.06 probably referred to aggregates that were not mass fractals but surface fractals [30]. Number fractal dimensions in the range of 0.67 to 3.92 [27] appear to be neither the mass nor the surface fractal dimension. A dimension for the aggregates greater than their embedding dimension, 3, has no physical meaning. In reexamining fractal dimensions, where $D > 3$, previously estimated for soil particle size distributions by using a number-based approach [31,32], Tyler and Wheatcraft [33] found that these were artifacts of the plotting algorithms and of assumptions of the density and size of the grains. The use of a mass-based approach was suggested to enable a consistent estimation of the fractal dimension of particle size distributions, although it was restricted to a narrow spectrum of natural soils.

Fractal analysis was also used to quantify soil fragmentation under various tillage methods and crop sequences at different times during the growing season [34]. The fractal dimension, D, calculated from aggregate size distribution for each treatment was used as an indicator of soil fragmentation and, consequently, of soil structure. In all tillage systems, the D value increased over autumn and winter, in the order of plow > chisel > disk > no-till, indicating an increased fragmentation of soil that became dominated by smaller aggregates. Formation of larger aggregates in soil increased during the growing season for all tillage systems, presumably because of the breakdown of plant residues by microorganisms and living plant roots, which provided aggregate-forming materials.

Corn appeared to increase formation of larger aggregates, whereas soybean decreased it [34].

A general theoretical framework for a fully self-consistent fractal representation of structured soil, as both a fragmented natural material and a porous medium, has recently been developed [35]. The fractal dimensions for both completely and incompletely fragmented soils and the domain of length scales across which fractal behavior occurred have been defined. A number of predictive equations that related the porosity, bulk density, and aggregate size distribution of soil structure to a characteristic fractal dimension have been derived [35]. The aggregate size distribution for a collection of unconnected aggregates (e.g., a completely fragmented soil), such as occurred in particle size analysis, was found to follow a power law expression involving a fractal dimension, D. This concept was then extended to describe a structured field soil with aggregates interconnected by solid material to form a porous system (incomplete fragmentation) for which equations were derived that involve a bulk fractal dimension, D_r, with $D < D_r < 3$, which represented the matrix and pore space of a medium that was not completely fragmented. In other words, D_r was the fractal dimension of the structured porous medium in nature and D was that of a collection of its aggregates. Considering the soil as a fractal, incompletely fragmented, and porous conducting network, a number of relationships were derived between soil water content, water potential, and hydraulic conductivity, expressed in terms of the fractal dimensions D and D_r [35]. All these concepts and equations were successively tested and validated experimentally with the limited available data on aggregate porosity, bulk density, and size distribution and with data on water properties of natural undisturbed soils [36].

Essentially all particle size distributions encountered in environmental samples were found [37] to follow a simple power law across a wide range of radii, expressed by

$$N(r) \propto r^{-\nu} \tag{10}$$

where N(r) is the number of particles per unit volume having a radius larger than a given radius, r, and ν is a fractional exponent, which can, in some cases, be interpreted as a fractal dimension. A range of $\nu = 2.5$ to 3.0 was observed for environmentally relevant particles [37]. Data of soil particle size distributions measured down to a radius of 20 nm, using both classical methods, such as sieving and sedimentation, and static and dynamic light scattering, followed the power law in Equation (10), with $\nu = 2.8$, over two to five decades of length scales for several soils investigated. This exponent was interpreted as the fractal dimension of an underlying structure [37]. Below the lower cutoff, typically around 50 to 100 nm, the number of particles remained almost constant, whereas above the upper cutoff, typically between 10 and 5000 μm, the distribution decreased much more steeply, possibly as a power law with an exponent $\nu = 5.4 \pm 0.2$, and clearly did not represent a fractal regime.

Pore size count data and areal porosity data, collected from image analysis of either photographs taken at several soil layer surfaces from several sites [38,39] or manually marked clear plastic sheets placed on soil surfaces at several depths and sites [40], were used to establish a statistical fractal dimension for soil macroporosity [41]. The average macroporosity dimension calculated across all sites and depths for a silt loam texture were consistent with values experimentally determined from soil texture [41]. Equations were also developed that were capable of predicting both matrix and macropore saturated conductivity by modifying the Marshall saturated hydraulic conductivity equation with the introduction of the fractal dimension of soil porosity [42]. A numerical method was developed for determining the hydraulic conductivity function using fractal concepts and information provided by the retention curve only, without the need of any hydraulic conductivity determinations [43].

The fractal approach was also applied to understand the possible underlying mechanisms in the formation of aggregates in Oxisols [44]. The higher surface fractal dimensions, D_s, determined for a gibbsite-rich Oxisol than for a goethite-rich Oxisol was attributed to the existence of stronger interaggregate bonds in the former soil. The finding that D_s increased with increasing soil organic carbon content was also attributed to an increase of interaggregate bonds, that is, to a rearrangement of aggregates to closer packings [44].

C. The Fractal Nature of Soil Humic Substances

It is well established that humic substances (HS) and their fractions, humic acid (HA), fulvic acid (FA), and humin, are a physically and chemically heterogenous mixture whose typical characteristics are the absence of a discrete structural organization and an intrinsic lack of order at the molecular level [45,46]. The extreme heterogeneity of substrates, intermediates, and biological and chemical processes involved in the genesis of HS is probably primarily responsible for the random assemblage of the building blocks, bridges, functional groups, and side chains that exist in humic macromolecules. These substances are also characterized by a wide variety of sizes and shapes, implying variable porosity and compactness of the molecular structures in both the solid and colloidal states. Humic particles are subject to complex aggregation and dispersion phenomena in aqueous media, which are of relevance to their physical behavior and chemical reactivity. The degree of irregularity and roughness of exposed surfaces of humic macromolecules has an important influence on the number, types, and availability of chemically and physically reactive sites, catalytic activity, adsorption capacity, and extent of interactions with mineral surfaces, metal ions, organic chemicals, plant roots, and microorganisms in soil.

Fractal geometry is thus expected to provide a powerful tool for the quantitative description of HS systems, of their formation and transformation processes, and of their physical and chemical behavior in soil. The major contribution that fractal geometry can make to an understanding of HS is in the general, simple, and succinct representation of complex data through the parameterization by a single number, that is, the fractal dimension. The major weakness of the fractal approach here is in the averaging out of any detailed properties of the molecules and in providing results and interpretations that can be valid only in an average or statistical sense for finite systems such as HS. An introductory review of the potential applications of the fractal concepts and methods to the study of HS has been provided recently [47]. Also, the first experimental proofs of the fractal nature of HS have recently been obtained, [48-54].

Water and D_2O solutions of two soil HAs at concentrations of 1 to 4 g L^{-1}, buffered at pH 5.0 and in the presence of 0.1 M NaCl, were analyzed by SANS after 48 h of standing at 11°C [48]. The measured scattering intensity, I(q), for both HAs obeyed a power law decay over one decade of the scattering vector, q, according to

$$I(q) \propto q^{-D} \tag{11}$$

Polydispersity was found not to affect the power law scattering observed, and thus the scattering depended only on the fractal dimension, D. D values calculated from the slope of the plot, log I(q)/log q, were 2.1 and 2.0 for the two HAs, and the same mathematically corrected value of $D = 2.3 \pm 0.1$ was obtained for both HAs [48]. Humic acid particles in solution were assumed to consist of building units (monomers) with a radial size ≤ 2.5 nm, aggregated into clusters with an average radius of 40 to 50 nm. A D value of 2.3 could be related either to a DLA model with D = 2.5 for colloidal aggregates, or more probably, to an RLA model yielding D = 2.1 [48]. In the latter case, partly charged carboxylate groups in HAs at pH 5 would provide a residual barrier among monomers which were thus excluded from the dense, central part of the cluster. Slow, reaction-limited processes would thus favor the development of an open, ramified structure of HA in these conditions.

A fractal dimension, D = 1.8, was measured by SANS for two HAs dissolved in D_2O at a concentration of 3.0 g L^{-1} at pH 5 and in 0.1 M NaCl, after storage at 4°C for 48h [49,50]. When the temperature was increased to 22°C, the fractal dimension of the HA particles in solution increased with time and reached a constant value of D = 2.35 after 60 h. At the low temperature, the initial aggregation was suggested to be a diffusion-limited process of the CLA type with D = 1.8 (D = 1.75 for the model). At 22°C, initial clusters appeared to restructure slowly to form more compact aggregates with D increasing to a value of 2.35 after 60 h, after which no further change was observed. Measurements of the electromotive force showed that observed restructuring was a proton-dependent process, which initially involved a fast release of protons followed by a slow uptake of protons. The process also led to the release of bound, structured water molecules from the HA clusters, indicating the formation of a relatively compact structure by a reaction-limited process. Humic acid was thus suggested to function as a sponge that released water as temperature increased [49].

Small-angle X-ray scattering technique has been used to investigate the fractal nature and measure the fractal dimension of a number of HAs and FAs from soil and other sources [51,52]. Experimental log-log plots of the intensity of the scattered X-rays, I(q), as a function of the scattering vector, q, were straight lines over a range of about two orders of magnitude of q for HAs and FAs irradiated either in the solid state or in solution. Two equations [11] have been used to evaluate which type of fractal are HAs and FAs. For a mass fractal, the exponent of the power law

$$I(q) \propto q^{-D_m} \tag{12}$$

is $D_m \leq 3$, whereas for a surface fractal, the exponent of the power law

$$I(q) \propto q^{-(6-D_s)} \tag{13}$$

is in the range $3 < 6 - D_s \leq 4$ [11]. Results showed that HAs and FAs in the solid state were surface fractals, whereas in solution they behaved as mass fractals. The calculated D_s for HAs and FAs in the solid state ranged between 2.2 and 2.8, whereas the D_m for two aquatic HAs gave two very different values, 2.5 and 1.6 [51].

The variation of the fractal dimension of a soil HA suspended in water at a concentration of 50 mg L^{-1} was studied as a function of pH [53]. The rapid, simple, and noninvasive turbidimetric technique, based on the measurement of the nonscattered, nonabsorbed, transmitted light by a conventional UV-vis spectrometer, was used in this study. Theoretical considerations showed the existence of a power law dependence for the turbidity, τ, on the wavelength, λ, with an exponent, β, that might be directly related to the fractal dimension of the system under observation. Under the experimental conditions used in this study, that is, aggregates in the micrometer size range, the value of the fractal dimension corresponded to the exponent, β, according to the following equation:

$$\beta = 4 - \gamma + (d \log\tau / d \log \lambda) \tag{14}$$

where γ is assumed constant and equal to the value of -0.2, which is normally used for proteins over the wavelength range of 400 to 800 nm [13]. Log-log plots of the turbidity, τ, versus the wavelength, λ, of water suspensions of soil HA at pH values of 3, 4, 5, and 6 were linear over the wavelength range (400-550 nm) examined, that is, the wavelength dependence of the turbidity conformed to a power law [53]. This result implied that HA particles exhibited a fractal nature, that is, HAs could be described as fractal objects (mass fractals), with a fractal dimension that could be calculated directly from the slope of the experimental plots by the use of Equation (14). Fractal dimensions of D = 2.71 and D = 2.73 were obtained for HA at pH 3 and 4, respectively [53]. This suggested a slow DLA-type process for the aggregation of colloidal HA. At pH 5, the value of D was lower (2.43) and close to the value obtained previously for similar systems [48], which suggested an RLA-model for the aggregation of HA particles. Finally, a much lower D value (1.34) was obtained for HA at pH 6 [53]. This indicated the occurrence of a profound modification of the shape and size of HA particles at this pH, possibly involving a CLA-type process similar to what was observed previously for HA at a temperature of 4°C [49]. The decreasing fractal dimension of HA with increasing pH from 3 to 6 was suggested to reflect the development of more expanded and open ramified structures from more compact, less porous structures, as previously observed by electron microscopy measurements [55].

The effects of pH, standing time, HA concentration, and ionic strength of the medium on the fractal dimension of HA have been examined for two additional soil HAs and a peat HA obtained from the International Humic Substances Society (IHSS) collection of reference and standard humic materials [54, N. Senesi et al., in preparation]. The wavelength dependence of the turbidity was confirmed as following a power law for the water suspensions of HAs in all the conditions examined. Granulometric analysis showed that the particle size distribution for HA fit a Gaussian curve (i.e., monodisperse system) and that the average size of suspended HA particles was large enough (one order of magnitude higher than the wavelength range used) to validate the use of Equation (14) for the calculation of the fractal dimension of HA.

The exponent, β, obtained experimentally from the log-log plot of τ versus λ for the three soil HAs in any condition of pH, standing time, concentration, and ionic strength, always resulted in $\beta \leq 3$ [54, N. Senesi et al., in preparation]. This implied the possible existence of a mass fractal of dimension, $D_m = \beta$, according to

$$\tau \propto \lambda^{D_m} \tag{15}$$

For the peat HA at low pH values (3, 4, and 5), the exponent, β, resulted in $3 < \beta < 4$, thus indicating the existence of a surface fractal of dimension $D_s = 6 - \beta$, according to

$$\tau \propto \lambda^{(6-D_s)} \tag{16}$$

The values of the exponent, β, of soil HAs in aqueous suspension, at a concentration of 30 mg L^{-1}, ranged from values close to 3 (nonfractal objects) for one HA sample at low pH values (3 and 4) and after short standing times (2 to 8 h) to values of about 1.21 at pH 7 after 24 h of standing. In general, the D_m

value decreased with increasing either the pH from 3 to 7 at any standing time or the time from 2 to 24 h at any pH. Although the trend was similar, D_m values were different from one HA to another, and the trend was more evident at low pH values and short standing times. Variations in D_m values, which indicated a change of the aggregation process model for HA particles, generally occurred above pH 5 after short standing times or even at pH 3 after long standing times. These results confirmed preliminary observations [53,54] and reflected the development of more porous, open, and expanded structures from more compact ones as the standing time or pH of the system increased. These trends were also confirmed by SEM observations of HA particles in the various conditions (N Senesi et al., in preparation).

A typical behavior was shown by the peat HA, which exhibited either a surface fractal or a nonfractal nature (D_s values ranging from 2.55 to about 3) at pH 3, 4, and 5 after any standing time and a mass fractal nature at pH 6 and 7, with D_m values much lower at pH 7 ($D_m < 2$) than at pH 6 (D_m between 2.50 to about 3) (N. Senesi et al., in preparation). The peat HA apparently evolved from a surface fractal regime to a mass fractal regime between pH 5 and 6. This was ascribed to an almost compact, space-filled structure with an irregular and rough surface that was assumed by peat HA particles at pH ≤ 5 and to a porous structure, possibly characterized by a rough surface, which was assumed at pH ≥ 6. This interpretation was supported by SEM observations of peat HA particles at various pH values (N. Senesi et al., in preparation).

An increase in ionic strength, obtained by the addition of NaCl at concentrations of 1, 5, and 10 mM to 30 mg L^{-1} suspensions of soil or peat HA at pH 3 to 7 and after a standing time of 4 h, produced an apparent decrease of the mass fractal dimension, D_m, of the HA particles (N. Senesi et al., in preparation). Such changes in D_m suggested a modification of the aggregation model for HA particles at different ionic strengths of the medium. The decrease in D_m with increasing pH, previously observed for HA suspensions in the absence of salt, was also apparent when the ionic strength was changed. The peat HA apparently maintained a mass fractal regime also at low pH values in the presence of salt in the solution. At all pH values, an increase in HA concentration from 30 to 40 mg L^{-1} produced a correspondent increase of the fractal dimension of the soil HA, but not of the peat HA (N. Senesi et al., in preparation).

The intrinsic physical and/or chemical reasons for the different fractal nature (mass or surface fractal) and fractal dimension exhibited by HS as a function of the nature, origin, state, and concentration of the sample and of the pH, standing time, temperature, and ionic strength of the medium is unknown. Further research is needed to provide a physical and chemical basis for the results of fractal analysis of HS and to elucidate at the molecular level the physical and chemical properties that determine the fractal behavior of HS. Further studies are also necessary to discover which new insights into the physical and chemical characteristics of HS may be provided by the proper interpretation of their fractal properties.

Although much effort has been directed in the last few years to the study of the fractal nature of HS, there are a number of key problems that still need to be considered in this field of research. Among these problems are (1) the exact definition of the kind and origin of HS samples to be subjected to fractal analysis; (2) the experimental conditions to be used, including the state (solid, solution, or suspension) and concentration of the sample; (3) the pH, temperature, and ionic strength, to be used; and (4) the consistency of the fractal dimension values obtained by different techniques, that is, scattering, turbidimetry, and microscopy.

A number of experimental constraints may influence the determination of the parameters that characterize a fractal structure by scattering and turbidity techniques, and these must be avoided or corrected. For instance, the presence of clusters of small size or a polydispersity of the particles forming the aggregate can, for various reasons, drastically affect the measurement of scattered intensity or turbidity. Multiple scattering processes, which may reduce the fractal dimension determined, can be avoided by using sufficiently thin samples. The amount of light transmission, that is, the ratio between the intensity of the transmitted beam and of the incident beam, must be at least of the order of 0.7. The sample concentration in the system should not, however, be excessive, to avoid internal scattering processes that may affect the measurement, especially of turbidity. Background corrections should be made by subtracting the contributions of both the solvent and incoherent scattering from the total scattered intensity.

Clarification and standardization is also needed for the appropriate use of fractal principles and methodology (e.g., equations, approximations and assumptions, models) when applied to the calculation and interpretation of the fractal dimension of HS samples from experimental data. In this respect, an important question to be addressed is the physical meaning of the fractal dimension obtained from experimental data, including the mass fractal versus the surface fractal concept.

IV. FRACTAL ANALYSIS OF PROTEINS, INCLUDING ENZYMES

A. Fractal Models of Protein Structure and Conformation

The secondary and tertiary structures of several proteins have extremely complex irregular forms that are beyond description by traditional geometries, that is, Euclidean and differential geometries. Fractal geometry has therefore been considered to be a potentially useful tool for the conformational analysis of proteins. Although the fractal dimension, D, is only a single parameter, it has been proven to be a useful indicator of protein conformation, as it provides a quantitative measure of the space-filling degree of a structure. Currently, several lines of evidence suggest that both the structure and conformation of proteins, including enzymes, are susceptible to fractal description.

On the basis of measurements of electron spin relaxation of low-spin Fe^{3+} in three hemoproteins, Stapleton et al. [56] deduced that they could be described as fractals with a dimension of $D = 1.65 \pm 0.04$, which agreed with the fractal dimension, 5/3, of a self-avoiding random walk (SAW) to which the protein backbone beared a marked resemblance. Experimental verification with X-ray data for myoglobin confirmed the value obtained for the fractal dimension, thus providing an excellent independent validation to the data obtained by the analysis of protein vibrations and spin relaxation [56].

Almost identical results (uncertainty of approximately 3% in both cases) were obtained for the fractal dimension of five ferric proteins calculated directly from X-ray structural data or measured by spin lattice relaxation [57]. Changes in protein structure as a function of solvent conditions could therefore be conveniently monitored by relaxation measurements. Proteins lack rigorous self-similarity, that is, dilatation invariance or repetition of the structure within some observable length scale. However, the concept of statistical self-similarity can be used for proteins when some statistical distribution function describing their properties (rather than the protein itself) exhibits self-similarity. On these bases, the fractal dimension, D, was calculated for several diamagnetic and paramagnetic native proteins (Table 1) from available X-ray structural data by drawing spheres of radius, R, about an arbitrary α-carbon on the polypeptide backbone and counting the number, n, of α-carbons to the sphere intersection [57]. The results, which defined the variations of n with R, were fitted to a power law:

$$n \propto R^D \tag{17}$$

This implied that the average end-to-end length R of a polymer chain, which is assumed to be unbranched, constituted a statistically self-similar property of the protein. Appropriate procedures were applied to eliminate the effect of the arbitrary choice of the origin and end effects. The resulting values of D among the different proteins varied by 20%, ranging from D = 1.76 (7/4) for lysozyme to D = 1.34 (4/3) for ferredoxin (Table 1), which was reasonable when compared to the theoretical result, 5/3, for a SAW in three-dimensional Euclidean space. In particular, the five hemoproteins examined had fractal dimensions close to 5/3, whereas ferredoxin differed from heme proteins by having a fractal dimension of D = 1.34, obtained from both X-ray data and relaxation data [57].

The fractal dimension of the protein chain and its significance in relation to the secondary and tertiary structure has been analyzed in 43 proteins with different secondary structure elements [58], selected according to Bernstein et al. [59] so as to cover the five structural classes of proteins. The fractal dimension of the protein chain was calculated from the slopes of the log-log plots of the length of

Table 1 Fractal Dimension of Some Selected Proteins (± Standard Error) (adapted from [57])

Protein	Fractal dimension
Myoglobin (ferric)	1.66 ± 0.04
Cytochrome *c* (ferric)	1.66 ± 0.05
Ferredoxin	1.34 ± 0.05
Hemoglobin (α, ferric)	1.64 ± 0.03
Hemoglobin (β, ferric)	1.62 ± 0.04
Cytochrome *c* 550 (ferrous)	1.69 ± 0.05
Cytochrome *c* 551 (ferric)	1.47 ± 0.04
Cytochrome *c* 2 (ferric)	1.71 ± 0.05
Cytochrome *b* 5 (ferric)	1.62 ± 0.06
Lysozyme	1.76 ± 0.05
Trypsin (Ca^{2+})	1.54 ± 0.05
Carboxypeptidase A (Zn^{2+})	1.56 ± 0.04
Papain	1.76 ± 0.06
Ribonuclease *s*	1.35 ± 0.05
Chymotrypsin α	1.36 ± 0.05
Chymotrypsin γ	1.35 ± 0.05
Rubredoxin (ferric)	1.49 ± 0.06

the protein molecule, L, as a function of the size of a length element, m. These plots consisted of two almost linear components in the respective ranges of $m \leq 10$ and $m > 10$. The folding of the protein thus appeared to retain the fractal nature within the entire range examined, with the smaller m reflecting the local (secondary) folding of the protein chain and the larger m reflecting the total (tertiary) folding. This implied that the folding of the protein chain was not controlled by a single rule for the local and total structures, but each type of structure seemed to be dictated by different rules. The mean values of the fractal dimensions of the 43 proteins for the short (D_s) and long segments (D_l) were $D_s = 1.34 \pm 0.05$ and $D_l = 1.95 \pm 0.17$. The smaller values of D_s were attributed to the more extended and/or swollen local conformation assumed by the protein molecule, possibly as the result of steric hindrance and repulsive forces, whereas the larger values of D_l were ascribed to the more compact total conformation of the globular protein, resulting from attractive forces. In conclusion, the fractal dimensions, D_s and D_l, presumably represented the complexity of local (secondary) (within the range of 10 amino acid residues) and total (tertiary) conformations, respectively. Steric hindrance and attractive forces appeared to be the major factors that determined the local and total conformations, respectively. Thus, the fractal dimension could be used as an indicator of the average strength of the forces acting in the protein molecule. The transition from D_s to D_l

was suggested to be related to the finite size of the α-helices or the β-sheet structures, which reflected the transition from a length scale of a secondary structure to a length scale of a tertiary structure [58]. Further, the proteins with α-helices showed the largest value of D_s, whereas the β-sheet structures showed the smallest values of D_s, which reflected the extended character of local (secondary) structure elements of β-sheet proteins compared to the closely packed character of α-helix proteins.

The total packing of the tertiary structure, however, appeared to be similar for the two classes of proteins, and hence, D_l was similar in both cases. No correlation was observed between the fractal dimensions, D_s, distributed within the range of 1.25 and 1.48, and D_l, ranging from 1.59 to 2.34 [58]. This implied that the total conformation of protein was determined regardless of the local structures of the protein, that is, the rule that dictated the formation of the total conformation was independent of the local conformation. Further, the fractal dimension, D_s, was independent of the chain length, whereas a negative correlation was observed between D_l and the chain length [58]. These findings confirmed the authors' suggestion that the shorter the chain length of the protein, the more unstable its tertiary structure, and that specific interactions are required to keep the tertiary structure stable. This implied that the fractal dimension of a protein with a short chain length should be greater than that of a protein with a longer chain length.

The calculation of the fractal dimension was refined and extended to 70 globular proteins by a method based on the statistical identification of the most common repeating motifs within the structure [60]. The fundamental relationship predicted that the number of monomer segments, N, of length, r, was related to the end-to-end length, R, of the polymer chain by the fractal dimension, D, according to

$$N = (R/r)^D = (R/r)^{1/\upsilon_F} \tag{18}$$

where the exponent, υ_F, is equal to the inverse Flory constant in polymer theory [61]. Fractal dimensions of proteins were estimated from the slope (1/D) in a log-log plot of Equation (18), where N was treated as the independent variable, and average values of R ($N \geq 2$) and r ($N = 1$) of the covalent polypeptide chain were derived from the backbone represented by α-carbon coordinates in a data bank [59]. Values of D ranged within the limits ($1 \leq D \leq 2$) of well-defined test structures and correlated principally with the dominant elements of secondary structures [60]. This implied that the fractal nature of polypeptide chains was determined primarily by the relative contributions of compact and extended secondary structures.

The excellent correlation between the fractal dimension, D, estimated from crystallographic coordinates of the α-carbon backbone atoms, and the spectral exponent, m, deduced from magnetic relaxation data [57], was refined and confirmed for four additional heme and iron-sulfur proteins [60]. Other factors that might affect the spectral exponents for these in proteins were identified and interpreted within the theoretical fractal model, which thus provided valuable predictions of the reported distribution of low-frequency vibrational modes of these proteins [60]. In conclusion, the fractal dimension of a protein, considered as a conformational index influenced by secondary structures and suprasecondary domains, could define the distribution of vibrational states associated with the collective motions of the polypeptide backbone [60].

Helman et al. [62] showed, however, that, for an accurate description of low-spin hemoproteins and ferredoxin, besides the fractal structure of the protein backbone (i.e., the polypeptide chain), the cross-connections (e.g., hydrogen bridges) between segments of the folded chain should be considered. A fractal model was proposed in which the protein was no longer described as a one-dimensional chain but as a more complex topological object. This fractal system was described [63] by the fracton dimensionality, d_{fr}, which differs in general from the classical fractal dimension, D, and reflects the topological aspects of the system. However, for a high enough density of bridges it resulted that d_{fr} = D, which was in agreement with previous experimental result [56,57].

A more refined model of a protein as a two-dimensional lattice of connectivity was subsequently proposed, and an approximate scaling law for connectivity was demonstrated, which explained the experimentally observed fracton dimension [64]. In this model, the dominant contributions to the low-frequency vibrational modes of a protein molecule, which determined the fracton dimension, were shown to derive not only from the main chain, but also from peculiar properties of interactions (cross-links) between the polypeptide chain, including hydrogen bonds and van der Waals and electrostatic forces [64]. The fracton dimension of a protein was therefore not a direct result of the geometric arrangement of the chain in space, but rather was determined by the bond structure of a protein. Hence, low-frequency vibrations "see" a protein as a less than two-dimensional lattice of bonds, despite the compact three-dimensional arrangement of a protein in space [65].

The turbidimetric technique was successfully applied to the fractal study of casein aggregates [13]. An average fractal dimension of $D = 2.33 \pm 0.03$ was obtained for casein aggregates from the power law dependence of turbidity on the wavelength. This value was in good agreement with D values determined from the power law dependence of either the hydrodynamic radius ($D = 2.29 \pm 0.07$) or the scattered light intensity ($D = 2.2 \pm 0.2$) on the scattering angle [13]. The values for the effective fractal dimension of casein micelles were between those predicted by the DLA model (D = 2.5) and the RLA model (D = 2.0) [66].

A fractal mechanism was proposed for the allosteric effect of an enzyme, that is, the conversion of protein conformation from one state to another induced by the attachment of excited substrate molecules [67]. An equation was derived that related the Hill coefficient, h, widely used as an index of cooperativity, to the fractal dimension, D, of the protein [67]. Thus the nonintegral character of the Hill coefficient might be considered a direct reflection of the fractal properties of proteins, and the h value might be regarded as a kind of fractal dimension. The values of the Hill coefficients were calculated from the values of fractal dimensions and related to the changes in protein conformation, which were, in turn, dominated by the substrate molecule and its conditions. The substrate molecules were depicted to randomly walk over the fractal networks of amino acid residues until they hit the active sites of the protein and could then be bound to these residues by hydrogen bonding and other electrostatic forces [68]. The authors concluded that the allosteric effects and the Hill coefficient are the reflection of the fractal properties of proteins and not of the number of binding sites on the protein molecule.

B. Fractal Surfaces of Proteins

The characteristic roughness and corrugation of protein surfaces are of extreme biological relevance in their function, including the association of different subunits; the recognition, diffusion, and binding of a ligand; and the release of products. The protein surface has been evaluated by rolling a sphere, usually with a radius of a water molecule, over it, or by summing the exposed surface area of the atoms modeled as spheres with radii determined from their known van der Waals properties. A powerful quantitative tool that allows the evaluation of the surface corrugation and roughness of proteins is fractal analysis. The major advantage of this approach is to provide a compact parameter that describes the degree of surface corrugation, that is, the fractal dimension of the surface, D_s. Protein surfaces may be analyzed at different levels of detail, thus providing either a single value of the fractal dimension or separate D_s values that are calculated for different portions of the surface. Because a scaling law cannot describe a local property, it is necessary for the value of the dimension to persist for reasonable variations in the length scale.

In fractal analysis, the surface is always measured by a probe, which can be a stick for a two-dimensional cross-section or a small sphere for a three-dimensional representation of the protein. The surface is then defined as the area of the probe times the number of probes required to cover the surface completely. Alternative calculations consider the contact surfaces of the probe with the protein or the surface defined by the center of the sphere. Different algorithms, however, may affect significantly the scaling properties measured; that is, the comparison between calculations of the fractal dimension employing different definitions for the surface is not trivial.

The fractal dimension, D_s, of the surface of lysozyme has been calculated by measuring the number of steps of length ε, $N(\varepsilon)$, that are required to walk around the length of the contour which closes nine different two dimensional cross-sections of the lysozyme surface [69]. Plots of N versus ε obeyed the power law:

$$N(\varepsilon) \propto \varepsilon^{-D_{cont}} \tag{19}$$

where the fractal dimension of the contour, D_{cont}, obtained from the slope of the straight line, is related to D_s by $D_{cont} = D_s - 1$. The fractal behavior was verified to extend over a wide range of ε (0.15 to 2.05 nm). Different cross-sections gave similar results; thus, the system was considered to be homogenous, and its surface was described by a single average surface dimension, $D_s = 2.17 \pm 0.02$. This result implied that the number of molecules that lysozyme could hold by adsorption (monolayer), N_{ads}, decreased with an increase in the cross-sectional area, σ, of the adsorbed species as

$$N_{ads} \propto \sigma^{-D_s/2} \tag{20}$$

A similar procedure was used to evaluate the fractal dimension of the contour of a two-dimensional projection of the surface of the enzyme [70]. Although the projection dimension was similar to that obtained from the cross-section, the former parameter was found not to correlate clearly with the real surface dimensionality.

A different approach was used by other authors in the evaluation of the fractal surface dimension of three proteins, including lysozyme [71]. The entire surface was examined simultaneously and D_s was obtained from the slope of the plot of the log of the surface area against the log probe size that was used to define the molecular surface according to

$$2 - D_s = d\log(A_s)/d\log(R) \tag{21}$$

where A_s and R are the molecular surface area and probe radius, respectively [1]. For probes with radii from 0.1 to 0.35 nm, corresponding to the approximate size of water molecules and side chains of studied proteins, an average value for D_s of approximately 2.4 was found for the three proteins examined.

The variation in D_s over the protein surfaces displayed as spherical projections was also calculated [71]. High fractal dimensions ($D_s > 2.5$) where obtained for highly irregular surface regions of lysozyme characterized by elevated reactivity, whereas lower D_s values, ranging from 2.3 to 2.5, were obtained for smoother surface regions, which were located near the active sites of the enzyme. The region of greatest surface roughness in superoxide dismutase corresponded to the dimer interface. Similarly, in concanavalin A, the most irregular surface regions corresponded to subunit interfaces. In general, the degree of irregularity varied across the protein surface. Regions involved in the formation of tight complexes and permanent binding (such as interfaces between subunits and, possibly, antibody combining regions) appeared to be more irregular than average ($D > 2.4$). The irregular surface apparently formed more stabilizing contacts, possibly by allowing a greater number of contacts in a given region than smoother surfaces, and increased the specificity. Surface regions of enzymes involved in the formation of transient complexes with ligands (such as the active sites) appeared to be smoother than average. To maintain catalytic efficiency, the active sites of an enzyme should not bind ligands strongly, as otherwise the release of the products, that is, the final stage of the enzymatic reaction, would be slow. Crystal packing and small differences in side-chain conformations were found to have no major effect on the fractal surface, which was shown to be relatively insensitive to the precise location and thermal fluctuations of individual atoms (within 0.1 to 0.2 nm) [71].

The local fractal dimension was also calculated by assigning average D_s values to each amino acid [72]. High local dimensions were found at interfaces for the satellite tobacco virus, whereas prealbumin showed small variations in the surface dimension at interfaces and low dimensions at active sites [72]. Different results were obtained for trypsin, which had an apparent global value of D_s (2.62 ± 0.01) lower than the D_s of the active site (2.80 ± 0.04) [65]. It was concluded that although binding was enhanced on a more corrugated site, a single parameter, such as the fractal dimension, might be insufficient to describe the behavior of the active site of the enzyme which involves very specific and complex interactions [65].

Global surface dimensions, D_s, calculated for several proteins were found to be similar and at most 2.2 (Table 2) [65]. Thus, values of the surface dimension

Table 2 Fractal Dimension of Protein Surfaces (adapted from [65])

Protein	Fractal dimension
Mouse immunoglobin A Fab	2.14
Bacterial serine protease A	2.09
Chicken lysozyme	2.12
Lysozyme	2.17
Lysozyme	2.19
Lysozyme	2.53
Subtilisin inibitor BPN complex dimer	2.11
α-Cobratoxin	2.13
Cytochrome *c* 3	2.12
Ribosomal protein L7/L12	2.13
Retinol binding protein	2.18
Prealbumin tetramer	2.21
Satellite tobacco necrosis virus	2.15
Trypsin	2.62

of proteins should have a physical basis and control some biologically relevant processes of proteins, for example, substrate diffusion to and along the surface. For example, molecules close to the surface were suggested to be captured at a rate that increased exponentially with D_S, whereas the slower the substrate migrated along the surface to the active site (surface diffusion), the larger was D_S [69]. In particular, a value of $D_S > 2$ was shown to accelerate diffusion to the surface, but to slow it down on the surface. Thus, an overall optimum between efficient diffusion to the surface and efficient trapping by the active sites might be expected when the protein surface has a dimension moderately larger than 2, for example, $D_S \approx 2.2$. The method used by Lewis and Rees [71] also produced values of D_S close to 2.2 for several proteins [72], whereas the exceptionally larger surface area obtained for some proteins [65,71] was considered an artifact of the calculation caused by the particular program used, which might generate surfaces in internal empty spaces or closed pockets that might not be exposed to the solvent [72].

More recently, the interdependence of the surface topography of proteins and bound water molecules has been studied by two methods: surface accessibility and fractal density measures [73]. A "surfractal" program has been used to quantitate the irregular surface topography of a protein with a fractal measure of the atomic density around each surface point. The calculated fractal atomic density (surfractal index) varied continuously, with higher values around surface points in deep grooves and with lower values around surface points at the ends of protrusions. A surfractal atomic density ≥ 2.2 was chosen to differentiate "deep groove" from "nongroove" solvent-accessible surfaces [73]. In contrast to the surface roughness functions previously used [e.g., 71], the surfractal function avoided undesiderable discontinuities and exhibited the linearity expected for a fractal measure, that is, the slope of the function (log of the number of neighbors versus log of the sphere radius) was fairly constant within the 0 to 1.0 nm boundaries used for the radius of the spheres [73]. The surfractal algorithm thus provided an objective, quantitative definition of regional surface shapes of proteins, and it might be useful for the automated identification of deep grooves as potential active drug-binding sites and of large surface protrusions as potential antigenic sites and mobile regions [73]. The combination of the surfractal dimension and measures of "access" radius provided complementary information on shape that was applicable to understanding many aspects of macromolecular structure and function of proteins. In particular, the shape dependence of water binding has important implications for understanding and controlling macromolecular interactions. The methods used and the results obtained were expected to be useful in interpreting structures, modeling bound solvent, and guiding the design of proteins and drugs [73].

The effect of the fractal nature of enzymes on the rates of enzymatic surface diffusion processes and reactions has been also studied [74]. A sequence of elementary reactions and transport processes occurring on the surface of immobilized enzymes has been simulated by modeling enzyme molecules as fractal objects. Rate multiplicity, which is generally attributed to complex reaction kinetics on a homogenous surface, was then tentatively related to the fractal nature of the surface, as an extension of the model applied to elementary reactions [74].

C. Fractal Dimension of Protein Aggregates

The aggregation of monomers and the formation of clusters is of central interest in biology and polymer and colloid chemistry, including the physical chemistry of biological macromolecules. Determination and interpretation of the fractal dimension of protein aggregates can provide insight into the structural basis and provide a better understanding of protein-protein interactions and aggregation in biomembranes, thus allowing a better assessment of their functional role.

Aggregates are characterized by a fractal dimension, D, according to

$$N \propto (r/R_o)^D \tag{22}$$

where N is the number of particles (monomers) inside a radius, r, from the center of the aggregate, and R_o is the monomer radius [1]. Thus, the characteristic radius, R_i, of an aggregate is related to the number of particles, i, in the cluster by

$$R_i = R_o i^{\beta} \tag{23}$$

with the cluster exponent, $\beta = 1/D$.

In a study of the heat-induced aggregation kinetics of immunoglobulin by quasielastic light scattering, the experimental results were found to be consistent with a scaling form of the cluster size distribution [75]. The cluster could be described by the fractal dimension of $D = 2.56 \pm 0.3$, a value significantly lower than the spatial dimension of $d = 3$. The results also showed that the aggregation process was irreversible and not diffusion limited [75].

Fluorescence-resonance energy transfer was used to determine the fractal dimension of the "coastline" of membrane protein aggregates [76]. The advantage of this approach was its accessibility to a wide range of systems and its possible use under physiological conditions. The disadvantages were that high levels of fluorescence energy transfer were difficult to achieve, and experimental conditions were restricted to measuring parameters varying by less than a decade, thus making it virtually impossible to measure multiple fractal dimensions. Two different experimental designs and theoretical analysis approaches—(1) energy transfer between proteins within a protein aggregate and (2) energy transfer from the lipid domain to the protein aggregate—were applied to experimental data for aggregates of bacteriorhodopsin reconstituted into phospholipid vesicles and aggregates of calcium-ATPase in sarcoplasmic reticulum [76]. Approach 2 was found to be preferable to approach 1, because the fractal dimension could not be independently determined with transfer within the aggregate. Further, energy transfer from lipids to protein provided a means of estimating the length of the "coastline" of the aggregate, and the fractal dimension could be directly determined from the slope of the log-log plot:

$$d \log E / d \log \alpha = D/2 \tag{24}$$

where E is the efficiency of the energy transfer and σ is the surface density of the acceptors. The fractal dimension, D = 1.6, obtained for bacteriorhodopsin was comparable with the D values of 1.56 for lattice animals [77] and 1.55 for chemical cluster-cluster aggregates [78]. This D value was consistent with favorable and relatively nonspecific protein-protein interactions, which probably produced extended aggregate structures. In contrast, the higher fractal dimension, D ≈ 1.8, estimated for calcium-ATPase was close to values obtained for percolation clusters [77], thus suggesting that purified ATPase vesicles might have such a high protein:lipid ratio that an almost totally contacted network (i.e., uniformly filled two-dimensional space) was formed. The approach used in this study [76] provided a means for assessing the nature of protein-protein interactions and aggregation in membranous systems.

Turbidity measurements [13,14] have been used to determine the fractal dimension of ramified clusters of α-elastin obtained by reversible aggregation (coacervation) of a nondispersed elastin solution upon increasing the temperature [79]. The obtained value of D = 2.24 compared well with D = 2.21 found for casein [13] and for RLA processes (see section II.C).

V. FRACTAL PATTERNS IN MICROBIAL GROWTH AND MORPHOLOGY

Important contributions have been made by fractal geometry to understanding the growth of inorganic systems in such processes as aggregation, cluster formation, and dendritic growth. Although biological growth generally leads to the formation of highly complex patterns, fractal concepts may, nevertheless, have a great potential for application to biology because of the wide variety of biological patterns that seem to be self-similar. This property implies that the global structure of an object may be complex, but that the fundamental growth concepts underlying the generation of the complex system may be simple [1].

The introduction of fractal analysis to microbiology may thus be of fundamental importance for the description of growth patterns and the development of morphology for a wide variety of microorganisms under a wide spectrum of growth conditions. These aspects are particularly important because of the high correlation that they have with pathogenicity, metabolic activity, enzyme production, and pigmentation. Fractal geometry is thus expected to provide a powerful tool for the geometric, pattern-oriented description and measure of irregularity of the complex structure of bacteria and mycelia, which have been usually described in empirical terms, such as diffuse, compact, smooth, rough, et cetera [e.g., 80].

Simple, nonequilibrium, probabilistic growth models resulting from computer simulations can also lead to complex structures that mimic certain types of biological morphologies [81]. These models result in structures in which the fractal dimension, D, is distinctly smaller than the Euclidean dimension, d, of the space or lattice in which the growth process is occurring. In other words, the mass of the object, M, scales with some characteristic length, l, which describes the overall size of the object, according to

$$M \propto l^{D} \tag{25}$$

where the fractal dimension, D, is not an integer and satisfies the condition, $D < d$ [1]. This mass-length scaling relationship is the basis of all methods used to measure the fractal dimension of real objects or objects generated in computer simulations.

Three mass-length scaling relationships are commonly used to measure the fractal dimension for computer simulations of nonequilibrium growth and aggregation processes [81]. These are

$$R_g \propto N^{(1/D_\beta)} \tag{26}$$

where R_g is the radius of gyration and N is the mass (number of particles or occupied lattice sites) of the growing objects;

$$M(l) \propto l^{D_\gamma} \tag{27}$$

where M(l) is the mass contained within the distance, l, measured from some occupied site (generally, l is measured from the initial growth site or "seed", i.e., the site of inoculation); and

$$C(r) \propto r^{(D_\alpha - d)} \tag{28}$$

where C(r) is the two-point density-density correlation function at a distance, r.

For all real and simulated systems, these relationships are obeyed only over a finite range of a length scale that is limited by upper and lower cutoffs. As a consequence, D_α, D_β, and D_γ are effective fractal dimensions that converge on a common value, D, in the asymptotic limit where the range of length scales becomes infinite. In general, $D < d$ for real fractal objects, and thus, the mean density of a fractal becomes smaller and smaller as the fractal grows larger and larger.

Although a variety of biological growth models that have been studied do not apparently lead to fractal structures, a model able to generate structures with a well-defined fractal geometry has been produced by a simple DLA process

[82]. This model has led to the development of a number of related models that are also able to generate fractal structures of interest in nonequilibrium growth processes. The DLA model is thus expected to provide new insight into the morphology of biological systems, the growth of which is limited by the diffusion of nutrients to or the diffusion of toxic metabolites away from the system.

Meakin [81] first developed a model in which growth occurred by the random occupation of vacant surface sites by a diffusion-limited process that assumed that the rate of growth (growth probability) depended on the local nutrient concentration, which diffused from a surrounding external source and was consumed by the growing system. In this case, a structure with a fractal dimension of about 1.7 was formed in two-dimensional ($d = 2$) simulations, and of about 2.5 in three dimensions ($d = 3$) [81,82]. Assuming that the growth probability, p, was proportional to some power, ε, of the local concentration, c, that is, $p \propto c^{\varepsilon}$, the fractal dimension generated by the model depended on the exponent, ε, and might vary from 1 to 2 or 3. Simulations of diffusion-controlled growth from one central site (inoculum) on a two-dimensional square lattice using growth exponents, ε, of 0.5, 1.0, and 2.0, generated typical clusters shown in Figure 1 [81]. In this simulation, the nutrient was supplied at the boundary, which maintained a fixed (steady state) concentration of nutrient in a circle of lattice sites surrounding the growing cluster. The effective fractal dimensions, D_{α} and D_{β}, of these clusters were estimated from Equations (28) and (26), respectively [81]. Simulations in similar conditions have also been obtained with two growth sites (i.e., seeds or inocula) separated by 10 lattice units (Figure 2). Simulations of diffusion-controlled growth with a point source of nutrient and a seed separated by 120 lattice units were also carried out (Figure 3) [81]. Further simulations were obtained using different conditions for the growth site(s) and nutrient source [81].

Two independent research groups [83,84] first reported on the unambiguous experimental identification of fractal growth processes in biological systems. *Serratia marcescens*, which is present everywhere as an inhabitant of plants, insects, animals, and inanimate materials, has been shown to form fractal colonies with a fractal dimension ranging from 1.7 to 1.8 when growing on a minimal-nutrient agar (Davis or Vogel-Bonner agar medium) after a week or longer incubation at 30°C. However, it formed a typical round colony not exhibiting fractal morphology on normal nutrient agar after 1 day of culture [83,85]. The fractal growth of *S. marcescens* on the agar surface was apparently related to the wetting activity of this bacterium. Three kinds of extracellular cyclic lipopeptides, named serrawettins, were found to be responsible for such wetting activity, and thus had a critical role in promoting rapid fractal spreading growth of *S. marcescens* [83,85]. Serrawettin-less mutants defective in the wetting activity, that is, unable to produce these exolipids, were all impaired in

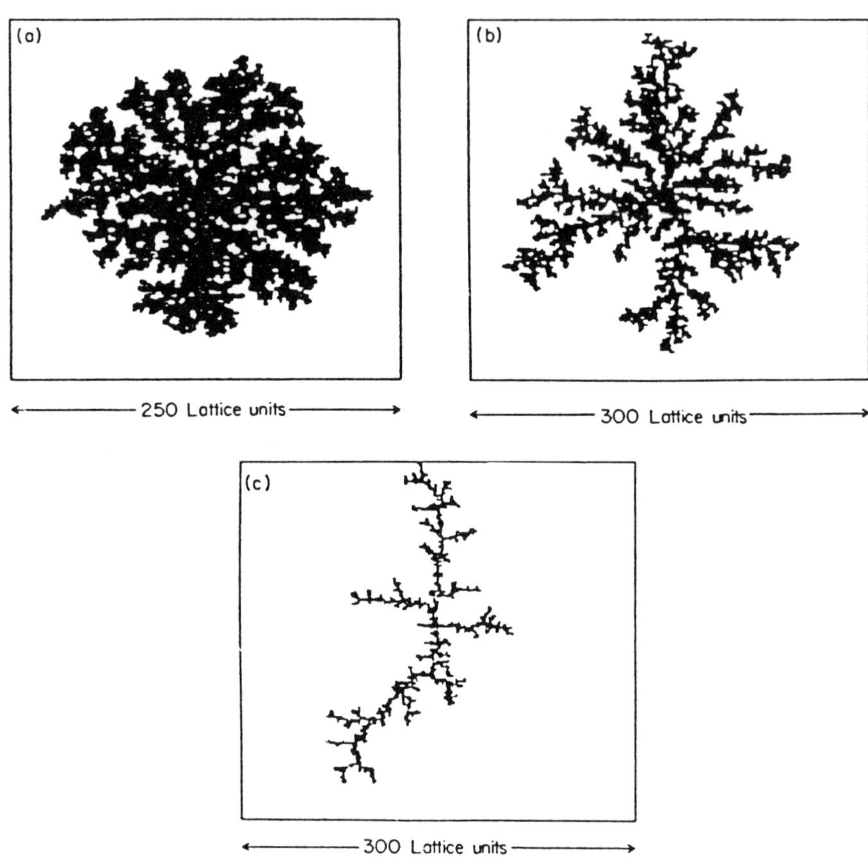

Figure 1 Two-dimensional simulations of typical clusters generated by a diffusion-limited process from one central site (inoculum), with growth exponents, ε, of (a) 0.5, (cluster of 12,000 sites), (b) 1.0 (cluster of 8000 sites), and (c) 2.0 (cluster of 2126 sites) (from Ref. 81).

forming a fractal giant colony, whereas they became effective when any serrawettin was supplied externally [85]. Fractal growth of *S. marcescens* was not observed when the wettability of the plates was increased by addition of a surfactant to the medium or when a peptone-glycerol agar with a strong wettability was used [83]. A longer incubation period was required for the development of a fractal colony on a poor nutrient medium, in contrast to the rapid, apparently nonfractal growth on a normal nutrient medium. The bacteria seemed, thus, to prefer a relatively low level of surface wettability, and minimal

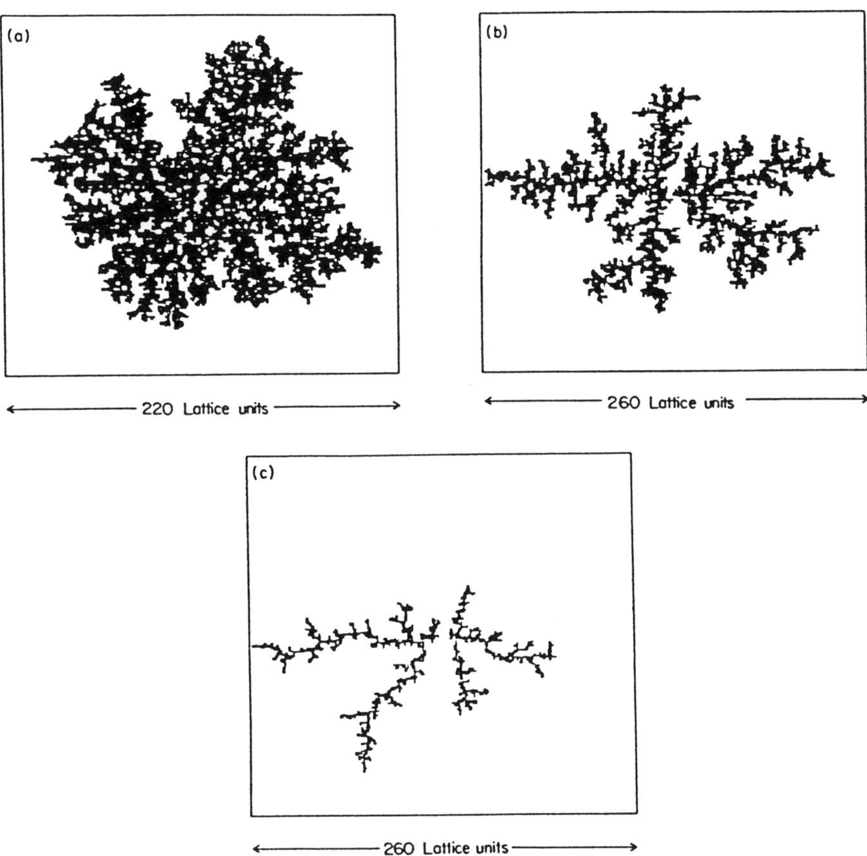

Figure 2 Two-dimensional simulations of a diffusion-limited process from two growth sites (seeds or inocula) separated by 10 lattice units obtained with growth exponents, ε, of (a) 0.5 (b) 1.0, and (c) 2.0 (from Ref. 81).

nutrient conditions, and they needed a relatively long period for fractal colony formation.

The importance of nutritional conditions and nutrient diffusion as critical factors for fractal colony growth of bacteria were confirmed by experiments developed on *Bacillus subtilis* [84, 86–88]. The growth of *B. subtilis* colonies on agar plates in the presence of low concentrations of a nutrient ($\leq$ 1 g L^{-1} of peptone) followed two-dimensional, random-branched, self-similar patterns that were governed by DLA processes in a concentration gradient of the nutrient. This implied a fractal growth of bacterial colonies that was strongly affected by both the conditions of the environment and the proliferative mode of the species.

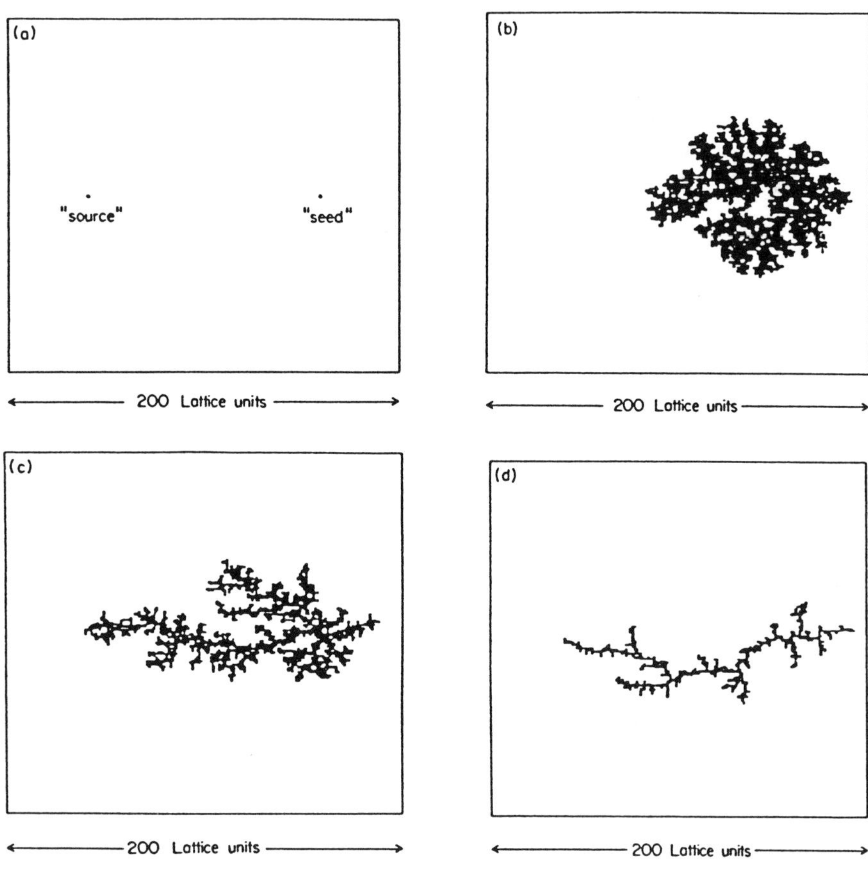

Figure 3 Two-dimensional simulations of a diffusion-limited process from a point source of nutrients and a seed separated by (a) 120 lattice units, obtained with growth exponents, ε, of (b) 0.5, and (c) 1.0, and (d) 2.0 (from Ref. 81).

B. subtilis was chosen as the sample bacterial species for a number of reasons. First, it is a rod-shaped, unicellular organism that proliferates through simple cell division, thus enabling a better study of the effects of environmental conditions, for example, nutrient concentration, moisture of the agar surface, and temperature. Second, the colony grows two-dimensionally on the surface of agar plates. The photos of bacterial colonies (Figure 4) obtained after incubation and growth for about 3 weeks on plates of dried agar containing concentrations ranging from 0 to 0.5 g L^{-1} of Bactopeptone as the nutrient were digitally analyzed by computer using an image scanner to calculate the fractal dimension by means of the box-counting method [86]. The average (14 samples) fractal

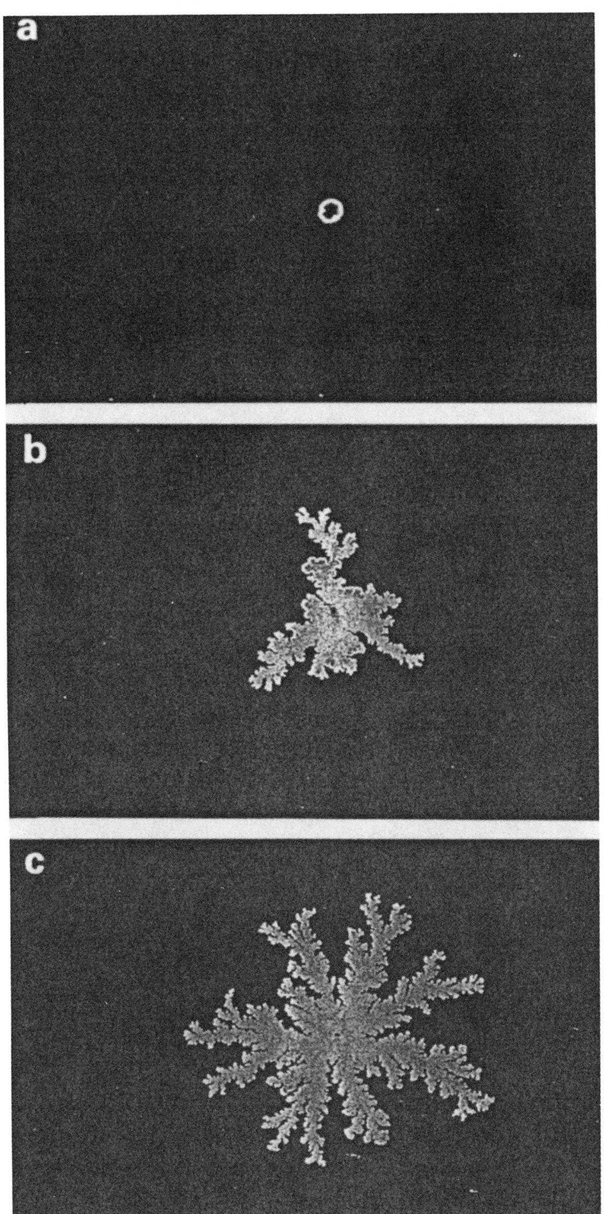

Figure 4 The effect of initial nutrient (peptone) concentrations of (a) 0 g L^{-1}, (b) 0.25 g L^{-1}, and (c) 0.5 g L^{-1} on the growth of colonies of *Bacillus subtilis* 3 weeks after inoculation (from Ref. 86).

Figure 5 A typical example of a DLA-like colony pattern ($D \approx 1.72$) of *Bacillus subtilis* incubated for 3 weeks after inoculation at 35°C on the surface of agar plates containing 1.0 g L^{-1} of peptone as nutrient (from Ref. 86).

dimension of the colony pattern obtained at an initial nutrient concentration of 1 g L^{-1} (Figure 5) was $D = 1.73 \pm 0.02$, which was close to that of two-dimensional DLA patterns.

Experimental evidence was also obtained for a number of effects that are inherent in DLA processes, including (1) the appearance of a screening effect of protruding main branches against inner ones with elapsing time from inoculation (Figure 6); (2) the repulsion between two neighboring colonies (Figure 7); and (3) the tendency of the colony to grow toward the nutrient (Figure 8) [86,87]. These results confirmed that the colony patterns of the organism were formed through a DLA process and could be described as fractals. To answer the question of whether nutrient or metabolite was responsible for the formation of fractal bacterial clusters, it was suggested that some nutrient might randomly concentrate into a colony and/or that some metabolites generated by the organism might diffuse away from it [84].

In a later study, the extracellular biosurfactant, surfactin, which is produced by some strains of *B. subtilis* and is structurally similar to serrawettins and active as a strong wetting agent, was suggested to promote the fractal morphogenesis of colonies of this bacterium [85]. Strains of *B. subtilis* that did not produce surfactin were unable to form fractal colonies. *Serratia rubidaea*, which shows spreading fractal growth, also was able to produce wetting agents, called rubiwettins, which are a mixture of hydroxy fatty acids [85].

Fractal structures were not formed when bacterial colonies of *B. subtilis* were grown on either dried (50°C for 40 min) agar in the presence of more nutri-

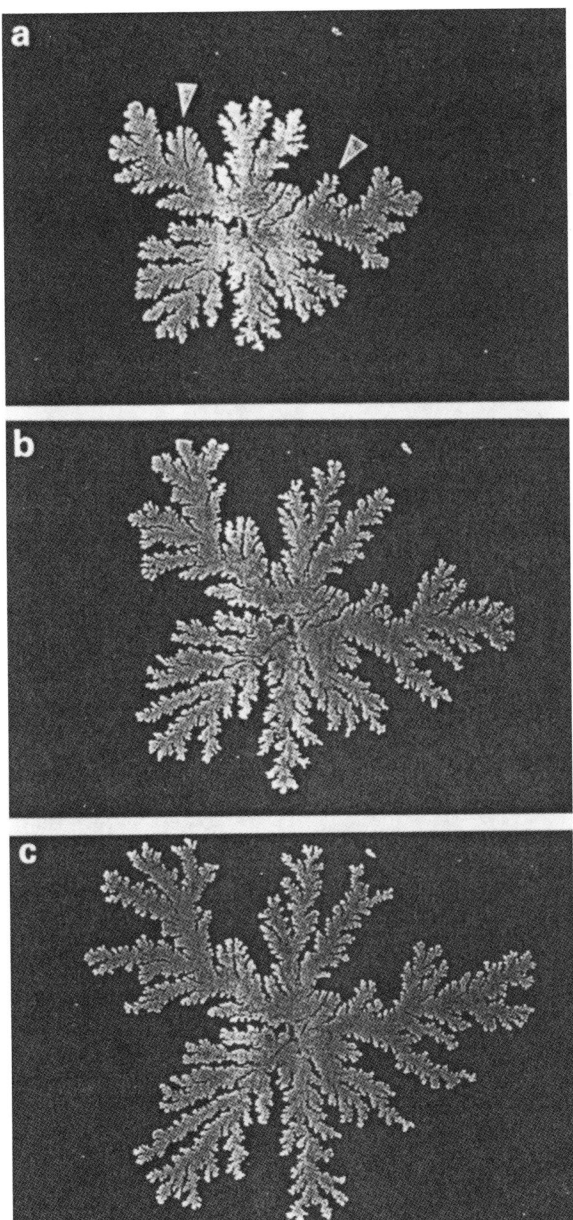

Figure 6 The growth of DLA-like colony patterns of *Bacillus subtilis* (a) 8, (b) 13, and (c) 19 days after inoculation, showing that some inner branches (pointed by triangles) stop growing (screening effect) after 8 days (from Ref. 86).

Figure 7 The repulsion between two neighboring colonies of *Bacillus subtilis* initially inoculated at two points 1 cm apart (from Ref. 86).

ent, that is, 15 g L^{-1} of peptone (Figure 9a), or on agar with a moist surface and containing 1 g L^{-1} of peptone (Figure 9b) [84]. In the former case, the colony pattern was smooth and almost round, resembling an Eden model cluster [88], whereas in the latter case, the bacterial colonies grew radially and isotropically with almost straight branches of the same lengths (dense radial morphology). On

Figure 8 The tendency of a colony of *Bacillus subtilis* to grow toward the nutrient (peptone) initially placed at the right side of the dish (from Ref. 86).

a

b

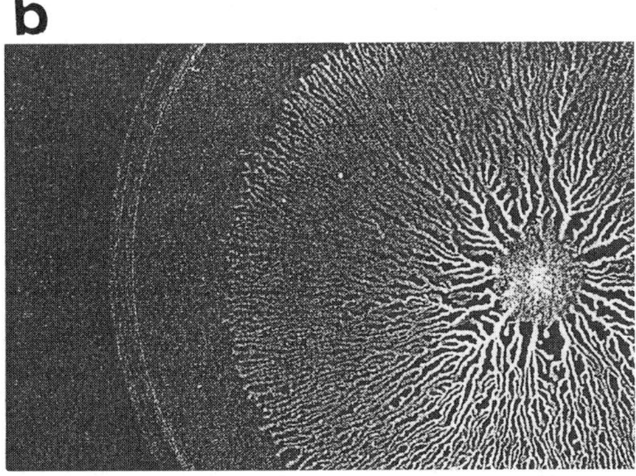

Figure 9 Colony patterns of *Bacillus subtilis* grown for (a) 9 days on a dried (50°C for 40 min) agar plate containing 15 g L^{-1} of peptone, and (b) grown for 1 day on an agar plate with a moist surface containing 1 g L^{-1} of peptone (from Ref. 84).

either a dried, but nutrient-rich, or a moist, but nutrient-poor agar plate, growth was fast, and it rapidly spread to cover the entire surface of the plate and did not allow the formation of fractal structures [84].

In many cases, the growing object itself may not be a fractal, but its surface may exhibit anisotropic fractal scaling. If the supply of nutrient is not limited, the stochastic sequence of the spontaneous multiplication of cells is expected to have a major role in the development of the surface of a colony, which may possibly follow a self-affine fractal scaling. The growth of *B. subtilis* and *Escherichia coli* on nutrient-rich agar plates for 3 to 6 days at 37°C resulted in the formation of apparently homogenous bulk colonies with dense branching morphology, which exhibited self-affine fractal surfaces, in agreement with the behavior of the surface of the Eden model clusters [89]. An effective fractal dimension of D ~ 1.22 and D ~ 1.26 was measured for the surfaces of *E. coli* and *B. subtilis*, respectively [89].

Several species of the family, *Enterobacteriaceae*, including various strains of *Proteus mirabilis*, *Citrobacter freundii*, *E. coli*, *Salmonella anatum*, *S. typhimurium*, and *Klebsiella pneumoniae*, were able to form fractal colonies after a relatively long incubation period in minimal-nutrient agar enriched with 0.4% glucose [90]. A fractal dimension of 1.7 to 1.8, similar to those of the DLA model, was measured for these colonies [90]. Growth models such as DLA, ballistic aggregation, and Eden have been applied to study fractal growth patterns of *Clostridium acetobutylicum* [91] and *Bacillus pumilus* [92,93]. An average fractal dimension of, respectively, 1.067 ± 0.015 and 1.30 ± 0.03, was measured for the ruggedness of the perimeter of these bacterial colonies [91–93].

In general, the type of mutant strains, that is, bacterium-specific factors, and some cultivation conditions, including surface tension, nutrient concentration, and moisture, appeared to constitute the most important factors affecting the type of bacterial proliferation and, in turn, the formation of fractal or nonfractal colony patterns. An evaluation of the intrinsic morphology or bacterial colonies could only be obtained within the limits of low nutrient concentrations and a low moisture content of agar, as only then would growth process inherent to a bacterial species produce a colony pattern possibly amenable to a fractal parameterization of its morphology.

Situations that induce a DLA-type fractal distribution of bacteria, that is, controlled by factors limiting the distribution and diffusion of nutrients and toxic metabolites, seem to be ubiquitous in nature [90]. Extended spreading fractal growth may be important for bacteria living on surfaces, in that it enables the bacteria to reach new surface environments efficiently, especially when the surfaces have fractal structures. Fractal structures can be shown to provide broader spaces for microbes than nonfractal surfaces [85]. The ability to occupy large surfaces in nature seems to be another important feature of many microbes that prefer to live on surfaces. Fractal structure of microbial niches and the small size of microbes might explain the extraordinary number of microbes and their ecological effects.

Fractal geometry has recently been used to probe the mechanisms of resource acquisition (foraging strategy) and the adaptation of morphology to the nutrient status of the substrate by a species of soil fungus [94]. Colonies of *Trichoderma viride* were grown over a homogenous substrate, that is, on humid cellophane sheets placed in petri dishes, inoculated with single germinated spores. The mechanism by which the fractal dimension was formed was through increased levels of branching. When the nutrient status of the substrate was poor, the colonies formed a low fractal-dimensional morphology by distributing as little hyphal mass as possible across a maximal area. When an elevated nutrient concentration was encountered, D increased toward 2, with the fungus filling the space as effectively as possible to exploit fully the substrate [94]. It was hypothesized that the fractal dimension reflected a compromise between exploitative and explorative growth forms [94].

Fractal analysis was also used as a tool for understanding the influence of a heterogenous distribution of resources (nutrients) on the morphological response of *T. viride* [95]. This aspect is of relevance when examining the ability of fungi to redistribute a patchily distributed nutrient resource in soil. Colonies were constrained to grow in two dimensions on cellophane in sterile petri dishes inoculated with single germinated spores of the fungus in a humid atmosphere with a discrete nutrient resource placed 2 cm away from the germling. Hyphal growth was sparse in the direction opposite to that of the nutrient source, whereas it formed a radially symmetric pattern on the other side of the colony. This suggested that the growth rate was not a local response to the nutrient status of the substrate. In fact, in the latter case, the hyphal profile should be elongated along the line joining nutrient source to inoculum site and not radially symmetrical about the point of inoculation. The fractal dimension of the colony structure that developed in the direction of the nutrient source did not differ significantly from 2, whereas that of the structure growing away from the nutrient source had a dimension significantly less than 2 [95]. However, an apparently greater amount of hyphal mass was measured in the direction away from the nutrient source, even though the space-filling efficiencies were greater in the direction toward the nutrient source. These results suggested that the space-filling capacity of the pattern adjusted to the heterogenous levels of nutrition and thus that the processes controlling branching, that is, space-filling efficiency, and the phenomenon of mass distribution were independent [95]. By enabling the quantitative separation of the phenomenon of mass distribution from space-filling efficiency, fractal analysis provides a powerful tool for understanding the morphological responses of fungal growth to nutrient status.

A fractal description has been provided for the space-filling characteristics and foraging strategies of two mycelial, saprotrophic basidiomycetes, *Phanerochaete velutina* and *Hyphaloma fasciculare*, as they extended into nonsterile soil from woody resource bases of varying nutrient status [96]. The commitment of mycelial biomass to exploration by these systems was also determined and related to their fractal geometry [96]. The possible mechanisms operating to generate fractal growth in these systems were also discussed [96]. The D values exhibited by foraging mycelial systems of *P. velutina* and *H. fasciculare* were found to depend on the potential of the nutrient base from which the mycelium was extending as a carbon/energy source [96]. Systems emanating from small-sized inocula had the lowest D values. Furthermore, as exploratory growth proceeded, mycelial growth became less complex, as represented by lower D values. This reduction in D values was suggested to be symptomatic of a biomass-conserving strategy through reabsorption of nutrients from autolyzed branches [96]. However, the degree of structural heterogeneity and branching, as described by the fractal dimension, was greater when the nutrient status of the resource base was high [96].

The foraging strategy employed by a fungus, which determines the balance between exploratory and exploitative growth as reflected by the fractal dimension, may depend on several factors, including its resource specificity, its tolerance to stressful microenvironmental conditions, its competitive ability, and the status of the resource base from which the mycelium is extending. Thus, the relatively high D value exhibited by *H. fasciculare*, which also produced highly branched, slow-extending systems with a greater commitment of mycelial biomass to exploration, reflected its short-range foraging strategy and could be associated with the unspecialized resource and microenvironmental requirements and high competitive ability of this fungus [96]. In contrast, the relatively low D value of *P. velutina*, the faster rate of its mycelial extension, and the smaller commitment to extra-resource mycelial biomass were indicative of greater resource specificity and, consequently, of the longer-range foraging strategy of this fungus [96]. The underlying processes generating the differences in structural complexity of the branching patterns of mycelial systems extending through a relatively hostile environment, that is, the soil, as described by their fractal dimension, were found to be congruent with hydraulic pressure-generated patterns [96].

Fractal concepts have also been applied to the description of growth patterns and aggregate formation for two other microbial species, a filamentous bacterium, *Streptomyces griseus*, and a fungus, *Ashbya gossypii* [97]. A cell was considered to be the smallest aggregating unit, and the colony was the aggregate under different sets of growth conditions (i.e., on filter paper and on agar). As mycelial structures could be mass fractals or surface fractals, two different box-counting methods were applied to evaluate the fractal dimension of the aggregates. The box-mass (BM) method was applied to the whole mass of the mycelium, and the box-surface (BS) method was applied to the surface of the system. These box-counting evaluations of digitized photographs of mycelial structures were performed by a standard computerized image analysis system. They led, respectively, to the evaluation of the box-mass dimension (D_{BM}) and the box-surface dimension (D_{BS}). The method consisted of covering the mycelium with a grid of cells (boxes) of side length, ε, and counting the number of boxes, $N_{box}(\varepsilon)$, that were intersected by the mycelium [97]. If the mycelium had a well-defined fractal dimension, D, the following power law was fulfilled:

$$N_{box}(\varepsilon) \propto \varepsilon^{-D} \tag{29}$$

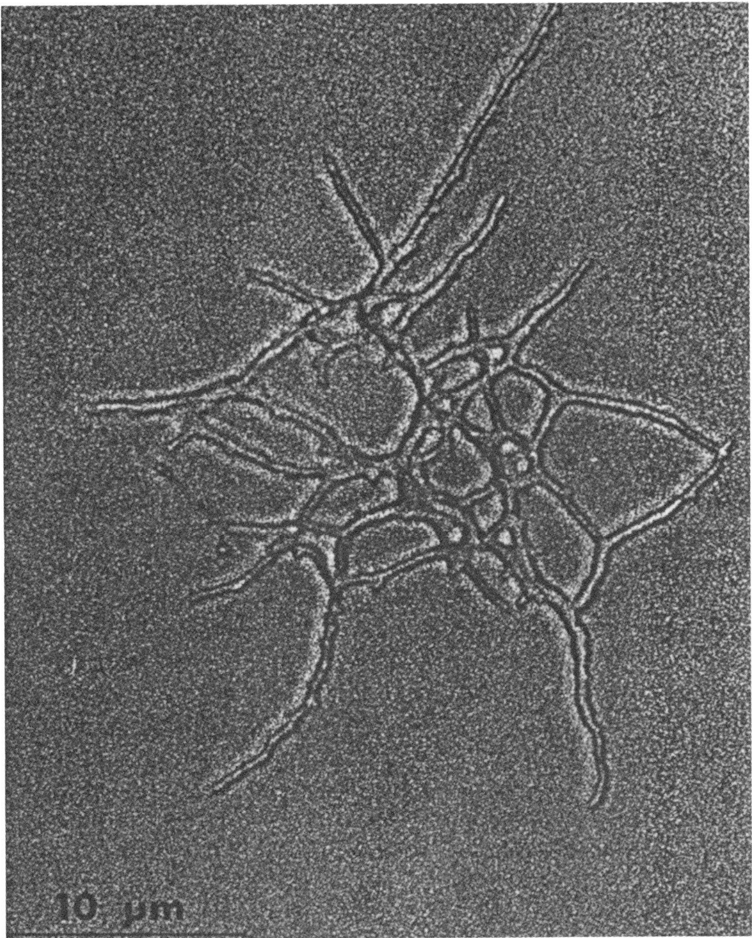

Figure 10 Mycelium of *Streptomyces griseus* grown for 16 h on filter paper (from Ref. 97).

where $N_{box}(\varepsilon)$ grew as ε^{-D}, within the ε range between the diameter of a single hypha, d_o (lower limit, or inner cutoff, ε_{min}), and the diameter, L, of the aggregate (upper limit, or outer cutoff, ε_{max}).

The mycelium of *S. griseus* (Figure 10), grown in d = 2 on filter paper in a petri dish, gave $D_{BM} = 1.49 \pm 0.01 \approx D_{BS} \approx 1.47 \pm 0.01$ over about a 0.5- to 6-µm (1 decade) length scale [97]. This implied that the aggregate was a self-similar mass fractal over this range of ε values. Consecutive photographs of the vegetative mycelium of *A. gossypii* (Figures 11 A–C), growing on a layer of

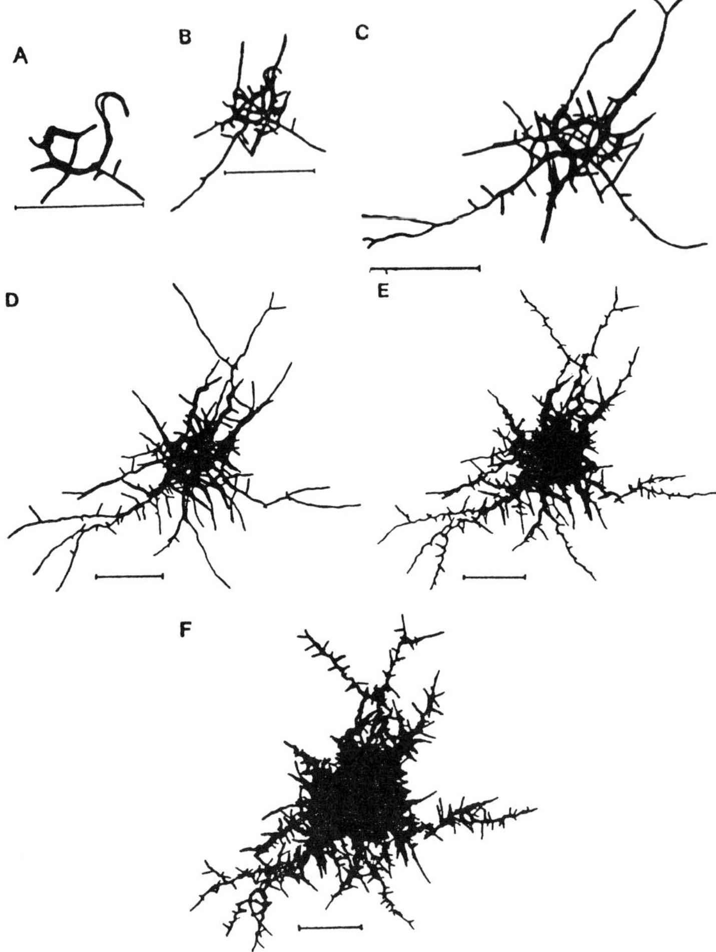

Figure 11 Mycelium of *Ashbya gossypii* grown on a 0.5-mm thick layer of agar after consecutive growth stages (from Ref. 97).

agar, 0.5mm thick, allowed the calculation of a fractal dimension, $D = D_{BM} = D_{BS}$ (mass fractal), ranging from 1.35 to 1.45 in earlier stages of colony growth. The hyphae of the young mycelium grew with branching on the agar layer in $d = 2$, with few hyphae growing into the agar layer and which were disregarded in the box counting. In the later stages of colony growth (Figures 11 D–F), the diameter of the mycelium increased very slowly, but more and more space-filling growth occurred, with a significant divergence between D_{BS} and D_{BM}. This indicated the transition from a mass fractal to a surface fractal. In the final stage

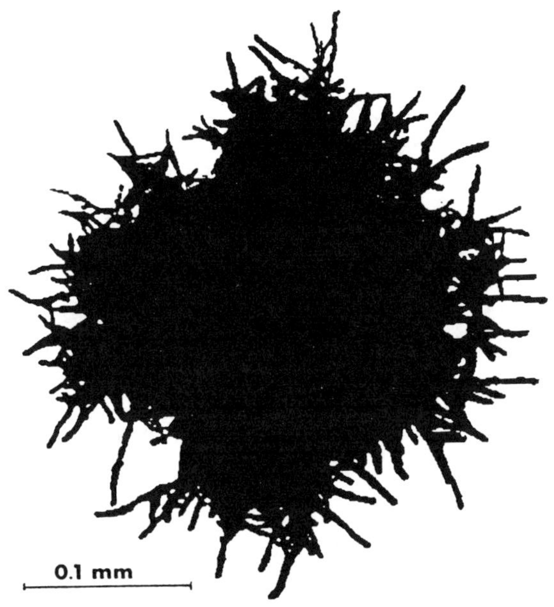

Figure 12 Mycelium of *Ashbya gossypii* grown for 60 h on a 0.5-mm thick layer of agar (from Ref. 97).

of growth (Figure 12), the colony of *A. gossypii* gave $D_{BS} = 1.45 \pm 0.01$ and $D_{BM} = 1.94 \pm 0.01$. In this aggregate, the hyphae grew over each other, into the agar, and into the air, thus completely filling the area in the center. The calculated D_{BM} value was thus close to the value of 2 that was expected for a planar photographic projection of a surface fractal. Consequently, the fractal dimension might serve as an important morphological characteristic of mycelial growth for a wide variety of filamentous microorganisms under a wide variety of growth conditions, for example, in the presence of chemical compounds that affect growth. An interesting practical application that was suggested was to establish a possible correlation of D with the production of antibiotics, to optimize fermentation processes that might be affected significantly by the shape of the producer organism [97].

A novel, rapid, and statistically reliable computerized image analysis method was introduced and described in detail, which could be usefully employed for quantifying the morphology of growth and calculating the fractal dimension of branching fungal colonies [98]. Biological variables or features of real fungal aggregates, which must be considered mathematically in obtaining reliable results, were discussed. A series of model and natural aggregates of known fractal dimension were analyzed to validate the method, that is, to ensure that the resulting fractal dimension values were accurate. The method was then applied to the analysis of branching of the white rot fungus, *Pycnoporus cinnabarinus*, which has been shown to be able to degrade organic aromatic pollutants. After incubation for 22 h on agar plates at 37°C, the fungal colonies displayed a filamentous branching fractal morphology with an estimated mean fractal dimension, $D = 1.67 \pm 0.05$ [98]. Branching growth for *P. cinnabarinus* was similar to that observed for other fungi, for example, *A. gossypii* [97] and *T. viride* [94].

Fungal colonies displaying a low degree of branching generally had lower D values compared with colonies showing more profuse branching [98]. These results implied that the fractal dimension is a useful parameter for quantifying the space-filling properties of the fungal colony and the self-similar (fractal) nature of the mass distribution (branching) of the hyphal network, which can be related to the efficiency of surface exploration (foraging growth) and exploitative (nutrient assimilation) growth mechanisms. Branching complexity was previously suggested to result from an equilibrium response between exploratory and exploitative growth [94]. This technique could then be used in studies of foraging ability and of underlying biological structures and biophysical processes inherent in the production of enzymes that are suggested to be responsible for the ability of white rot fungi to degrade successfully a wide range of aromatic compounds of environmetal concern.

Several techniques for measuring fractal dimensions were applied to colonies of three actinomycetes with different colony morphology and/or space-filling properties. *Nocardia asteroides*, *N. otitidiscaviarum*, and *Rhodococcus erythropolis* were examined at regular growth intervals during incubation on agar plates at 25°C [99]. The objective was to assess, on a comparative basis, the suitability of various techniques for the quantitative fractal characterization of colony morphology. Although box-count methods were superior to methods based on the use of calipers, values of D obtained by different methods were found to be equally valid but not directly comparable. Therefore, it is important that the methodology used for measurement and data analysis using fractal techniques be carefully selected and always stated. Within the limits of growth observed, *N. asteroides* behaved as a mass fractal, *R. erythropolis* was a surface fractal, and *N. otitidiscaviarum* showed a transition from mass to surface fractal [99]. The fractal dimension increased with the size and complexity of the colonies of the filamentous *N. asteroides* and *N. otitidiscaviarum*, but decreased slightly with the nonfilamentous colonies of *R. erythropolis* [99]. The increases of D values during colony growth measured in this study [99] agreed with findings for colonies of *S. griseus* and *A. gossypii* [97], *T. viride* [94], *Sordaria macrospora* [100], and *Phanerochaete magnoliae* [101]. The D values slightly above 1 measured for *R. erythropolis* were similar to those obtained for colonies of *C. acetobutylicum*, which had a similar nonfilamentous shape [91].

An apparent fractal dimension ranging from 2.55 ± 0.17 to 2.76 ± 0.14 has been obtained for three yeast strains of *Saccharomyces cerevisiae* by comparing the release of reducing sugar, that is, sucrose hydrolysis activity, against the diameter of yeast flocs cultivated on a sucrose medium in a continuous reactor [102]. However, larger fractal dimensions, which were close to the Euclidean value of the embedding dimension, 3, were obtained for flocs of *S. cerevisiae* when the floc dry weight was compared with the floc volume [102]. In general, the D values measured for yeast flocs of *S. cerevisiae* were much higher than those reported for other microbial aggregates and suggested the existence of dense and compact aggregates with a highly rugose surface, which was in good agreement with the Eden model for particle-cluster aggregation phenomena [102].

Most biological aggregates obtained from natural environments (e.g., marine "snow") and bioreactors (e.g., bioflocculated microbial aggregates generated by the activated sludge process) had fractal dimensions smaller than those measured for inorganic colloidal aggregates, such as clays, alum, and ferric oxihydroxides [103]. Biological aggregates in wastewater treatment bioreactors, known as activated sludge flocs, have also been shown to possess scaling properties and to develop fractal structures [103–105].

Physical characteristics of activated sludge flocs are known to have an important role in the performance of the activated sludge process, which is one of the major techniques of biological wastewater treatment [104]. The most important physical characteristic of activated sludge flocs is probably floc structure, which has a substantial influence on the mechanism of substrate transfer and on the flow through the floc itself. The fractal characterization of activated sludge flocs should thus enable the introduction of the laws of probability into the study of their irregular structure and provide a fractal model of floc structure. The fractal approach thus represents an improved alternative to the simplistic and approximate assumption of uniformity currently suggested and applied to structure models of activated sludge flocs [103,104].

Activated sludge flocs had fractal dimensions generally ranging from 1.4 to 2.07, whereas flocs containing a large number of filamentous microorganisms had a fractal dimension close to 1.0 [105]. The fractal nature of microbial aggregates was found to be a function of the type of microorganism and the conditions used to develop aggregates [106,107]. Different values of the fractal dimension were obtained as a result of different mechanisms of aggregate formation [106,107]. The magnitude of the fractal dimension for a collection of microbial aggregates, that is, a fractal classification of microbial aggregates, could also provide a useful reference for identifying the coagulation mechanisms and for calculating settling rates of fractal aggregates.

Aggregates (0.4–1.4 nm) of *Zoogloea ramigera* that developed in rotating test tubes were both surface and mass fractals, with fractal dimensions of D_S =

1.69 ± 0.11 and $D_m = 1.79 \pm 0.28$, respectively [106]. When these bacteria were grown in a laboratory benchtop fermentor, the aggregates maintained their surface fractal characteristics ($D_s = 1.78 \pm 0.11$) but lost their mass fractality ($D_m = 2.99 \pm 0.36$) [106]. Aggregates of the yeast *S. cerevisae*, grown in rotating test tubes were also fractals, but they had average fractal dimensions higher ($D_s = 1.92 \pm 0.08$ and $D_m = 2.66 \pm 0.34$) that bacterial aggregates grown under physically identical conditions [106]. These results suggested that scaling relationships based on fractal geometry might be more useful than equations derived from Euclidean geometry for quantifying the effects of different fluid mechanical conditions on aggregate morphology, density, and porosity [106].

Coagulation equations and self-similar size distribution analyses used to describe coagulation were modified to incorporate the fractal geometry of aggregates [107]. For example, aggregates of *Z. ramigera* developed in the rotating tubes were found to be formed by a differential sedimentation mechanism, for which D values between 1.6 and 2.3 are expected, rather than by shear coagulation, for which values of D greater than 2.4 apply [107].

Ligninolytic fungi, such as selected strains of white rot fungi possessing extracellular enzymes with an active lignocellulose degrading capacity, for example, *P. cinnabarinus*, showed considerable industrial potential for the treatment of a wide range of recalcitrant organic pollutants. The use of spore inocula to initiate the processing of a particular pollutant presented the advantage of a uniform germination phase, but the disadvantages of a long lag phase and reduced germination efficiency in environments containing high pollutant levels. In contrast, the use of a filamentous mycelial inoculum reduced the problem of an extended lag phase, with growth proceeding readily even at high levels of pollutants in the medium. The preparation and quantification of a uniform inoculum composed of homogenous hyphal fragments was difficult to obtain [108].

However, a new method based on the concepts of fractal geometry was proposed to obtain a quantitative relationship between frequency and size distribution of mycelial fragments of *P. cinnabarinus* and to measure the effects of different homogenizer speeds, which correspond to different shear rates, and duration of shearing on the distribution of mycelial fragments [108]. In this analysis, the D value was used to measure the mycelial fragmentation distribution function, that is, to describe the degree of fragmentation for a set of mycelial fragments, and not a fractal shape. The fractal dimension, D, of mycelial inocula subjected to different shear processes provided a descriptive summary of a complex system, that is, the D value provided information on the shear history that produced the fine-particle system. The decrease in the D value as the shear rate and duration increased (up to a threshold value) indicated a production of a more fragmented and more uniform inoculum characterized by a lower population diversity [108]. Further developments of this method could provide useful relationships between D, biomass growth rates and enzymatic activity for the optimization of inoculum preparations suitable for the design and operation of filamentous organisms in bioreactors that process various materials containing recalcitrant organic pollutants [108].

VI. FRACTAL HABITATS AND ORGANISM DYNAMICS

Fractal surface areas of vegetation and roughness of soil pores measured at different resolutions of scale for organisms of various size may be used to evaluate the habitable space available to species or individuals of that size. This concept has been used to relate body size to population density in fractal habitats [109,110].

Similar to coral reefs, plants surfaces were found to be fractal, thus influencing the type of arthropods (mainly insects) living on them [109]. Estimation of the fractal dimension, D, for arbitrarily positioned transects across the surface of a selection of plants ranged from 1.28 to 1.79, with a mean value of $D \approx 1.5$ [109]. Plants that apparently had a more complex growth form had a higher fractal dimension. The fractal nature of plant surfaces has implications for the animals living on them. For example, assuming that animals perceive and use vegetation surfaces in a way proportional to their body length, it was calculated that for a homogenous fractal surface having transects with $D = 1.5$ the area perceived by animals 3-mm long was up to an order of magnitude greater than the same area perceived by animals 30-mm long [109]. The increase in available space for animals of smaller body length was then combined with the way in which metabolic rate scales with body length to make predictions about the distribution of body lengths of animals living on vegetation [109]. As population densities are inversely proportional to metabolic rates [111] and the metabolic rate scales as the 0.75th power of the body weight, it was derived that the number of arthropods, N(l), of size, l, living on a vegetation of surface fractal dimension, D_s, scales as

$$N(l) \propto l^{-(D_s+0.25)} \tag{30}$$

With all other parameters being equal, a decrease of an order of magnitude in body length resulted in a 178-fold increase in the density of individuals. This increase in density was combined with the expected increase in the available surface area for a given decrease in body length to predict relative numbers of individual animals of different body lengths living on the surface of vegetation. For an order-of-magnitude decrease in body length, such calculations predicted an increase of between 560-fold and 1780-fold in the number of individual animals. Although data on the number of individual animals of different body lengths collected from the surface of vegetation are scarce, the relative numbers of small and large individual arthropods collected from vegetation were broadly consistent with theoretical predictions based on the fractal nature of vegetation and individual rates of resource utilization [109]. Calculations could be reversed to predict the fractal dimension of the surface of vegetation, given the mean distribution of individual body lengths of animals collected on the surface. This calculation predicted a fractal dimension of D = 2.78 for the surface of vegetation [109].

The fractal dimension as a measure of the roughness of soil pore walls has been utilized to estimate the available pore area for microarthropods of different sizes and to predict a relationship between the relative abundance of various size classes and body size of soil microarthropod communities [110]. Photographic images of thin sections of soil were computer scanned, and only intersected soil pores actually habitable to microarthropods, that is, pores having an area ≥ 0.003 mm^2, were analyzed as two-dimensional patches. The fractal dimension of the patch perimeter, D_p, corresponded to the slope of the log-log plot of the calculated perimeter, P, and the area, A, used to estimate the boundary dimension of each patch according to [110]

$$P \propto A^{D_p/2} \tag{31}$$

The corresponding pore surface dimensions, D_s, calculated by adding 1 to D_p [1], varied only slightly around 2.32. Accordingly, a decrease of an order of magnitude of body length would increase available habitat space approximately four times and result in a fourfold increase in density of individuals in a given pore area [110].

An important factor in population dynamics of soil microorganisms and of the dynamics of cycling and transport of nutrients via the soil microbial population is the persistence of the smallest prey species, which, in turn, depends on the existence of refuge sites in soil, that is, niches of small-scale structure accessible to bacterial species but excluding predatory protozoa. Soil structure is thus expected to have a large impact on temporal dynamics of microbial populations in soil. As bacteria tend to reside on the surface of pore walls where they may be subjected to predation by protozoa, fractal data of soil structure (discussed in section III) should allow the prediction of the magnitude of the area of refuge sites for prey species of various size and the calculation of the fraction of the potential habitat of the prey species that is inaccessible to predators [95]. For example, in a soil of fractal dimension, D = 2.36, half of the potential habitable area of the bacteria has been estimated to be inaccessible to predation by protozoa [95].

Although microbial movement in soil is mainly affected by mass flow and fluid films, the fractal nature of soil structure probably also has an important influence on the spatial dynamics of microbes in soil, that is, structure will influence small microbes differently from large ones in a tortuous soil. By combining fractal data of soil structure with measurements of the diffusion rate to body size, it has been estimated that, despite their low mobility, large species could move more rapidly than expected through soil because the effect of constricting pore necks is to limit the effective tortuosity, thus permitting a greater mobility to an otherwise more slowly moving, large predator [88]. Fractal geometry may thus provide a valuable theoretical and quantitative framework with which to understand the role of soil structure in the spatio temporal dynamics of microorganisms in soil and microbial processes involved in nutrient cycling in soil [95].

VII. CONCLUSIONS AND RECOMMENDATIONS

Fractal geometry can provide invaluable insights into the mechanism whereby spatial organization may influence the interaction between soil structure, including pore space and surface features, and biotic processes and spatial and temporal dynamics of microorganisms in soil.

Further investigations are needed on the variability of fractal dimensions of habitable pore surfaces in time and space, particularly in organic soil layers where seasonal changes and high spatial variability resulting from comminution and decomposition of plant debris occur. Data on the rate and extent of changes of the fractal dimensions may provide answers to whether various microbial communities actually equilibrate with the geometric features of their habitats. Extension of the applications of fractal geometry is also expected to probe further the mechanisms of resource acquisition by microbial species in soil and to help in understanding the dependence of growth and morphology of different microbial strains on nutrient concentrations and culture conditions.

The fractal nature and properties of microbial aggregates, as functions of the type of microorganism and the fluid mechanical environment existing in different soil conditions, are crucial themes for further research that aims at providing a better understanding of attachment mechanisms of microorganisms to surfaces, such as suspended organic and inorganic particles and other microorganisms. Evaluation of the fractal dimension of microbial aggregates may also help to elucidate the effects of various factors on bioflocculation, coagulation, and settling rate mechanisms, as well as on the morphology, size, density, and porosity of aggregates in the soil system.

In conclusion, the major contribution that fractal geometry can provide to a better understanding of soils consists of a unifying theoretical and quantitative model of the soil ecosystem. The fractal soil model should incorporate (1) soil physical properties that govern the supply of oxygen, water, and nutrients to plant and microorganisms; (2) chemical and biochemical processes that involve organic and inorganic soil colloids and surfaces; and (3) microbial population dynamics and biological processes, including nutrient cycling in soil. The major weakness of the fractal approach is in the averaging out of any detailed property of the system. In other words, fractal theory provides results that are valid only in an average or statistical sense for finite, highly irregular, and heterogenous systems, such as soil physical, chemical, and biological components and their properties and processes.

ACKNOWLEDGMENT

The author wishes to thank Dr. Federica R. Rizzi for her valued suggestions and help in the preparation of this paper.

REFERENCES

1. Mandelbrot, B. B. 1982. The fractal geometry of nature. W. H. Freeman, San Francisco.
2. Pfeifer, P., and M. Obert. 1989. Fractals: basic concepts and terminology, p. 11–40. *In* D. Avnir (ed.), The fractal approach to heterogenous chemistry: surfaces, colloids, polymers. John Wiley & Sons, Chichester, England.
3. Pfeifer, P. 1988. Fractals in surface science: scattering and thermodynamics of adsorbed films, p. 283–305. *In* R. Vanselow and R. Howe (ed.), Chemistry and physics of solid surfaces, Vol. 7. Springer, Berlin.
4. Meakin, P. 1983. Diffusion controlled cluster formation in two, three and four dimension. Phys. Rev. A. 27:604–607.
5. Meakin, P. 1989. Simulations of aggregation processes, p. 131–158. *In* D. Avnir (ed.), The fractal approach to heterogenous chemistry: surfaces, colloids, polymers. John Wiley & Sons, Chichester, England.
6. Brown, W. D., and R. C. Ball. 1985. Computer simulation of chemically limited aggregation. J. Phys. A. 18:L517–L521.
7. Farin, D., and D. Avnir. 1989. The fractal nature of molecule-surface interactions and reactions, p. 271–291. *In* D. Avnir (ed.), The fractal approach to heterogeneous chemistry: surfaces, colloids, polymers. John Wiley & Sons, Chichester, England.
8. Farin, D., and D. Avnir. 1987. Reactive fractal surfaces. J. Phys. Chem. 91:5517–5521.
9. Havlin, S. 1989. Molecular diffusion and reactions, p. 251–266. *In* D. Avnir (ed.), The fractal approach to heterogeneous chemistry: surfaces, colloids, polymers. John Wiley & Sons, Chichester, England.
10. Farin, D., and D. Avnir. 1988. The reaction dimension in catalysis on dispersed metals. J. Am. Chem. Soc. 110:2039–2045.
11. Schmidt, P. W. 1989. Use of scattering to determine the fractal dimension, p. 67–78. *In* D. Avnir (ed.), The fractal approach to heterogeneous chemistry: surfaces, colloids, polymers. John Wiley & Sons, Chichester, England.
12. Martin, J. E., and J. Hurd. 1987. Scattering from fractal. J. Appl. Crystal. 20:61–78.

13. Horne, D. S. 1987. Determination of the fractal dimension using turbidimetric techniques. Application to aggregating protein systems. Faraday Discuss. Chem. Soc. 83:259–270.
14. Teixeira, J. 1986. Experimental methods for studying fractal aggregates, p. 145–162. *In* H. E. Stanley and N. Ostrowsky (ed.), On growth and form. M. Nijhoff, Dordrecht, Netherlands.
15. Weitz, D. A., and M. Oliveira. 1984. Fractal structures formed by kinetic aggregation of aqueous gold colloids. Phys. Rev. Lett. 52:1433–1436.
16. Burrough, P. A. 1983. Multiscale source of spatial variation in soil. I. The application of fractal concepts to nested levels of soil variation. J. Soil Sci. 34:577–597.
17. Burrough, P. A. 1983. Multiscale source of spatial variation in soil. II. A non-Brownian fractal model and its application in soil survey. J. Soil Sci. 34:599–620.
18. Armstrong, A. C. 1986. On the fractal dimension of some transient soil properties. J. Soil Sci. 37:641–652.
19. Folorunso, O. A., C. E. Puente, D. E. Rolston, and J. E. Pinzón. 1994. Statistical and fractal evaluation of the spatial characteristics of soil surface strength. Soil Sci. Soc. Am. J. 58:284–294.
20. Ahl, C., and J. Niemeyer. 1989. The fractal dimension of the pore-volume inside soils. Z. Pflanz. Bodenk. 152:457–458.
21. Bartoli, F., R. Philippy, M. Doirisse, S. Niquet, and M. Dubuit. 1991. Structure and self-similarity in silty and sand soils: the fractal approach. J. Soil. Sci. 42:167–185.
22. Bartoli, F., G. Burtin, R. Philippy, and F. Gras. 1993. Influence of fir root zone on soil structure in a 23 m forest transect: the fractal approach. Geoderma 56:67–85.
23. Young, I. M., and J. W. Crawford. 1991. The fractal structures of soil aggregates: its measurements and interpretation. J. Soil Sci. 42:187–192.
24. Toledo, P. G., R. A. Novy, H. T. Davis, and L. E. Scriven. 1990. Hydraulic conductivity of porous media at low water content. Soil Sci. Soc. Am. J. 54:673–679.
25. Perfect, E., and B. D. Kay. 1991. Fractal theory applied to soil aggregation. Soil Sci. Soc. Am. J. 55:1552–1558.
26. Rasiah, V., B. D. Kay, and E. Perfect. 1992. Evaluation of selected factors influencing aggregate fragmentation using fractal theory. Can. J. Soil Sci. 72:97–106.
27. Perfect, E., V. Rasiah, and B. D. Kay. 1992. Fractal dimensions of soil aggregate-size distributions calculated by number and mass. Soil Sci. Soc. Am. J. 56:1407–1409.
28. Perfect, E., B. D. Kay, and V. Rasiah. 1993. Multifractal model for soil aggregate fragmentation. Soil Sci. Soc. Am. J. 57:896–900.
29. Rasiah, V., B. D. Kay, and E. Perfect. 1993. New mass-based model for estimating fractal dimension of soil aggregates. Soil Sci. Soc. Am. J. 57:891–895.
30. McBratney, A. B. 1993. Comments on "Fractal dimensions of soil aggregate-size distributions calculated by number and mass." Soil Sci. Soc. Am. J. 57:1393.
31. Tyler, S. W., and S. W. Wheatcraft. 1989. Application of fractal mathematics to soil water retention estimation. Soil Sci. Soc. Am. J. 53:987–996.
32. Tyler, S. W., and S. W. Wheatcraft. 1990. Fractal processes in soil water retention. Water Resour. Res. 26:1047–1054.
33. Tyler, S. W., and S. W. Wheatcraft. 1992. Fractal scaling of soil particle-size distributions: analysis and limitations. Soil. Sci. Soc. Am. J. 56:362–370.

34. Eghball, B., L. N. Mielke, G. Calvo, and W. W. Wilhelm. 1993. The fractal description of soil fragmentation for various tillage methods and crop sequences. Soil Sci. Soc. Am. J. 57:1337–1341.
35. Rieu, M., and G. Sposito. 1991. Fractal fragmentation, soil porosity and soil water properties: I. Theory. Soil Sci. Soc. Am. J. 55:1231–1238.
36. Rieu, M., and G. Sposito. 1991. Fractal fragmentation, soil porosity and soil water properties: II. Applications. Soil Sci. Soc. Am. J. 55:1239–1244.
37. Borkovec, Q. W. N., and H. Sticher. 1993. On particle-size distributions in soils. Soil Sci. Soc. Am. J. 57:883–890.
38. Edwards, W. M., L. D. Norton, and C. E. Remond. 1988. Characterizing macropores that affect infiltration into nontilled soil. Soil Sci. Soc. Am. J. 52:483–487.
39. Edwards, W. M., M. J. Shipitalo, and L. D. Norton. 1988. Contributions of macroporosity to infiltration into a continuous corn no-tilled watershed: Implications for contaminant movement. J. Contam. Hydrol. 3:193–205.
40. Logsdon, S. D., R. R. Allmaras, L. Wu, J. B. Swan, and G. W. Randall. 1990. Macroporosity and its relation to saturated hydraulic conductivity under different tillage practices. Soil Sci. Soc. Am. J. 54:1096–1101.
41. Brakensiek, D. L., W. J. Rawls, S. D. Logsdon, and W. M. Edwards. 1992. Fractal description of macroporosity. Soil Sci. Soc. Am. J. 55:1721–1723.
42. Rawls, W. J., D. L. Brakensiek, and S. D. Logsdon. 1993. Predicting saturated hydraulic conductivity utilizing fractal principles. Soil Sci. Soc. Am. J. 57:1193–1197.
43. Shepard, J. S. 1993. Using of fractal model to compute the hydraulic conductivity function. Soil Sci. Soc. Am. J. 57:300–306.
44. Bartoli, F., R. Philippy, and G. Burtin. 1992. Influence of organic matter on aggregation in oxisols rich in gibbsite or in goethite. I. Structures: the fractal approach. Geoderma 54:231–257.
45. Hayes M. H. B., P. MacCarthy, R. L. Malcolm, and R. S. Swift. 1989. The search for structure: setting the scene, p. 3–30. *In* M. H. B. Hayes, P. MacCarthy, R. L. Malcolm, and R. S. Swift (ed.), Humic substances II: In search for structure. John Wiley & Sons, Chichester, England.
46. Hayes M. H. B., P. MacCarthy, R. L. Malcolm, and R. S. Swift. 1989. Structures of humic substances: the emergence of "forms," p. 689–731. *In* M. H. B. Hayes, P. MacCarthy, R. L. Malcolm, and R. S. Swift (ed.), Humic substances II: In search for structure. John Wiley & Sons, Chichester, England.
47. Senesi, N. 1994. The fractal approach to the study of humic substances, p. 3–41. *In* N. Senesi and T. M. Miano (ed.), Humic substances in the global environment and implications on human health. Elsevier, Amsterdam.
48. Österberg, R., and K. Mortensen. 1992. Fractal dimension of humic acids, a small angle neutron scattering study. Eur. Biophys. J. 21:163–167.
49. Österberg, R., L. Szajdak, and K. Mortensen. 1994. Temperature-dependent restructuring of fractal humic acids: a proton-dependent process. Environ. Int. 20: 77–80.
50. Österberg, R., and K. Mortensen. 1994. Fractal geometry of humic acids. Temperature-dependent restructuring studied by small-angle neutron scattering, p. 127–132.

In N. Senesi and T. M. Miano (ed.), Humic substances in the global environment and implications on human health. Elsevier, Amsterdam.

51. Rice, J., and J. S. Lin. 1993. Fractal nature of humic materials. Environ. Sci. Technol. 27:413–414.
52. Rice, J., and J. S. Lin. 1994. Fractal dimension of humic materials, p. 115–120. *In* N. Senesi and T. M. Miano (ed.), Humic substances in the global environment and implications on human health. Elsevier, Amsterdam.
53. Senesi, N., G. F. Lorusso, T. M. Miano, G. Maggipinto, F. R. Rizzi, and V. Capozzi. 1994. The fractal dimension of humic substances as a function of pH by turbidity measurements, p. 121–126. *In* N. Senesi and T. M. Miano (ed.), Humic substances in the global environment and implications on human health. Elsevier, Amsterdam.
54. Senesi, N., F. R. Rizzi, P. Dellino, P. Acquafredda, G. Maggipinto, and G. F. Lorusso. 1994. The fractal morphology of soil humic acids. Trans. 15th World Cong. Soil Sci. 3b:81–82.
55. Chen, Y., and M. Schnitzer. 1989. Sizes and shapes of humic substances by electron microscopy, p. 621–638. *In* M. H. B. Hayes, P. MacCarthy, R. L. Malcolm, and R. S. Swift (ed.), Humic substances II: In search for structure. John Wiley & Sons, Chichester, England.
56. Stapleton, H. J., J. P. Allen, C. P. Flynn, D. G. Stinson, and S. R. Kurtz. 1980. Fractal form of proteins. Phys. Rev. Lett. 45:1456–1459.
57. Allen, J. P., J. T. Colvin, D. G. Stinson, C. P. Flynn, and H. J. Stapleton. 1982. Protein conformation from electron spin relaxation data. Biophys. J. 38:299–310.
58. Isogai, Y., and T. Itoh. 1984. Fractal analysis of tertiary structure of protein molecule. J. Phys. Soc. Japan 53:2162–2171.
59. Bernstein, F. C., T. F. Koetzle, G. J. B. Williams, E. F. Meyer, Jr., M. D. Brice, J. R. Rodgers, O. Vennard, T. Shimanouchi, M. Tasumi. 1977. The protein data bank: a computer-based archival file for macromolecular structures. J. Mol. Biol. 112:535–542.
60. Wagner, G. C., J. T. Covin, J. P. Allen, and H. J. Stapleton. 1985. Fractal models of protein structure, dynamics and magnetic relaxation. J. Am. Chem. Soc. 107:5589–5594.
61. Flory, P. J. 1953. Principles of polymer chemistry. Cornell University Press, Ithaca, New York.
62. Helman, J. S., A. Coniglio, and C. Tsallis. 1984. Fractons and the fractal structure of proteins. Phys. Rev. Lett. 53:1195–1197.
63. Alexander, S., and R. Orbach. 1982. Density of states on fractals: fracton. J. Phys. Lett. 43:L625–L631.
64. Elber, R., and M. Karplus. 1986. Low-frequency modes in proteins: use of the effective-medium approximation to interpret the fractal dimension observed in electron-spin relaxation measurement. Phys. Rev. Lett. 56:394–398.
65. Elber, R. 1989. Fractal analysis of protein, p. 407–423. *In* D. Avnir (ed.), The fractal approach to heterogenous chemistry: surfaces, colloids, polymers. John Wiley & Sons, Chichester, England.
66. Kolb M. and R. Jullien. 1984. Chemically limited versus diffusion limited aggregation. J. Phys. Lett. 45:L971–L977.

67. Li, H., S. Chen, and H. M. Zhao. 1990. Fractal mechanisms for the allosteric effects of proteins and enzymes. Biophys. J. 58:1313–1320.
68. Li, H. Q., and H. M. Zhao. 1990. Fractal theory and investigation of enzyme model. Nature J. 13:326–331.
69. Pfeifer, P., U. Welz, and H. Wippermann. 1985. Fractal surface dimension of proteins: lysozyme. Chem. Phys. Lett. 113:535–540.
70. Farin, D., A. Volpert, and D. Avnir. 1985. Determination of adsorption conformation from surface resolution analysis. J. Am. Chem. Soc. 107:3368–3370.
71. Lewis, M., and D. C. Rees. 1985. Fractal surfaces of proteins. Science 230:1163–1165.
72. Aqvist, J., and O. Tapia, 1987. Surface fractality as a guide for studying protein-protein interactions. J. Mol. Graph. 5:30–34.
73. Kuhn, L. A., M. A. Siani, M. E. Pique, C. L. Fischer, E. D. Getzoff, and J. A. Tainer. 1992. The interdependence of protein surface topography and bound water molecules revealed by surface accessibility and fractal density measures. J. Mol. Biol. 228:13–22.
74. Sharma, A., U. P. Shinde, and B. D. Kulcarni. 1990. Effect of fractal nature on enzymatic reactions. Biotechnol. Lett. 12:737–742.
75. Feder, J., T. Jossang, and E. Rosenqvist. 1984. Scaling behavior and cluster fractal dimension determined by light scattering from aggregation proteins. Phys. Rev. Lett. 53:1403–1406.
76. Dewey, T. C., and M. M. Datta. 1989. Determination of the fractal dimension of membrane protein aggregates using fluorescence energy transfer. Biophys. J. 56: 415–420.
77. Stauffer, D. 1979. Scaling theory of percolation cluster. Phys. Rep. 54:1–74.
78. Jullien, R., and M. Kolb. 1984. Hierarchical model for chemically limited cluster-cluster aggregation. J. Physiol. A17:L639–L643.
79. Tamburro, A. M., and V. Guantieri. 1991. Classical and fractal description of elastin structure, p. 391–400. *In* C. Rossi and E. Tiezzi (ed.), Ecological physical chemistry, Proc. Int. Workshop, 8–12 November 1990, Siena, Italy. Elsevier, Amsterdam.
80. Belisle, J. T., and P. J. Brennan. 1989. Chemical basis of rough and smooth variation in mycobacteria. J. Bacteriol. 171:3465–3470.
81. Meakin, P. 1986. A new model for biological pattern formation. J. Theor. Biol. 118: 101–113.
82. Witten, T. A., and L. M. Sander. 1981. Diffusion-limited aggregation, a kinetic critical phenomenon. Phys. Rev. Lett. 47:1400–1403.
83. Matsuyama, T., M. Sogawa, and Y. Nakagawa. 1989. Fractal spreading growth of *Serratia marcescens* which produces surface active exolipids. FEMS Microbiol. Lett. 61:243–246.
84. Fujikawa, H., and M. Matsushita. 1989. Fractal growth of *Bacillus subtilis* on agar plates. J. Phys. Soc. Japan 58:3875–3878.
85. Matsuyama, T., and M. Matsushita. 1993. Fractal morphogenesis by a bacterial cell population. Crit. Rev. Microbiol. 19:117–135.
86. Matsushita, M., and H. Fujikawa. 1990. Diffusion-limited growth in bacterial colony formation. Physica A. 168:498–506.

87. Fujikawa, H., and M. Matsushita. 1991. Bacterial fractal growth in the concentration field of nutrient. J. Phys. Soc. Japan 60:88–94.
88. Fujikawa, H. 1994. Diversity of the growth patterns of *Bacillus subtilis* colonies on agar plates. FEMS Microbiol. Ecol. 13:159–168.
89. Vicsek, T., M. Cserzo, and V. K. Horvath. 1990. Self-affine growth of bacterial colonies. Physica A. 167:315–321.
90. Matsuyama, T., and M. Matsushita. 1992. Self-similar colony morphogenesis by gram-negative rods as the experimental model of fractal growth by a cell population. Appl. Environ. Microbiol. 58:1227–1232.
91. Markx, G. H., and C. L. Davey. 1990. Applications of fractal geometry. Binary 2: 169–175.
92. Schindler, J., and T. Rataj. 1992. Fractal geometry and growth models of a bacillus colony. Binary 4:66–72.
93. Schindler, J. 1993. Dynamics of bacillus colony growth. Trends Microbiol. 1:333–338.
94. Ritz, K., and J. W. Crawford. 1991. Quantification of the fractal nature of colonies of *Trichoderma viride*. Mycol. Res. 94:1138–1141.
95. Crawford, J. W., K. Ritz, and I. M. Young. 1993. Quantification of fungal morphology, gaseous transport and microbial dynamics in soil: an integrated framework utilizing fractal geometry. Geoderma 56:157–172.
96. Bolton, R. G., and L. Boddy. 1993. Characterization of the spatial aspects of foraging mycelial cord systems using fractal geometry. Mycol. Res. 97:762–768.
97. Obert, M., P. Pfeifer, and M. Sernetz. 1990. Microbial growth patterns described by fractal geometry. J. Bacteriol. 172:1180–1185.
98. Jones, C. L., G. T. Lonergan, and D. E. Mainwaring. 1993. A rapid method for the fractal analysis of fungal colony growth using image processing. Binary 5:171–180.
99. Soddell, J. A., and R. J. Seviour. 1994. A comparison of methods for determining the fractal dimensions of colonies of filamentous bacteria. Binary 6:21–31.
100. Obert, M., M. Sernetz, and U. Neushulz. 1991. Comparison of different microbial growth patterns described by fractal geometry, p. 293–306. *In* H. O. Peitgen, J. M. Henriques, and L. F. Penedo (ed.), Fractals in the fundamental and applied sciences. Elsevier, Amsterdam.
101. Ainsworth, A. M., and A. D. M. Rayner. 1991. Ontogenetic stages from coenocyte to basidiome and their relation to phenoloxidase activity and colonization processes in *Phanerochaete magnoliae*. Mycol. Res. 95:1414–1422.
102. Fontana, A., C. Bore, C. Ghommidh, and J. P. Guiraud. 1992. Structure and sucrose hydrolysis activity of *Saccharomyces cerevisiae* aggregates. Biotechnol. Bioeng. 40: 475–482.
103. Li, D. H., and J. J. Ganczarczyk. 1989. Fractal geometry of particle aggregates generated in water and wastewater treatment processes. Environ. Sci. Technol. 23: 1385–1389.
104. Li, D. H., and J. J. Ganczarczyk. 1990. Structure of activated sludge flocs. Biotechnol. Bioeng. 35:57–65.
105. Logan, B. E., and D. B. Wilkinson. 1990. Fractal geometry of marine snow and other biological aggregates. Limnol. Oceanogr. 35:130–136.

106. Logan, B. E., and D. B. Wilkinson. 1991. Fractal dimension and porosities of *Zoogloea ramigera* and *Saccharomyces cerevisae* aggregates. Biotechnol. Bioeng. 38:389–396.

107. Jlang, Q., and E. Logan. 1991. Fractal dimensions of aggregates determined from steady-state size distributions. Environ. Sci. Technol. 25:2031–2038.

108. Cameron, L. J., G. T. Lonergan, and D. E. Mainwaring. 1993. Mycelial fragment size distribution: an analysis based on fractal geometry. Appl. Microbiol. Biotechnol. 39:242–249.

109. Morse, D. R., J. H. Lawton, M. M. Dodson, and M. H. Williamson. 1985. Fractal dimension of vegetation and the distribution of arthropod body lengths. Nature 314: 731–733.

110. Kampichler, C., and M. Hauser. 1993. Roughness of soil pore surface and its effects on available habitat space of microarthropods. Geoderma 56:223–232.

111. Peters, R. H. 1983. The ecological implications of body size. Cambridge University, Press, Cambridge.

11

Field Implementation of In Situ Bioremediation: Key Physicochemical and Biological Factors

Richard J. F. Bewley Dames & Moore, Manchester, England

I. INTRODUCTION

Bioremediation is the application of microbial processes to convert environmental contaminants into harmless substances [1]. As a treatment technology for contaminated land, it involves adjusting or controlling the physical, chemical, and biological properties of the soil environment to achieve the conditions under which such conversions are brought about.

In situ bioremediation essentially refers to the application of such treatment to the contaminated environment without displacing the contaminant from its source. In the strict sense, it therefore excludes not only commonly applied bioremedial practices involving excavation of soil and its placement in treatment beds, windrows, biopiles [2–6], or soil slurry reactors [7], but also pump-and-treat technologies in which the contaminant is removed from the saturated zone in an aqueous medium and transferred to an above ground treatment system. The latter could include processes such as percolating filtration, aerated lagoons, activated sludge treatment [8], and suspended or attached growth systems [9–12]. In this sense, engineered wetland systems for treatment of water contaminated from another source (e.g., municipal wastewater, storm water runoff, coal mine drainage, and certain industrial wastes [13]) could also be viewed as an ex situ rather than an in situ treatment.

This chapter is not intended to provide a comprehensive coverage of all full-scale in situ treatments, but rather to focus on the key issues concerning the development and implementation of this technology that relate to the physicochemical and biological properties of the soil environment and the potential for their amelioration. As such, key laboratory and pilot-scale studies that have provided a significant insight into these factors will be considered, together with examples of ex situ treatment where these cover issues of relevance to both types of bioremediation systems. It should be noted that there are actually few published accounts of case studies that represent truly in situ treatment, and in a number of instances, at least for the saturated zone, a combination of both in situ and aboveground treatments has been involved [e.g., 14–17]. In many of such systems the treatment is primarily ex situ, but the reinjection of partially treated ground water, enriched with nutrients as well as contaminant-degrading microorganisms, into the subsurface will promote truly in situ treatment.

II. STRATEGIES FOR IN SITU BIOREMEDIATION

The reasons for the persistence of environmental contaminants in soil can be divided into those that relate to the chemical or physical properties of the contaminant and those that relate to the environment.

The chemical properties of contaminants and their susceptibility to degradation will be dependent upon whether appropriate enzymes have been acquired by specific microorganisms during the evolutionary process. The key requirements are for such enzymes to accept the structural configuration of the molecule as a substrate and the ability of the compound to induce or depress the appropriate catabolic enzymes present [18]. In some cases, biodegradability may be enhanced by partial preoxidation [19].

The physical properties of the contaminant that inhibit biodegradation are often related to its inherent bioavailability, most notably solubility, and the physical nature of the matrix in which it is distributed. The rapidity of degradation of a potentially degradable contaminant, such as naphthalene, is likely to be very different if it is distributed within the soil in a mass of tarry material compared with its occurrence in, for example, creosote or a mineral oil. In the case of the former, the low bioavailability is attributable not so much to solubility but rather to the relatively impenetrable mass of the contaminant matrix and the low surface area: volume ratio available for microbial attack.

The concentration of the contaminant will impose limits to biodegradation potential at both ends of the scale. Inhibition of degradation of the contaminant resulting from toxicity of high concentrations present may be a primary reason for persistence of the contaminant, as evidenced by reductions in the size of populations of microorganisms, for example, as noted where high levels of gasoline were present [20], or the occurrence of high concentrations of what are known to be potentially degradable compounds, for example, phenols at coal gasification sites even in near-surface environments [21]. Conversely, there may be an absence of degradation below a given threshold where the contaminant concentration is too low to support growth [22–24].

Strategies for overcoming these specific issues that relate to contaminant properties may be less easy to devise for in situ treatment than those that address the physicochemical and biological factors relating to the environment. Application of surfactants has been used for enhancement of bioavailability in full-scale bioremediation projects as a means of dispersing the mass of the contaminant itself [2], as well as in promoting the partitioning of the contaminant into the aqueous phase in the subsurface systems, a mechanism that is discussed later. Such application may require some form of physical disturbance to promote the distribution of the surfactant throughout the contaminated medium, for example, through rotorvation of treatment beds [2]. This may not be readily achievable in situ except for a near-surface distribution. Similarly, for high concentrations of the contaminant, dilution of the soil with a bulking agent, for example, organic mulch, wood or bark chippings [1], may assist in reducing the levels of the contaminant by its redistribution through the matrix. Apart from land farming systems (i.e., without excavation of the contaminated soil), this cannot readily be effected for in situ systems, except through mobilization of the contaminant into the aqueous phase and subsequent aboveground dilution before treatment.

The strategic objectives of in situ treatment are, in most cases, focused on ameliorating the physicochemical properties of the environment rather than those associated with the contaminant itself. The essential features of all such systems are based on providing to the contaminated environment oxygen or an alternative electron acceptor, together with various other chemical supplements, to address nutrient imbalance, promote cometabolism, or change some other property of the soil.

Contamination may be present within the near-surface horizons, at lower depths within the vadose zone, or within the saturated zone. It may be adsorbed on the surface of mineral particles and organic matter or be associated predominantly within the aqueous phase, where it may occur in a truly dissolved, emulsified, or immiscible phase. Volatile contaminants will also partition into the gaseous phase within the interstitial spaces between soil particles. An understanding of the distribution of the contaminant, both on a macro- and microscale, is essential for the selection of the most suitable in situ scheme. A variety of configurations can be engineered according to the contaminant distribution, and examples of these are shown in Figure 1 for the vadose zone and in Figures 2 and 3 for the saturated zone. There is clearly a wide range of permutations available that involve not only combinations of ex situ and in situ systems but also combinations of techniques for each of these. In general however, key forms of in situ treatment can be divided into the following categories.

(a) IN SITU LAND FARMING

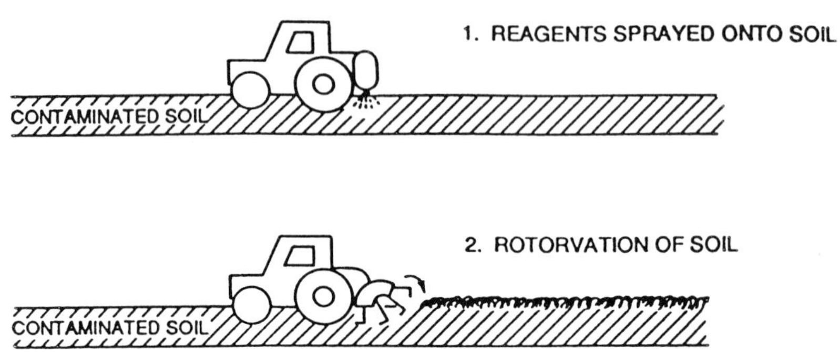

(b) SOIL VAPOUR EXTRACTION

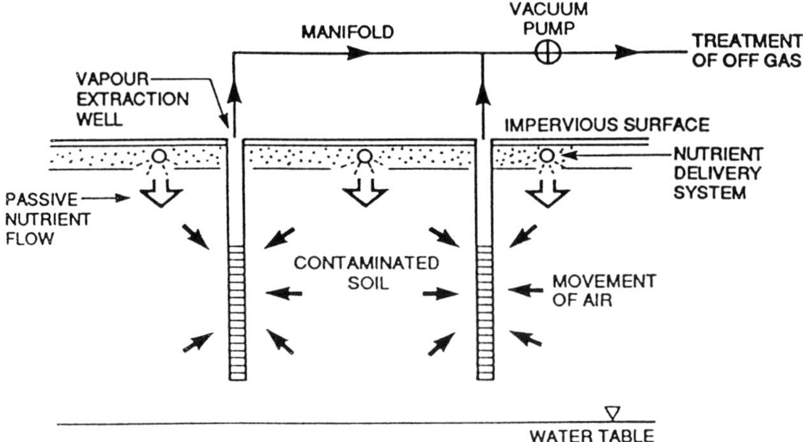

Figure 1 Various approaches for in situ treatment of vadose zone. Figure 1(d) reproduced from Rainwater et al., 1991 [30], with kind permission, Technomic Publishing Co.

A. Vadose Zone

In Situ Land Farming

This technique is applicable to contamination that is distributed within the plow layer. It represents the earliest type of process used for the land farming of oil sludge or oil waste [25,26] and takes advantage of the fact that nearly 90% of all soil microorganisms are concentrated in the upper 25 cm of surface soil [27]. These processes were somewhat at variance from a conventional understanding

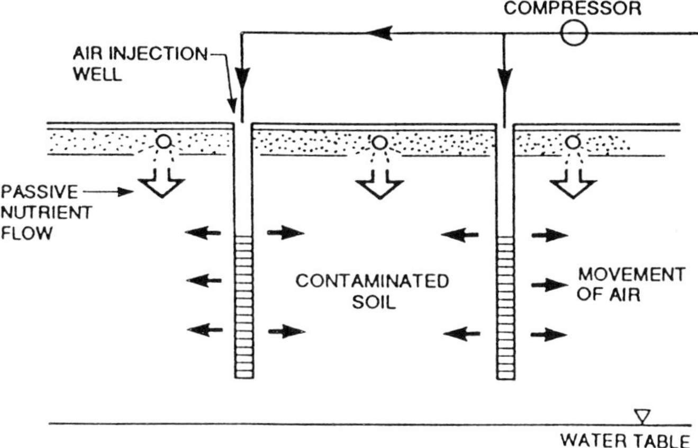

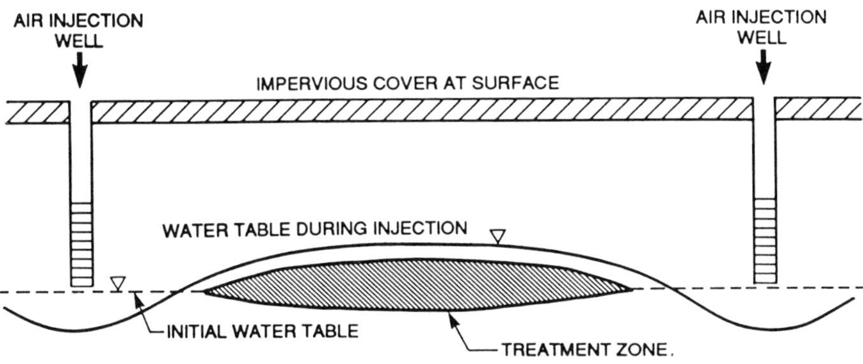

of bioremediation as a treatment technology, for here the soil is used almost as a bioreactor, that is, as a medium for the purposeful disposal and degradation of a waste product. The basic principle remains the same, however; that is, the periodic rotorvation of the upper layer of soil to achieve the mass transfer of oxygen to the sites of microbial activity for effecting degradation of the contaminant. Addition of nutrients can be effected through conventional agricultural spraying equipment. The process is clearly limited by the depth at which rotorvation can

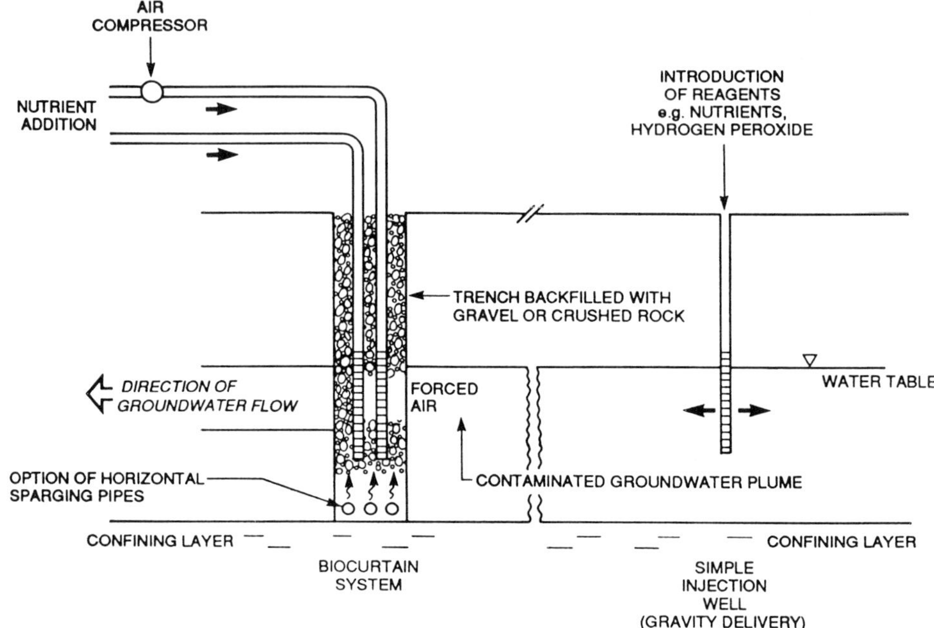

Figure 2 Two approaches for passive in situ treatment of the saturated zone.

be achieved: using conventional equipment, this is usually approximately 600 mm [15]. Successful application of the technology can therefore be achieved in sites where the contamination is present in the surface layers, for example, in some railroad facilities where it has been used to treat various hydrocarbon residues [28]. Disadvantages include the potential for odor generation, losses of volatiles, leaching, and run off [27].

Soil Vapor Extraction

This process involves the application of a vacuum to the contaminated zone via a vent well to produce a flow of air through the vadose zone [29]. This continual supply of oxygen allows biodegradation to proceed at a relatively rapid rate (promoted also by passive nutrient application), while volatile organics will partition into the gaseous phase and be recovered by conventional physical means, with the off-gas being treated at the surface.

One example of a variation on this process involves installation of a set of horizontal pipes for vacuum extraction within a gravel layer and injecting air

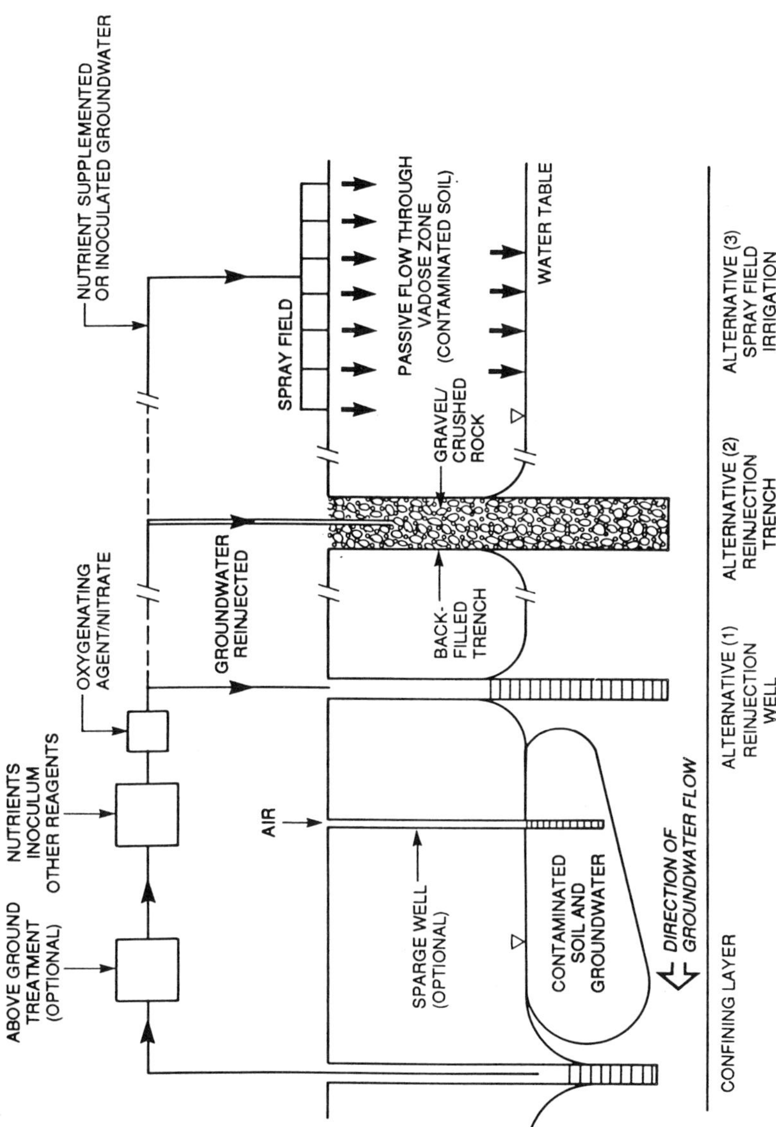

Figure 3 Three alternative approaches for dynamic treatment of the saturated zone.

vertically into the underlying soil via a standard well pipe (Anonymous, Hazardous Waste Consultant, March/April 1989, p. 1-14 to 1-16).

Bioventing

The objective of inducing an air stream through the contaminated vadose zone may also be achieved by injecting rather than withdrawing air from the soil. This system has the advantage of obviating the requirement for treatment of off-gas, although monitoring of the treatments may be more complex [29].

Hydrodynamic Control

In this approach, cyclic movement of the water table is induced by injecting air through two or more wells below an impervious surface [30]. This also provides oxygen to the unsaturated zone. The effect of this action is to redistribute microorganisms and cause alternate contact of the organic contaminants with the partially filled pore space in the vadose zone. Downward migration of contaminants in the smaller pores to the area of greatest biological activity may also be promoted.

B. Saturated Zone

Treatments for the saturated zone generally have to address contamination that may be distributed both within the aqueous phase (i.e., ground water) and that which is adsorbed onto the soil surface.

Passive systems may involve simple application of nutrients and an oxygenating agent into the contaminated subsurface [31] at one or more point sources or through a network of air sparging/nutrient delivery wells (Figure 2). The latter may be combined with venting systems [32].

A further development of such systems involves the use of a "biocurtain" sparge barrier in which the contaminated ground water passes through a zone where biological activity is maximized, for example, a line of sparge points [32] or a vertical trench backfilled with pea gravel and with horizontally installed air injector tubes at its base [33]. In this case, some other oxygenating agent, for example, hydrogen peroxide, or oxygen as microbubbles may be added, together with nutrients, et cetera. Such a system could also involve induced ground water flow rather than relying solely on passive migration. The installation of cutoff walls with low hydraulic conductivity (e.g., sealable joint sheet piling) on either side of the zone of maximal biological activity enables the ground water plume to be channeled through the zone. This treatment option is referred to as the funnel and gate system [34]. The use of such systems is particularly appropriate where the contaminant is primarily in the aqueous phase within the saturated zone. Where, however, a significant fraction of the contamination is also adsorbed onto the solid phase, then there is a requirement for a completely in situ scheme to effect bioremediation throughout the entire contaminated zone.

Dynamic systems for in situ bioremediation may take several forms, but all generally involve the removal of ground water from the area hydraulically downgradient of the contamination source, supplementation with chemical reagents (primarily nutrients and oxygenating agents and, in some cases, inocula), and reinjection via a series of wells or a trench system backfilled with gravel or crushed rock (infiltration gallery) hydraulically upgradient of the source area [18,31]. Although biodegradation then occurs in situ, such systems also allow for further treatment ex situ [14–16], which may include physical and chemical as well as biological processes.

In some instances, the implementation of a dynamic in situ treatment system for treatment of contamination in the saturated zone may be combined with treatment of the vadose zone by allowing the recirculated water to infiltrate passively over an irrigation spray field and percolate through the vadose zone [35]. Such treatments are clearly limited to situations where the substratum is highly permeable to avoid the possibility of ponding or waterlogging the underlying soil. There may also be significant health and safety limitations imposed on such a process if volatile contamination is present. Alternatively, if the contamination in the vadose zone is close to the surface, it may be possible to address this by ex situ treatment while in situ treatment is simultaneously being carried out in the saturated zone below [36].

In summary, there are a variety of potential options for in situ bioremediation, but all involve basically similar concepts. The selection of the most appropriate option is necessarily site specific. However, a recent modeling study evaluated the performance, over a 1-year clean-up, of 19 different remediation scenarios for the biodegradation and transport of naphthalene present in the saturated zone, with the objective being to effect the greatest reduction in naphthalene concentration without increasing the volume of contaminated ground water in the aquifer [37]. The results indicated that in situ bioremediation and in situ bioremediation/pump-and-treat systems should perform better than pump-and-treat systems for sand and gravel aquifers contaminated with biodegradable hydrocarbons. There were clear benefits to be gained through (1) minimizing the distance for pulling the plume toward extraction wells or the distance through which the dissolved oxygen needs to be circulated to reach the contaminated zone, (2) injecting as high a concentration of dissolved oxygen as possible, and (3) designing the well field to produce convergent flow toward a central location and to minimize divergent flow and upgradient and intraplume stagnation areas.

The range of possible options for bioremediation will clearly be limited, however, by the distribution of the contaminant and the nature of the subsurface environment. Such limitations and the scope for manipulation of the parameters concerned are discussed in the following section.

III. PHYSICOCHEMICAL FACTORS AND THEIR POTENTIAL FOR OPTIMIZATION

A. Soil Physical Factors

Soil structural and textural characteristics are often the primary obstacles to the development of an appropriate bioremediation strategy, as they will control the effectiveness of delivery of the key supplements, such as oxygen, nutrients, and inocula, to the areas where the contaminant is distributed. For ex situ bioremediation, amelioration of soil properties may be achieved within a suitably engineered treatment bed, by the addition of organic mulches, or by using specialized rotorvation equipment [2–4, 15]. The only form of in situ treatment where this may potentially be achievable concerns in in situ land farming, where additives, such as bark, wood chips, or other organic mulches [1], are plowed into the upper layer of the soil to improve textural and structural properties. The benefits of adding such supplements have been demonstrated in lysimeter experiments performed by the TNO (Netherlands Organization for Applied Scientific Research) using a loamy soil with a high silt content (>20%) [4]. The depth of the soil that was contaminated with cutting oils was 0.4 to 0.5 m, and over the course of 1 year, 95% reductions in oil contamination were obtained. When drainage was improved by the addition of bark or by allowing plants to develop, reductions from 2300 mg kg^{-1} to 700 and 1140 mg kg^{-1}, respectively, were obtained over a 42-day period compared with only 1650 mg kg^{-1} through the addition of fertilizer and 2250 mg kg^{-1} in the control. For in situ bioremediation of soil, however, such treatment is limited by the depth to which rotorvation can be effected, as discussed above. Also, where high concentrations of clays are present, even the addition of significant quantities of an organic material may have little effect on the overall structure and accessibility of the contaminated to microbial attack.

For other forms of in situ treatment requiring delivery of air or liquid throughout the depth of the contaminated zone, the physical characteristics of the soil, in particular, particle size, will influence hydraulic conductivity (K), the parameter often taken as the yardstick for assessing the applicability of in situ bioremediation to a given medium. K is defined by Darcy's law [38] in the equation, $Q = KIA$, where Q = flow rate (length/time), K = hydraulic conductivity (length/time), I = hydraulic gradient (dimensionless), and A = cross-sectional area ($length^2$).

Where soil bioventing systems are employed to enhance bioremediation within the vadose zone, increasing the density of the well network will be required with increasing proportion of finer-grain particle sizes. However, where hydraulic conductivities fall below 0.001 cm s^{-1}, such systems may not be appropriate [39]. A similar limitation occurs for ground water recirculation systems. The lower limit of K values would typically be representative of a fine sand [38], that is, a grain diameter of approximately 0.1 mm.

In addition to the difficulties imposed on bioremediation by soil structure and texture, the nature of the subsurface environment in terms of its physical heterogeneity often leads to significant difficulties, not only in the development of an appropriate remedial treatment but in monitoring its overall effectiveness. Soil from contaminated industrial sites, in many cases, consists of "made ground" that can be composed of materials as diverse as rubble, slag, clay, concrete, cobbles, paper, and timber [40].

The effect of these differences in physical structure is to provide a network of different migration pathways for the contaminant and also to create a vast range of microenvironments with different physicochemical properties. For each microsite where the contaminant is distributed, the relative importance of the various physical and chemical properties controlling contaminant degradation will be different. Therefore, tailoring an optimal treatment process necessarily has to be a compromise that is based on the average condition across the contaminated zone as a whole.

The practical difficulties of treating heterogeneous environments of this nature are mainly concerned with channeling and uneven treatment [39] and also with effective monitoring of the treatment process. Where the variation in levels of contamination is very high, as is often the case [1,2], and deviates significantly from that of a normal distribution, the sampling requirements and statistical criteria for establishing that effective reductions in contaminant levels have occurred through bioremediation may be highly complex.

The influence of soil structural and textural properties on other soil physical characteristics, such as moisture content and temperature, that will directly affect microbial activity is well established [41]. Degradation rates are generally highest at soil tensions between 0 and 100 kPa [42]. At 10 and 33 kPa, for example, rapid mineralization of methyl parathion occurred with the formation of bound residues [43]. As the soil became drier, changes in the ratio of the degradation products were observed (e.g., the concentration of *p*-nitrophenol in relation to *p*-aminophenol increased) and mineralization was significantly slower in dry soil (1500 kPa).

Where in situ land farming is used as the method for bioremediation or where bioventing of the unsaturated zone occurs, irrigation (e.g., sprinklers or subirrigation) may be used to increase soil moisture content, with application being made frequently in relatively small amounts to avoid leaching [42]. Excess water may be removed by using either a surface drainage system of open ditches or buried pipe drains into which water from abstraction wells is passed to lower the water table. The ameliorating effect of mulches on soil properties and promoting drainage has already been discussed. Synthetic substances may be used to enhance soil water content in arid regions, for example, urea formaldehyde and amorphous sodium hydrosilicates [44].

Mulches and other soil amendments may also be used for the purpose of enhancing soil temperature where land farming is carried out, as for ex situ treatment. Where softwood bark was mixed into a mound of chlorophenol contaminated soil in a 1:2 ratio by volume, temperatures of 5 to 15°C above ambient temperature were recorded [45], the highest temperatures being measured closest to the surface (32°C in mid-summer and 15°C in October). This system reduced concentrations of chlorophenols from 212 to 30 mg kg^{-1} in four summer months. Although direct incorporation of residues into the soil being land farmed in situ may not have as great an influence on temperature amelioration as in the case of an ex situ soil treatment bed, the effect may not be insignificant, particularly when taken with concomitant improvements in structural properties, as described above.

Application of heat to accelerate microbial degradation of contaminants in situ, either through raising the temperature of air used in bioventing systems or that of ground water in recirculation systems, is an attractive concept, particularly if there is a source of waste heat on site that can be used for such purposes [46]. In other instances, it is likely to be a costly procedure. For the in situ remediation of a German refinery site contaminated with hydrocarbons, recharge ground water was heated to 25°C. For an 18°C average ground water temperature, it was estimated that a daily degradation rate of 1.5 g of hydrocarbons m^{-3} was achieved [14]. There could be instances, however, where too high a temperature increase may slightly inhibit degradation. In laboratory studies of degradation of oily sludge in Norwegian soils, an optimal temperature of 18°C was observed, with distinctly slower rates at 24°C as well as well as 12°C [47].

B. Electron Acceptors

Options Available for In Situ Treatment

Although there may be instances where strictly anaerobic conditions may be appropriate for removal of contaminants, in the majority of cases there is a requirement for provision of either oxygen itself or an oxygen-containing compound (usually nitrate) to act as an electron acceptor. Oxygen limitation is indeed often the rate-limiting factor for biodegradation [48]. Depending on the distribution of the contamination, there may be a requirement for oxygen to be supplied to the vadose, saturated, or both zones of the subsurface.

Methods for introduction of oxygen into the subsurface that have been used for in situ treatment systems include venting for the vadose zone and air sparging, injection of oxygen-saturated water or hydrogen peroxide, or nitrate addition for the saturated zone. Tables 1 and 2 (from Ref. 49) provide a comparison of both cost and treatment effectiveness for each of these various options for two

Table 1 Estimated Cost-Effectiveness Comparison of Oxygenation Methods, Assuming a High Degree of Contamination

System	Flow rate (Ls^{-1})	Oxygen (kg day^{-1})	Site treated (%)	System Utilization efficiency (%)	Treatment time (days)	Contaminant traeatment cost[a] (\$ kg^{-1})
Air sparging	15 wells at 0.94 (2 cfm)	2.7	41	1 (sparg) 70 (DO)	1716	199
Water injection	4.4 (70 gpm)	3.6	85	50	1580	221
Venting[b]	75 (160 gpm)	1814	72	5	132	30
Peroxide	4.4 (70 gpm)	86	95	15	330	144
Nitrate[c]	4.4 (7.6 Ls^{-1} or 120 gpm recovery)	96	85	13	335	170

[a] Costs quoted are from original source and have not been adjusted for inflation.
[b] Vent system costs include a vapor-phase control system (catalytic oxidizer).
[c] Nitrate capital control costs include significant measures for control of off-site migration.
Information reproduced from R. A. Brown and J. R. Crosbie [49] with permission of the Hazardous Materials Control Research Institute and converted into SI units.

contamination scenarios. Both scenarios are based on an area of contamination that is 76 m by 30 m involving approximately 1900 L of petroleum hydrocarbon fuel in a permeable sand.

In the first scenario, there is significant contamination in the sorbed and dissolved phase, and the majority of the contaminant in the sorbed phase is at or above the water table. Based on cost performance, venting is the best method

Table 2 Estimated Cost/Performance of Oxygenation Systems for a Low Degree of Contamination (Dissolved Phase Only)[a]

System	Oxygen delivered (kg day^{-1})	Treatment time (days)	Treatment cost (\$ kg^{-1})
Air sparging	2.7	180	258
Water injection	3.6	330	692
Venting	Not applicable	—	—
Peroxide	86	180	295
Nitrate	96	240	366

[a] See footnotes for Table 1.
Information reproduced from R. A. Brown and J. R. Crosbie [49] with permission of the Hazardous Materials Control Research Institute and converted into SI units.

(though applicable only to the vadose zone), followed by peroxide, nitrate, air sparging, and water injection. Based on treatment effectiveness (percentage of the site treated), the order is peroxide, nitrate, water injection, venting, and air sparging (Table 1).

However, if it is assumed that there is no soil contamination above the water table, that the concentrations in soil are below 100 ppm, and that the contamination is primarily in the dissolved phase, then air sparging is both the most cost-effective treatment and the one with the shortest treatment time, followed by peroxide (same treatment time), nitrate, and water injection (Table 2).

For both scenarios, it is assumed that each of the techniques being considered is equally technically feasible and will, in fact, deliver the theoretical amount of oxygen. In reality, site constraints and physicochemical properties will determine the applicability and effectiveness of each particular method.

Oxygen

For the vadose zone, the application of air injection or vapor extraction systems are often the most suitable means of achieving the required mass transfer of oxygen. In one example of this, bioventing (using air withdrawal) was used to treat soil contaminated by a spill of 100,000 L of JP-4 jet fuel at Hill Air Force Base, near Ogden, Utah [29], where hydrocarbon concentrations averaging 1500 mg kg^{-1} were detected. Based upon the stoichiometric oxygen requirement for hexane mineralization, the total amounts of JP-4 volatilized and biodegraded were 47,000 and 15,000 kg, respectively, over a 14-month period, as estimated from measurements of oxygen and hydrocarbon concentrations in the off-gas during extraction. As the treatment progressed, the concentration of volatiles declined, and by the end of the treatment period, approximately 4 kg day^{-1} of hydrocarbons were being volatilized compared with 96 kg day^{-1} being degraded. The authors reported a further study at Tyndall Air Force Base in Florida, where a more controlled experimental design was carried out to assess relative volatilization and biodegradation rates for a soil contaminated with up to 20,000 mg kg^{-1} of JP-4. In this example, dewatering was employed to maintain the water depth at approximately 1.6 m, and much higher rates of biodegradation and volatilization were recorded. In one treatment cell, 25 kg of hydrocarbon had been volatilized, and the same amount had been lost through biodegradation over a 120-day period. A decline in biodegradation rates, apparently unrelated to the disappearance of substrate was also observed, possibly as a result of the onset of winter temperatures.

The use of vapor-venting techniques for physical remediation of gasoline contamination is well established, although even with such lighter hydrocarbon fractions, biodegradation may make a significant contribution to total contaminant removal. At a former service station facility in Massachusetts, a soil-venting system was installed to address gasoline hydrocarbons in the vadose zone that had arisen from underground storage tanks. The rate of carbon dioxide production in the effluent soil vapor was observed to parallel the concentrations of methane and total petroleum hydrocarbons present in the vapor phase, the 11% concentration of carbon dioxide that was observed initially being over 300 times its normal atmospheric level [49].

Direct injection of air may be extended to the saturated zone where oxygen is introduced into the ground water by air sparging and allowed to diffuse outward from the well bore. The process is limited in terms of the amount of oxygen that can be supplied (8 to 12 mg L^{-1} at normal temperatures) [50]. The use of pure oxygen in place of air allows the introduction of maximum oxygen levels of 40 to 50 mg L^{-1}, but it has significant cost and handling implications [50]. Also, the hydrostatic pressure of shallow aquifers is essentially atmospheric pressure so that degassing will occur immediately [51]. The amount of oxygen supplied by sparging is determined by the rate of water flow, which in turn is a function of the ground water gradient, the hydraulic conductivity, and the surface area of the formation affected by the well bore [49]. Table 3 (taken from Ref. 49) gives the amount of oxygen supplied from a single well, by sparging with either air or oxygen, in aquifers of differing hydraulic conductivity and for various hydraulic gradients.

The application of sparged air was reported in one of the earliest documented examples of bioremediation for the cleanup of residual gasoline contamination in a fractured dolomitic aquifer near Ambler, Pennsylvania, following a pipeline fracture that had released 300,000 L of gasoline [52; also Raymond et al., 1975, 1976, cited in Ref. 53]. Following physical recovery of approximately 63% of the spilled gasoline, nutrient addition was initiated with air sparging. Ten months after the nutrient addition had been completed, gasoline present in the ground water was below the level of detection. In a further study (Raymond et al., cited in Ref. 53) a similar approach was adopted for a gasoline-contaminated sandy aquifer in Millville, New Jersey, but as the result

Table 3 Oxygen Supplied by Sparging (Single Well)[a] (kg day^{-1})

Hydraulic conductivity (cm s^{-1})	Hydraulic gradient (m/m)					
	(High) 0.1		(Medium) 0.01		(Low) 0.001	
	Air	Oxygen	Air	Oxygen	Air	Oxygen
5×10^{-1} (gravel)	2.7	14	0.27	1.4	0.027	0.14
5×10^{-3} (medium sand)	0.03	0.14	2.7×10^{-3}	1.4×10^{-2}	2.7×10^{-4}	1.4×10^{-3}
5×10^{-6} (silt)	2.7×10^{-5}	1.4×10^{-4}	2.7×10^{-6}	1.4×10^{-5}	2.7×10^{-7}	1.4×10^{-6}

[a] Assumes a 9.1 m (30 ft) saturated thickness and a lateral influence of 0.91 m (3 ft) for the well.

Information reproduced from R. A. Brown and J. R. Crosbie [49] with permission of the Hazardous Materials Control Research Institute and converted into SI units.

of the rapid utilization of dissolved oxygen around the injection points, oxygen was not sufficiently well distributed throughout the treatment zone, and although product was removed, concentrations of gasoline in the soil did not decrease. Phenol formation, attributed to the lack of sufficient oxygen, was also reported, and was removed after further aeration.

A bioremediation program for a gasoline spill was also initiated for a fractured dolomite formation after recovering 333,000 L from a reported loss of 507,000 L [54]. Air-sparging wells only had a localized effect and reduced contamination levels from 5 to 8 ppm to 2.5 ppm during the 12-month nutrient addition phase. However, after a further 6 months after termination of nutrient addition, the hydrocarbon level had decreased to concentrations below 500 ppb. The authors described a further example of the application of air-sparging techniques to bioremediation of soil and ground water, following recovery of 70,000 L of gasoline as free product from an estimated 114,000 L that had leaked from an underground storage tank. In this case, hydrocarbons were also measured in the sorbed phase at 0.9 to 2.4 m above the water table, where contaminant levels of 2000 to 3000 ppm were detected. Ground water contaminant concentrations were reduced from 30 to 40 ppm to below 1 ppm over a 10-month period, during which time periodic nutrient addition and continuous air-sparging through 14 wells had taken place. Soil concentrations were reduced to below 50 ppm.

Another example of the use of air-sparging was reported for gasoline contamination in a fractured red-brown shale and siltstone aquifer, where more than 15 mg L^{-1} hydrocarbons were present in the dissolved phase [49,55]. The strategy adopted involved reinfiltration of abstracted ground water supplemented with a nutrient mixture via an infiltration gallery (with air-stripping to remove volatiles) and air-sparging in the peripheral regions of the plume (outside the gallery). During the initial 11 months of treatment, a 50 to 85% reduction in contamination was observed. The reduction stabilized at this level, however, and was unsuccessful in addressing the contamination adsorbed in the fractures of the bedrock. There were significant maintenance costs associated with the mechanical cleaning of the thick biomass growth that fouled and plugged the sparge wells. The program had to be completed by increasing concentrations of oxygen supplied to the trickle feed and application of dilute hydrogen peroxide (100 mg L^{-1}).

The use of aeration was also employed as part of the cleanup procedure for the New Jersey Biocraft Laboratories site [38] where ground water was contaminated with approximately 85 mg L^{-1} acetone and 55 mg L^{-1} methylene chloride and had a COD of 648 mg L^{-1}. The system here involved (1) collection of the contaminated plume from a downgradient trench, (2) surface treatment involving a dual system of two aeration and two sludge settling tanks, each tank being of 20,000-L capacity, and (3) reintroduction of the treated water via two infiltration trenches. In this example, a series of nine injection wells at 9-m spacings, based on an assumed 4.5-m radius of influence of each aeration point, were established over the site area. Air was continuously injected at a pressure of 28 to 62 kN m^{-2}. Between August 1981 and June 1985, approximately 95% of the site had been remediated. It was reported that 60% of the degradation took place in the surface reactors and 40% in situ.

These cases demonstrate that applications of air-sparging techniques have met with mixed success in the field. Another method that has been proposed for the introduction of air to the saturated zone involves the use of colloidal gas aphrons [56]. These consist of gas bubbles of the order of 25 μm in diameter encapsulated in thin surfactant films that are of sufficient tenacity to withstand being coalesced, even if the bubbles are pressed together. Most water-soluble surfactants can be used for such purposes. Their small size and relative stability allow these microbubbles to remain in suspension, so that they can flow through channels in a permeable aquifer. In laboratory tests, a combination of colloidal gas aphrons, *Pseudomonas putida*, and nutrients injected into a saturated sand microcosm degraded more than 60% of phenol present during a 24-h period. Using a conservative estimate that approximately 15 to 20% of the injected oxygen can be dissolved in the flowing ground water, although experiments have indicated that as much as 20 and 30% may be retained in the saturated zone [33], and assuming that 10% would be required to degrade the surfactant, then approximately 5 to 10% would be available for biodegradation of the contaminant. There have not been any major reports concerning the application of the technology for full-scale bioremediation. One possible disadvantage is that injection of the microbubbles has been shown to decrease hydraulic conductivity, possibly as a result of partially blocking the flow paths in the saturated soil matrix [33].

Hydrogen Peroxide

The limitations of adding air or oxygen directly to the saturated zone, as imposed by solubility constraints, has stimulated interest in the application of hydrogen peroxide. When it is considered that one molecule of oxygen is generated for every two molecules of hydrogen peroxide, and its miscibility making it an ideal reagent for use in the saturated zone, the immediate advantages are apparent in comparison with direct aeration. Concentrations of hydrogen peroxide up to 100 or 200 mg L^{-1} can be added without toxic effect, although this can apparently be increased to even 1000 mg L^{-1} with proper acclimation [51,57]. Tolerance appears to be site specific; in one instance, 200 mg L^{-1} completely inhibited microbial activity in samples taken from a particular aquifer [48]. The degradation of hydrogen peroxide to harmless end products would appear to obviate problems concerning the fate of the substance itself, although its reactions with soil organic matter may potentially result in the formation of a wide range of typically polar, water soluble compounds [58] that have both significant mobility and oxygen demand. The key limitation of hydrogen peroxide is its lack of stability and rapid decomposition upon contact with the soil. Decomposition not only reduces its efficiency and cost-effectiveness, but also may cause blockage of pores [57]. In columns of sandy loam soil, concentrations of hydrogen peroxide of 100 mg L^{-1} or more significantly reduced permeability by over 50%, either through oxygen bubbles or precipitation of oxidation products [59]. Decomposition in the saturated zone is known to be catalyzed by the presence of iron [51], and several potential additives were evaluated in the laboratory for their effect in promoting stability, including phosphate buffer, Calgon, sodium hexametaphosphate, urea, uric acid, acetanilide, and sodium tripolyphoshate. Only the latter significantly prolonged the life of hydrogen peroxide in a sandy loam or a sandy clay loam, although a greater effect was achieved in sterilized systems, indicating a biological mechanism of decomposition through microbial catalase.

The relative effects of abiotic and biological decomposition of hydrogen peroxide were examined in detail during a full-scale research project initiated at Eglin Air Force Base, Florida, where an estimated 75,000 to 95,000 L of jet fuel had contaminated approximately 3,060,000 L of soil and shallow ground water [60]. Of this, over 26,000 L had been recovered using skimmer pumps. Four abstraction wells, a ground water treatment unit, and gravel-filled infiltration galleries were established for the addition of hydrogen peroxide and nutrients. Laboratory studies indicated that when added to a slurry consisting of a 1:3 ratio of soil:deionized water, the half-life of hydrogen peroxide was 4 h, but when added to the same mixture following pretreatment with a nutrient mixture containing 800 ppm orthophosphate (as a stabilizer in the presence of iron), this could be extended to 12 h. However, when hydrogen peroxide concentrations were measured in a monitoring well located 0.5 m from the infiltration gallery, a half-life of between 1 and 2 h was observed following supplementation of feed water to the gallery, even though 90% of the iron had been removed from the ground water before reinjection. Even when the gallery was conditioned with phosphate-supplemented feed water for 6 days before repeating the field experiment, no significant improvement in stability was observed. Experiments in flasks indicated that sterilization reduced decomposition rates of hydrogen peroxide to undetectable levels. It was concluded that the action of microbial catalase was responsible. Further field testing indicated that decomposition of hydrogen peroxide in gallery material was negligible when fresh tap water was used rather than recirculated ground water. It was concluded that the solution to unwanted decomposition of hydrogen peroxide was to eliminate microbial activity in the infiltration gallery.

Despite these problems, the use of hydrogen peroxide as the primary source of oxygen has been reported in a number of full-scale in situ bioremediation projects. As already noted, there have been examples where it was used following air-sparging, sometimes as a commercially supplied formulation in combination with microbial nutrients. In a system where residual (gasoline) hydrocarbon levels in ground water of 2.5 ppm hydrocarbons had been achieved from initial contamination levels of 15 ppm after 20 months of air-sparging, the application of hydrogen peroxide and microbial nutrients resulted in further reductions to below detectable levels in 8 of the 12 monitoring wells and to 200 to 1200 ppb in the remaining four, 6 months later [54]. Contamination by residual fuels and solvents (consisting of a 45% aromatic/55% alkane mixture) following recovery of free product was also treated in situ by a similar nutrient-peroxide formulation. This involved a reduction in contaminant levels in ground water from 22 to 45 ppm to below 10 ppb over a 2.5-month period. Subsequent treatment using activated carbon over a 3.5-month period was used as a polishing process to reduce concentrations to below 10 ppb [54].

Hydrogen peroxide was also used in another example involving contamination arising from a leak of 3400 L of leaded gasoline that was impacting a municipal water supply. Phase separated product (100 L) had been recovered using a water table depression pump, with excavation of soils containing an estimated further 190 L. Hydrogen peroxide was added to recovered water after it had been passed through an air stripper, before reintroduction via a reinfiltration gallery [31]. It was reported that 99% of the residual bound and dissolved hydrocarbons in the soil was degraded over 18 months of operating the system and that this represented biodegradation of over 2500 L of hydrocarbons, compared with only 190 L removed by air stripping of the abstracted water.

Hydrogen peroxide was also used for an in situ pilot project at Kelly Air Force Base, Texas [61] where concentrations were stepped up on a weekly basis from 100 to 500 mg L^{-1}. The initial data gave inconclusive results concerning bioremediation in soil. However there was a decrease in hydrocarbons in aquifer material over a 2-month period, and there was a slight decrease or no change in concentrations in ground water in two wells but a large increase in concentration in a third. It was suggested that this may have been the result of microbial emulsification, although it should be noted that different analytical methods were used on each occasion. After approximately 8 months of treatment, however, a GC scan for total hydrocarbons indicated an overall quantitative reduction in organics present in the ground water [62]. Whereas a decrease in tetrachloroethylene and trichloroethylene was also observed (from an average of 4 to 0.93 mg L^{-1}), there was also an increase in the decomposition product, *trans*-1,2-dichloroethylene, from 0.03 to 1.4 mg L^{-1}.

As an oxidizing agent, the action of hydrogen peroxide may be to increase the susceptibility of certain contaminants to microbial degradation. This was suggested for a number of the key components of creosote by liquid culture studies performed as part of the development of a bioremedial treatment for contamination at Blekholmstorget, in central Stockholm [16]. The design and implementation of the full-scale treatment is interesting, inasmuch as it involved both aboveground treatment of abstracted ground water in reactor tanks, where hydrogen peroxide was added (together with nutrients and inocula), as well as further aeration, before reinfiltration and passage through both the vadose and saturated zones, where the contamination was distributed. A schematic diagram is shown in Figure 4. Concentrations of creosote at eight sampling locations were reduced by an average of 64% over a 4-month period from a mean starting concentration of 7903 mg kg^{-1}.

Nitrate as an Electron Acceptor

As well as having associated technical problems, hydrogen peroxide is a relatively expensive reagent, and the search for an alternative soluble oxygen source has led investigators to examine the possibility of using anaerobic respiration, and in particular, nitrate as a terminal electron acceptor, rather than attempting to satisfy oxygen demands through the achievement of strictly aerobic conditions within the aquifer.

A range of hydrocarbons [63] and their derivatives [64] can potentially be degraded under anaerobic conditions. These include compounds that frequently occur as primary contaminating substances. A mixed population of microorganisms was shown to degrade phenol in anaerobic liquid culture in the presence of nitrate, the latter being reduced mainly to gaseous nitrogen [65]. Similarly, anaerobic oxidation of *p*-cresol with reduction of nitrate to molecular nitrogen has been observed [66]. Such laboratory studies, which indicated the potential for anaerobic degradation of phenols, have been supported by field evidence. Investigations of attenuation of a plume of coal tar derivatives in ground water indicated that disappearance of the contaminants could not be explained simply by sorption or dilution alone, and it was concluded that anaerobic degradation of phenolic compounds was primarily responsible [67].

For aromatic hydrocarbons without oxygen-containing functional groups, experimental evidence for degradation with nitrate as an electron acceptor is somewhat inconsistent. Rapid mineralization of toluene and *m*-xylene took place in a laboratory aquifer column operated in the absence of molecular oxygen and with nitrate as the electron acceptor [68]. Although organisms adapted to utilizing *m*-xylene were able to degrade toluene, benzaldehyde, benzoate, *m*-toluyaldehyde, *m*-toluate, *m*-cresol, *p*-cresol, and *p*-hydroxybenzoate, they were unable to metabolize benzene, naphthalene, methylcyclohexane, 1,3-dimethylcyclohexane [69]. In nitrate-amended ground water samples contaminated with a gasoline spill, indigenous microorganisms caused a 95% loss of benzene, toluene, and xylene during 95 days, compared with an unamended control, with removal exceeding 98% when phosphate was added [70]. The addition of acetate and methanol decreased the reduction in benzene, toluene, and xylene, although the addition of the trace elements, iron and molybdenum, increased the reduction

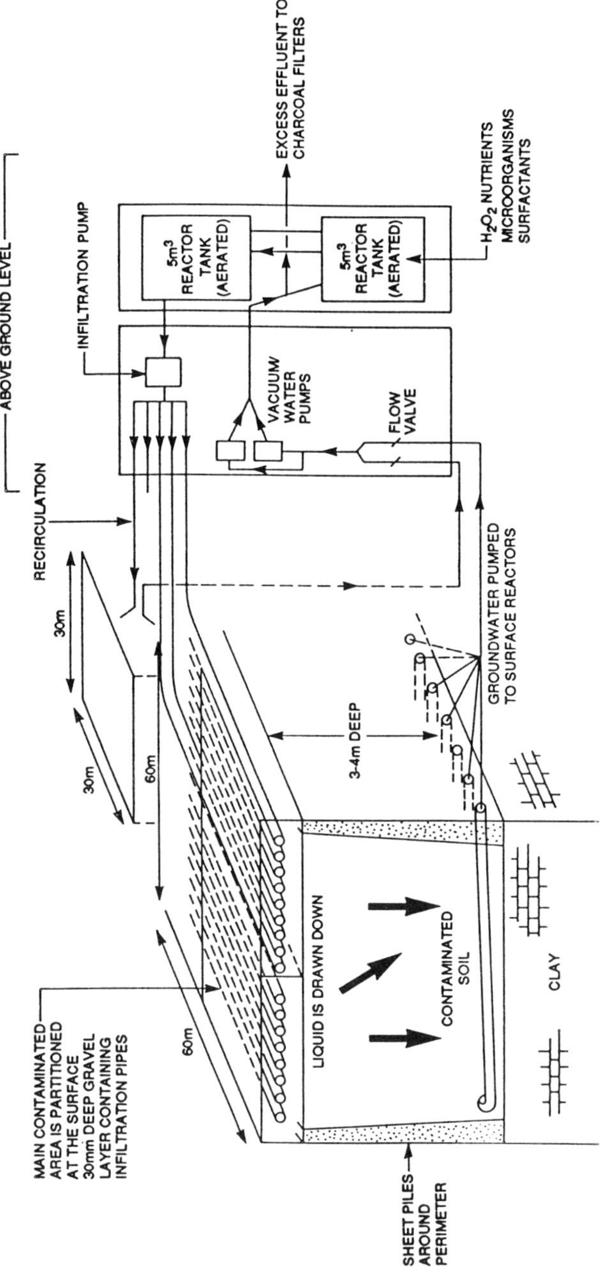

Figure 4 An in situ biotreatment system for creosote-contaminated soil and groundwater at Blekholmstorget in Sweden. From Ellis *et al.*, 1990 [16]. Reproduced with kind permission, Selper Ltd.

of the aromatics during the early stages. In laboratory studies using contaminated ground water samples, degradation of benzene, toluene, ethyl benzene, and the *p*- and *m*- isomers of xylene occurred under anaerobic conditions only when nitrate was added [48], and as before, the extent of degradation was greater when inorganic nutrients were also provided. No degradation of *o*-xylene occurred, however. In soil slurry experiments, mineralization of naphthalene also occurred under denitrifying conditions [71]. Close to the aqueous saturation level of 50 ppm naphthalene, approximately 90% of the naphthalene was mineralized within 50 days.

Whereas these and other studies using soil/water slurries [72] have indicated the degradation of aromatic compounds under denitrifying conditions, other laboratory investigations have produced negative results, at least for hydrocarbons. For example, in a liquid medium seeded with sewage effluent and maintained under anaerobic conditions in the presence of nitrate, no utilization of ethyl benzene, chlorinated benzenes, or naphthalene occurred after 11 weeks of incubation [73].

To examine the feasibility of using nitrate as an electron acceptor in the biorestoration of an aquifer contaminated with JP-4 jet fuel, microcosms were prepared using core samples taken from uncontaminated and contaminated areas of the aquifer under anaerobic sampling conditions and supplemented with nitrate, other nutrients, and aromatic hydrocarbons [74]. These were then incubated under a nitrogen atmosphere at 12°C. In the case of the uncontaminated core, no observable lag period occurred before degradation of toluene, although there was a 30-day lag before degradation of xylenes, ethyl benzene, and 1,2,4-trimethylbenzene, and there was little evidence for degradation of benzene. However, in the contaminated aquifer microcosms, both extended lag phases and decreased biodegradation rates were observed, with no significant degradation of benzene, ethyl benzene, or *o*-xylene, even after 6 months, despite the occurrence of active denitrification. Incubation of uncontaminated cores supplemented with either benzene or *m*-xylene indicated that the basal rate of denitrification was inhibited in their presence. It was concluded that the longer lag observed for the contaminated aquifer microcosms was attributable to either preferential metabolism of indigenous carbon or the inhibitory effects of these aromatic compounds or their metabolites.

After this laboratory evaluation, a field study was conducted that involved a combination of aerobic and anaerobic processes. An infiltration gallery (9 m × 9 m) was installed above a section of the aquifer that contained 2.65 m^3 of JP-4 [75]. A total of 11,400 m^3 of ground water was recirculated through the infiltration gallery per week to create a hydraulic mound above the contaminated zone. Over a 40-day period, recirculation of ground water initially containing only ambient levels of oxygen and nitrate reduced concentrations of benzene from 760 to less than 1 $\mu g\ L^{-1}$, toluene from 4500 to 17 $\mu g\ L^{-1}$, ethyl benzene from

840 to 44 μg L^{-1}, *m*- and *p*-xylenes from 2600 to 490 μg L^{-1}, and *o*-xylene from 1400 to 260 μg L^{-1}. Average core concentrations of benzene were reduced from 0.84 to 0.032 mg kg^{-1}, toluene from 33 to 0.13 mg kg^{-1}, ethyl benzene from 18 to 0.36 mg kg^{-1}, *m*- and *p*-xylene from 58 to 7.4 mg kg^{-1}, and *o*-xylene from 26 to 3.2 mg kg^{-1}. These reductions were achieved through hydraulic flooding (resulting in mixing and dilution) and some aerobic degradation. However, the reductions achieved with benzene and toluene were greater than what could be achieved simply by dilution. This would be consistent with the observation that these monoaromatics will readily degrade aerobically under appropriate conditions.

Ground water was then supplemented with nitrate (10 mg L^{-1} nitrogen) and with other nutrients and recirculated for 76 days. The final concentrations of contaminants in the ground water were <1 μg L^{-1} for benzene and toluene, 6 μg L^{-1} for ethyl benzene, and 20 to 40 μg L^{-1} for the xylene isomers. The respective core concentrations for benzene, toluene, ethyl benzene, *m*- and *p*-xylenes, and *o*-xylene were 0.017, 0.036, 0.019, 0.059, and 0.27 mg kg^{-1} respectively. The greater degree of recalcitrance observed with o-xylene isomer was consistent with the observations in the microcosm study [74]. Reductions during both the initial recirculation phase and the nitrate application phase were also observed for the total quantity of JP-4 fuel (as measured by GC analysis), although a significant residue remained. The total mass of JP-4 within the contaminated zone was reduced from an estimated 2250 kg to 1970 kg during the initial phase, and to 1420 kg over the second phase of treatment. Reductions in other hydrocarbon classes (alkanes, cycloalkanes), as well as in the alkyl benzenes, were also observed during the nitrate-addition phase. Evidence for active denitrification was provided by decreased nitrate concentrations across the investigation area, transient nitrite formation, and enrichment of denitrifier populations. Approximately 10 times the amount of nitrate was consumed than could be accounted for stoichiometrically by the oxidation of benzene, toluene, ethyl benzene, and xylenes alone, so that other carbon sources present were apparently being metabolized.

The application of nitrate as an electron acceptor was reported in the biorestoration of an aquifer contaminated with predominantly aromatic hydrocarbons in the upper Rhine valley [76] after removal of product by pumping had occurred. In this field demonstration, some aerobic degradation was also involved as ground water was abstracted downgradient of the contaminated area, and aerated and filtered through sand before reinjection. To achieve a flushing effect, uncontaminated ground water from a deeper aquifer was also reinjected upgradient from the first injection system. In addition to surface treatment, including removal of iron and methane, various supplements were added to the infiltration water, including ammonia and phosphate as nutrient sources and nitrate (at 300 mg L^{-1}) as an electron acceptor. During the 300 days of operation, approximately 50,000 kg of nitrate were utilized and there was evidence of gaseous nitrogen evolution in samples recovered from the flushing water. The treatment was accelerated by raising the temperature of the flushing water from 12 to 13°C to 22°C. Over a period of 2 years, aliphatic hydrocarbons were reduced from 2 to 0.1 mg L^{-1}, and complete elimination of benzene occurred after 6 months. Toluene and xylenes required a longer period, although from starting concentrations of around 2 to 3 mg L^{-1}, xylenes reduced to 0.07 mg L^{-1} after 2 years of operation.

A similar, full-scale bioremediation system utilizing both aerobic and anaerobic treatment has been described with ground water being abstracted from 16 wells and then reinjected at a rate of 400 m^3 h^{-1} via a system of infiltration trenches, with a 40 to 60 day residence time in the remediation area [14]. The ground water temperature was raised prior to reinjection. At an average ground water temperature of 18°C, the consumption rate of nitrate was 100 mg L^{-1} for a mean residence time of 50 days. Between January and August 1992, it was estimated that approximately 50,000 kg of hydrocarbons were mineralized from the consumption of 17,000 kg of oxygen and 200,000 kg of nitrate. Stripping of volatile hydrocarbons was estimated to remove 30,000 kg $year^{-1}$. It was reported in 1993 that a 5-year operational period (from commencement in 1991) was expected.

Such field applications concerning nitrate application have therefore not involved solely degradation under anaerobic conditions. In the case of the upper Rhine valley example [76], calculations undertaken indicated that the quantity of dissolved oxygen present in the ground water in situ would be insufficient for degradation of the reported concentrations of hydrocarbons present. This information, together with the observed reduction in nitrate concentrations, and the reported evolution of nitrogen gas from aquifer samples, provided circumstantial evidence that degradation of the various hydrocarbons (including benzene) had occurred under anaerobic conditions, with nitrate as the electron acceptor. In the case of benzene, this would apparently contradict those laboratory studies that indicated no significant degradation under denitrifying conditions [69,74]. However, saturation of the water with oxygen prior to reinfiltration may have been sufficient to initiate the sequence of benzene degradation, with complete mineralization being effected under denitrifying conditions after the oxygen had been depleted [74].

The application of nitrate as an electron acceptor was also used for the in situ treatment of pentachlorophenol (PCP) at a window and door manufacturing facility [77]. Laboratory bench-scale testing on soil cores indicated that approximately 10% of the PCP could be removed from the soil by leaching, and up to 85% could be removed by either aerobic or anaerobic degradation. A treatment system was established involving recycling of the ground water through both the saturated and unsaturated zone via recovery wells and recharge trenches. Aboveground treatment involved removal of soluble metals (particularly iron) by oxidation with potassium permanganate and filtration, removal of other organic constituents, using ozonation, and finally tertiary treatment using adsorption on activated carbon. Soil borings taken from the saturated zone showed a 94% reduction in the concentration of PCP from 25,100 to 1550 μg kg^{-1} over a 6-month period. Within the immediate path of the reinjected water from the recharge trench, reductions of 98, 98, and 51% in PCP concentrations were obtained at respective depths of 1.2 to 1.8 m, 9.1 to 9.8 m, and 9.8 to 11 m. These reductions were attributed to desorption and biological transformation, based on preliminary laboratory studies. Evidence for the latter was the continuous release of chloride from the breakdown of PCP in the ground water. An increase in alkalinity was also observed.

Based on observations that, at least for some aquifer systems, anaerobic nitrate respiration is a potential mechanism for the removal of hydrocarbons and other contaminants, it is perhaps surprising that there have not been more field examples of the application of nitrate to in situ bioremediation technology. It should, however, be noted that nitrate itself is an important ground water contaminant, and there could be potential regulatory problems in its application.

Other Anaerobic Systems

There is little information concerning the use of sulfate as an alternative electron acceptor to nitrate in field examples of bioremediation. The range of contaminants that can be metabolized under sulfate-reducing conditions may be somewhat less than under nitrate-reducing conditions, although fewer studies appear to have been carried out in this area [64]. A recent laboratory study involving highly reduced sediments from San Diego Bay, California, incubated under anaerobic conditions, demonstrated that benzene could be oxidized with sulfate serving as an electron acceptor [78]. The results also suggested that oxidation to carbon dioxide took place without production of extracellular intermediates. In samples from an anoxic aquifer, there was almost complete degradation of phenol and benzoate but not the haloaromatic substances tested, although dehalogenation did occur following biological removal of the sulfate [79]. Stimulation of methanogenesis without removal of sulfate did not result in dehalogenation, indicating inhibition of the activity of dehalogenating microorganisms by high concentrations of sulfate.

Degradation mechanisms involving fermentative as opposed to anoxic systems have been described for a number of important contaminant groupings, particularly halogenated organics [80]. Laboratory studies of aquifer material obtained from a site contaminated by an aviation gasoline spill and which in some areas was actively methanogenic, indicated relatively rapid biological removal of benzene, toluene, and xylenes under both aerobic and anaerobic (methanogenic) conditions [81]. Although the actual rate of disappearance of the aromatic hydrocarbons in the field did not compare well with the laboratory studies under aerobic conditions, the kinetics of anaerobic removal in the laboratory were similar to the kinetics in the field in the anaerobic area of the plume. This was probably a reflection of the limitation of mass transport of oxygen in the field under aerobic conditions, whereas methanogenic processes were not limited by the availability of oxygen.

The anaerobic microcosms were also supplemented with 1,1,1-trichloroethane (TCA) and trichloroethylene (TCE), and after 95 weeks of incubation, 1,1-dichloroethylene (1, 1-DCE) and 1,1-dichloroethane (1,1-DCA) were identified. Under anaerobic conditions, chlorinated hydrocarbons can be reductively dechlorinated (e.g., tetrachloroethylene (PCE) to TCE to DCE to vinyl chloride and ultimately to ethylene [82]), but as successive removal of chlorine occurs at a progressively slower rate, vinyl chloride may accumulate [83].

The application of fermentative systems for true in situ bioremediation has not been widely reported, probably as the result of the difficulties in chemically inducing, maintaining, and controlling anaerobic conditions in the subsurface [32]. For near-surface areas of contamination this is sometimes achievable through flooding the soil and incorporating organic material, an approach that has produced positive results, at least in laboratory or pilot testing, for various chlorinated pesticides [84,85], as discussed below.

Some pilot plant systems for treating the saturated zone at depth have been described, but they have involved treatment in surface systems rather than actual treatment in situ. One such example was the use of a lysimeter system consisting of soil mixed with granular activated carbon [86] above a sand base. Ground water contaminated with several thousand parts per billion of 1,1,1-trichloroethane was passed through the pilot plant at flow rates of over 189 L day^{-1} and was successfully treated by an anaerobic consortium to concentrations below the detection limit of 20 ppb. No other halogenated hydrocarbons were detected in the effluent. Such verification is an important requirement in the overall assessment of the efficacy of anaerobic treatment of chlorinated hydrocarbons, given the potential for the accumulation of halogenated intermediates under such conditions.

Under aerobic conditions, however, methanotrophic bacteria may effect cometabolic mineralization of such halogenated intermediates in the presence of methane [87,88]. This suggests the possibility of a combined anaerobic/aerobic approach for a treatment strategy. An interesting aboveground pilot-scale system for the treatment of ground water contaminated with both nitrate and chlorinated aliphatic compounds has been described (D. J. Warrelmann and V. Schulz-Berendt, Preprints, International Symposium, Soil Decontamination using Biological Processes, Karlsruhe/D 1992, p. 806–810). The system consists of solid-bed reactors comprising three biological stages. In the first of these, denitrification occurs to reduce the high concentrations of nitrate present. In the second, anaerobic stage, the redox potential is below -200 mV to allow microbial sulfate reduction and methanogenesis to occur. Under these conditions, anaerobic degradation of chlorinated hydrocarbons occurs using sucrose as a cosubstrate. The anaerobic stage is then followed by an aerobic stage in which methanotrophic conditions are created by supplying 1 to 2% (v/v) methane. This allows for cometabolism of intermediate chlorinated hydrocarbons that may be produced as a result of partial degradation in the anaerobic stage. Biofiltration units are used to treat off-gas from the reactor systems. Although such a system is essentially one of pump and treat, rather than in situ treatment, it raises the possibility of using combinations of in situ and ex situ systems for treatment of chlorinated hydrocarbons by alternately anaerobic degradation and aerobic methanotrophic cometabolism.

C. Nutrients and Cosubstrates

Most contaminants are utilized by microorganisms as carbon sources. As in agricultural systems [89], nitrogen and then phosphorus are the most important limiting nutrients, as they are required primarily for the formation of amino acids, proteins, nucleotides, and vitamins in the case of nitrogen, and nucleic acids and coenzymes in the case of phosphorus. Sulfur, essential for some amino acids and coenzymes, may also be limiting, as well as trace amounts of minerals, particularly potassium, cobalt, molybdenum, magnesium, calcium, and iron. Although these are often available as a result of the dissolution of minerals, they are consumed in the degradation process and may therefore be depleted when concentrations of organics are high [50].

In most cases, however, deficiency of trace elements is not a major problem, especially in contaminated sites where such elements are often present as associated contaminants, for example, as catalysts and additives in fuels, in the fill material commonly associated with industrial sites [90,91] or as residues in coal and coal ash, such as may occur at railroad facilities [28]. Laboratory studies that evaluated the factors involved in the degradation of oil sludge in soil indicated only a marginal response to the addition of trace elements [92].

The importance of nutrient additions in the enhancement of the degradation of hydrocarbons has been highlighted in a number of reviews [93–98]. There are instances where additions of inorganic nutrients have resulted in a significant stimulatory effect on biodegradation relatively soon after application [92] and others where the effect occurred only after several months had elapsed. Such a delay in the onset of a stimulatory response with added nutrients, including ammonium sulfate, was observed in laboratory studies of oil degradation [99], and similar effects have been observed in field plots treated with various oil residues, where in one particular instance, the stimulatory effect of fertilizer was not observed until approximately half the oil had been degraded [25]. Similarly, in a further field plot study concerning the degradation of crude oil, significant reductions in saturated hydrocarbons compared to the control were observed after 66 and 308 days in fertilizer-amended soil, but not after 12 days. Aromatics and heterocyclics did not appear to be significantly affected [100].

In some soils, the effects of nutrient addition on degradation of contaminants are particularly clear-cut. In a field experiment starting with levels of between 2 and 3.5% oil, respective reductions in oil content of 4, 9, 22, and 26% were obtained in a previously uncultivated soil over a 9-month period following addition of 0, 200, 400, and 600 kg N ha^{-1} (as a 25-3-6 N-P-K fertilizer). In a previously cultivated, more organic-rich soil, the corresponding reductions were 10, 15, 39, and 45% [47]. After 24 months, there was little difference between the two fields for corresponding fertilizer application rates, the respective mean reductions being 35, 50, 74, and 83%. In contrast, during another field experiment when waste oil (21 L/m^{-2}) was mixed into the top 15 cm of soil, degradation over the course of 1 year was not affected by application of four levels of fertilizer [101].

In some sites, addition of nutrients may actually repress the degradation of contaminants. The potential for bioremediation was examined in a soil contaminated with a residue consisting predominantly of branched aliphatic, high molecular weight aromatic and polar compounds. Not only did the addition of mixtures of various fertilizers, consisting of ammonium, potassium and sodium phosphates, nitrates, or urea, not increase the rate of mineralization of the hydrocarbons, but some treatments actually inhibited the process relative to the control [20,59]. The inhibitory effect of such additions was also observed for the mineralization of glucose and it was suggested that the microbiota in this particular site had become adapted to oligotrophic conditions.

Determination of the requirement for and the application of the optimal quantities and types of nutrients are therefore key elements in a bioremediation strategy. It is possible to develop optimal carbon:nitrogen ratios based on the content of these elements in microbial cells and the probable efficiency of conversion of the carbonaceous substrate into carbon dioxide, water, and microbial biomass. For the decomposition of 100 units of substrate carbon, it is necessary to provide 1 to 2, 3 to 4, and 3 to 6 units of nitrogen for bacteria, fungi, and actinomycetes, respectively [41]. For hydrocarbons, a C:N:P ratio of 300:10:1 has been proposed, based on one-third of the substrate being converted to cell mass and two-thirds to carbon dioxide [6]. In reality, the optimal carbon:nitrogen:phosphorus ratio for bioremediation often deviates substantially from such ratios [92]; for waste oils and sludges for example, optimal C:N ratios ranging from 9:1 to 200:1 have been reported [97]. Examples of carbon:nitrogen ratios reported in studies on the degradation of contaminants are provided in Table 4. Comparisons of different contaminating substances added to a particular soil indicate differing optimal C:N ratios, which are not attributable just to differences in the indigenous levels of carbon and nitrogen alone. In laboratory tests, where a refinery sludge was added to soil samples, a greater degree of biodegradation was observed with a C:N ratio of 9:1 compared with no nutrient addition (C:N ratio 78:1) or with a C:NPK ratio of 40:1. With a petrochemical sludge of lower nitrogen content, adjusting the C:N ratio from 380:1 to 124:1 also increased the rate of decomposition, but this effect was reduced with a 23:1 ratio [104].

There are several important reasons for this variability and deviation from theoretically derived optimal C:N ratios. First, the degree of recalcitrance of contaminant residues will vary substantially, particularly for petroleum compounds [92], which range from readily degradable gasoline, kerosene, and diesel to lubricating oil residues and heavy-end crude oil fractions [105]. The extent to which such residues can be mineralized to carbon dioxide and water or incorporated into cell biomass will similarly vary. Second, the availability of the organic carbon in the soil will vary, not just that which is part of the contaminant, but also that which occurs as natural soil organic matter. In some instances, therefore, it may be more appropriate to estimate nitrogen requirements upon the basis of the organic carbon content of the contaminant itself rather than on the total soil organic carbon. Such may be the case where the contaminant is present in a soil with a low organic matter content or where the organic carbon content consists of highly humified material. In other cases, if the soil contains a high content of readily degradable carbonaceous substrate, the total organic carbon content may be a more useful parameter for estimating nitrogen requirements. Availability of the contaminant will itself, however, decrease with increasing soil organic content and influence that which is available for microbial consumption [71]. Finally, the availability of the nutrients themselves, both the indigenous levels of nitrogen, phosphorus, and other essential elements together with the nutrients being applied, will vary according to both their chemical nature and soil physicochemical properties.

In conclusion, theoretical determinations of optimal C:N ratios provide a useful starting point for estimating nutrient requirements, but for practical purposes, investigations using, for example, experiments in microcosms are advisable for deriving nutrient application rates. The requirements for nutrient supplementation may also vary substantially across the treatment area. In some (pristine) aquifer samples, inorganic nutrient additions resulted in a rapid adaptation to, and higher rate of metabolism of several xenobiotics, including ethylene dibromide, *p*-nitrophenol, phenol, and toluene, whereas in others, such additions had little to no effect [106].

There are a number of possible forms in which nutrients may be supplied. The use of ammonium and potassium phosphates, nitrates, and sulfates have been reported in several examples, although comparisons of the relative effects of different anionic or cationic compositions of the nutrient salts for particular

Table 4 Examples of Carbon:Nitrogen Ratios Used in Studies on Oil Degradation in Soil Systems

Description of system	C:N:P ratio used or which was most effective	Contaminant reductions achieved by optimal C:N ratio	Contaminant reductions achieved in other treatments tested	Nutrients used	Ref.
Oil sludge amended laboratory microcosms	C:N 60:1; C:P 800:1	33% loss in hydrocarbons over 0–131 days, 60% loss over 131–285 days, with sludge loadings on days 0 and 131. Overall percentage biodegradation 45% over 0–285 days.	Comparative losses for control, 26% over 0–131 days, 35% over 131–285 days, and 30% over 0–285 days. For C:N ratio of 300:1 and C:P ratio of 4000:1, respective losses of 28, 54, and 40%. For C:N ratio of 15:1 and C:P ratio of 200:1, respective losses of 27, 49, and 37%.	NH_4NO_3,K_2HPO_4	[92]
Lysimeter columns, sand/soil mixture amended with oil sludge	C:N 200:1; C:P:K 2640:1:1	20 to 24% biodegradation of 5% w/w hydrocarbons in 120 days.	10% biodegradation in control without fertilizers.	Urea, urea-paraffin adduct, or urea formaldehyde with octylphosphate or triple superphosphate (all at same C:N ratio) with KCl	[102]
Laboratory microcosms amended with Bachaquero gas oil	C:N:P 100:10:1	60% loss in alkanes, 56% loss in aromatics, and 40% loss	38% loss in alkanes, 36% loss in aromatics, and 17% loss	NH_4NO_3,K_2HPO_4	[103]

		in total fatty acids after 42 days of incubation.	in total fatty acids in soil without fertilizer		
Respirometer studies on soil amended with refinery sludge (78:1 C:N ratio)	C:N 9:1	80% biodegradation as determined by residual hydrocarbons, and 82% as determined by CO_2 evolution after 161 days.	Percent biodegradation obtained was 63% for control, 76% for C:P (only) ratio of 40:1, 72% for C:NPK ratio of 40:1, as determined by residual hydrocarbons and 60, 71 and 74% respectively, as determined by CO_2 evolution after 161 days.	$Ca(NO_3)_2 \cdot 4H_2O$, KCl, $CaH_4(PO_4)_2 \cdot H_2O$	[104]
Respirometer studies on soil amended with petrochemical sludge (380:1 C:N ratio)	C:N 124:1	50% biodegradation as determined by residual hydrocarbons, and 18% as determined by CO_2 evolution after 161 days.	Percent biodegradation obtained was 34% for control, 43% for C:P (only) ratio of 124:1, and 43% for C:N (only) ratio of 23:1, as determined by residual hydrocarbons and 15, 13, and 13%, respectively, as determined by CO_2 evolution after 161 days.	$Ca(NO_3)_2 \cdot 4H_2O$, KCl, $CaH_4(PO_4)_2 \cdot H_2O$	[104]

Table 4 Continued

Description of system	C:N:P ratio used or which was most effective	Contaminant reductions achieved by optimal C:N ratio	Contaminant reductions achieved in other treatments tested	Nutrients used	Ref.
Field plots, Alberta Canada, amended with crude petroleum	C:N 10:1	Reduction in the saturate fraction of crude from 59–60% of the total 12 days after fertilizer added to 46% of the total after 60 days and 38–39% after 308 days (no significant difference in proportions of aromatics or heterocyclics).	Saturate fraction in soil not supplemented with fertilizer reduced from 63 to 51% after 66 days and 55% after 308 days.	Urea-phosphate ($N-P_2O_5-K$, 27:27:0)	[100]
Field plots (6 m^2), Norway, using previously uncultivated soil, low organic matter content	C:N approximately 130:1, estimated from application rate of 600 kg N ha^{-1}, assuming soil density 1.6 T m^{-3}, mixing depth 0.25 m and initial oil content of 2.3% of which 85% is C.	Reduction in oil content of 26% in 9 months and 83% after 12 months.	Reductions in oil content obtained with 0, 200, and 400 kg N ha^{-1} after 9 months were, respectively, 4, 9, and 22%, and 34, 54, and 75% after 24 months. (200 and 400 kg N ha^{-1} estimated as approximately 400:1 and 200:1 C:N ratios, respectively).	Unspecified fertilizer; 25-3-6 NPK	[47]

As above, but using previously cultivated soil with higher organic matter content	C:N approximately 200:1 from 600 kg N ha^{-1}, estimated as above with initial oil content of 3.4%	Reduction in oil content of 45% after 9 months and 82% after 24 months.	Reductions obtained with 0, 200, and 400 kg N ha^{-1} after 9 months were 10, 15, and 39%, and 37, 46, and 73% after 24 months, respectively, (200 and 400 kg N ha^{-1} estimated as approximately 600:1 and 300:1 C:N ratios, respectively.)	Unspecified fertilizer; 25-3-6 NPK	[47]
In situ treatment of soil; refinery site	70:5:1 C:N:P	Mean oil concentrations in soil reduced from 185 ± 48 mg kg^{-1} to 26 ± 6 mg kg^{-1} over 15 weeks (2- to 6-m depth).	None specified.	Unspecified organic and inorganic nutrient mixture	[15]

sites are few. It has been reported that in one example, ammonium sulfate resulted in greater growth of the microbiota responsible for contaminant degradation than ammonium nitrate [93]. Urea has been used as a nitrogen source, often in combination with inorganic, nitrogen salts [3,15]. The relative effect of urea may be influenced by soil pH, as the pH of recipient soils increases as ammonium is produced [41]. For in situ treatment, particular consideration needs to be given to potential problems arising from precipitation and adsorption effects. In column studies, ready adsorption of ammonium in a sandy loam and sandy clay loam was observed compared with an acid-washed sand as a result of cation exchange binding, whereas nitrate was freely transported in all soils. Phosphate was also retarded in the natural soils, and as a result of its precipitation, could cause significant pore blockage, severely limiting in situ treatment (e.g., 960 mg of H_2PO_4 reduced soil conductivity by almost 70%). Addition of sodium tripolyphosphate and, to a lesser extent, sodium hexametaphosphate decreased the extent of phosphate removal, and in one soil tested, even released bound phosphate. However, both additives resulted in disruption of the crumb structure of the soil, which could cause potential problems in the field [59].

As well as suppression of degradation through an inhibitory action, for example ammonia or nitrate toxicity [92], excess nutrient application may also result in unacceptable leaching from the zone of application. Based on the latter consideration, interest has focused on the use of slow-release fertilizers. Application of a ureic nitrogen source to soil in column experiments indicated that over a 20-day period, only 2.8 mg of mineral nitrogen was eluted from urea-formaldehyde as the nitrogen source compared with 38.6 mg from urea and 29.9 mg from a urea-paraffin adduct [102]. For long-term in situ biodegradation of contaminants by land farming, application of a slow-release nutrient formulation may be appropriate, particularly if there is concern over the impact of losses of nutrients through leaching.

Conversely, a slow-release approach may not be applicable where more rapid reductions is contaminant concentrations are necessary. In the development of a treatment strategy for an oil refinery site contaminated with 2% oil residues arising from bombing during World War II, there was little effect of adding a granular fertilizer on contaminant concentrations after 8 to 9 weeks compared with a formulation with similar elements incorporated into a liquid fertilizer [3]. The insolubility of such material also causes problems for in situ treatment where it is necessary to supply the nutrients to a ground water recirculation system.

For stimulation of the degradation of petroleum hydrocarbons in seawater, oleophilic fertilizers have been developed that include paraffinized urea and octylphosphate [107] and a paraffin-supported fertilizer containing $MgNH_4PO_4$ as the active ingredient [108]. As a result of their lipophilic nature, such fertilizers will partition into the organic phase and stimulate biodegradation at the interface, whereas conventional fertilizers would be subject to dissolution in the aqueous phase and therefore be of limited effectiveness. The use of similar oleophilic fertilizers has attracted much recent interest, particularly concerning their application to marine environments (sea or shoreline) for treatment of oil spills, most notably that of the *Exxon Valdez* [109–111]. Oil degradation was significantly enhanced using the oleophilic fertilizer, with visual evidence of disappearance: the supposition that this was primarily biological in nature was supported by observations of a reduction in the octadecane:phytane ratio (the former being readily degradable, the latter being biologically recalcitrant [110]). As other spills had indicated, however, that phytane was less recalcitrant than originally thought, further studies used hopane (a C30 pentacyclic triterpane) as an internal conserved standard, as this was known to be resistant to biodegradation over several years and of low solubility and volatility [111]. Comparisons of the ratios of other hydrocarbons to hopane confirmed that the reductions observed were likely to have been biological. The rates of biodegradation were dependent on the concentration of nitrogen within the shoreline, oil loading, and the extent to which natural degradation had already occurred. In the case of subsurface oil, treatment with a combination of oleophilic and slow-release fertilizer resulted in greater oil degradation than an untreated reference beach [110].

The effectiveness of application of oleophilic fertilizers to in situ soil and/or ground water systems, is likely to be more limited, apart from the obvious example of intertidal zones, and has not been widely reported. There may, however, be a use for such nutrients in treating localized areas of contamination in situ, for example, in the "smear zone," where fluctuations in the water table have resulted in product being distributed across the boundary of the vadose and saturated zones. Column studies indicated little difference in the extent of degradation of oil sludge when nutrients were supplied as urea plus superphosphate, the oleophilic urea-paraffin adduct plus octylphosphate, or the slow-release urea formaldehyde plus superphosphate [102]. In addition to leaching more nitrogen, one other disadvantage of the oleophilic fertilizer was a greater degree of total organic carbon elution compared with the other treatments, which the authors attributed to detergent properties of the octylphosphate.

The relative merits of adding an organic nutrient source are more complex. The purpose of applying organic material as part of a bioremediation treatment may be threefold: to provide an alternative source of nitrogen, phosphorus, vitamins, or trace elements to or as a supplement to a mineral nutrient application; to provide a supplemental source of carbon and achieve an overall stimulation of microbial activity; and to promote cometabolism. In many cases, however, there is often evidence that such additions may inhibit degradation through provision of a preferential carbon source [e.g., 106].

The inclusion of organic substrates, such as molasses or yeast extract, has been reported in examples of successful full-scale bioremediation projects [3].

Whereas molasses is utilized primarily as a carbon source, and any benefits arising from its application are probably the result of an overall stimulation of microbial activity, the application of yeast extract may accelerate degradation through a number of mechanisms. Yeast extract has the advantage of containing both macro- and micronutrients in concentrations similar to those of the microbial cell, and as a solid rather than a liquid supplement, it provides the opportunity for a slower release of nutrients with less potential for leaching from the soil and contamination of ground water or adjacent surface water courses.

Pot experiments examining the effects of various supplements to soil amended with fuel or waste oil indicated that the addition of brewer's yeast at 0.25 or 0.75% w/w (95% water content) accelerated the rate of consumption of oil by two- to tenfold [112]. This effect was far greater than that resulting from the addition of inocula or mineral fertilization and was also observed when a further six soils, "naturally" contaminated with various concentrations of oil arising from sites of accidental spillage and from a refinery, were also supplemented with 1% fresh brewer's yeast. A separate study involving the development of a treatment strategy for soil at an oil refinery contaminated by bombing in World War II tested the effects of the addition of a series of concentrations of brewer's yeast or yeast extract [3]. There was a progressive enhancement of the extent of hydrocarbon degradation with increasing concentration of brewer's yeast added to the basal medium consisting of a urea/mineral salts mixture. The maximum reduction in concentration of oil hydrocarbons and associated polar breakdown products occurred when 0.75% brewer's yeast was added: from 16,300 to 7600 mg kg^{-1} in a 14-week period compared with 14,500 mg kg^{-1} in a control. Brewer's yeast was therefore included as one of the supplements (together with inorganic nutrients) in the field application of the treatment. Concentrations of total petroleum hydrocarbons were reduced by this treatment from 22,600 to 2000 mg kg^{-1} over approximately 450 days: the treatment time was prolonged as a result of the chemical nature of the residue, which consisted of a high proportion of higher boiling point fractions and branched cyclic alkanes. It is not known whether the beneficial effect of the yeast was as a result of the particular form in which the nitrogen was supplied, the provision of vitamins and growth factors, a general stimulation of microbial activity, or a cometabolic effect.

Yeast extract or waste brewer's yeast may also be used for in situ treatments involving direct land farming of the soil, where it may be applied as a fine suspension using conventional agricultural spraying equipment. For in situ treatment of the saturated zone using ground water recirculation systems or even passive nutrient application systems with direct injection to the water table, application of a suspended rather than a dissolved nutrient has obvious limitations in terms of treating the less permeable strata, and could result in localized blockage of injection wells through the buildup of excessive microbial biomass.

It should also be noted that the use of yeast extract does not always have a beneficial effect on the degradation of hydrocarbons. In one set of microcosm experiments [92], a slight stimulation of carbon dioxide production was observed after addition of yeast extract, but the extent of biodegradation of the hydrocarbons was not affected. The same set of experiments also evaluated the addition of 1% sewage sludge, which had the effect of decreasing both carbon dioxide evolution and biodegradation of oil sludge in treated samples, an effect that may be attributable to the availability of more degradable substrates. Similarly, both sewage sludge and cow manure inhibited biodegradation of parathion and fonofos in soil [113]. On the other hand it was reported that the addition of composted sewage sludge to soil enhanced the biodegradation of 1,2,3-trichlorobenzene [114]. Experimental work examining the degradation of crude oil in soil at low temperatures indicated that the addition of freshly prepared birch sawdust reduced hydrocarbon mineralization (as measured by carbon dioxide evolution), by the provision of a more readily available carbon source [115]. As with the yeast extract, however, in situ applications of such material would probably be limited to land farming type scenarios. Not only would there be physical problems in applying such material to ground water, particularly in its distribution through the contaminated medium, but the addition of readily available carbonaceous substrates is likely to create an unacceptably high biochemical oxygen demand.

Incorporation of various plant residues appears to have a stimulatory effect on the degradation of various organochlorine insecticides in submerged soils. In laboratory studies, degradation of DDT to DDD was stimulated by the addition of 0.5% organic matter as rice straw, cellulose, or glucose [84]. In a soil already rich in organic matter, addition of a further 0.5% organic matter as rice straw or cellulose did not significantly affect the rate of degradation of three isomers of BHC, but glucose caused a retardation. However in a clay soil, all four BHC isomers were degraded at a faster rate when 0.5% glucose, cellulose, or rice straw were added, the degree of stimulation being greatest with the latter, probably due to a high content of available carbon and nitrogen.

The degradation of toxaphene occurred more rapidly in laboratory studies in (flooded) soil amended with alfalfa than in soil amended with or without nutrient broth [85]. A similar degree of degradation was achieved with amendments of cotton gin waste, although a higher concentration was required (25% cotton gin waste as compared with 1% alfalfa meal), and composted cotton gin waste was not effective. A field experiment to test the potential for anaerobic degradation of toxaphene in drainage ditch sediment amended with steer manure was inconclusive as the result of input of further fresh residues of toxaphene during the experimental period. However, changes in the GC peak profile in the organic-amended compared with the nonamended area suggested that biotransformations were occurring.

The role of organic supplements in biodegradation is, therefore, not clearly understood, and the potential advantages of their addition may be offset by the possible inhibition of the degradation process. Such issues need to be carefully evaluated on a case-by-case basis. A further effect of adding an organic waste material, such as a manure or sewage sludge, is the potential for the coupling of the contaminant to the material through binding mechanisms similar to those that occur with humus [116]. This would minimize leachability of the contaminant but, at the same time, reduce its bioavailability and increase its overall persistence in the soil system.

One of the postulated advantages of adding an organic source of carbon is the provision of a potential substrate for cometabolic transformations of the contaminants present. More specific applications of cometabolism in biodegradation studies have been through the use of analog enrichment; for example, the degradation of certain polychlorinated biphenyls (PCBs) have been stimulated by the application of biphenyl [117]. In practice, however, there has not been widespread reporting of the use of such techniques on a field scale, possibly as the result of both limited applications in terms of the total number of PCB congeners that can be successfully treated using this approach and potential regulatory concerns over the introduction to the environment of the analog itself. A similar environmental concern probably limits the use of systems where trichloroethylene (TCE) is cometabolized by strains of *Pseudomonas putida* in the presence of phenol or toluene [118–120].

A potentially more promising (at least in terms of environmental acceptability) approach using cometabolism for the in situ bioremediation of TCE and other chlorinated aliphatic solvents is through the application of methane. The methane monooxygenase enzyme complex possessed by methanotrophic bacteria has been shown to be capable of oxidizing TCE and other chlorinated ethenes in laboratory experiments [87,88] and was subsequently incorporated into a patent for a treatment process using a packed column system (J. T. Wilson and B. H. Wilson, U.S. Patent No. 4,713,343; 1987). Such a system involves the stimulation of the methanotrophic population through exposure to 0.6% methane, although it was reported that other low molecular weight alkanes such as ethane, *n*-propane, and *n*-butane could also be used. This concept has been extended from such reactor systems, which are applicable to pump-and-treat systems, to a truly in situ approach. The establishment of a test zone involving a shallow semiconfined aquifer, which could be artificially contaminated with a series of chlorinated aliphatic hydrocarbons [121], has allowed a detailed evaluation of this process [122]. In the first stage, biostimulation of methanotrophs was achieved through the addition of methane and oxygen, which were pulse injected to avoid biofouling, to achieve injected concentrations of 5.9 and 20.8 mg L^{-1} respectively. Rapid decreases in the concentrations of methane and dissolved oxygen were achieved between 200 and 430 h (methane levels decreased below detection at the end of this period), with a mass ratio of oxygen to methane consumed of 2.5, which was significantly lower than the ratio required for complete methane oxidation. Biotransformation studies were then conducted, first involving continuous injection of TCE over a 3-month period to achieve steady-state conditions, during which time-continuous pulse injection of methane and oxygen was also carried out. Reductions in TCE were normalized using bromide as a conservative tracer to account for nonbiotic losses, and it was estimated that 25 to 30% of the TCE had been degraded. The experiments were further extended to monitor the transformations of other chlorinated ethenes, including *cis*- and *trans*-dichloroethylene (DCE) and, later, vinyl chloride. The degree of transformation ranged from 90 to 95% for vinyl chloride, 80 to 90% for *trans*-DCE, 45 to 55% for *cis*-DCE, and 20 to 30% for TCE; that is, the less chlorinated compounds were transformed more. The limitation of the process was considered to be the amounts of methane and oxygen that could be delivered to the aquifer, therefore limiting the growth of the methanotrophic population responsible for carrying out these transformations. However, the addition of higher concentrations of methane may inhibit the process. Model simulations, similar to the field observations, indicated that when greater concentrations of methane were added, competition for the enzyme between methane and the chlorinated hydrocarbon countered the potential benefits conferred by the development of a higher population of the methanotrophs [123]. Laboratory experiments by other workers using a bioreactor system also indicated a reduced rate of TCE degradation at higher methane concentrations, although degradation of *cis*-DCE was less affected [82].

D. Bioavailability and Chemical Enhancement

The chemical supplements provided in most bioremedial treatments involve the addition of nutrients, cosubstrates, or sometimes other reagents to address one or more physical or chemical factors of the soil environment that limit microbial activity. Another group of reagents may be added for the purpose of addressing another issue, that is, the low bioavailability of the contaminant. The success or failure of full-scale applications of such reagents has not been widely reported, but a review of a number of laboratory investigations provides some indication as to potential pitfalls.

The production of surfactants is one of the mechanisms used by microorganisms to utilize insoluble compounds, such as petroleum hydrocarbons, by both promoting attachment of the organism to the contaminant and the solubilization of the contaminant [124]. Different strategies are used by different microorganisms. It should also be noted that emulsification of oil, for example, in the case of *Acinetobacter calcoaceticus* RAG-1 through production of emulsan, may be a mechanism with which to achieve detachment of the cell from the oil droplet once the utilizable hydrocarbon has been depleted [125].

Applications of synthetic surfactants or dispersants used for the emulsification of oil to experimental systems produced conflicting results. Based on oxygen uptake measurements (after establishing that the test surfactants themselves did not contribute significantly), concentrations of up to 7% (in oil) of the surfactants, BDOWH X and BDOWE 8, appeared to enhance the degradation of crude oil [126]. A similar enhancement of oxygen uptake was noted in sand columns spiked with oil with the dispersant Corexit 7664 [127]. Another comparison of four dispersants indicated that all supported microbial growth, but the one with the poorest emulsifying capacity (Sugee 2, Handy Chemical Ltd.) was the only one that actually stimulated biodegradation [128].

There is laboratory evidence of direct enhancement of the degradation rates of individual hydrocarbons through surfactant application. The use of one or more nonionic surfactants of the Neodol group supplied by Shell Development Co., consisting of ethoxylated alcohols, resulted in the solubilization of *n*-decane and *n*-tetradecane in micelles, and reduced the doubling time of two alkane-degrading bacteria during exponential growth on the hydrocarbon substrate by a factor of at least five compared with surfactant-free systems. Application of surfactant also increased the rate of hydrocarbon consumption (as measured by GC analysis) in the case of one of the strains tested, and for a second strain, degradation of the two hydrocarbons occurred only in its presence [129].

Surfactants may also promote biodegradation of one compound while inhibiting that of another. Triton X-100 inhibited the mineralization of hexadecane dissolved in 2,2,4,4,6,8,8-heptamethylnonane by *Arthrobacter* sp. by preventing adherence of the bacteria to the heptamethylnonane-water interface [130]. However, the same surfactant apparently increased the rate and extent of naphthalene degradation, when the latter was dissolved in the same solvent. This was probably as a result of greater partitioning of naphthalene into the aqueous phase. The authors postulated that, when no longer attached to the interface, the *Arthrobacter* sp. was probably free of the spatial and nutrient limitations associated with the solvent-water interface, so that degradation of naphthalene was enhanced.

The efficacy of a surfactant in promoting desorption of contaminants from soil surfaces will vary according to soil type. The nonionic surfactants, Alfonic 810-60 and Novel II 1412-56, promoted desorption of phenanthrene from a mineral but not from an organic soil [131]. However, both surfactants, at 10 $\mu g\ g^{-1}$, significantly enhanced the extent of the degradation of phenanthrene in both the mineral and organic soils, the stimulatory effect being more pronounced in the latter. As with phenanthrene, neither surfactant increased the desorption of biphenyl in the organic soil; however, biodegradation was enhanced by Alfonic 810-60 at 100 $\mu g\ g^{-1}$. It appears, therefore, that low concentrations of some surfactants may increase the mineralization of sorbed aromatic contaminants, even if the apparent degree of desorption does not appear to be appreciable.

Conversely, even where significant desorption can be achieved, surfactant toxicity problems and interference with contaminant degradation can occur. Degradation of PAHs was inhibited by sodium dodecyl sulfate because the surfactant was preferred as a growth substrate, whereas various other nonionic surfactants with an average ethoxylate chain length of 9 to 12 monomers were toxic to PAH-degrading cultures [132]. Toxicity decreased with increasing hydrophilicity (increasing ethoxylate chain length), and the nontoxic surfactants, Sapogenat T-300 (Hoechst AG, Frankfurt) and Tegopren 5851 (Goldschmidt AG, Essen) promoted degradation of mixtures of fluorene, phenanthrene, anthracene, fluoranthene, and pyrene. In the presence of 2.0 mM of Tegopren 5851, only 14.6% of the PAH mixture remained after incubation compared with 61.8% without added surfactant.

Both Triton X-100 and Tensoxid S50 (Tensia, Liege, Belgium) were relatively effective in desorbing Aroclor 1242 from sand. However, the growth of several bacteria capable of metabolizing a number of PCB congeners were inhibited to varying degrees in the presence of Tensoxid S50. Triton X-100 was also inhibitory to bacterial growth, but less so than Tensoxid S50, particularly for two of the isolates tested, H850 and BTL333. Degradation of the relatively easily degradable 2,3-dichlorobiphenyl by both these isolates was not significantly affected by the presence of 0.5% Triton X-100. However, the degradation of other PCB congeners having a greater degree of chlorination was considerably reduced. It was suggested that in such cases, the surfactant provided a more readily available carbon source [133].

Extraction of PCBs from soil using surfactants may, in fact, be considered as a separate treatment strategy from that of bioremediation [134]. Indeed, for a number of hydrophobic contaminants, the application of surfactant-assisted flushing to the saturated zone can be regarded as an in situ treatment technology [38,135], in which aboveground treatment of the recovered contaminant may [136] or may not include a biological element. In both cases however, careful evaluation of potential physicochemical interactions is required before scale up, as the application of some surfactants may lead to a drastic reduction in permeability [137]. Even where degradation of the surfactant does not inhibit that of the contaminant, a further problem may be the oxygen demand created. The surfactants, Brij 30/Brij 35 at a concentration of 0.1%, promoted degradation of PAH in coke waste, but laboratory data indicated that the oxygen demand associated with the surfactant overwhelmed the delivery of oxygen, even using the maximum feasible hydrogen peroxide concentration [138]. A further practical difficulty in application was the form of the surfactant as a waxy solid.

Despite these potential problems, there are reports of application of surfactants and other compounds used to desorb contaminants from soil to field scale in situ bioremediation projects. As part of the formulation of an in situ treatment strategy for oil-contaminated soil at a refinery site adjacent to the Rhine river, over 50 compounds were screened in shake flasks for their efficacy in contaminant removal from soil, and the most promising were subjected to both permeability tests and soil column leach tests [15]. Cyanamer P70 (Cyanamid), a descalant was selected, as it removed a relatively high proportion of the oil contaminant (47%) from the soil, while maintaining soil permeability, even at concentrations in excess of 1000 mg L^{-1}. This compound was an antiscalant, a polyacrylate used to control scale formation and precipitation in cooling water and boiler systems. The compound was subsequently used in a field-scale in situ treatment, together with nutrients and microbial inocula. Ground water was abstracted from a central well to an aerator tank and then piped back via a system of gravel infiltration trenches. Over a 15-week period, concentrations of hydrocarbons in soil from the saturated zone were reduced from a mean of approximately 200 mg kg^{-1} to below 50 mg kg^{-1}.

A similar procedure was adopted in the development of a treatment system for a creosote-contaminated site in Sweden [16]. In soil column tests, 22 and 25% of creosote in the soil could be leached with the surfactants, ethylan CD916 and ethylan BCP (Lankro), respectively, compared with only 9% when water alone was used. Tank leaching tests (whereby 10 kg of soil was leached with 30 L of liquid over across a surface of 1000 cm^2) achieved a 26.4% reduction in concentrations of creosote in soil after 4 days of leaching with 0.5% ethylan BCP, compared with a reduction of only 10.2% with water alone.

A further example of the use of nonionic surfactants (unnamed) in the bioremediation of aromatic, polycyclic, and aromatic compounds has been described for a former asphalt production site [139]. An increase in the solubility of PAH by a factor of 2100 to 3100 was reported with mobilization of both high and low molecular weight compounds. The treatment system involved both a surface bioreactor and stimulation of in situ bioremediation by adding nutrients containing nitrate as an electron acceptor.

The application of surfactants or other compounds capable of promoting emulsification or desorbing contaminants from the solid phase could, therefore, potentially make a significant contribution to the degradation process in certain instances, but the literature clearly indicates that a very careful screening program is necessary. The general criteria for successful application of a surfactant are an absence of toxicity, and lack of interference with both the biodegradation process and the soil properties. In addition, other considerations of site and contaminant properties are necessary before application, as factors such as hardness, alkalinity, and pH will all influence performance [136].

The use of solvents for solubilizing contaminants has also been investigated. Shake flask studies indicated that 1% methanol was effective in promoting the solubility of several PAHs, and the application of 0.5% propanol to flushing water from an in situ test field treatment of gasoline and diesel fuel enhanced recovery of hydrocarbons. Although the presence of the alcohol did not appear to have any adverse influence on the degradation of aromatic hydrocarbons, the authors considered that the use of alcohols for in situ bioremediation was precluded by decreases in soil porosity resulting from biomass or gas production [140].

E. Other Chemical Factors

The pH of soil may potentially influence biodegradation rates, both directly and indirectly. The optimum pH range for decomposition of most waste residues is from 6.5 to 8.5, although indirect effects of pH in promoting or reducing nutrient availability may also be of importance when these are key limiting factors, particularly in the case of phosphorus [141], or for reduction or enhancement of heavy metal availability [42]. It has been suggested that because changes in soil pH may alter the relative activities of bacteria and fungi, this could potentially result in different biochemical degradation pathways, for example, formation of mutagenic arene oxides from the degradation of polyaromatic hydrocarbons if fungal activity predominates [141].

Increasing the pH of an acid Podzol from 4.5 to 7.4 almost doubled the degradation rate of both alkanes and aromatic hydrocarbons [103]. Further increase to pH 8.5 resulted in a reduction in the extent of degradation. Similarly, the extent of oil sludge biodegradation was increased when the soil pH was raised from 5 to 7, although in this instance, a further increase in degradation was observed at pH 7.8 [92]. Where a land farming approach is used, ameliorating acid soil conditions can be achieved through liming, that is, the addition to the soil of any calcium- or calcium plus magnesium-containing compound that is capable of raising the soil pH [42, 141]. Selection of the most suitable reagent will depend on the degree to which the material can be finely ground, if supplied in solid form, the relative costs, and the neutralizing value. The latter is often expressed as the calcium carbonate equivalent (CCE), in which the neutralizing value of $CaCO_3$ is set at 100. CCE values for various liming materials are 100 for calcitic limestone ($CaCO_3$, 100% purity), 89 for dolomitic limestone (65% $CaCO_3$ plus 20% $MgCO_3$, 87% purity), 151 for quicklime (CaO, 85% purity), 120 for slaked lime ($Ca(OH)_2$, 85% purity), 50 for marl ($CaCO_3$, 50% purity), and between 75 and 90 for blast-furnace slag ($CaSi_2O_3$) [42].

When there is a requirement to lower the soil pH, this may be achieved through the addition of elemental sulfur, sulfuric acid, liquid ammonium polysulfide, or aluminum and iron sulfates [141].

Adjustment of the pH of the saturated zone may also be desirable, although altering the pH of a large quantity of ground water may be difficult and could have major cost implications. The selection of a suitable liming reagent will also be influenced, in this case, by the requirement for a soluble compound. As with the unsaturated zone, the amount of reagent required will be highly dependent on the buffering capacity of the soil and will need to be determined experimentally on a case-by-case basis. The pH of the saturated zone could be of particular importance if nitrate as an electron acceptor is to be used. The addition of nitrate to soil that did not contain an actively denitrifying population inhibited methanol degradation caused by the accumulation of nitrite, whereas when the pH was raised above pH 6 such inhibition was relieved (W. G. Wilson, Master's thesis, Virginia Polytechnic Institute and State University, Blacksburg, Virginia, 1986 [cited in Ref. 142]).

There are few detailed reports of the adjustment of pH for full-scale in situ projects. Ground water contaminated with ethylene glycol from leakage of a surface storage lagoon was adjusted from a mean pH of around 4.5 to 6 to 7 (together with nutrient addition) after abstraction, and before aboveground treatment in an aerated bioreactor [143]. The treated ground water was reinjected, both via reinjection wells and through surface application, to achieve flushing of contaminated soil. The amount of biodegradation that actually occurred in situ as distinct from the aboveground system is not known, although concentrations of ethylene glycol were reduced from over 1000 ppm to below the limit of detection in all production wells at the end of the treatment period, which consisted of a 14-day operational phase, a 3-month monitoring phase, and a 9-month maintenance program. During the latter period, pH control was also affected by the addition of lime to the soil surface in the lagoon and contaminant plume areas.

The application of reagents to alter other soil chemical properties as part of field-scale in situ bioremediation has not been widely reported. It is possible that there may be substances within the soil environment, including both organic and inorganic contaminants, that could inhibit the activities of the microbiota responsible for contaminant degradation. Although the effects of contaminants on microbe-mediated processes in soil have received considerable attention in recent years, it is difficult to draw too many overall conclusions, as the relative effects on microbial activity of such substances, particularly heavy metals, are largely a function of the physicochemical properties of the specific soil involved [144]. Certain processes that are mediated by a narrow group of microbial species, such as nitrifiers [145] or vanillin-degrading organisms [146,147], sometimes appear to be more sensitive to pollutant-induced environmental stress, for example, acid rain, than processes mediated by a wide range of soil microorganisms, such as ammonifiers [145] or organisms of degrading glucose [147,148]. Similarly, it is possible to speculate, therefore, that degradation processes involving contaminants that can be metabolized by a wide range of soil microorganisms, such as petroleum hydrocarbons [18], may be less sensitive to environmental stress than those involving compounds, such as certain halogenated aromatics, that are metabolized by a more restricted group of species.

There are a number of examples in the literature that indicate that the type and concentration of clay minerals and organic matter will mitigate the toxicity of heavy metal and acidic contaminants toward microbe-mediated processes [e.g., 144,147–150]. The addition of clay or organic matter to in situ treatment for the sole purpose of reducing the inhibitory effect of metals (or nontarget organic compounds) on contaminant degradation has not been widely reported. Although the beneficial effects of organic matter in promoting structural and moisture retentive properties have been discussed, the possible inhibitory effects of providing an alternative carbon source as well as reductions in contaminant availability need to be considered [116]. Addition of clay may result in conditions that are unfavorable for effective, aerobic biodegradation systems. If heavy metals are a potential problem, the application of lime to maintain the soil at neutral to slightly alkaline pH levels may be sufficient to mitigate their effect on microbial activity.

IV. BIOLOGICAL FACTORS

Having optimized as far as possible the various physical and chemical factors of the soil environment and, where possible, of the contaminating compound itself, the question arises as to whether there is any requirement for the addition of microorganisms as part of the bioremediation strategy. The application of inocula clearly involves a significant number of protocols and procedures, not least of which is to ensure that the organisms added are not pathogenic [151]. The degree to which bioaugmentation is necessary, or at least advantageous, has been discussed in a number of reviews [e.g., 1,40,94,96,124,152]. It has been postulated that there are six general circumstances where bioaugmentation may confer advantages above just stimulation of the natural microbiota: (1) for the treatment of highly recalcitrant compounds that are degraded slowly, even under optimal conditions by the natural microbiota; (2) for the treatment of contaminants present in relatively high or very low concentrations; (3) for the treatment of sites contaminated by very recent spillages; (4) for the treatment of environments where a particular physicochemical characteristic inhibits the activity of the indigenous microbiota; (5) for faster and more effective cleanup through an accelerated rate of degradation (or reduced lag phase) or achievement of lower residual contaminant levels; and (6) for an enhancement of the overall reliability of the treatment process [40]. For particular situations, there may be special circumstances where bioaugmentation is advantageous; for example, it has been suggested that for wetlands and marshes, augmentation of the biomass of oil degraders through introduced bacteria may assist in changes to the consistency of the oil, promoting emulsification and reducing the adverse ecological impact on fauna and rooted vegetation [152].

There are also instances where an organism may be the prime element in a biotreatment strategy by virtue of having a particular degradative capacity that encompasses a range of polymeric compounds. The most notable example of this concerns white rot fungi, in particular *Phanerochaete chrysosporium*. The lignin peroxidase system of this organism has been shown to be capable of degrading a wide range of organic compounds, including PAHs, PCBs, PCP, and various organochlorine pesticides, in soil, and the potential application for bioremediation has been discussed in a recent review [153]. One of the conclusions arising from this review is that *P. chrysosporium* is not a typical soil organism, and the soil environment must be modified accordingly with substantial quantities of a suitable growth substrate, such as wood chips, straw, corn cobs, etcetera, to provide it with a competitive advantage. Such systems lend themselves more to an ex situ approach involving excavation of the soil and treatment in an appropriately designed bed. An in situ approach was adopted for a field trial in which the plots were sterilized by fumigation before amendation with wood chips and peat [154]. Treatment with *P. chrysosporium* or *P. sordida* reduced PCP concentrations of 250 to 400 mg kg^{-1} by 88 to 91%. For the saturated zone, an alternative approach may be to attempt to mobilize the contaminant in the aqueous phase, recover the ground water, and treat it in a bioreactor wherein the fungus is maintained under controlled conditions. A rotating biological contact system for utilizing *P. chrysosporium* in such a treatment system has been patented (H. M. Chang, T. W. Joyce, T. K. Kirk, and V. B. Huynh, U.S. Patent No. 4,554,075; 1985). Such applications, however, are essentially pump-and-treat rather than truly in situ systems, and though fungi clearly have an enormous potential role in bioremediation, there is little information concerning their application to in situ cleanup.

There are few comparative studies of the relative effects on contaminant degradation with or without inoculation in full-scale bioremediation projects or even on a pilot scale, at least for in situ treatment. The various laboratory studies undertaken to develop bioremediation strategies or as pure research have indicated varying results of bioaugmentation that involved inoculation of either indigenous or nonindigenous organisms ranging from no effect to accelerated decomposition. The general conclusion appears to be that for compounds that are naturally occurring or which are relatively easily degradable, such as petroleum hydrocarbons, the results of bioaugmentation are not particularly dramatic. This finding is similar for marine and terrestrial systems [124] or where commercially available mixed bacterial cultures have been compared with mixed populations of bacteria from activated sludge in aerobic batch conditions [155].

In a comparison of the extent of degradation of petroleum hydrocarbons and associated breakdown products in refinery soil contaminated by weathered oil and inoculated separately with 25 strains of indigenous bacteria, only one (unidentified) isolate significantly reduced concentrations below that of the addition of only nutrients [3]. Even here, the difference in mean percentage degradation after 7 weeks was only 11%. Similarly, inoculating test soils contaminated by light fuel oil or heavy waste oil with indigenous bacteria (isolated by growth on aliphatic hydrocarbons) resulted in no significant difference in degradation of the oils compared with treatments without inoculation [112]. In contrast, the addition of selected autochthonous soil bacteria to a soil contaminated with tank bottom residue consisting of a reportedly aromatic hydrocarbon residue increased degradation by 22% over that in uninoculated soil or soil supplemented with fertilizer [156]. The authors used bacteria maintained in a nutrient-supplemented (commercially produced) vermiculite and attributed the reason for the more positive response compared with previous investigations [112] to the greater degree of hydrocarbon contamination in their study (10% hydrocarbons compared with 0.5%). It is also possible, however, that the nutrient effect of the bacteria themselves may have been a contributory factor.

There is more evidence for enhancement of the rate or extent of biodegradation by bioaugmentation in microcosm or field studies with less readily degradable contaminants, such as halogenated compounds, which, in contrast to petroleum hydrocarbons, involve chemical substituents that are not normally encountered in pristine environments [157]: for example, with pentachlorophenol [158,159], 2- and 3-chlorobenzoic acid [160], 2,4,5-T [161], and in a number of other instances [35]. Laboratory microcosm experiments are often a necessary preliminary stage in the development of a bioremediation strategy, at least for the more recalcitrant compounds. However, where the benefits of bioaugmentation have been identified in such studies, the implications for in situ treatment are not always clear. It has been postulated that laboratory studies may overestimate or underestimate these benefits, where the inoculated organisms are unable to function as effectively against the rigors of the prevailing physicochemical conditions in situ in comparison with optimal conditions in the microcosm; alternatively, particularly hardy strains may enhance degradation in the field [40]. When comparative studies are made of the effects of bioaugmentation in systems of increasing complexity, leading up to field-scale implementation, conflicting results are often obtained. In shake flask experiments, inoculation of oiled beach material with a mixed bacterial community of indigenous microorganisms or a single strain resulted in both an enhancement of carbon dioxide evolution and a two- to threefold increase in the degradation of phenanthrene. When the medium was replaced by a simulated tidal cycle, however, there was no significant enhancement of carbon dioxide evolution or the rate of extent or (crude oil) degradation compared with inorganic nutrient addition alone. This was attributed to several possibilities, such as the washout of cells and lack of adhesion capability of the inocula or simply the rapid enhancement of the activities of the indigenous microbiota through nutrient stimulation [162]. After a series of laboratory tests on 10 commercial products, consisting of eight bacterial inocula, one dispersant, and one fertilizer formulation, two of the microbial products (produced by Sybron, Inc., and ERI-Waste Microbes, Inc.) that had resulted in significantly greater alkane degradation [163] were tested in an in situ field experiment [164]. This involved measuring degradation of alkanes in a set of four randomized blocks on a beach contaminated 16 months earlier with the *Exxon Valdez* oil spill. Each block consisted of a control plot, a mineral nutrient plot, and two plots receiving mineral nutrients together with each of the two products. No significant differences were obtained ($P > 0.05$) in the weight of the residual oil or alkane hydrocarbon profile over the 27-day test period. One important issue in the interpretation of such results, however, concerned once again the considerable variability of the data, which may have confounded the effect of a genuine stimulatory response. Based on such considerations, the authors conducted a power analysis, which indicated that the experimental error variance was too high reliably to detect differences of the type observed in the experiment; that is, the results were inconclusive, rather than negative. However, the somewhat limited time frame for the experiment and the weathered nature of the oil were also noted as being potential contributory factors to the lack of a clear response.

A rather longer in situ field trial was undertaken over a 308-day period in north-central Alberta, Canada, which involved soil plots consisting of a control, and treatments with crude oil, oil and bacteria, oil and fertilizer, and oil, bacteria, and fertilizer [100]. Significant reductions in saturated hydrocarbons compared with the control were obtained after 66 and 308 days, where both bacteria and fertilizer or fertilizer alone had been applied, although there were only slight differences between these latter treatments. The bacteria-amended soil did show an increased utilization of the higher C20 to C25 chain-length compounds (often considered to be toward the top end of the range of hydrocarbons that are readily degradable [95,96,97] both after 66 and 308 days. Even here, however, the effect was not observed in all replicates.

In contrast to the relatively easily degradable saturate fraction (at least for the *n*-alkanes), there is greater evidence for the possible benefits of bioaugmentation in the field in the case of the more recalcitrant PAHs. A field demonstration compared the efficacy of an inoculum of indigenous bacteria with that of a commercial inoculum (Microbe Masters, Inc., Baton Rouge) in degrading PAHs introduced through the addition of 4% oil and grease to the soil [165]. A supplement of mineral nutrients was added to each treatment. After a 14-day acclimation period, the indigenous consortium was capable of degrading PAHs at rates similar to that of the commercial culture: in this example therefore, the sole advantage of the commercial inoculum was a reduction in the onset of degradation. For full-scale projects, however, particularly where bioremediation is competing against other disposal options, reductions in treatment time may be a critical factor in gaining acceptance of the technology.

Although this particular example did not include a further treatment to compare the effect of nutrient stimulation alone with the other treatments, bioaugmentation with either an indigenous strain (*P. fluorescens*) or an isolate from a culture collection (*P. putida*, ref. NCIB 9816) was compared with nutrient stimulation alone together with an unamended control) as part of a field-scale implementation of the remediation of a coal gasification site at Blackburn, U.K. [2,40]. In a prototype treatment bed, only the treatment containing *P. putida* resulted in a statistically significant reduction in the concentration of total PAHs (a mean reduction of 5000 mg kg^{-1} over an 8-week period), whereas no reductions were obtained either with nutrients and surfactant alone or in the unamended control. Reductions in two of the individual PAHs (pyrene and fluoranthene) were observed for the treatment with the indigenous *P. fluorescens*, but the reduction in total PAH content was not significant. Subsequent full-scale treatment involved application of both strains of bacteria.

Although both the previous examples were not continued for long periods, they illustrate that, at least in the short term, application of specific isolates may reduce the overall treatment time. Such organisms have been referred to as vanguard organisms in bioremedial treatment technology [166,167], based on account of the hypothesis that they are responsible for catalyzing initial, rate limiting biochemical reactions such as ring cleavage. It is only necessary for such inoculated organisms to have a temporary competitive advantage, if the degradation process is then completed by an active, indigenous microbiota [168]. Cometabolism (cooxidation) has been proposed as a possible mechanism for degradation of the higher ringed PAHs [169]: if this is the case, then an overall enhancement of total microbial activity, brought about by the introduction of large numbers of vanguard organisms, together with nutrient supplements, would also accelerate the degradation process if both substrate and cosubstrate are readily available.

In situ degradation of PCBs has also been the subject of field trials involving bioaugmentation. After a laboratory study which achieved a 50% reduction in PCB concentration in 100 days, an in situ approach was adopted for a field trial involving a drag strip, the top 15 cm of which was contaminated with a weathered Aroclor 1242 [134, 170]. Part of the area was covered with a tent and inoculated with a bacterial isolate, LB400, a bacterial species that had demonstrated PCB degradation capacity in the laboratory. Up to 25% degradation of the PCB was observed in the near-surface layer; the lesser degree of degradation was attributed, at least partly, to poor survival of LB400 under field conditions.

Several full-scale or partial in situ bioremediation projects have used bacterial inocula, although their overall contribution to the total degradation process has not been quantified. In an oil contaminated soil where inocula of indigenous bacteria were included together with nutrients and surfactant as part of a ground water recirculation scheme, mean oil concentrations in soil were reduced by

86% from 185 to 26 mg kg^{-1} over 15 weeks [15]. Mean soil concentrations of creosote (measured as the sum of 10 PAHs and dibenzofuran) were also reduced by 64% over 4 months using a recirculation system through both the vadose and saturated zone, where microorganisms were included together with nutrient supplements [16]. In the cleanup of a spill of crude oil in an area of 16,000 m^2 contaminated to a depth of 0.5 m, 23 kg of Hydrobac, a commercially available culture, blended with 1900 L of warm water, was sprayed over the area and rototilled into the soil, after the addition of 18 kg of a nonionic surfactant (Polybac E) and 180 kg of a nutrient formulation (Polybac N) [38,171]. The oil content of the soil was reduced by 66% from August 5 to September 10. The treatment was repeated after 6 weeks and continued until crude oil residues were no longer observed. The costs of biodecontamination were $16,000 [171]. The same inoculum (Hydrobac) was also used for a partial in situ treatment of a formaldehyde spill from a railroad car that contaminated the rail ballast, the underlying soil, and an adjacent ditch. In this example a 75,000 L holding tank was seeded with inoculum, nutrients, and surfactant, and the contents were mixed by aeration, sprayed onto the ballast, and allowed to percolate through the contaminated ballast and soil before being collected in the drainage ditch and recirculated to the holding tank. The degree to which treatment occurred in situ, as compared with within the holding tank, and the relative contribution of the inoculum to the overall degradation process are not known, although reductions in formaldehyde concentrations from 750 to 250 mg L^{-1} were achieved in the leachate within 1 week. After 3 weeks, the formaldehyde concentration had declined to below 1 mg L^{-1} [171].

In some instances, in situ biological degradation through addition of inocula is used as a polishing system after removal of the bulk of the contamination by other methods. A mixture of various (unspecified) chemical compounds in ground water was treated via physical and chemical processes, which included clarification and precipitation, adsorption on granular, activated carbon, and sparging, before reinjection and surface flushing. When concentrations of the compounds had declined from initial levels of up to 10,000 ppm to below 1000 ppm, hydrocarbon-degrading bacteria and nutrients were added before reinjection. After initiation of this program, the total chemical concentration in the ground water was reduced to below 1 ppm, and soil concentrations of the contaminants were reduced by 5- to 10-fold over a 2-month period. The contribution of the commercial inoculum to the overall degradation was not known, although it did not appear to perform significantly better in laboratory studies when compared with the indigenous microbiota [172].

In another case, over 375,000 L of a mixture of ethylene glycol, propyl acetate, and other unspecified organic compounds were released into the soil after a transportation accident [173]. After surficial cleanup that involved the excavation of contaminated soil, a recovery system was initiated for the ground water, which was treated by clarification, aeration, and granular, activated carbon. Once the concentrations of the three key contaminants present had been reduced to below 200 ppm, water was inoculated with bacterial cultures, nutrients, and air before reinjection. This was considered by the authors to have accelerated the treatment, with reductions of ethylene glycol and propyl acetate to below 50 mg L^{-1} and of total concentrations of spilled contaminants to below 100 mg L^{-1}. In another example [173], air-stripping was used to treat easily volatilized hydrocarbon contaminants and other organic chemicals at an industrial facility, and the recovered water was inoculated with hydrocarbon-degrading bacteria. The treatment was apparently so effective that treatment with granular, activated carbon proved unnecessary. Once acceptable concentrations of the contaminants were achieved, the inoculated water was reinjected to achieve in situ biodegradation in the soil, although the influence of the commercial inoculum was not quantified. Other examples where inoculation of specific microbial strains has taken place in pump-and-treat systems for ground water remediation have been cited [42,53].

In conclusion, there has been little definitive evidence of full-scale bioremediation being accomplished in situ primarily as the result of the application of specific inocula as compared with stimulation of the activities of the indigenous microbiota. However, there is circumstantial evidence that bioaugmentation may significantly enhance the overall effectiveness of the cleanup process in the case of less easily degradable compounds, even if this occurs only through a reduction in the lag phase. The use of introduced organisms poses a number of potential hurdles that need to be surmounted, particularly mass transfer and the requirement of the inocula to move efficiently through soil pores to reach the target compound, acclimation to the physicochemical properties of the particular soil environment (a problem confounded by inherent heterogeneity), and competitive ability with the preadapted indigenous microbiota [23,40,152]. Moreover, it has been noted that because many of the key bacteria involved in bioaugmentation are gram negative, nonspore formers, their storage in large quantities may result in difficulties in maintaining viability [98]. This is less of a problem in the case of fungal spores, although there is probably greater potential for the application of fungi to ex situ rather than in situ bioremediation.

V. LIMITATIONS AND POTENTIAL CONCERNS

Bioremediation of contaminated soil and ground water may be effected by a manipulation of the physical, chemical, and biological factors specific to the environment within which the contaminant is distributed. The practicality of altering the key limiting factors will determine the success of bioremediation as a treatment strategy, and this in turn is often greatly influenced by soil heterogeneity. Although soil in general is heterogeneous, contaminated soil often tends to be particularly so, especially when "made ground" is involved, including for example, concrete and brick rubble, slag, timber, and putrescible waste. This in turn creates a complex distribution pattern for the contaminant(s) of concern. In situations where a particular organism is to be introduced, differences in the physical and chemical characteristics of the soil will affect its performance directly through effects on its physiology and indirectly through shifts in competitive advantage with the naturally established microbiota from one microsite to the next [40].

From a practical standpoint, environmental heterogeneity imposes problems in terms of monitoring and validating the treatment process and of assessing whether bioremediation has actually been achieved. The greater the degree of environmental variability, the greater the number of samples required during the monitoring process, and complex statistical analysis may be required, especially if the contaminant concentration pattern does not follow a normal distribution. Sampling (especially for in situ treatment) and chemical analysis make a significant contribution to the overall cost of bioremediation, in terms of the monitoring itself (e.g., construction of monitoring wells) and the laboratory expenses. Where such costs are excessive, they could seriously reduce the overall cost-competitiveness of the process in relation to alternative approaches.

In situ systems of bioremediation suffer most from variability-related problems, and ex situ processes offer greater opportunities to reduce soil heterogeneity and the potential for greater control of the treatment process. A significant degree of mixing and homogenization may be achieved by processing the soil prior to incorporation in a treatment bed [e.g., 2,15,16] or by treating the soil in a slurry reactor system [7]. The variability that exists in both the vadose and saturated zones often makes it difficult to prove that the loss of contaminant has occurred strictly as the result of bioremediation, rather than by physical or chemical processes. A review of the extent to which successful in situ bioremediation has been proved indicates that the evidence often appears to be circumstantial and that for most accidental spills and contaminated sites of unknown history, attempts to derive quantitative mass balances are futile [174]. This review categorized a number of investigations into those suggesting or those successfully proving in situ biodegradation had occurred. Where bioremediation had been proved in soil, the evidence cited was the enhanced loss of the contaminant, in one case when two white rot fungi were added [154], and in two cases when fertilizers were added, together with an observed increase in total microbial numbers [175] or numbers of contaminant-degrading bacteria [25]. In the case of ground water, two examples of proof were cited. In the first of these, the cometabolic transformation of chlorinated ethenes was demonstrated in a shallow confined aquifer where mass balance determinations were made possible by artificially contaminating the ground water [122]. In the second, mineralization of naphthalene and phenanthrene occurred in sediments in an aquifer contaminated by coal tar residues, but not in adjacent, pristine areas [176]. Increased numbers of bacteria and their protozoal predators were observed only in the case of the PAH-contaminated environment. It is notable that all of these studies cited as evidence for successful in situ biodegradation [174] refer primarily to pilot-scale examples of bioremediation rather than to full-scale implementation. There has yet to be a full-scale in situ bioremediation project where sufficient controls have been established to allow a definitive demonstration of contaminant removal by a biological mechanism. Inasmuch as the establishment of suitable controls with appropriate replication would be a major cost factor, it is not difficult to understand why there is a paucity of such information.

The primary objective of any remedial system is to reduce the concentration of the contaminants to an acceptable and safe target level (i.e., established or approved by a regulatory agency). Hence, the extent to which this reduction is achieved by biological, chemical, or physical means is often solely of academic interest. Assuming that it is possible to achieve the cleanup target, a criterion that is of equal or more importance is that the process should not result in any untoward side effects in the form of potential hazards. Such hazards may relate to any of the supplements added, for example, the use of nitrate as a nutrient or electron acceptor in sensitive aquifers, potential eutrophication arising from the addition of nitrogenous and phosphate materials, potentially toxic effects of surfactants on aquatic fauna, problems arising from increased chemical oxygen demand, et cetera. Many of these hazards can be controlled through judicious application and engineering procedures to contain the contaminated plume, for example, through hydraulic controls.

One particular issue of concern in bioremediation is the potential accumulation of intermediates. Different intermediates may result from different electron acceptors (e.g., denitrifying, sulfate reducing, and methanogenic conditions), at least for some isomers of hydroxybenzoate [177]. In some cases, the production of the intermediate is well established as part of the degradation pathway, and where the intermediate is relatively easily degradable, it can be tracked as part of the monitoring program, for example, degradation of acetone as an intermediate of the degradation of isopropanol [17]. In other cases, the potential intermediates arising from the degradation of a particular contaminant may be less well understood.

Intermediates, including dihydrodiols, diols, diones, and carboxylic acid PAH derivatives, have been identified from the degradation of both lower and higher molecular weight PAHs (ranging from two to five fused benzene rings) in laboratory studies [e.g., 178, 179]. However, in soil or sediment systems, the presence of an active microbial community rather than a single organism may result in somewhat different situation. For example, the concentrations of intermediates in sediments enriched with ^{14}C-labeled PAHs never rose to a significant fraction of the total ^{14}C present [180], and in a further microcosm study where the degradation of PAHs with three or more rings was investigated [168], it was similarly concluded that following initial ring cleavage, degradation by indigenous microorganisms proceeded relatively rapidly with little or no accumulation of intermediates. A key issue here would appear to be the existence of an active indigenous microbiota.

Even with contaminants whose degradability is well established, there may be a potential for intermediates of greater polarity than the parent compound to be produced. Not only are these more mobile, and therefore more liable to be leached from the environment, but they may also be inhibitory to further biodegradation and so limit the effectiveness of the process itself. Fatty acids, particularly short chain ones, produced as intermediates from the degradation of crude oil reduced the rate and extent of the mineralization of the oil, and together with the crude oil were shown to have a synergistic toxic effect [181].

Although the evidence from field studies is limited, it does tend to support the view that, at least as far as hydrocarbons and lower-ringed PAHs are concerned, toxicity arising from the accumulation of intermediates appears to be transitory in a well-managed system. The disappearance of hydrocarbons from soils in lysimeter units supplemented with one of three oils—jet fuel, heating oil, or diesel oil—was measured periodically in untreated and treated systems (the latter involved nutrient and lime addition, the former involved water only [182]). At the same time, a Microtox assay was used to assess the relative toxicity of the residue which was compared with a similar but uncontaminated soil. This bioassay is based on monitoring changes in light emission by the luminescent *Photobacterium phosphoreum* when exposed to a toxic substance [183]. The degree of toxicity is measured as the effective concentration of a test sample that causes a 50% decrease in light output (EC_{50}), so that the greater the EC_{50} value, the lower the toxicity. The results are shown in Figure 5. The jet fuel, which was the least persistent of the three oils tested, was rapidly detoxified and EC_{50} values returned to those of the control in 2 weeks for the treated soil and after 6 weeks for the untreated soil. For the soil supplemented with heating oil (of intermediate persistence), both the treated and untreated soils increased initially in toxicity, the degree of toxicity being slightly higher in the treated soil. However, the subsequent decline in toxicity in the treated soil occurred much sooner (after 6 weeks) than in the untreated soil (12 weeks), and after 20 weeks toxicity values were the same as in the uncontaminated soil. A similar pattern was observed in soil contaminated with diesel oil, the slowest of the three oils to degrade. Even though there was an initial increase in toxicity, it was only in the treated soil that toxicity declined to background soil levels after 20 weeks, whereas in the untreated soil, there was still significant toxicity present. These data were supported by the results of plant studies that demonstrated that the bioremediation partially restored the capacity of the soil to support seed germination and

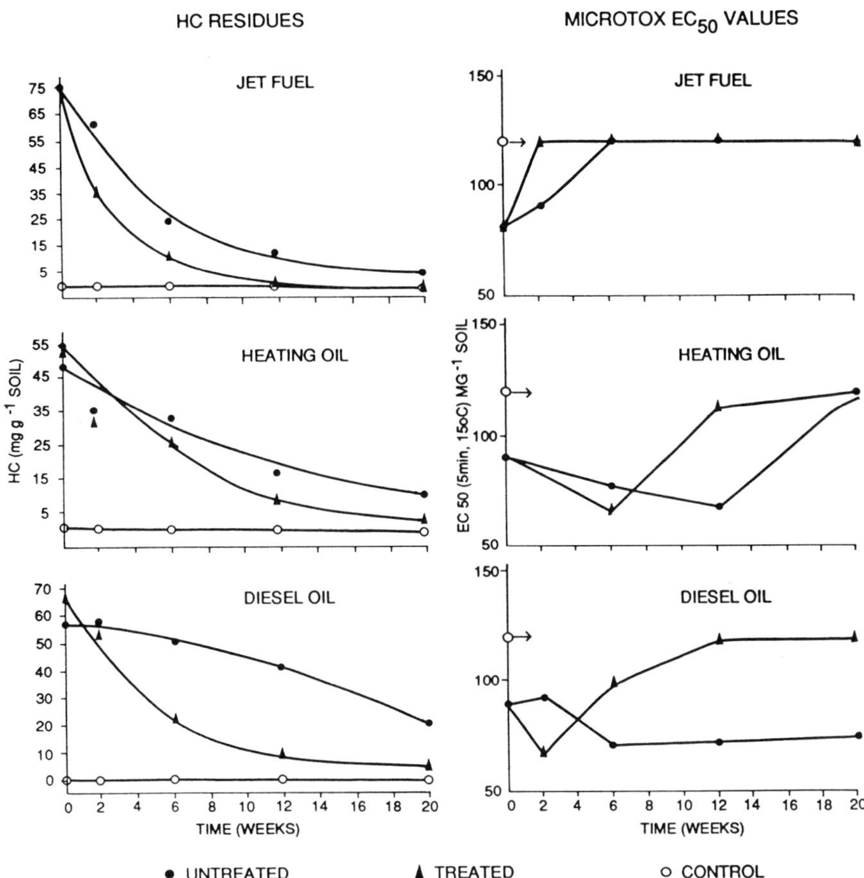

Figure 5 Changes in hydrocarbon residues and Microtox EC_{50} values with time in oil-supplemented lysimeters. Redrawn from Soil Biology & Biochemistry 22:501–505. X. Wang and R. Bartha, Effects of Bioremediation on Residues, Activity and Toxicity in Soil Contaminated by Fuel Spills [181], copyright 1990; with kind permission from Elsevier Science Ltd., The Boulevard, Langford Lane, Kidlington, OX5 1GB, England.

plant growth; for example, after 4 weeks the percent germination of rye grass was 0% in the untreated soil, 68% in the treated soil, and 80% in the control. With soybean, the respective values were 6, 67, and 80%.

In the diesel oil-amended soil, which probably contained the highest level of aromatics, all of the PAHs in the treatment receiving nutrient and lime addition were present at less than 5% of their initial concentrations after 12 weeks; apart from methyl pyrene, 10% of which remained. In contrast, greater persistence of aromatics and PAHs (up to 32.5% of the initial concentrations) were observed in the untreated soil, that is, which received water only [184]. Mutagenicity tests using the Ames test were also performed on extracts from the treated soil. As with the Microtox data, there was an initial increase in toxicity as measured by mutagenicity, which was thought to correspond to the hydroxylation of the PAHs, but this was followed by a decline to background levels over the subsequent 20-week period.

Although these studies were conducted only at the lysimeter scale, they demonstrate that for petroleum hydrocarbons, including associated PAHs of up to four rings, the toxic effects arising from the products of biodegradation appear to be only transitory. Moreover, in actively treated soils, where nutrients and lime were added to enhance microbial activity, the toxicity of soil extracts was reduced to background levels, in contrast to untreated systems. Further investigations are necessary to determine whether this is also true for the higher-ringed PAHs or other contaminants. However, the conclusions from these studies indicate that there is a greater likelihood of toxic intermediates persisting longer when a contaminant remains untreated rather than when it is treated by active bioremediation. This may, in some cases, strengthen the argument for implementing an active bioremediation system to treat a spillage of contaminant rather than relying on the slower processes of natural attenuation.

One example that indicates the potentially detrimental effects of the production of intermediates is illustrated by a microcosm study with a genetically modified microorganism (GMMO). There have been no reports of in situ treatment using GMMOs [185]. Although the subject of the potential risks arising from release of GMMOs is outside the scope of this review, the introduction of a GMMO with a particular biodegradative ability may influence the activity of the indigenous microbiota through chemical changes in the soil environment [186]. For example, when a genetically modified *P. putida* PP0301 (pRO 103) was introduced to soil enriched with 500 mg kg^{-1} of 2,4-dichlorophenoxyacetate (2,4-D), there was an accumulation of the metabolite, 2,4-dichlorophenol (2,4-DCP), together with a significant decrease in fungal propagules and the rate of carbon dioxide evolution [187]. Demonstration of the toxicity of 2,4-DCP toward fungi, both in situ and in vitro, supported the conclusion that production of this intermediate was responsible for such changes. There is therefore a need to evaluate such potential effects, not only related to the introduction of a GMMO, but also in terms of intermediate production in general, particularly for halogenated aromatics and other xenobiotics, as part of protocols for the development of bioremediation strategies.

The greater the degree to which a contaminant is recalcitrant in nature and the greater the requirement for a biological element with the overall strategy, that is, use of bioaugmentation techniques, the greater are the potential pitfalls involved. Failure of inocula to achieve in situ what they can perform in axenic culture has been attributed to the contaminant occurring at too low a concentration to support growth, the presence of inhibiting substances, the effects of predation by protozoa, utilization of alternate substances, and failure to move through the soil pore spaces [23]. New tools for tracking the specific organisms involved in bioremediation should greatly improve monitoring capabilities; for example, gene probe technology provides accurate determination of specific catabolic bacteria and their response to changing environmental conditions [188].

The development of microorganisms with the capacity for degrading particular compounds and which are also capable of functioning at environmental extremes has a significant potential for in situ treatment. One such example is demonstrated by a laboratory study where transfer of the TOL plasmid, pWWO, from the mesophile, *P. putida* PaW1, to the naturally occurring psychotroph, *P. putida* Q5, allowed the latter to degrade 1000 mg L^{-1} toluate at temperatures as low as 0°C [189]. At the same time, there is a requirement for developing better systems for introducing and distributing nutrients, inocula, and oxygen, or other electron acceptors, to the subsurface and integrating in situ bioremediation more closely with physical and chemical treatment technologies.

Finally, successful implementation of bioremediation will require an understanding of site heterogeneity and the accommodation of this factor within the overall strategy for cleanup.

DISCLAIMER

The views expressed in this paper are those of the author and do not necessarily represent those of Dames & Moore.

REFERENCES

1. Bewley, R. J. F. 1992. Bioremediation of contaminated ground, p. 270–284. *In* J. F. Rees (ed.), Contaminated land treatment technologies. Elsevier, London.
2. Bewley, R., B. Ellis, P. Theile, I. Viney, and J. Rees. 1989. Microbial clean-up of contaminated soil. Chem. Ind. 23:778–783.
3. Bewley, R. J. F., B. Ellis, and J. F. Rees. 1990. Development of a microbiological treatment for restoration of oil contaminated soil. Land Deg. Rehabil. 2:1–11.
4. de Kreuk, J. F., and G. J. Annokkee. 1988. Applied biotechnology for decontamination of polluted soils; possibilities and problems, p. 679–686. *In* K. Wolf, W. J. van den Brink, and F. J. Colon (ed.), Contaminated soil '88. Kluwer Academic, Dordrecht.
5. Heely, D. A., E. S. Werk, and R. G. Kowalski. 1994. Bioremediation and reuse of soils containing no. 5 fuel oil in New England using an above ground treatment

cell: a case study, p. 163–173. *In* D. L. Wise and D. J. Trantolo (ed.), Remediation of hazardous waste contaminated soils. Marcel Dekker, New York.

6. Fan, C. Y., and A. N. Tafuri. 1994. Engineering application of biooxidation process for treating petroleum-contaminated soil, p. 373–401. *In* D. L. Wise and D. J. Trantolo (ed.), Remediation of hazardous waste contaminated soils. Marcel Dekker, New York.
7. League, J. 1991. Liquid solids contacts (LSC) or how to bug hazardous waste to death, p. 19–24. In H. M. Freeman and P. R. Sferra (ed.), Innovative hazardous waste treatment technology, Vol. 3, Biological processes. Technomic Publishing, Lancaster, Pennsylvania.
8. Chivers, G. E. 1984. The treatment of wastes from the petrochemical industry, p. 130–190. *In* D. Barnes, C. F. Forster, and S. E. Hrudey, Petroleum and organic chemical industries. Pitman, Boston.
9. Cherminsoff, P. N. 1987. Biotechnology: treating industrial/municipal wastes and wastewater. Pollut. Eng.:74–77.
10. Troy, M. A., H. A. Minnigh, and W. O. Pipes. 1988. Aerobic microbiological remediation of contaminated groundwater, p. 524–536. *In* J.C. Evans (ed.), Toxic and hazardous wastes. Proceedings of the Nineteenth Mid-Atlantic Industrial Waste Conference.
11. Parkin, G. F., and C. R. Calabria. 1986. Principles of bioreclamation of contaminated ground water and leachates, p. 150–173. *In* D. Lorenzen, R. A. Conway, L. P. Jackson, A. Hamza, C. L. Perket, and W. J. Lacy (ed.), Hazardous and industrial solid waste testing and disposal, Vol. 6. ASTM, Philadelphia, Pennsylvania.
12. van der Hoek, J. P., L. G. C. M. Urlings, and C. M. Grobben. 1989. Biological removal of polycyclic aromatic hydrocarbons, benzene, toluene, ethylbenzene, xylene and phenolic compounds from heavily contaminated ground water and soil. Environ. Tech. Letters. 10:185–194.
13. Bandyopadhyay, S., S. K. Bhattacharya, and P. Majumdar. 1994. Engineering aspects of bioremediation, p. 55–75. *In* D. L. Wise and D. J. Trantolo (ed.), Remediation of hazardous waste contaminated soils. Marcel Dekker, New York.
14. Battermann, G., R. Fried, M. Meier-Lohr, and P. Werner. 1993. Pilot and large scale experiences in the in situ bioremediation of a refinery site polluted with hydrocarbons, p. 1135–1144. *In* F. Arendt, G. J. Annokkee, R. Bosman, and W. J. van den Brink (ed.), Contaminated soil '93. Kluwer Academic, Dordrecht.
15. Ellis, B., M. T. Balba, and P. Theile. 1990. Bioremediation of oil contaminated land. Environ. Tech. 11:443–455.
16. Ellis, B., P. Harold, and H. Kronberg. 1991. Bioremediation of a creosote contaminated site. Environ. Tech. 12:447–459.
17. Flathman, P. E., and G. D. Githens. 1985. In-situ biological treatment of isopropanol, acetone, and tetrahydrofuran in the soil/groundwater environment, p. 173–185. *In* E. K. Nyer (ed.), Groundwater treatment technology. Van Nostrand Reinhold, New York.
18. Balba, M. T., and R. J. F. Bewley. 1991. Organic contaminants and microorganisms, p. 237–274. *In* K. C. Jones (ed.), Organic contaminants in the environment. Elsevier, London.

19. Carberry, J. B. 1994. Enhancement of bioremediation by partial preoxidation, p. 543–597. *In* D. L. Wise and D. J. Trantolo (ed.), Remediation of hazardous waste contaminated soils. Marcel Dekker, New York.
20. Morgan, P., and R. J. Watkinson. 1990. Assessment of the potential for in situ biotreatment of hydrocarbon-contamainated soils. Water Sci. Tech. 22:63–68.
21. Department of the Environment. 1987. Problems arising from the redevelopment of gas works and similar sites (2nd ed). Her Majesty's Stationery Office, London.
22. Alexander, M. 1985. Biodegradation of organic chemicals. Environ. Sci. Technol. 18:106–111.
23. Goldstein, R. M., L. M. Mallory, and M. Alexander. 1985. Reasons for possible failure of inoculation to enhance biodegradation. Appl. Environ. Microbiol. 50:977–983.
24. Borden, R. C., M. D. Lee, J. M. Thomas, P. B. Bedient, and C. H. Ward. 1989. *In situ* measurement and numerical simulation of oxygen limited biotransformation. Ground Water Monit. Rev. 9:83–91.
25. Raymond, R. L., J. O. Hudson, and W. V. Jamison. 1976. Oil degradation in soil. Appl. Environ. Microbiol. 31:522–535.
26. CONCAWE. 1980. Sludge farming: a technique for the disposal of oily refinery wastes. CONCAWE Report No. 3/80. CONCAWE, The Hague.
27. Marks, R. E., Y. B. Acar, and R. J. Gale. 1994. In situ remediation of contaminated soils containing hazardous mixed wastes by bio-electrokinetic remediation and other competitive technologies, p. 405–436. *In* D. L. Wise and D. J. Trantolo (ed.), Remediation of hazardous waste contaminated soils. Marcel Dekker, New York.
28. Troy, M. A., L. B. Allen, and D. E. Jerger. 1994. Bioremediation of petroleum-contaminated soils at railroad facilities, p. 227–243. *In* D. L. Wise and D. J. Trantolo (ed.), Remediation of hazardous waste contaminated soils. Marcel Dekker, New York.
29. Hinchee, R. E., R. N. Miller, and R. R. Dupont. 1991. Enhanced biodegradation of petroleum hydrocarbons: an air-based *in situ* process, p. 177–183. *In* H. M. Freeman and P. F. Sferra (ed.), Innovative hazardous waste treatment technology, Vol. 3, Biological processes. Technomic Publishing, Lancaster, Pennsylvania.
30. Rainwater, K., M. P. Mayfield, C. E. Heintz, and B. J. Claborn. 1991. Cyclic vertical water table movement for enhancement of *in situ* biodegradation of diesel fuel, p. 123–130. *In* H. M. Freeman and P. F. Sferra (ed.), Innovative hazardous waste treatment technology, Vol. 3, Biological processes. Technomic Publishing, Lancaster,
31. Wilson, S. B., and R. A. Brown. 1989. *In situ* bioreclamation: a cost effective technology to remediate subsurface organic contamination. Ground Water Monit. Rev. 9:173–179.
32. Leahy, M. C., and R. A. Brown 1994. Bioremediation: optimizing results. Chem Eng.:108–116.
33. Michelsen, D., and M. Lotfi. 1991. Oxygen microbubble injection for *in situ* bioremediation: possible field scenario, p. 131–142. *In* H. M. Freeman and P. F. Sferra (ed.), Innovative hazardous waste treatment technology, Vol. 3, Biological processes. Technomic Publishing, Lancaster, Pennsylvania.

34. Starr, R. C., and J. A. Cherry. 1994. *In situ* remediation of contaminated groundwater: the funnel and gate system. Ground Water: 465–476.
35. Morgan, P., and R. J. Watkinson. 1989. Microbiological methods for the clean up of soil and groundwater contaminated with halogented organic compounds. FEMS Microbiol. Rev. 63:277–300.
36. Bewley, R. J. F., and J. G. Alexander. 1993. Technical factors in the design of a programme for *ex situ* and *in situ* bioremediation of a former oil distribution terminal, p. 1147–1148. *In* F. Arendt, G. J. Annokkee, R. Bosman, and W. J. van den Brink (ed.), Contaminated soil '93. Kluwer Academic, Dordrecht.
37. Marquis, S.A., Jr., and D. Dineen. 1994. Comparison between pump and treat, biorestoration and biorestoration/pump and treat combined: lessons from computer modeling. Ground Water Monit. Rev. 14:105–119.
38. Hazardous Materials Control Research Institute (HMCRI). 1987. Hazardous materials *in situ* stabilisation. HMCRI, Silver Spring, Maryland.
39. Ghassemi, M. 1988. Innovative *in situ* treatment technologies for clean up of contaminated sites. J. Haz. Mat. 17:189–206.
40. Bewley, R. J. F. 1992. Bioremediation and waste management, p. 33–45. *In* D.E.S. Stewart-Tull and M. Sussman (ed), The release of genetically modified microorganisms. Plenum Press, New York.
41. Alexander, M. 1977. Introduction to soil microbiology. John Wiley & Sons, New York.
42. Chambers, C. D., J. Wills, S. Giti-Pour, J. L. Zieleniewski, J. F. Rickabaugh, M. I. Mecca, B. Pasin, R. C. Sims, J. E. McLean, R. Mahmood, R. R. Dupont, and K. Wagner. 1991. *In situ* treatment of hazardous waste contaminated soils. (2nd ed.). Noyes Data Corporation, Park Ridge, New Jersey.
43. Ou, L. T., P. S. C. Rao, and J. M. Davidson. 1983. Methyl parathion degradation in soil: influence of soil-water tension. Soil Biol. Biochem. 15:211–215.
44. Nimah, M. H., J. Ryan, and M. Chaudry. 1983. Effect of synthetic conditioners on soil water retention, hydraulic conductivity, porosity and aggregation. Soil Sci. Soc. Am. J. 47:742–745.
45. Valo, R., and M. Salkinoja-Salonen. 1986. Bioreclamation of chlorophenol-contaminated soil by composting. Appl. Microbiol. Biotechnol. 25:68–75.
46. Satijn, H. M. C., and P. A. de Boks. 1988. Biorestoration, a technique for remedial action on industrial sites, p. 745–753. *In* K. Wolf, W. J. van den Brink, and F. J. Colon (ed.), Contaminated soil '88. Kluwer Academic, Dordrecht.
47. Sandvik, S., A. Lode, and T. A. Pedersen. 1986. Biodegradation of oily sludge in Norwegian soils. Appl. Microbiol. Biotechnol. 23:297–301.
48. Morgan, P., S. T. Lewis, and R. J. Watkinson. 1993. Biodegradation of benzene, toluene, ethylbenzene and xylenes in gas-condensate-contaminated ground water. Environ. Pollut. 82:181–190.
49. Brown, R. A., and J. A. Crosbie. 1990. Oxygen sources for *in situ* bioremediation, p. 22–28. In Biotreatment. Hazardous Materials Control Research Institute Silver Spring, Maryland.
50. Rainwater, K., and R. J. Scholze. 1991. *In situ* biodegradation for treatment of contaminated soil and groundwater, p. 107–121. *In* H. M. Freeman and P. F. Sferra

(ed.), Innovative hazardous waste treatment technology, Vol. 3, Biological processes. Technomic Publishing, Lancaster, Pennsylvania.

51. Dragun, J. 1988. The soil chemistry of hazardous materials. Hazardous Materials Control Research Institute, Silver Spring, Maryland.
52. Jamison, V. W., R. L. Raymond, and J. O. Judson, Jr. 1975. Biodegradation of high octane gasoline in groundwater. Devel. Ind. Microbiol. 16:305–312.
53. Lee, M. D., J. M. Thomas, R. C. Borden, P. B. Bedient, J. T. Wilson, and C. H. Ward. 1988. Biorestoration of aquifers contaminated with organic compounds. Crit. Rev. Environ. Control. 18:29–89.
54. Brown, R.A., R. D. Norris, and G. R. Brubaker. 1985. Aquifer restoration with enhanced bioreclamation. Pollut. Eng. 17:25–28.
55. Yaniga, P. M., and W. Smith. 1987. Aquifer restoration via accelerated in situ biodegradation of organic contaminants, p. 1–6. *In In situ* treatment Hazardous Materials Control Research Institute, Silver Spring, Maryland.
56. Michelsen, D. L., D. A. Wallis, and F. Sebba. 1984. In situ biological oxidation of hazardous organics. Environ. Prog. 3:103–107.
57. Thomas, J. M., and C. H. Ward. 1989. In situ biorestoration of organic contaminants in the subsurface. Environ. Sci. Technol. 23:760–766
58. Griffith, S. M., and M. Schnitzer. 1977. Organic compounds formed by the hydrogen peroxide oxidation of soils. Can. J. Soil Sci. 57:223–231.
59. Morgan, P., and R. J. Watkinson. 1992. Factors limiting the supply and efficiency of nutrient and oxygen supplements for the *in situ* biotreatment of contaminated soil and groundwater. Water Res. 26:73–78.
60. Spain, J. C., J. D. Milligan, D. C. Downey, and J. K. Slaughter. 1989. Excessive bacterial decomposition of H_2O_2 during enhanced biodegradation Ground Water 27: 163–167.
61. Heyse, E., S. C. James, and R. Wetzel. 1986. *In situ* biodegradation of aquifer contaminants at Kelly Air Force Base. Environ. Prog. 5:207–211.
62. Wetzel, R. S., D. H. Davidson, C. M. Durst, and D. J. Sarno. 1987. Effectiveness of *in situ* biological treatment of contaminated groundwater and soils at Kelly Air Force Base, Texas, p. 128–133. *In In situ* treatment. Hazardous Materials Control Research Institute, Silver Spring, Maryland.
63. Bertrand, J. C., P. Caumette, G. Mille, M. Gilewicz, and M. Denis. 1989. Anaerobic biodegradation of hydrocarbons. Sci. Prog. (Oxford) 73:333–350.
64. Sleat, R., and J. P. Robinson. 1984. The bacteriology of anaerobic degradation of aromatic compounds. J. Appl. Bacteriol. 57:381–394.
65. Bakker, G. 1977. Anaerobic degradation of aromatic compounds in the presence of nitrate. FEMS Lett. 1:103–110.
66. Bossert, I. D., and L. Y. Young. 1986. Anaerobic oxidation of *p*-cresol by a denitrifying bacterium. Appl. Environ. Microbiol. 52:1117–1122.
67. Ehrlich, G. G., D. F. Goerlitz, E. M. Godsy, and M. F. Hult. 1982. Degradation of phenolic contaminants in ground water by anaerobic bacteria: St. Louis Park, Minnesota. Ground Water 20:703–710.
68. Zeyer, J., E. P. Kuhn, and R. P. Schwarzenbach. 1986. Rapid microbial mineralization of toluene and 1,3-dimethylbenzene in the absence of molecular oxygen. Appl. Environ. Microbiol. 52:944–947.

69. Kuhn, E. M., J. Zeyer, P. Eicher, and R. P. Schwarzenbach. 1988. Anaerobic degradation of alkylated benzenes in denitrifying laboratory aquifer columns. Appl. Environ. Microbiol. 54:490–496.

70. Gersberg, R. M., W. J. Dawsey, and M. D. Bradley. 1991. Biodegradation of monoaromatic hydrocarbons in groundwater under denitrifying conditions Bull. Environ. Contam. Toxicol. 47:230–237.

71. Al-Bashir, B., T. Cseh, R. Leduc, and R. Samson. 1990. Effect of soil/contaminant interactions on the biodegradation of naphthalene in flooded soils under denitrifying conditions. Appl. Microbiol. Biotechnol. 34:414–419.

72. Milhelic, J. R., and R. G. Luthy. 1991. Sorption and microbial degradation of naphthalene in soil-water suspensions under denitrification conditions Environ. Sci. Technol. 25:169–177.

73. Bouwer, E. J., and P. L. McCarty. 1983. Transformations of halogenated organic compounds under denitrification conditions. Appl. Environ. Microbiol. 45:1295–1299.

74. Hutchins, S. R., G. W. Sewell, D. A. Kovacs, and G. A. Smith. 1991. Biodegradation of aromatic hydrocarbons by aquifer microorganisms under denitrifying conditions. Environ. Sci. Technol. 25:68–76.

75. Hutchins, S. R., W. C. Downs, J. T. Wilson, G. B. Smith, D. A. Kovacs, D. D. Fine, R. H. Douglass, and D. J. Hendrix. 1991. Effect of nitrate addition on biorestoration of fuel contaminated aquifer: field demonstration. Ground Water 29:571–580.

76. Werner, P. 1985. A new way for the decontamination of polluted aquifers by biodegradation. Water Supply 3:41–47.

77. Campbell, J. R., J. K. Fu, and R. O'Toole. 1989. Biodegradation of PCP contaminated soils using *in situ* subsurface bioreclamation, p. 17–27. *In* Biotreatment. Hazardous Materials Control Research Institute, Silver Spring, Maryland.

78. Lovley, D., J. D. Coates, J. C. Woodward, and E. J. Phillips. 1995. Benzene oxidation coupled to sulfate reduction. Appl. Environ. Microbiol. 61:953–958.

79. Gibson, S., and J. M. Suflita. 1986. Extrapolation of biodegradation results to groundwater aquifers: reductive dehalogenation of aromatic compounds. Appl. Environ. Microbiol. 52:681–688.

80. Rittman, B. E., D. Jackson, and S. L. Storck. 1988. Potential for the treatment of hazardous organic chemicals with biological processes, p. 15–64. *In* D. L. Wise (ed.), Biotreatment systems, Volume 3. CRC Press, Boca Raton, Florida.

81. Wilson, B. H., J. T. Wilson, D. H. Kampbell, B. E. Bledsoe, and J. M. Armstrong. 1990. Biotransformation of monoaromatic and chlorinated hydrocarbons at an aviation gasoline spill site. Geomicrobiol. J. 8:225–240.

82. Leahy, M. C., M. Findlay, and S. Fogel. 1989. Biodegradation of chlorinated aliphatics by a methanotrophic consortium in a biological reactor, p. 3–9. *In* Biotreatment. Hazardous Materials Control Research Institute, Silver Spring, Maryland.

83. Freedman, D. L., and J. M. Gossett. 1989. Biological reductive dechlorination of tetrachloroethylene and trichloroethylene to ethylene under methanogenic conditions. Appl. Environ. Microbiol. 55:2144–2151.

84. Castro, T. F., and T. Yoshida. 1974. Effect of organic matter on the biodegradation of some organochlorine insecticides in submerged soils. Soil Sci. Plant Nutr. 20: 363–370.
85. Mirsatari, S. G., M. M. McChesney, A. C. Craigmill, W. L. Winterlin, and J. N. Seiber. 1987. Anaerobic microbial dechlorination: an approach to on-site treatment of toxaphene-contaminated soil. J. Environ. Sci. Health B22:663–690.
86. Boyer, J.D., R. C. Ahlert, and D. S. Kosson. 1988. Pilot plant demonstration of *in situ* biodegradation of 1,1,1-trichloroethane. J. Water Pollut. Control Fed. 60:1843–1849.
87. Fogel, M. M., A. R. Taddeo, and S. Fogel. 1986. Biodegradation of chlorinated ethenes by a methane-utilizing mixed culture. Appl. Environ. Microbiol. 51:720–724.
88. Wilson, J. T., and B. H. Wilson. 1985. Biotransformation of chloroethylene in soil. Appl. Environ. Microbiol. 49:242–243.
89. Young, C. S., and R. G. Burns. 1993. Detection, survival and activity of bacteria added to soil, p. 1–63. *In* J. M. Bollag and G. Stotzky (ed.), Soil biochemistry, Vol. 8. Marcel Dekker, New York.
90. Department of the Environment. 1990. Pilot survey of potentially contaminated land in Cheshire—a methodolgy for identifying potentially contaminated sites. Her Majesty's Stationery Office, London.
91. Interdepartmental Committee on the Redevelopment of Contaminated Land. 1987. Guidance on the assessment and redevelopment of contaminated land. ICRCL 59/83 (2nd ed.). Her Majesty's Stationery Office, London.
92. Dibble, J. T., and R. Bartha. 1979. Effect of environmental parameters on the biodegradation of oil sludge. Appl. Environ. Microbiol. 37:729–739.
93. Lee, M. D., and C. H. Ward. 1985. Biological methods for the restoration of contaminated aquifers. Env. Toxicol. Chem. 4:743–750.
94. Atlas, R. M. 1977. Stimulated petroleum biodegradation. Crit. Rev. Microbiol. 5: 371–386.
95. Bossert, I., and R. Bartha. 1984. The fate of petroleum in soil ecosystems, p. 435–473. *In* R. M. Atlas (ed.), Petroleum microbiology. Macmillan, New York.
96. Atlas, R. M. 1981. Microbial degradation of petroleum hydrocarbons: an environmental perspective. Microbiol. Rev. 45:180–209.
97. Morgan, P., and R. J. Watkinson. 1989. Hydrocarbon degradation in soils and methods for soil biotreatment. Crit. Rev. Biotechnol. 8:305–333.
98. Bartha, R. 1986. Biotechnology of petroleum pollutant biodegradation. Microb. Ecol. 12:155–172.
99. Odu, C. T. I. 1978. The effect of nutrient application and aeration on oil degradation in soil. Environ. Pollut. 15:235–240.
100. Jobson, A., M. McLaughlin, F. D. Cook, and D. W. S. Westlake. Effect of amendments on the microbial utilization of oil applied to soil. Appl. Microbiol. 27:166–171.
101. Watts, J. R., J. C. Corey, and K. W. McLeod. 1982. Land application studies of industrial waste oils. Environ. Pollut. (ser. A) 28:165–175.
102. Dibble, J. T., and R. Bartha 1979. Leaching aspects of oil sludge biodegradation in soil. Soil Sci. 127:365–370.

103. Verstraete, W., R. Vanloocke, R. de Borger, and A. Verlinde. 1976. Modelling of the breakdown and the mobilization of hydrocarbons in unsaturated soil layers, p. 99–112. *In* J. M. Sharpley and A. M. Kaplan (ed.) Proceedings of the Third International Biodegradation Symposium. Applied Science Publishers, London.
104. Brown, K. W., K. C. Donnelly, and L. E. Duel, Jr. 1983. Effects of mineral nutrients, sludge application rate, and application frequency on biodegradation of two oily sludges. Microb. Ecol. 9:363–373.
105. Butt, J. A., D. F. Duckworth, and S. G. Perry. 1985. Characterisation of spilled oil samples: purpose, sampling, analysis and interpretation. John Wiley & Sons, Chichester.
106. Swindoll, C. M., C. M. Aelion, and F. K. Pfaender. 1988. Influence of inorganic and organic nutrients on aerobic biodegradation and on the adaptation response of subsurface microbial communities. Appl. Environ. Microbiol. 54:212–217.
107. Atlas, R. M., and R. Bartha. 1973. Stimulated biodegradation of oil slicks using oleophilic fertilizers. Environ. Sci. Technol. 7:538–541.
108. Olivieri, R., P. Bacchin, A. Robertiello, N. Oddo, L. Degen, and A. Tonolo. 1976. Microbial degradation of oil spills enhanced by a slow-release fertiliser. Appl. Environ. Microbiol. 31:629–634.
109. Swannell, R. P. J., and I. M. Head 1994. Bioremediation comes of age. Nature 368: 396–397.
110. Pritchard, P. H., and C. F. Costa. 1991. EPA's Alaska oil spill bioremediation project. Environ. Sci. Technol. 25:372–379.
111. Bragg, J. R., R. C. Prince, E. J. Harner, and R. M. Atlas. 1994. Effectiveness of bioremediation for the *Exxon Valdez* oil spill. Nature 368:413–418.
112. Lehtomaki, M., and S. Niemala. 1975. Improving microbial degradation of oil in soil. Ambio 4:126–129.
113. Lichtenstein, E. P., T. T. Liang, and M. K. Koeppe. 1982. Effects of fertilizers, captafol, and atrazine on the fate and translocation of [^{14}C] fonfos and [^{14}C] parathion in a soil plant microcosm. J. Agric. Food Chem. 30:871–878.
114. Marinucci, A. C., and R. Bartha. 1979. Biodegradation of 1,2,3,-and 1,2,4-trichlorobenzene in soil and liquid enrichment culture. Appl. Environ. Microbiol. 38: 811–817.
115. Loynachan, T. E. 1978. Low temperature mineralization of crude oil in soil. J. Environ. Qual. 7:494–500.
116. Bollag, J.-M. 1983. Cross-coupling of humus constituents and xenobiotic substances, p. 127–141. *In* R. F. Christman and E. T. Gjessing (ed.), Aquatic and terrestial humic materials. Ann Arbor Science Publishers, Ann Arbor, Michigan.
117. Focht, D. D. 1993. Microbial degradation of chlorinated biphenyls, p. 341–407. *In* J.-M. Bollag and G. Stotzky (ed.), Soil biochemistry, Vol. 8. Marcel Dekker, New York.
118. Nelson, M., S. O. Montgomery, E. J. O'Neill, and P. H. Pritchard. 1986. Aerobic metabolism of trichloroethylene by a bacterial isolate. Appl. Environ. Microbiol. 52: 383–384.
119. Nelson, M., S. O. Montgomery, W. R. Mahaffey, and P. H. Pritchard. 1987. Biodegradation of trichloroethylene and involvement of an aromatic biodegradative pathway. Appl. Environ. Microbiol. 53:949–954.

120. Nelson, M., S. O. Montgomery, and P. H. Pritchard. 1988. Trichloroethylene metabolism by microorganisms that degrade aromatic compounds. Appl. Environ. Microbiol. 54:604–606.
121. Roberts, P. V., G. D. Hopkins, D. M. Mackay, and L. Semprini. 1990. A field evaluation of *in situ* biodegradation of chlorinated ethenes: Part 1, methodology and field site characterization. Ground Water 28:591–604.
122. Semprini, L., P. V. Roberts, G. D. Hopkins, and P. L. McCarty. 1990. A field evaluation of *in situ* biodegradation of chlorinated ethenes: Part 2, results of biostimulation and biotransformation experiments. Ground Water 28:715–727.
123. Semprini, L., and P. L. McCarty. 1992. Comparison between model simulations and field results for *in situ* biorestoration of chlorinated aliphatics: Part 2, cometabolic transformations. Ground Water 30:37–44.
124. Prince, R. 1993. Petroleum spill bioremediation in marine environments. Crit. Rev. Microbiol. 19:217–242.
125. Rosenberg, E., R. Legmann, A. Kushmaro, R. Taube, E. Adler, and E. Z. Ron. 1992. Petroleum bioremediation—a multiphase problem. Biodegradation 3:337–350.
126. Robichaux, T. J., and H. Nugent Myrick. 1972. Chemical enhancement of the biodegradation of crude oil pollutants. J. Petrol. Technol. 24:16–20.
127. Bloom, S. A. 1970. An oil dispersant's effect on the microflora of beach sand. J. Mar. Biol. Assoc. U.K. 50:919–923.
128. Mulkins-Phillips, G. J., and J. E. Stewart. 1974. Effect of four dispersants on biodegradation and growth of bacteria on crude oil. Appl. Microbiol. 28:547–552.
129. Bury, S. J., and C. A. Miller. 1993. Effect of micellar solubilization on biodegradation rates of hydrocarbons. Environ. Sci. Technol. 27:104–110.
130. Efroymson, R. A., and M. Alexander. 1991. Biodegradation by an *Arthrobacter* species of hydrocarbons partitioned into an organic solvent. Appl. Environ. Microbiol. 57:1441–1447.
131. Aronstein, B. N., Y. M. Calvillo, and M. Alexander. 1991. Effect of surfactants at low concentrations on the desorption and biodegradation of sorbed aromatic compounds in soil. Environ. Sci. Technol. 25:1728–1731.
132. Tiehm, A. 1994. Degradation of polycyclic aromatic hydrocarbons in the presence of synthetic surfactants. Appl. Environ. Microbiol. 60:258–263.
133. Viney, I., and R. J. F. Bewley. 1990. Preliminary studies on the development of a microbiological treatment for polychlorinated biphenyls. Arch. Environ. Contam. Toxicol. 19:789–796.
134. McDermott, J. B., R. Unterman, M. J. Brennan, R. E. Brooks, D. P. Mobley, C. C. Schwartz, and D. K. Dietrich. 1989. Two strategies for PCB soil remediation: biodegradation and surfactant extraction. Environ. Prog. 8:46–51.
135. Rixey, W. G. 1994. Enhanced *in situ* soil-flushing fundamentals, p. 691–717. *In* D. L. Wise and D. J. Trantolo (ed.), Remediation of hazardous waste contaminated soils. Marcel Dekker, New York.
136. Kimball, S. L. 1994. The use of surfactants to enhance pump-and-treat processes for *in situ* soil remediation, p. 633–643. *In* D. L. Wise and D. J. Trantolo (ed.), Remediation of hazardous waste contaminated soils. Marcel Dekker, New York.
137. Sobisch, T., G. Reinisch, H. Hubner, A. Karg, and H. Neibelschutz. 1993. The potential use of surfactants in the field of physicochemical and biological clean up,

p. 1453–1454. *In* F. Arendt, G. J. Annokkee, R. Bosman, and W. J. van den Brink (ed.), Contaminated soil '93. Kluwer Academic, Dordrecht.

138. Sanseverino, J., D. A. Graves, M. E. Leavitt, S. K. Gupta, and R. G. Luthy. 1994. Surfactant-enhanced bioremediation of polynuclear aromatic hydrocarbons in coke waste, p. 345–372. *In* D. L. Wise and D. J. Trantolo (ed.), Remediation of hazardous waste contaminated soils. Marcel Dekker, New York.
139. Reusing, G. 1993. *In-situ* biorestoration of soil contaminated by mineral oil, PAHs and phenols in combination with extraction using non-ionic surfactants, p. 1433–1434. *In* F. Arendt, G. J. Annokkee, R. Bosman, and W. J. van den Brink (ed.), Contaminated soil '93. Kluwer Academic, Dordrecht.
140. Werner, P., and H. J. Brauch. 1988. Aspects on the *in situ* and on site removal of hydrocarbons from contaminated sites by biodegradation, p. 695–704. *In* K. Wolf, W. J. van den Brink, and F. J. Colon (ed.), Contaminated soil '88. Kluwer Academic, Dordrecht.
141. Dupont, R., R. C. Sims, J. L. Sims, and D. L. Sorensen. 1988. Biological treatment of hazardous waste-contaminated soils, p. 23–94. *In* D. L. Wise (ed.), Biotreatment systems, Vol. 2. CRC Press, Boca Raton, Florida.
142. Morris, M. S., and J. T. Novak. 1989. Mechanisms responsible for the biodegradation of organic compounds in the subsurface. J. Haz. Mat. 22:393–406.
143. Flathman, P. E., D. E. Jerger, and L. S. Bottomley. 1989. Remediation of contaminated ground water using biological techniques. Ground Water Monit. Rev. 9:105–119.
144. Babich, H., and G. Stotzky. 1980. Enviromental factors that influence the toxicity of heavy metal and gaseous pollutants to microorganisms. Crit. Rev. Microbiol. 8:99–145.
145. Bewley, R. J. F., and G. Stotzky. 1983. Effects of cadmium and simulated acid rain on ammonification and nitrification in soil. Arch. Environ. Contam. Toxicol. 12: 285–291.
146. Bewley, R. J. F., and D. Parkinson. 1986. Monitoring the impact of acid deposition on the soil microbiota, using glucose and vanillin decomposition. Water Air Soil Pollut. 27:57–68.
147. Bewley, R. J. F., and G. Stotzky. 1984. Degradation of vanillin in soil-clay mixtures treated with simulated acid rain. Soil Sci. 137:415–418.
148. Bewley, R. J. F., and G. Stotzky. 1983. Simulated acid rain (H_2SO_4) and microbial activity in soil. Soil Biol. Biochem. 15:425–429.
149. Bewley, R. J. F., and G. Stotzky. 1983. Effects of cadmium and zinc on microbial activity in soil; influence of clay minerals. Part I: Metals added individually. Sci. Total Environ. 31:41–55.
150. Bewley, R. J. F., and G. Stotzky. 1983. Effects of cadmium and zinc on microbial activity in soil; influence of clay minerals. Part II: Metals added simultaneously. Sci. Total Environ. 31:57–69.
151. Fletcher, R. D. 1994. Practical considerations during bioremediation, p. 39–53. *In* D. L. Wise and D. J. Trantolo (ed.), Remediation of hazardous waste contaminated soils. Marcel Dekker, New York.
152. Pritchard, P. H. 1992. Use of inoculation in bioremediation. Curr. Opin. Biotechnol. 3:232–243.

153. Bumpus, J. A. 1993. White rot fungi and their potential use in soil bioremediation processes, p. 65–100. *In* J.-M. Bollag and G. Stotzky (ed.), Soil biochemistry, Vol. 8. Marcel Dekker, New York.

154. Lamar, R. T., and D. M. Dietrich. 1990. *In situ* depletion of pentachlorophenol from contaminated soil by *Phanerochaete chrysosporium*. Appl. Environ. Microbiol. 56: 3093–3100.

155. Dott, W., D. Feidieker, P. Kampfer, H. Schleibinger, and S. Strechel. 1989. Comparison of autochthonous bacteria and commercially available cultures with respect to their effectiveness in fuel oil degradation. J. Ind. Microbiol. 4:365–374.

156. Vecchioli, G. I., M. T. Del Panno, and M. T. Painceira. 1990. Use of selected autochthonous soil bacteria to enhance degradation of hydrocarbons in soil. Environ. Pollut. 67:249–258.

157. Hardman, D. 1991. Microbial pollution control: a technology in its infancy. Chem. Ind. (London): 244–246.

158. Edgehill, R. U., and R. K. Finn. 1983. Microbial treatment of soil to remove pentachlorophenol. Appl. Environ. Microbiol. 45:1122–1125.

159. Crawford, R. L., and W. W. Mohn. 1985. Microbiological removal of pentachlorophenol from soil using a *Flavobacterium*. Enzyme Microb. Technol. 7: 617–620.

160. Ellis, B., and R. J. F. Bewley. 1990. Biotreatment of contaminated land, p. 231–240. In P. Howsam (ed.), Microbiology in civil engineering. E. & F.N. Spon, London.

161. Chatterjee, D. K., J. J. Kilbane, and A. M. Chakrabarty. 1982. Biodegradation of 2,4,5-trichlorophenoxyacetic acid in soil by a pure culture of *Pseudomonas cepacia*. Appl. Environ. Microbiol. 44:514–516.

162. Mueller, J. G., S. M. Resnick, M. E. Shelton, and P. H. Pritchard. 1992. Effect of inoculation on the biodegradation of weathered Prudhoe Bay crude oil. J. Ind. Microbiol. 10:95–102.

163. Venosa, A. D., J. R. Haines, W. Nisamaneepong, R. Govind, S. Pradhan, and B. Siddique. 1991. Efficacy of commercial inocula in enhancing oil biodegradation in closed laboratory reactors. J. Ind. Microbiol. 10:13–23.

164. Venosa, A. D., J. R. Haines, and D. M. Allen. 1992. Efficacy of commercial inocula in enhancing biodegradation of weathered crude oil contaminating a Prince William Sound beach. J. Ind. Microbiol. 10:1–11.

165. Catallo, W. J., and R. J. Portier. 1992. Use of indigenous and adapted microbial assemblages in the removal of organic chemicals from soils and sediments. Water Sci. Tech. 25:229–237.

166. Bewley, R. J. F., R. Sleat, and J. F. Rees. 1991. Waste treatment and pollution clean up, p. 507–519. *In* V. Moses and R. E. Cape (ed.), Biotechnology: The science and the business. Harwood Academic, London.

167. Bewley, R. J. F., and P. Theile. 1988. Decontamination of a coal gasification site through application of vanguard microorganisms, p. 739–743. *In* K. Wolf, W. J. van den Brink, and F. J. Colon (ed.), Contaminated soil '88. Kluwer Academic, Dordrecht.

168. Heitkamp, M., J. P. Freeman, D. W. Miller, and C. E. Cerniglia. 1988. Pyrene degradation by a *Mycobacterium* sp.: identification of ring oxidation and ring fission products. Appl. Environ. Microbiol. 54:2556–2565.

169. Keck, J., R. C. Sims, M. Coover, K. Park, and B. Symons. 1989. Evidence for cooxidation of polynuclear aromatic hydrocarbons in soil. Water Res. 23:1467–1476.

170. Unterman, R. 1990. Bacterial treatment of PCB-contaminated soils, p. 17–18. *In* Bioremediation '88: Hazardous waste treatment by genetically engineered or adapted organisms. Hazardous Materials Control Research Institute, Silver Spring, Maryland.

171. Kretschek, A., and M. Krupka. 1987. Biodegradation as a method of hazardous waste treatment in soil and subsurface environments, p. 39–45. *In* In situ treatment. Hazardous Materials Control Research Institute, Silver Spring, Maryland.

172. Ohneck, R. J., and G. L. Gardner. 1982. Restoration of an aquifer contaminated by an accidental spill of organic chemicals. Ground Water Monit. Rev. 2:50–53.

173. Quince, J. R., and G. L. Gardner. 1982. Recovery and treatment of contaminated ground water: Part II. Ground Water Monit. Rev. 2:18–25.

174. Madsen, E. L. 1991. Determining *in situ* biodegradation. Environ. Sci. Technol. 25: 1663–1673.

175. Westlake, D. W. S., A. M. Jobson, and F. D. Cook. 1978. *In situ* degradation of oil in a soil of the boreal region of the Northwest Territories. Can. J. Microbiol. 24: 254–260.

176. Madsen, E. L., J. L. Sinclair, and W. C. Ghiorse. 1991. *In situ* biodegradation: microbiological patterns in a contaminated aquifer. Science 252:830–833.

177. Kuhn, E. P., J. M. Suflita, M. D. Rivera, and L. Y. Young. 1989. Influence of alternate electron acceptors on the metabolic fate of hydroxybenzoate isomers in anoxic aquifer slurries. Appl. Environ. Microbiol. 55:590–598.

178. Evans, W. C., H. N. Fernley, and E. Griffiths. 1965. Oxidative metabolism of phenanthrene and anthracene by soil pseudomonads. Biochem. J. 95:819–831.

179. Gibson, D. T., D. M. Jerina, H. Yagi, and H. J. C. Yeh. 1975. Oxidation of the carcinogens benzo(a)pyrene and benzo(a)anthracene to dihydrodiols by a bacterium. Science 189:295–297.

180. Herbes, S. E., and L. R. Schwall. 1978. Microbial transformation of polycyclic aromatic hydrocarbons in pristine and petroleum-contaminated sediments. Appl. Environ. Microbiol. 35:306–316.

181. Atlas, R. M., and R. Bartha. 1973. Inhibition by fatty acids of the biodegradation of petroleum. Antonie van Leeuwenhoek J. Microbiol. Serol. 39:257–271.

182. Wang, X., and R. Bartha. 1990. Effects of bioremediation on residues, activity and toxicity in soil contaminated by fuel spills. Soil Biol. Biochem. 22:501–505.

183. Bulich, A. A., and D. L. Isenberg. 1980. Use of the luminescent bacterial system for the rapid assessment of aquatic toxicology. Adv. Instrum. 35:35–40.

184. Wang, X., X. Yu, and R. Bartha. 1990. Effect of bioremediation on polycyclic aromatic hydrocarbon residues in soil. Environ. Sci. Technol. 24:1086–1089.

185. Santas, P., and R. Santas. 1994. Status of sea-borne bioremediation technologies, p. 459–479. *In* D. L. Wide and D. J. Trantolo (ed.), Remediation of hazardous waste contaminated soils. Marcel Dekker, New York.

186. Doyle, J. D., G. Stotzky, G. McClung, and C. W. Hendricks. 1995. Effects of genetically engineered microorganisms on microbial populations and processes in

natural habitats. *In* S. L. Neidleman and A. I. Laskin (ed.), Advances in applied microbiology. Academic Press, New York.

187. Short, K. A., J. D. Doyle, R. J. King, R. J. Seidler, G. Stotzky, and R. H. Olsen. 1991. Effects of 2,4-dichlorophenol, a metabolite of a genetically engineered bacterium, and 2,4-dichlorophenoxyacetate on some microorganism-mediated ecological processes in soil. Appl. Eviron. Microbiol. 57:412–418.
188. Sanseverino, J., J. T. Fleming, A. Heitzer, B. M. Applegate, and G. S. Sayler. 1994. Applications of environmental biotechnology to bioremediation, p. 97–123. *In* D. L. Wise and D. J. Trantolo (ed.), Remediation of hazardous waste contaminated soils. Marcel Dekker, New York.
189. Kolenc, R. J., W. E. Inniss, B. R. Glick, C. W. Robinson, and C. I. Mayfield. 1988. Transfer and expression of mesophilic plasmid-remediated degradative capacity in a psychotrophic bacterium. Appl. Environ. Microbiol. 54:638–641.

Index